HANDBOOK OF METAHEURISTIC ALGORITHMS

Uncertainty, Computational Techniques, and Decision Intelligence Book Series

Series Editors

Tofigh Allahviranloo, PhD
Faculty of Engineering and Natural Sciences, Istinye University, Istanbul, Turkey

Narsis A. Kiani, PhD
Algorithmic Dynamics Lab, Department of Oncology-Pathology & Center of Molecular Medicine, Karolinska Institute, Stockholm, Sweden

Witold Pedrycz, PhD
Department of Electrical and Computer Engineering, University of Alberta, Alberta, Canada

Volumes in Series

- Sahar Tahvili, Leo Hatvani, Artificial Intelligence Methods for Optimization of the Software Testing Process (2022)
- Farhad Lotfi, Masoud Sanei, Ali Hosseinzadeh, Sadegh Niroomand, Ali Mahmoodirad, Uncertainty in Data Envelopment Analysis (2023)
- Chun-Wei Tsai, Ming-Chao Chiang, Handbook of Metaheuristic Algorithms (2023)
- Stanislaw Raczynski, Reachable Sets of Dynamic Systems (2023)

For more information about the UCTDI series, please visit: https://www.elsevier.com/books-and-journals/book-series/uncertainty-computational-techniques-and-decision-intelligence

HANDBOOK OF METAHEURISTIC ALGORITHMS

From Fundamental Theories to Advanced Applications

CHUN-WEI TSAI
Department of Computer Science and Engineering
National Sun Yat-sen University
Kaohsiung, Taiwan, ROC

MING-CHAO CHIANG
Department of Computer Science and Engineering
National Sun Yat-sen University
Kaohsiung, Taiwan, ROC

ACADEMIC PRESS
An imprint of Elsevier

ELSEVIER

Notices

Knowledge and best practice in this field are constantly changing. As new research and experience
broaden our understanding, changes in research methods, professional practices, or medical
treatment may become necessary.

Practitioners and researchers must always rely on their own experience and knowledge in
evaluating and using any information, methods, compounds, or experiments described herein. In
using such information or methods they should be mindful of their own safety and the safety of
others, including parties for whom they have a professional responsibility.

To the fullest extent of the law, neither the Publisher nor the authors, contributors, or editors,
assume any liability for any injury and/or damage to persons or property as a matter of products
liability, negligence or otherwise, or from any use or operation of any methods, products,
instructions, or ideas contained in the material herein.

ISBN: 978-0-443-19108-4

For information on all Academic Press publications
visit our website at https://www.elsevier.com/books-and-journals

Publisher: Mara Conner
Editorial Project Manager: John Leonard
Production Project Manager: Sajana Devasi P K
Cover Designer: Greg Harris

Typeset by VTeX

Working together
to grow libraries in
developing countries

www.elsevier.com • www.bookaid.org

To our wives, our parents, and the next generation.

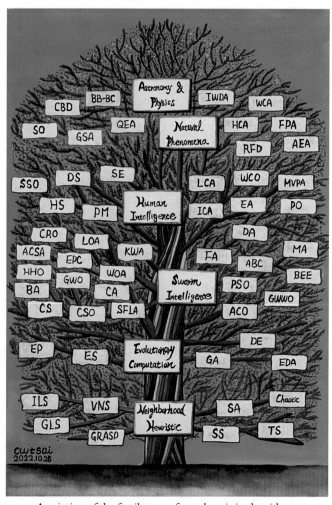

A painting of the family tree of metaheuristic algorithms.

Contents

PART 1 Fundamentals

PART 2 Advanced technologies

All the source code described in this book can be found in https://github.com/cwtsaiai/metaheuristics_2023/.

List of figures

List of tables

List of algorithms

List of listings

About the authors

Chun-Wei Tsai (1978–)

Chun-Wei Tsai received his PhD degree in Computer Science and Engineering from National Sun Yat-sen University, Kaohsiung, Taiwan in 2009. He was a postdoctoral fellow with the Department of Electrical Engineering, National Cheng Kung University, Tainan, Taiwan before joining the Department of Applied Geoinformatics and the Department of Information Technology, Chia Nan University of Pharmacy & Science, Tainan, Taiwan in 2010 and 2012, respectively. He joined the Department of Computer Science and Information Engineering, National Ilan University, Yilan, Taiwan and the Department of Computer Science and Engineering, National Chung-Hsing University, Taichung, Taiwan, in 2014 and 2017, respectively. In 2019, he joined the faculty of the Department of Computer Science and Engineering, National Sun Yat-sen University, Kaohsiung, Taiwan, where he is currently an Assistant Professor. He has more than 20 years of experience in metaheuristic algorithms and their applications. He has served as the Secretary-General of the Taiwan Association of Cloud Computing from 2018 to 2021; as Associate Editor for the *Journal of Internet Technology*, *IEEE Access*, *IET Networks*, and the *IEEE Internet of Things Journal* since 2014, 2017, 2018, and 2020, respectively; and as Content Manager for *IEEE Sensors Alert* since 2021. He has also been a member of the Editorial Board of the *Elsevier Journal of Network and Computer Applications* and *Elsevier ICT Express* since 2017 and 2021, respectively. His research interests include computational intelligence, data mining, cloud computing, and internet of things.

Ming-Chao Chiang (1956–)

Ming-Chao Chiang received his BS degree in Management Science from National Chiao Tung University, Hsinchu, Taiwan in 1978 and his MS, M.Phil., and PhD degrees in Computer Science from Columbia University, New York, NY, USA in 1991, 1998, and 1998, respectively. He has over 12 years of experience in the software industry encompassing a wide variety of roles and responsibilities in both large and start-up companies in Taiwan and the USA before joining the faculty of the Department of Computer Science and Engineering, National Sun Yat-sen University,

Kaohsiung, Taiwan in 2003, where he is currently a Professor. His research interests include image processing, evolutionary computation, and system software.

Preface

Our very first memory of a metaheuristic algorithm, the genetic algorithm (GA), in the early 2000s, is quite vivid. Many other metaheuristic algorithms, including simulated annealing (SA), tabu search (TS), ant colony optimization (ACO), particle swarm optimization (PSO), and differential evolution (DE), mushroomed at the end of the 1980s and continued to thrive, to become a big family nowadays. This finding led us to embark on a challenging but rewarding adventure to this new world that has its own system and rules; however, almost everything can be broken and innovation is a norm in this world. About 10 years ago, we came across the idea of writing a book to pass our knowledge of metaheuristic algorithms on to whoever are interested in this research domain for one reason: We found that many students have a hard time realizing some of the metaheuristic algorithms based on descriptions given in reference books or research papers. Although they may be able to find source code for some of the metaheuristic algorithms on the Internet, the coding styles and explanations are generally quite different. It would normally take students a lot of time and effort to adapt themselves to these different coding styles to figure out how to realize them. Apparently, there exists a gap between theory and implementation in this research domain. To narrow down this gap, we first present a unified framework for metaheuristics (UFM) and then use it to describe well-known metaheuristic algorithms and their variants. To make it easier for the audience of this book to understand how the metaheuristic algorithms and advanced technologies discussed herein are realized, the source code developed by the authors from scratch based on the UFM is included to show how SA, TS, GA, ACO, PSO, DE, parallel metaheuristic algorithms, hybrid metaheuristics, local search, and some other relevant advanced technologies are implemented in C++ and Python, where C++ is the programming language of our choice for the examples described in this book, while Python is the programming language suggested by one of the reviewers of the proposal of this book. We are not particularly good at Python, and it took us about a month to finish the implementation.

In the very beginning, the aim was to reduce the learning curve of the audience by developing an integrated library. Different versions of this library even have nicknames such as the USS Enterprise (NCC-1701) and Space Battleship Yamato in our groups. With the first version of the library, only about 10 to 20 instructions are needed to realize a new metaheuristic

algorithm for solving a new optimization problem. This all benefits from
the characteristics of an object-oriented programming language (namely,
encapsulation, inheritance, and polymorphism) in general and the template
of C++ in particular. Although this library is very charming in the sense
that it makes it very easy to realize most metaheuristic algorithms for solving various optimization problems—be it *discrete* or *continuous*—students
have difficulty understanding how metaheuristic algorithms work. From
the perspective of teaching, this library is not useful for students who do
not know what a metaheuristic algorithm is. So, although we still have this
library today, the students in our groups are not encouraged to use it. Our
teaching and learning experience shows that the best way to learn about a
new metaheuristic algorithm is to first understand the theory (algorithm)
and then implement it. The realization process helps us to fully understand
the pros and cons of metaheuristic algorithms. However, the learning performance of some students was still not as good as expected, even though
some of the implementation was simplified in the second version of the
book. For example, in the second version, the getopt function was used
to simplify the parsing of the command line options, but students who did
not have enough experience with C++ or other relevant programming languages were still struggling. We soon realized that theory and programming
should not be an obstacle for students and researchers who are interested
in entering this research domain. The price that must be paid to write a
more flexible program is certainly that it will be more complicated as well.
Instead of making the implementation more flexible, and thus more complicated, in this third version, the aim is to make implementation as simple
and thus as easy to understand as possible, so chances are that the audience
of this book will have no difficulty understanding not only the theory but
also the implementation of a set of representative metaheuristic algorithms,
to make it easier for them to enter this research domain.

　　The ultimate goal of this book is to share with the audience our experience and know-how on metaheuristic algorithms from the ground up,
that is, from the basic ideas to advanced technologies, even for readers who
have no background knowledge in artificial intelligence or machine learning. This book is roughly divided into three parts: fundamentals, advanced
technologies, and the appendix.

- For readers who are new to this area, we suggest reading the book from
  the very beginning to the very end chapter by chapter. In addition to
  the implementation in C++ described in the main text, the imple-

mentation in Python can be found in Appendix B: Implementation in Python.

- For readers who already have some background and experience in metaheuristic algorithms, we suggest reading Chapters 3 and 4 first to get familiar with the writing and programming styles of this book and then selecting an algorithm in Chapters 5–10 that is of interest as the second starting point before jumping to chapters in the second part of the book (advanced technologies) to find the knowledge that is of interest.

- For readers who already are quite familiar with most metaheuristic algorithms, we suggest starting with Chapter 4 and then jumping to the chapter that is of interest to find the needed information. Moreover, brief reviews of SA, TS, GA, ACO, PSO, and DE are provided in the end of Chapters 5–10, and discussions on advanced technologies in the beginning of Chapters 11–20 may provide some useful information.

After reading this book, it is expected that the audience can not only realize most of the metaheuristic algorithms, but also use them to solve real-world problems. It is also hoped that this book can be used by students and researchers as a reference for self-study to enter this research domain or by teachers as a reference or textbook for a course. This book does not cover all metaheuristic algorithms; however, it provides insight into the metaheuristic algorithms and it is expected that people who read the whole book will have no difficulty understanding metaheuristic algorithms not covered in this book. The strategy on which this book is based is like the famous proverb: "Give a man a fish, and you feed him for a day; teach a man to fish, and you feed him for a lifetime." Finally, it is hoped that this will lead the audience of this book to come up with valuable ideas and solutions in using metaheuristic algorithms. Last but not least, we would like to take this opportunity to thank Mara E. Conner, John Leonard, and Sajana Devasi P K at Elsevier for their support in the publication of this book. Special thanks go to the anonymous reviewers who suggested that this book should also provide (1) the implementation in Python so that readers who are not familiar with C++ can also benefit from this book, (2) brief reviews of SA, TS, GA, ACO, PSO, and DE, and (3) discussions on advanced applications of metaheuristic algorithms. The days of the week writing this book were the days we enjoyed the most, because we forgot about almost everything else, like children playing a video game all night. We may write another book to cover more metaheuristic algorithms presented after 2000 in the near future. It sounds very exciting and is worthy of being looked forward to, isn't it?

Chun-Wei Tsai and Ming-Chao Chiang
Sizihwan
October 2022

PART ONE

Fundamentals

CHAPTER ONE

Introduction

Optimization problems are ubiquitous. In general, in optimization problems the goal is to find the best solution from a large number of possible solutions[1] in a given solution space. One way to solve an optimization problem is to first *find* good solutions; then *estimate* these solutions; and finally use information thus obtained to *make a decision* between looking for other possible solutions[2] or simply announcing that the solution for the problem in question has been found. This technique for solving a problem can be traced back to "the crude life of our savage ancestors in the remote past." For instance, using rocks or animal bones as weapons for hunting large beasts seems to be a better solution than fighting with savage beasts bare-handed. The lesson we learn from the above is that problems have been solved as follows: (1) find possible solutions, (2) evaluate possible solutions to obtain qualities, and (3) compare qualities of solutions we have and determine whether or not to repeat the process until an appropriate solution is found. Although optimization problems may be different, a similar process can be used to find an applicable solution, thus making our life much easier.

The two types of problems that we often confront in our daily lives are: (1) those that can be solved intuitively and (2) those that cannot be solved intuitively. The first type is, of course, very simple and can often be solved intuitively or reflexively. In our daily lives, we pull our hands back immediately when our fingers touch fire neglectfully. Since this example has only two possible solutions—either pull our hands back or keep touching the fire—we have to quickly make a good decision without thinking for a long time to avoid being in jeopardy. In contrast to the problems that can be solved intuitively, many complex and/or large-scale optimization problems we face in our daily lives cannot be solved intuitively because all these problems have a common characteristic—a very large solution space that is composed of a huge number of candidate solutions for the optimization problem in question—and there is simply no way to know which one of these solutions is the best intuitively, because it is like looking for a needle in a haystack. Although the best solution can be pinpointed from a large

---

[1] Possible solutions are also referred to as feasible or candidate solutions.
[2] Solutions that are either around the current location or in other regions in the solution space.

number of possible solutions by checking each of the solutions one by one, it may take an unreasonable amount of time for a complex optimization problem.

A good example is the so-called traveling salesman problem (TSP) (Applegate et al., 2007; Lin, 1965). This well-known complex optimization problem is defined as follows: Given a set of cities and the distance between each pair of cities, find the shortest path to visit each city and then return to the city of departure. This kind of problem can often be found in our daily lives in different forms. For instance, a truck driver of an express home delivery service typically has to drive to the place of dispatch in the departure city c_1 to pick up the goods before they are delivered to different customers in cities c_2, c_3, and c_4. Finally, the truck driver has to drive back to city c_1. Table 1.1 shows the three possible distinct paths r_1, r_2, and r_3.

Table 1.1 Possible routes for cities c_1, c_2, c_3, and c_4.

Path	Departure city		Second city		Third city		Fourth city		Destination city
r_1	c_1	→	c_2	→	c_3	→	c_4	→	c_1
r_2	c_1	→	c_3	→	c_2	→	c_4	→	c_1
r_3	c_1	→	c_3	→	c_4	→	c_2	→	c_1

Fig. 1.1 shows graphically the delivery order of cities c_2, c_3, and c_4, which may affect the efficiency of a truck driver in terms of the total delivery time or cost. Finding the shortest or lowest-cost route is a critical problem for such a company. It seems that this problem can be easily handled. Nevertheless, we will soon realize that there are in fact a total of $(n-1)!/2$ possible routes if we always depart from and return to city c_1, where n is the number of cities.

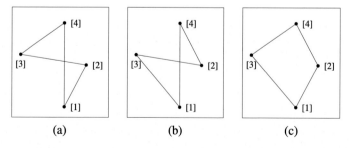

(a)   (b)   (c)

Figure 1.1 Different paths for a truck driver in the case of four cities. (a) Path 1 (r_1). (b) Path 2 (r_2). (c) Path 3 (r_3).

Table 1.2 shows that as the number of cities increases, the number of possible routes will grow explosively. There are 12 possible routes for five cities, 60 possible routes for six cities, and so forth. For such a complex problem, despite the fact that intuition cannot guarantee that the best solution will be found, it is reasonable for human beings to give it a try, especially when in a hopeless tangle.

Table 1.2 Number of possible routes for an *n*-city TSP.

Number of cities	Number of possible routes
4	3
5	12
6	60
7	360
8	2520
9	20,160
10	181,440
11	1,814,400

Now the following question arises: Can modern computers find the best solution of a TSP in a reasonable time? Eventually, all we know is that the advance of computer technology has witnessed many "impossibles" being made "possibles" in many different fields. The answer is yes and no! Computers are indeed powerful and useful for computations that have to be repeated again and again, such as checking for a large number of possible routes in a short time. However, letting the computer list all possible routes of a TSP before picking the best route may not be a feasible strategy, even with the fastest computer in the world today, when the number of cities is high. The reason is simple. With a "large" problem, the number of possible routes may become an astronomical number. In this case, even modern computers may not be able to list all of them in "a reasonable time," let alone check all of them for the best solution, again, in a reasonable time. This TSP example demonstrates that using only the computational ability of a computer is not guaranteed to solve complex problems that we confront efficiently. Accordingly, modern computer systems have typically been designed for integration with smart or intelligent methods to make it possible for them to solve complex problems in a reasonable time. Deep Blue developed by IBM that beats Kasparov in a chess game (Silver, 2012) and AlphaGo developed by Google (Silver et al., 2016) that beats Lee Sedol in a Go game are two representative examples to show the possibilities of computers with artificial intelligence. AlphaGo, in addition to deep learning algorithms, uses the so-called minimax and alpha-beta pruning methods to reduce computations that are essentially irrelevant to make it possible for such a system to find a "good decision" in a Go game in a reasonable time. This illustrates the importance of artificial intelligence methods in modern computer systems.

To date, computers have been successfully applied to not only chess but also many other aspects in our daily lives, such as weather analysis (Jain, 1991), seismic analysis (Mavko et al., 2009), customer analysis (Punj and Stewart, 1983), and even DNA analysis (Lander et al., 1987). Similar to the aforementioned applications, we can now enjoy 3D Avatar in a cinema as a result of the advance of computer technology, such as much faster computing power, much larger storage space, much smaller devices, much more accurate predictions, and much more accurate services. These results make us believe that we may be able to solve complex problems in a reasonable time in the forthcoming future by using better hardware, software, systems, and algorithms.

1.1. Why metaheuristic algorithms

Now, let us move back to the question if we can find the best solution from all feasible solutions of a complex optimization problem efficiently via a search algorithm. The reality is that modern computers are incapable of solving all optimization problems to their full generality because most optimization problems (Blum and Roli, 2003) are complex and large-scale problems that are either NP-complete or NP-hard. For most complex optimization problems, modern computers are incapable of finding the best solution using exhaustive search (ES) because the number of possible candidate solutions is simply way too large to be checked in a reasonable time. As long as it is impossible to check all solutions in the solution space of a complex optimization problem in a reasonable time, metaheuristic algorithms[3] (Glover and Kochenberger, 2003) provide a possible solution to the problem in the sense that an approximate, or even optimal, solution can be found in a reasonable time.

Since metaheuristics provide an efficient way (in terms of computation time) to search for an approximate solution of a complex problem,

---

[3] Throughout this book, we will use the term "metaheuristics" as a plural noun and the term "metaheuristic" as an adjective.

several metaheuristic algorithms have been presented since the 1960s or even earlier, namely, simulated annealing (SA) (Černý, 1982; Kirkpatrick et al., 1983), tabu search (TS) (Glover, 1986, 1989, 1990a; Glover and Laguna, 1997), genetic algorithm (GA) (Goldberg, 1989; Holland, 1962, 1975), differential evolution (DE) (Storn and Price, 1997), ant colony optimization (ACO) (Dorigo and Gambardella, 1997; Dorigo et al., 1996; Dorigo and Stützle, 2004), and particle swarm optimization (PSO) (Engelbrecht, 2006; Kennedy and Eberhart, 1995). Successful results and approaches using metaheuristics can also be found in many studies (Blum and Roli, 2003; Glover and Kochenberger, 2003). Because these metaheuristics are designed and developed for different purposes, none of them beats all the others for all kinds of optimization problems (Wolpert and Macready, 1997). This implies that each of the aforementioned metaheuristic approaches has its pros and cons and thus may perform well for some optimization problems but not for others.

A critical question is, how do we choose an applicable metaheuristic algorithm for an optimization problem in question? This can be summarized as a research issue on how these metaheuristics are implemented and compared for an optimization problem. For beginners in this research domain, a lot of time will be spent on programming. The main research issues in the implementation of metaheuristics can be summarized as follows:

1. Implementation of metaheuristics: It is generally difficult for beginners to realize a metaheuristic algorithm described in a research paper from scratch although it is well known that it is important for knowing the pros or cons of a metaheuristic algorithm.

2. Implementation of improved versions: Another research issue in the implementation of metaheuristics is that several studies attempt to enhance the performance of metaheuristics, especially the "add, remove, and modify operators" of a metaheuristic algorithm. Even though the difference between the improved version and the original version of a metaheuristic algorithm is very small, we usually still have to reimplement quite a few parts of the code.

3. Comparison between metaheuristics: In this situation, researchers usually want to ensure that all metaheuristics can be compared based on exactly the same criteria—not only consistent measurements but also the same simulation environment. Thus, the most important thing is to modularize the "common" parts of metaheuristics. We can then employ these common modules (or library) to make all the comparisons as fair as possible.

4. Combination of different metaheuristics: In this situation, the simplest and easiest way to implement a combination of different metaheuristics is apparently to reuse the implementation of the original metaheuristics. This is not that simple usually because the implementation of each operator generally depends on other relevant operators of the same metaheuristic algorithm. These characteristics explain that it is very important to have a "common interface" for every operator of metaheuristics.

5. Applying metaheuristics to different optimization problems: Another important issue for most studies on metaheuristics is the way the problem in question is "represented," because different optimization problems may require different representations for their solution. In addition to the representation of different optimization problems, most of the operators also need to be modified.

These observations show that some breaches in programming are simply because the loose coupling issues are not taken into account. This can be attributed to three major reasons: (1) Most implementations of a metaheuristic algorithm simply ignore the fact that its operators will be used by other metaheuristics or can be used to solve other optimization problems. (2) The terms used for the parameters of the same thing of different metaheuristics may be different. For example, the term chromosome is used for GA while the term particle is used for PSO to represent the same thing, i.e., the solution. (3) The design concepts of different metaheuristics may differ, meaning that their implementations are usually very different, especially for the beginners in this research domain.

Our observation shows that from the *theoretical perspective*, to simplify the implementation of metaheuristics, the very first thing that we have to do is to come up with a unified framework for most of the metaheuristics so that all of them can be included in the same framework. To provide a better way to implement metaheuristics and their variants from the *practical perspective*, the basic idea to implement a unified framework for metaheuristics will be presented in this book. How to implement this framework and how to use it to develop more efficient and effective metaheuristics will then be presented. This book will also provide a discussion on advanced research issues for simulation results, such as diversification vs. intensification, convergence trends, and statistical analysis. As noted previously, this book begins with the fundamentals of metaheuristics from the perspectives of theory and practice by using a novel unified framework for metaheuristics. The details of this framework, its applications, and advanced research issues will then be given. In summary, the ultimate goal of this book is to facilitate the lives of beginners in this research domain.

For instructors: This book can be used as a textbook for a brief introduction to metaheuristics, to support new students in this research domain. This means that after reading this book, most students will understand not only the basic ideas of this discipline but also how the algorithms are implemented to solve optimization problems of interest.

For researchers and students: This book is also suitable for self-study, especially for graduate students wanting to build background knowledge in this field by themselves. Of course, for students who have some rudimentary knowledge of this field, this book can also be used to enhance their programming skills, especially for implementing metaheuristics.

1.2. Organization of this book

This book consists of 20 chapters and is divided into two parts, as shown in Fig. 1.2. The first part is aimed at well-known optimization problems in which metaheuristics play an important role. Chapters 1–4 are focused on the basic ideas of optimization problems and traditional methods for solving these problems. Chapters 5–10 describe in detail six well-known metaheuristic algorithms—namely, SA, TS, GA, ACO, PSO, and DE—for solving the problems given in Chapter 2. The second part is aimed at some advanced technologies to enhance the performance of metaheuristic algorithms. It begins with solution encoding and initialization, followed by some ways to redesign the essential operators of a metaheuristic algorithm in Chapters 11–13. Chapters 14–18 focus on ways to redesign the whole search strategy of a metaheuristic algorithm. Chapter 19 shows some interesting and promising applications using metaheuristic algorithms. The last chapter, Chapter 20, gives a brief review on the development of metaheuristic algorithms in recent years and points out some future research directions. Appendix A describes some additional technologies to show how the performance of a metaheuristic algorithm is measured and analyzed. Appendix B gives examples to show how a metaheuristic algorithm is implemented in Python.

To make the life of the audience of this book easier, as shown in Fig. 1.3, the presentation of most sections will follow the same order—a brief introduction, the main contents, the simulation results, and a further discussion. The arrow lines from Section 2 to Section 1 and from Section 3 to Section 2 indicate that these parts are the suggested learning path to understand

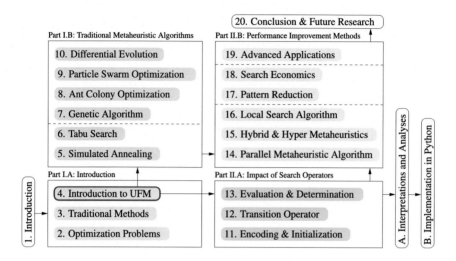

Figure 1.2 Organization of this book.

the deep meaning behind the search algorithms described in each part. The book can be read chapter by chapter and for each chapter section by section. You can always move back to the previous step to better understand the prerequisite knowledge of the current step.

Figure 1.3 The structure of each section in this book.

Section 1 is an introduction, the focus of which is on the topic to be presented in that section, which may contain the basic ideas, motivation, pseudocode, simple examples, and so forth, so that the audience of the book can quickly understand the topic of each section. In other words, this section is aimed at helping the audience of the book build up the prerequisite knowledge and background, e.g., an introduction to SA in Section 5.1. Section 2 provides the main contents, the focus of which is either on a detailed implementation of the metaheuristic algorithm or on an in-depth introduction. For instance, in Section 5.2, a description of the implementation of SA is given. Section 3 shows the simulation results of the metaheuristic algorithm discussed in the first two sections. Section 4 provides a further discussion, the focus of which is on the related work of the metaheuristic algorithm discussed in the first two sections and issues that a researcher needs to take into account and on pointing out some promising research issues. Note that the focus of the first part is on the fundamentals of various metaheuristic algorithms; therefore, Section 1 is typically just a brief introduction to make the audience more familiar with the metaheuristic algorithm to be discussed in the chapter, while Section 4 is a brief discussion on some related work to make the audience aware of other possible solutions and research directions. Since the book describes some advanced technologies, we suggest that you skim it over for the first reading, perhaps coming up with some questions and ideas for the moment, and then come back to read it again once you finish reading the entire book. The answers to most of the questions with which you came up earlier will suddenly become clarified. In contrast to the first part, the focus of the second part is on advanced technologies; thus, Section 1 is on related work because we strongly believe that you should now already have enough background, questions, and ideas in the metaheuristic algorithms and are able to learn many new technologies. On the other hand, Section 4 in the second part will be nothing more than a summary of promising research issues.

CHAPTER TWO

Optimization problems

Optimization problems have been part of our daily lives for years. That is why we have been constantly looking for applicable methods for solving these problems. Before we proceed to develop a useful method for solving optimization problems, we have to be aware of concepts that are relevant to the problem in question, such as *objective function*, *constraint*, and *solution*. This information makes it possible to precisely define the optimization problem. This chapter will begin with the definition of an optimization problem in Section 2.1, followed by some examples to illustrate the combinatorial and continuous optimization problems in Sections 2.2 and 2.3, respectively. Finally, a summary of the chapter is given in Section 2.4.

2.1. Problem definition

A clear and precise definition is generally needed to describe the characteristics of a problem before we proceed to develop a search algorithm for solving it. A typical definition for the optimization problem similar to those given by some studies (Blum and Roli, 2003; Grötschel et al., 1981; Papadimitriou and Steiglitz, 1982) is as follows.

Definition 1 (Optimization problem $\mathbb{P}$). An optimization problem $\mathbb{P}$ is to find the optimal value, possibly subject to some constraints, out of all possible solutions; that is,

$$\underset{s\in\mathbf{A}}{\mathrm{opt}}f(s), \qquad \text{subject to } \forall c_i(s) \odot b_i, \ i = 1, 2, \ldots, m,$$

where:
- opt is either min (for minimization) or max (for maximization),
- s is a candidate solution,
- $\mathbf{A}$ and $\mathbf{B}$ are the domain and codomain of the problem $\mathbb{P}$, namely, $\mathbf{A}$ is the set of all possible solutions and $\mathbf{B}$ is the set of all possible outcomes of the objective function,
- $f(s) : \mathbf{A} \to \mathbf{B}$ is the objective function,
- $c_i(s) \odot b_i$ is the ith constraint, and
- $\odot$ is $<$, $>$, $=$, $\leq$, or $\geq$.

Handbook of Metaheuristic Algorithms
https://doi.org/10.1016/B978-0-44-319108-4.00015-0

The optimal solution and optimal value of the optimization problem $\mathbb{P}$ are defined as follows.

Definition 2 (Optimal solution s^*). The optimal solution is a solution, out of all feasible candidate solutions of the optimization problem $\mathbb{P}$, that gives the optimal value:

$$f(s^*) = \underset{s\in\mathbf{A}}{\mathrm{opt}}f(s), \forall c_i(s) \odot b_i, \; i = 1, 2, \ldots, m.$$

Definition 3 (Optimal value f^*). If the optimal solution s^* for the problem $\mathbb{P}$ exists, then the optimal value f^* is defined as

$$f^* = f(s^*).$$

Now that we have Definition 1, the minimization problem of minimizing $f(s)$ subject to some constraints can be defined as

$$\underset{s\in\mathbf{A}}{\min}f(s), \qquad \text{subject to } \forall c_i(s) \odot b_i, \tag{2.1}$$

while the maximization problem of maximizing $f(s)$ subject to some constraints can be defined as

$$\underset{s\in\mathbf{A}}{\max}f(s), \qquad \text{subject to } \forall c_i(s) \odot b_i. \tag{2.2}$$

Other optimization problems can be defined similarly. These examples indicate that Definition 1 will be useful for defining most of the optimization problems we are facing. With the definition of optimization problems at hand, we can now think about how to find the solution of any optimization problem both efficiently and effectively. The optimization problems we are facing can be classified into two categories based on the variable type of the solution space (Papadimitriou and Steiglitz, 1982), namely, solutions encoded as *discrete* variables and solutions encoded as *continuous* variables. The former is referred to as the combinatorial optimization problem (COP), while the latter is referred to as the continuous optimization problem. This classification is important when using metaheuristics to solve an optimization problem because it has something to do with not only the way the solutions are encoded but also the way the search operators are designed for the search processes.

For a COP (Blum and Roli, 2003), we are looking for the best solution from a finite set—usually either a set of integer numbers or a subset, a permutation, or even a graph structure of something. For a continuous optimization problem (Boyd and Vandenberghe, 2004), unlike the discrete problem, we are looking for a set of real numbers that not only satisfy all the given constraints but also give the best solution. Note that the solution space and feasible set in mathematical optimization represent the set of all possible solutions for the problem in question (see also https://en.wikipedia.org/wiki/Feasible_region).

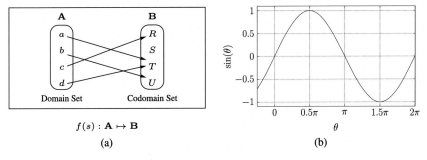

Figure 2.1 Two examples for illustrating the definition of optimization problems from the perspectives of (a) set and (b) function.

Fig. 2.1 gives two examples to show the relationships between the solution s, the optimization function $f()$, the domain $\mathbf{A}$, and the codomain $\mathbf{B}$, which can be mapped to the *discrete* and *continuous* optimization problems, respectively. As shown in Fig. 2.1(a), the domain is $\mathbf{A} = \{a, b, c, d\}$, while the codomain is $\mathbf{B} = \{R, S, T, U\}$. Given $\mathbf{A}$, $\mathbf{B}$, the objective function $f()$, and the possible solution s, we can easily understand solutions that map elements in the domain to elements in the codomain. For example, if $s = a$, the function $f()$ will map $a \in \mathbf{A}$ to $T \in \mathbf{B}$. Unlike the example given in Fig. 2.1(a), which is aimed to explain the optimization problem from the perspective of set, the example given in Fig. 2.1(b) is aimed to explain the optimization problem from the perspective of function. This example shows that a sine function is used as the objective function $f()$, the domain $\mathbf{A}$ is the whole x-axis, and the codomain $\mathbf{B}$ is part of the y-axis. It can be easily seen from Table 2.1 how a θ in the domain $\mathbf{A}$ is mapped to $\sin(\theta)$ in the codomain $\mathbf{B}$; for example, $\theta = 0.50\pi$ is mapped to $\sin(\theta) = 1.00$. From these two examples, it can be easily seen that the number of possible candidate solutions of a continuous optimization problem depends on the precision. If the interval between two different candidate solutions of a continuous optimization problem can be infinitesimal, then the number of candidate solutions in the solution space of such a problem will be uncountably infinite.

Table 2.1 Relation between θ and $\sin(\theta)$.

θ	0.00	0.25π	0.50π	0.75π	1.00π	1.25π	1.50π	1.75π	2.0π
$\sin(\theta)$	0.00	0.70	1.00	0.70	0.00	−0.70	−1.00	−0.70	0.0

2.2. Combinatorial optimization problems

The goal of COPs is to find the optimal solution from a finite set or a countably infinite set of solutions. The possible solutions of a COP are generally "discrete" or can be discretized.

2.2.1 The one-max and 0-1 knapsack problems

The one-max problem (Schaffer and Eshelman, 1991) is a very simple COP, which can be regarded as a touchstone to measure the performance of a new search algorithm.

Definition 4 (The one-max problem $\mathbb{P}_1$). Given a binary string[1] $s = (s_1, s_2, \ldots, s_n)$, the one-max problem is to find the solution that maximizes the number of ones in s:

$$\max_{s \in \mathbf{A}} f(s) = \sum_{i=1}^{n} s_i, \qquad \text{subject to } s_i \in \{0, 1\}.$$

The optimal solution of this problem is that all the subsolutions assume the value 1; i.e., $s_i = 1$ for all i. For instance, the optimal solution for $n = 4$ is $s^* = (1111)$. As shown in Fig. 2.2, the objective value of a possible solution $s = (0111)$ can be easily calculated. It is a count of the number of ones in the solution s, as the objective function $f(s) = \sum_{i=1}^{n} s_i$ described in Definition 4 states; that is, $f(s) = f((0111)) = 0 + 1 + 1 + 1 = 3$.

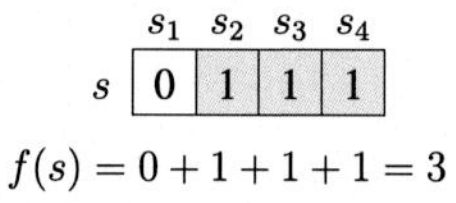

$$f(s) = 0 + 1 + 1 + 1 = 3$$

Figure 2.2 How the objective value of a one-max problem is calculated.

Fig. 2.3 shows a possible landscape of the solution space of a one-max problem of size $n = 4$. If a greedy search algorithm is used and is allowed to

---

[1] It can also be represented as a vector, a sequence, a tuple, or anything else that makes sense and can be handled efficiently.

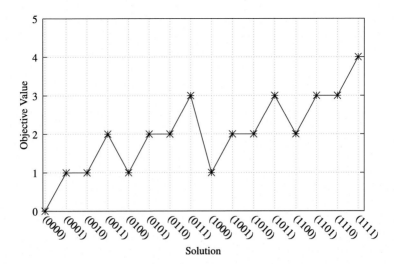

Figure 2.3 A possible landscape of the solution space of a one-max problem of size $n = 4$.

randomly add one to or subtract one from the current solution s to create the next possible solution v for solving the one-max problem, that is, it is allowed to move one and only one step to either the left or the right of the current solution in the landscape of the solution space described in Fig. 2.3, this makes it very difficult for the greedy search algorithm to find the optimal solution s^*. For instance, provided that the current solution is $s = (0111)$, the greedy search algorithm will move to either $v = (0110)$ or $v = (1000)$, none of which is superior to the current solution, so the search process will eventually get stuck at the current solution in this case. Apparently, without knowledge of the landscape of the solution space, the search process will easily get stuck in the peaks of this solution space. Hence, most researchers prefer using the one-max problem as an example because it is easy to implement and also because it can be used to prove if a new concept for a search algorithm is correct.

Of course, the one-max problem can be easily transformed to other optimization problems. A good example is the 0-1 knapsack problem, for which the value of each bit s_i in a bit string s also needs to be determined. The main difference between the one-max problem and the 0-1 knapsack problem is that for the 0-1 knapsack problem a value and a weight are associated with each subsolution s_i and a feasible solution must satisfy the constraint weight W. The 0-1 knapsack problem can be defined as follows.

Definition 5 (The 0-1 knapsack problem $\mathbb{P}_2$). Given a binary string $s = (s_1, s_2, \ldots, s_n)$ and a maximum weight capacity W, the 0-1 knapsack problem is to maximize the value of the objective function $f(s)$ under the constraint W:

$$\max_{s \in A} f(s) = \sum_{i=1}^{n} s_i v_i, \qquad \text{subject to } w(s) = \sum_{i=1}^{n} s_i w_i \leq W, s_i \in \{0, 1\},$$

where:
- v_i is a value associated with s_i and
- w_i is a weight associated with s_i.

Let us take a closer look at these two problems. It can be easily seen that the development of a new search algorithm for solving the one-max problem can be easily adapted to solve the 0-1 knapsack problem by adding v, w, and W and modifying the objective function, as described in Definition 5.

$$s \quad \begin{array}{|c|c|c|c|} \hline s_1 & s_2 & s_3 & s_4 \\ \hline 0 & 1 & 1 & 1 \\ \hline \end{array}$$

$$v \quad \begin{array}{|c|c|c|c|} \hline v_1 & v_2 & v_3 & v_4 \\ \hline 1 & 2 & 3 & 4 \\ \hline \end{array}$$

$$w \quad \begin{array}{|c|c|c|c|} \hline w_1 & w_2 & w_3 & w_4 \\ \hline 4 & 3 & 2 & 1 \\ \hline \end{array}$$

$$f(s) = (0 \times 1) + (1 \times 2) + (1 \times 3) + (1 \times 4) = 9$$
$$w(s) = (0 \times 4) + (1 \times 3) + (1 \times 2) + (1 \times 1) = 6$$

Figure 2.4 How the objective value of the 0-1 knapsack problem is calculated.

As shown in Fig. 2.4, suppose there are four items s_i, for $1 \leq i \leq 4$, each of which is associated with a weight w_i and a value v_i. Further suppose that the maximum weight capacity W of the sack is 7. For this example, a possible solution is $s = (0111)$ that satisfies the given constraint W because the total weight w is 6, which is less than $W = 7$. Although the way the solution to the 0-1 knapsack problem is encoded is exactly the same as that for the one-max problem, the objective functions and thus the objective values are different. The objective function for the one-max problem is defined as $\sum_{i=1}^{n} s_i$ and thus the objective value is calculated as $\sum_{i=1}^{n} s_i = 0+1+1+1 = 3$, while the objective function for the 0-1 knapsack problem is defined as $\sum_{i=1}^{n} s_i v_i$ and thus the objective value is calculated as $\sum_{i=1}^{n} s_i v_i = 0 \times 1 + 1 \times 2 + 1 \times 3 + 1 \times 4 = 9$.

2.2.2 The B2D and deceptive problems

With a minor modification, the solution space of the one-max problem can be simplified as the solution space of another optimization problem. Let us call the modified version of the one-max problem the binary to decimal (B2D) problem. One way to define the B2D problem is as follows.

Definition 6 (The B2D-1 problem $\mathbb{P}_3$). The B2D-1 problem is to maximize the value of the objective function of a binary string $s = (s_1, s_2, \ldots, s_n)$:

$$\max_{s \in \mathbf{A}} f(s) = \text{B2D}(s), \qquad \text{subject to } s_i \in \{0, 1\},$$

where:
- the B2D function will return the decimal value of s.

As shown in Fig. 2.5, the solution space is much simpler than that in Fig. 2.3, so it is expected that even a simple greedy algorithm will not get stuck at some solutions in finding the best solution of this problem during the convergence process, which is $s^* = (1111)$ and which gives the objective value 15, i.e., $f(s^*) = 15$.

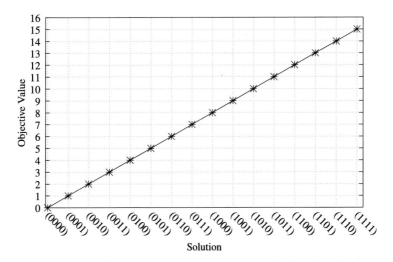

Figure 2.5 A possible landscape of the solution space of a B2D-1 problem.

Fig. 2.5 provides a possible landscape to show the relationship between candidate solutions that assumes that the search algorithm in question can only generate a new candidate solution v by moving the current solution s to the left or to the right by decreasing or increasing B2D(s) by 1. This implies that there are only two possible next states (candidate solutions) that can be generated from the current solution except for solutions (0000) and (1111), which can only be moved to the right and to the left, respectively. For example, the next candidate solution that can be generated from the current solution (0010) is either (0001) or (0011). These two candidate solutions (0001) and (0011) are the nearest neighbors of the current solution (0010) compared with other candidate solutions because both of them need only one hop from solution (0010), but not the others.

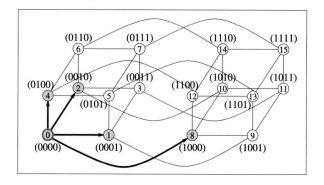

Figure 2.6 Another landscape of the solution space of a B2D-1 problem.

If another search algorithm can generate a new candidate solution by randomly inverting (flipping) one of the subsolutions of the current solution, the number of possible states of the new candidate solution will be n, where n is the number of subsolutions. As shown in Fig. 2.6, there are four possible states for the new candidate solution of each solution. In this example, if the current solution is (0000), then the new candidate solution can be (0001), (0010), (0100), or (1000). This example shows that the landscape of a solution space will be affected by the problem definition, solution encoding, and search algorithm.

To better understand the performance of a search algorithm, we can make a minor change to the objective function of this problem and call it B2D-2 (Goldberg, 1989; Michalewicz, 1996).

Definition 7 (The B2D-2 problem $\mathbb{P}_4$). The B2D-2 problem again can be defined as a problem that aims to maximize the value of the objective function of a binary string $s = (s_1, s_2, \ldots, s_n)$ except that the problem is now deceptive:

$$\max_{s \in \mathbf{A}} f(s) = |\text{B2D}(s) - 2^{n-2}|, \qquad \text{subject to } s_i \in \{0, 1\}, n > 2,$$

where:
* the B2D function will return the decimal value of s.

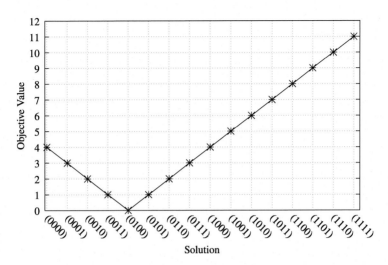

Figure 2.7 The landscape of the solution space of a B2D-2 problem (a deceptive problem).

As shown in Fig. 2.7, the minimum value of the candidate solutions is 0, while the maximum value (the global optimal solution) of the candidate solutions is 11. The objective value of the solution $s = (0000)$ is 4, but it is a local optimum as far as this example is concerned, for if the greedy algorithm starts with a solution that falls in the range of $s = (0000)$ to $s = (0100)$ and is allowed to either add one to or subtract one from the current solution, then it will get stuck at this local optimum. A similar example can also be found in (Michalewicz, 1996), which was used to explain the differences between hill climbing (a greedy search algorithm) and simulated annealing (a metaheuristic algorithm). This kind of optimization problem is often referred to as a deceptive problem (Goldberg, 1989), which is typically used to test whether a search algorithm is capable of escaping local optima or not. As summarized in (Michalewicz, 1996), for hill climbing and simulated annealing, the former will fall into a local optimum if it starts with particular values, but the latter has the chance of escaping the local optimum.

2.2.3 The traveling salesman problem (TSP)

The traveling salesman problem (TSP) (Applegate et al., 2007; Lin, 1965) is a very traditional COP. Since the number of permutations of the routing order for the n given cities of the TSP is $(n-1)!/2$, it is impossible in practice to calculate all the permutations in a reasonable amount of time, especially when the number of cities is large. With these concerns and constraints as well as Definition 1 in mind, we can then define the TSP as follows.

Definition 8 (The traveling salesman problem $\mathbb{P}_5$). Given n cities and the distance between each pair of cities, seek a tour (a closed path) with minimum distance that visits each city in sequence once and only once and then returns directly to the first city. Formally,

$$\min_{s\in c_\pi}f(s) = \left[\sum_{i=1}^{n-1} d(c_{\pi(i)}, c_{\pi(i+1)})\right] + d(c_{\pi(n)}, c_{\pi(1)}),$$

where $c_\pi = \{\langle c_{\pi(1)}, c_{\pi(2)}, \ldots, c_{\pi(n)}\rangle\}$, that is, all permutations of the n cities.

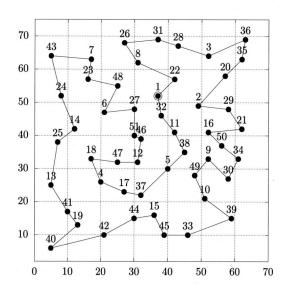

Figure 2.8 The optimal solution of eil51.

As shown in Fig. 2.8, the benchmark eil51 (Reinelt, 1991; TSPLIB, 1995) has 51 cities, where the first city[2] is denoted by ⊚ while the other cities are denoted by •. The optimal tour is denoted by the closed path 1 → 22 → 8 → 26 → 31 → 28 → 3 → 36 → 35 → 20 → 2 → 29 → 21

[2] The first city is the city where the tour starts.

$\rightarrow$ 16 $\rightarrow$ 50 $\rightarrow$ 34 $\rightarrow$ 30 $\rightarrow$ 9 $\rightarrow$ 49 $\rightarrow$ 10 $\rightarrow$ 39 $\rightarrow$ 33 $\rightarrow$ 45 $\rightarrow$ 15 $\rightarrow$ 44 $\rightarrow$ 42 $\rightarrow$ 40 $\rightarrow$ 19 $\rightarrow$ 41 $\rightarrow$ 13 $\rightarrow$ 25 $\rightarrow$ 14 $\rightarrow$ 24 $\rightarrow$ 43 $\rightarrow$ 7 $\rightarrow$ 23 $\rightarrow$ 48 $\rightarrow$ 6 $\rightarrow$ 27 $\rightarrow$ 51 $\rightarrow$ 46 $\rightarrow$ 12 $\rightarrow$ 47 $\rightarrow$ 18 $\rightarrow$ 4 $\rightarrow$ 17 $\rightarrow$ 37 $\rightarrow$ 5 $\rightarrow$ 38 $\rightarrow$ 11 $\rightarrow$ 32 $\rightarrow$ 1. That is, the optimal tour starts at city 1, visits each city in sequence, and then returns to city 1. As defined above, each solution is a permutation of the given cities, and the objective function is the overall distance traveled. In this case, the overall distance of the optimal tour for eil51 is 429.983, meaning that it is the shortest distance possible. It is interesting to note that according to (Reinelt, 2007), the overall distance of the optimal tour for eil51 is 426. This difference is because in (Reinelt, 1995, 1991), the Euclidean distance between each pair of cities is computed by rounding the real distance up to the nearest integer by the "nint()" function. In summary, as one of the most well-known COPs that can be easily transformed to other optimization problems (e.g., the network packet routing problem) as well as an NP-hard problem, the TSP has been widely used as a yardstick to measure the performance of search algorithms.

2.3. Continuous optimization problems

Unlike the COP, the possible solutions for a continuous optimization problem are typically "uncountably infinite." This means that the number of solutions in the solution space is tantamount to the number of real values in the given space, that is, infinite. The domain for the continuous optimization problem is normally a continuous space, be it a single-objective optimization problem (SOP)[3] or a multi-objective optimization problem (MOP).

2.3.1 The single-objective optimization problem

Based on Definition 1, we can also define an SOP, which is typically in the form of a test function with its solution encoded as a set of real numbers.

Definition 9 (The single-objective optimization problem $\mathbb{P}_6$). Given a function and a set of constraints, the SOP is to find the optimal value, subject to the constraints, out of all the solutions of the function:

$$\operatorname*{opt}_{s \in \mathbf{R}^n} f(s), \qquad \text{subject to } c_i(s) \odot b_i, \ i = 1, 2, \ldots, m,$$

where:

[3] It is also referred to as a single-objective real-parameter or global function optimization problem.

- $\mathbf{R}^n$ and $\mathbf{R}$ are the domain and codomain, respectively,
- $f(s) : \mathbf{R}^n \rightarrow \mathbf{R}$ is the objective function to be optimized,
- $c_i(s) : \mathbf{R}^n \rightarrow \mathbf{R} \odot b_i$, $i = 1, 2, \ldots, m$, are the constraints, and
- opt and $\odot$ are as given in Definition 1.

Now that we have Definition 9, we can proceed to describe optimization problems with n variables, which can be regarded as searching for a solution in an n-dimensional solution space. The Ackley function[4] (Ackley, 1987; Bäck, 1996) can be used as an example of the continuous optimization problem.

Definition 10 (The Ackley optimization problem $\mathbb{P}_7$). We have

$$\min_{s \in \mathbf{R}^n} f(s) = -20 \exp\left(-0.2\sqrt{\frac{1}{n}\sum_{i=1}^n s_i^2}\right) -$$

$$\exp\left(\frac{1}{n}\sum_{i=1}^n \cos(2\pi s_i)\right) + 20 + e,$$

$$\text{subject to } -30 \le s_i \le 30, \ i = 1, 2, \ldots, n.$$

It can be easily seen from Definition 10 that the Ackley function is evaluated in the hypercube $s_i \in [-30, 30]$, $i = 1, 2, \ldots, n$. The global optimum (minimum) of the Ackley function is $f(s^*) = 0$ located at $s^* = (0, 0, \ldots, 0)$. For instance, for $n = 2$, the global optimum (minimum) is $f(s^*) = 0$ located at $s^* = (0, 0)$. Fig. 2.9 shows that this function has many local optima, which makes it hard for the search algorithm to find the global optimum. In addition to the Ackley function, sphere, Rosenbrock, Rastrigin, and many other well-known functions can also be found on Wikipedia (at https://en.wikipedia.org/wiki/Test_functions_for_optimization) that can be used to help you understand the performance of a metaheuristic algorithm. More detailed and advanced information of recent trends in the single-objective optimization problem research field can be found in (Jamil and Yang, 2013; Suganthan et al., 2005; Wu et al., 2017).

2.3.2 The multi-objective optimization problem

Unlike the SOP, the MOP takes into account two or more objectives at the same time for the optimization problem in question, which may even

---

[4] See also http://www.sfu.ca/~ssurjano/ackley.html.

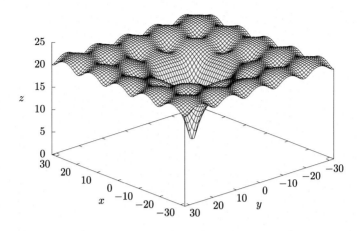

Figure 2.9 The landscape of the solution space of the Ackley function.

contradict each other. The multi-objective optimization problem, however, is more suitable to represent problems in the real world. For example, when buying a new car, the price and safety normally conflict with each other. Like the SOP, we can also define the MOP based on Definition 1.

Definition 11 (The multi-objective optimization problem $\mathbb{P}_8$). Given a set of functions and a set of constraints, the MOP is to find the optimal value or a set of optimal values (also called Pareto front), subject to the constraints, out of all possible solutions of these functions:

$$\operatorname*{opt}_{s \in \mathbf{R}^n}(f_1(s), f_2(s), \ldots, f_k(s)), \qquad \text{subject to } c_i(s) \odot b_i, \ i = 1, 2, \ldots, m,$$

where:
- $\mathbf{R}^n$ and $\mathbf{R}$ are the domain and codomain, respectively,
- $f_j(s) : \mathbf{R}^n \to \mathbf{R}$, $j = 1, 2, \ldots, k$, are the objective functions to be optimized,
- $c_i(s) : \mathbf{R}^n \to \mathbf{R}$, $i = 1, 2, \ldots, m$, are the constraints, and
- opt and $\odot$ are as given in Definition 1.

Here, we will use the Schaffer min-min global optimization problem[5] as an example to explain the characteristics of an MOP.

---

[5] See also https://en.wikipedia.org/wiki/Test_functions_for_optimization.

Definition 12 (The Schaffer min-min global optimization problem $\mathbb{P}_9$). We have

$$\min_{s \in \mathbf{R}^n} \begin{cases} f_1(s) = s^2, \\ f_2(s) = (s-2)^2, \end{cases} \qquad \text{subject to } s \in [-10^3, 10^3].$$

There typically exist no solutions that minimize all the objective functions of an MOP simultaneously. This means that a solution s_a may not be *better* than (dominate) another solution s_b for all objective functions because some of the objective functions of the MOP are in conflict.

Definition 13 (Domination). If we say that a solution s_a dominates another solution s_b, this means that $f_i(s_a) \leq f_i(s_b)$, $\forall i \in \{1, 2, \ldots, n\}$. In case $f_i(s_a) \leq f_i(s_b)$, $\neg \forall i \in \{1, 2, \ldots, n\}$, the solutions s_a and s_b cannot dominate each other.

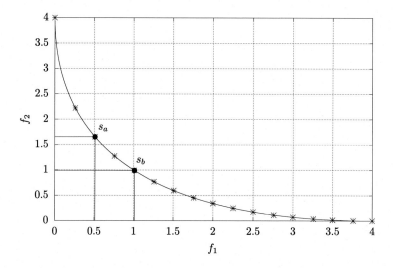

Figure 2.10 The Pareto front of the Schaffer min-min function.

To solve this kind of optimization problem, we normally want to find a set of solutions (also called the Pareto front) that do not dominate each other. As shown in Fig. 2.10, "$*$" represents solutions that are not dominated by all the other solutions. If we can find a sufficient number of such solutions, we can then draw a curve to approximate the Pareto front of such a problem. For example, the solutions s_a and s_b cannot dominate each other because $f_2(s_b) < f_2(s_a)$ while $f_1(s_a) < f_1(s_b)$. This means that s_b is a better solution as far as f_2 is concerned whereas s_a is a better solution as far as f_1 is concerned.

2.4. Summary

The very first thing to do to enter a new research domain is to understand the assumptions, conditions, and objectives of the problem. With these definitions at hand, it is easier to understand the definitions of a new optimization problem, which in turn makes it easier to enter a new research domain. Once we are capable of understanding the definitions of optimization problems, the next step is simply to use this knowledge to develop an applicable search algorithm to solve these optimization problems.

CHAPTER THREE

Traditional methods

In this chapter, we will turn our attention to two traditional methods for optimization problems, namely, the exhaustive search (ES) and hill climbing (HC) search methods, and use them for the one-max problem $\mathbb{P}_1$ and the B2D-2 problem $\mathbb{P}_4$ to demonstrate how they work. These two traditional methods are essentially the epitome of the early studies for solving optimization problems, and both of them are still used in modern computer systems. The basic ideas of these algorithms and how they are implemented are the main focus of this chapter. Their pros and cons will, of course, also be discussed based on simulation results.

3.1. Exhaustive search (ES)

3.1.1 The basic idea of ES

As mentioned in Chapter 2, many complex optimization problems exist in our daily lives. An intuitive method for solving these complex optimization problems is the so-called ES.[1] The basic idea of ES is to check *all* the candidate solutions in the solution space of the problem in question; therefore, it is always capable of finding the best solution for the problem in question. To explain how ES solves an optimization problem, we will use it to solve the one-max problem in this section. As shown in Fig. 3.1, the search strategy of ES can be summarized as follows: check the first possible solution x_1 in the solution space, then check the second possible solution x_2, then check the third x_3, and so on, until all possible solutions in the solution space are checked. The landscape of this example, as depicted in Fig. 3.1, shows that there are in total 16 possible solutions in the solution space because the number of feasible solutions for a one-max problem is 2^n, where n is the number of bits in a solution s. The symbols $\mathbb{S}$ and $\mathbb{E}$ in the figure indicate the starting point and the end point of the convergence process, respectively. In this example, the number of bits is $n = 4$; therefore, the number of feasible solutions is $2^4 = 16$. Fig. 3.1 further shows that there

---

[1] Exhaustive search is also referred to as full search or brute force.

are three local optima (x_4, x_8, and x_{12}) and one global optimum (x_{16}) in the solution space. Since ES will check all 16 possible solutions, it is guaranteed to find the best solution.

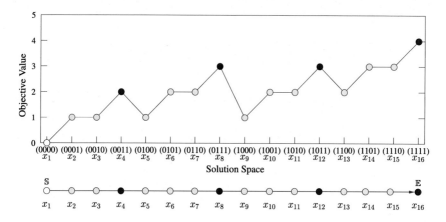

Figure 3.1 The search strategy of ES.

Algorithm 1 Exhaustive search.

1 Set the initial solution $s = (0, 0, \ldots, 0)$	I
2 $f_s = $ Evaluate(s)	E
3 $v = s$	
4 **While** the termination criterion is not met	
5 $\quad$ $v = $ GenNext(v)	T
6 $\quad$ $f_v = $ Evaluate(v)	E
7 $\quad$ If f_v is better than f_s	D
8 $\quad\quad$ $s = v$	
9 $\quad\quad$ $f_s = f_v$	
10 $\quad$ EndIf	
11 **End**	
12 **Output** s	O

As lines 1–2 of Algorithm 1 show, ES will start with the initial solution $s = (0, 0, \ldots, 0) = x_1$, followed by the function call Evaluate() to calculate the objective value of the initial solution s. As far as this algorithm is concerned, the function Evaluate() is used to calculate the objective value of a solution while the function GenNext() is used to generate the next candidate solution. For a one-max problem, the next candidate solution of (0000) is obviously (0001). The main loop of ES will check all candidate solutions one by one, as shown in lines 4–11. It will first call GenNext() to generate the next candidate solution v, as shown in line 5, and then call

Evaluate() to calculate the objective value f_v of v, as shown in line 6. As far as this algorithm is concerned, the objective value denotes the number of 1's in the solution, as stated in Definition 4. If the objective value f_v of the new candidate solution v is better than the objective value f_s of the old candidate solution s, then the values of s and f_s will be replaced by the values of v and f_v, respectively, as shown in lines 7–10. This process (lines 4–11) will be repeated until all the candidate solutions are checked. In this way, ES is guaranteed to find the optimal solution for the problem in question.[2]

3.1.2 Implementation of ES for the one-max problem

ES is very simple and thus very easy to implement for not only the one-max problem but also many other problems. The hard part for the implementation of the one-max problem is, instead, to generate all the possible solutions for a huge one-max problem, say, of size 100 or even larger. For a one-max problem of size $n = 100$, the number of candidate solutions is $2^n = 2^{100} \approx 1.2676506002 \times 10^{30}$. For $n = 1000$, the number of candidate solutions that ES has to search is $2^{1000} \approx 1.0715086072 \times 10^{301}$, which implies that the basic data types, such as `unsigned int`, `unsigned long`, and even `unsigned long long`, of most popular programming languages are simply not large enough to accommodate all these solutions.

The implementation of ES described herein is aimed to make it capable of checking all the possible candidate solutions of a one-max problem of any size. As shown in Fig. 3.2, for the one-max problem, the modules `main.cpp`, `search.h`, and `search.cpp` of ES correspond to the source files `es_main.cpp`, `es_search.h`, and `es_search.cpp`, respectively, residing in the directory `src/c++/0-es/1-onemax/`. The coding style of this program can be summarized as follows: The module "`es_main.cpp`" is used as the driver of the ES program, which is divided into two parts: the interface named "`es_search.h`" and the implementation named "`es_search.cpp`." This coding style will also be used for other (meta)heuristic algorithms in this book, meaning that the module "`main.cpp`" is responsible for invoking the search

---

[2] Keep in mind that the icon or icons to the far right of each line in the description of all the algorithms indicate what kind of operator it is. In brief, $\boxed{\text{I}}$ indicates it is an initialization operator; $\boxed{\text{T}}$ indicates it is a transition operator; $\boxed{\text{E}}$ indicates it is an evaluation operator; $\boxed{\text{D}}$ indicates it is a determination operator; and $\boxed{\text{O}}$ indicates it is an output operator. Moreover, a combo of these operators, such as $\boxed{\text{T}}\!\mapsto\!\boxed{\text{E}}$, indicates that it is a transition and evaluation operator, that is, an operator that performs first the transition function and then the evaluation function.

process by calling the *search* function defined in "search.cpp" using the interface specified in "search.h."

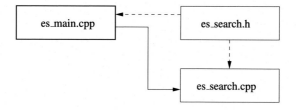

Figure 3.2 Relationship between modules of ES.

Note that most of the programs described in this book are written in C++ and are complied and tested on Linux, and the commands to compile, recompile, and run this particular program are as follows:

```
$ make clean dep all
$ ./es 10
```

The style of the above commands is also used for other programs of metaheuristic algorithms described in this book. The first command in this example is used to (1) clean up all the files generated in the previous run of this command, (2) generate dependencies (see GNU make for more details), and (3) recompile the program. The second command is used to run the search algorithm. In this example, the ES will be run for a one-max problem of size 10 bits. Now, let us move back to the implementation of ES. The driver of ES, named es_main.cpp, is given in Listing 3.1. Line 3 of Listing 3.1 shows that es_main.cpp will include es_search.h first. Lines 12–15 show that this program will exit immediately after printing a message to show how the program should be invoked if an invalid size for the one-max problem is provided by the user of the program; otherwise, it will call the es_search() function in line 22 to search for the optimal solution of the one-max problem by checking all the candidate solutions in the solution space. As noted previously, the number of candidate solutions is 2^n, where n is the size of the one-max problem, that is, the number you enter when starting the program. Lines 21 and 23 of this program are used to keep track of the computation time of ES by recording the start and end times of the search process of ES, respectively. After that, the computation time of ES is calculated and displayed, as shown in lines 25–26.

Listing 3.1: src/c++/0-es/1-onemax/es_main.cpp: The driver of ES.

```
 1 #include <iostream>
 2 #include <ctime>
 3 #include "es_search.h"
 4
 5 using std::cerr;
 6 using std::endl;
 7 using std::atoi;
 8 using std::clock;
 9
10 int main(int argc, char *argv[])
11 {
12     if (argc != 2 || atoi(argv[1]) <= 0) {
13         cerr << "Usage: ./es <n> where <n> is a positive integer." << endl;
14         return 1;
15     }
16
17     const int n = atoi(argv[1]);
18
19     cerr << "# number of bits: " << n << endl;
20
21     const clock_t begin = clock();
22     const int rc = es_search(n);
23     const clock_t end = clock();
24
25     const double cpu_time = static_cast<double>(end - begin) / CLOCKS_PER_SEC;
26     cerr << "# CPU time used: " << cpu_time << " seconds." << endl;
27
28     return rc;
29 }
```

The interface of ES, named es_search.h, is given in Listing 3.2. The main purpose of this module is to specify the interface to the es_search() function defined in the implementation of ES, named es_search.cpp. In this case, as shown in line 4, one and only one function—namely, es_search()—is declared to be called by the driver of ES.

Listing 3.2: src/c++/0-es/1-onemax/es_search.h: The interface of ES.

```
1 #ifndef __ES_SEARCH_H_INCLUDED__
2 #define __ES_SEARCH_H_INCLUDED__
3
4 extern int es_search(int n);
5
6 #endif
```

The implementation of ES, named es_search.cpp, is given in Listing 3.3. Apparently, after including all the necessary interfaces in lines 1–4, the code in lines 6–9 continues with the declarations using std::cout, using std::cerr, using std::endl, and using std::string to ease the use of functions defined in the namespace "std" of the standard library. That is, after the declarations, the function std::cout can be invoked by the name `cout` instead of the long, fully qualified name `std::cout`. The other functions can be invoked similarly.[3] The transition, evaluation, and determination operators are then declared in lines 13–15, for the purpose of generating a new solution, evaluating the objective value of the new solution, and determining if the new solution is better than the best solution found so far.

Listing 3.3: src/c++/0-es/1-onemax/es_search.cpp: The implementation of ES.

```
1 #include <iostream>
2 #include <string>
3 #include <gmpxx.h>
4 #include "es_search.h"
5
6 using std::cout;
7 using std::cerr;
8 using std::endl;
9 using std::string;
10
11 using solution = string;
12
13 static inline solution transit(const solution& s);
14 static inline int evaluate(const solution& s);
15 static inline void determine(int fv, const solution& v, int& fs, solution& s);
16
17 int
18 es_search(int n)
19 {
20     // 0. initialization
21     solution s(n, '0');
22     int fs = evaluate(s);
23     cout << fs << " # " << s << endl;
24     // 1. transition
25     solution v = s;
26     while ((v = transit(v)).size() != 0) {
27         // 2. evaluation
28         int fv = evaluate(v);
29         // 3. determination
30         determine(fv, v, fs, s);
31         cout << fs << " # " << v << endl;
32     }
33     cerr << "# " << fs << endl;
34     return 0;
35 }
36
37 static inline string
38 transit(const string& s)
39 {
40     mpz_class v = mpz_class(s, 2) + 1;
41     if (v < 0 || v > (mpz_class(1) << s.size()) - 1)
42         return string();
43     string t = v.get_str(2);
```

---

[3] However, in order to save space, in what follows in this book, we will use the so-called using directive "using namespace std" instead of the so-called using declaration that we have just used from time to time.

```
44      t.insert(0, s.size() - t.size(), '0');
45      return t;
46  }
47
48  static inline int
49  evaluate(const solution &s)
50  {
51      return count(s.begin(), s.end(), '1');
52  }
53
54  static inline void
55  determine(int fv, const solution& v, int& fs, solution& s)
56  {
57      if (fv > fs) {
58          fs = fv;
59          s = v;
60      }
61  }
```

Now, it is time to move on to the main search procedure named es_search() (lines 17–35), which can be divided into four parts: initialization, transition, evaluation, and determination. First comes the initialization in lines 21–22 in which the variables s and fs that represent the best solution found so far and the objective value of the best solution found so far, respectively, are defined and initialized. The variable s is initialized to "0000"[4] for a one-max problem of size 4; that is, the number of 0's required to initialize the initial solution s depends on the size of the one-max problem. The initial objective value fs is set equal to the value returned by the evaluate() function. Next come the variables v and fv in lines 25 and 28 that are defined to keep track of, respectively, the next candidate solution, denoted v, and the objective value of v, denoted f_v, in Algorithm 1. In lines 26–32, the transition, evaluation, and determination operators are performed repeatedly until all candidate solutions are generated and checked.

The transition operator defined in lines 37–46 is called by es_search() from line 26. It is responsible for generating the next solution v from the current solution v; that is, v = v + 1.[5] More precisely, line 40 of the transit() operator will first convert the current solution s to a multiple precision (MP) integer by instantiating an instance of the class mpz_class by calling its constructor that takes as input two parameters, the first of which is a string encoding the current solution while the second is an integer specifying the base. It is 2 in this case because the first parameter (the input string)

---

[4] That is, a string of four 0's in this case because the solution is encoded as a string.

[5] Note that the variable v plays two different roles here, as the current solution and the next (or new) solution. This is exactly the same as v described in Algorithm 1.

encodes a binary number.[6] A one is then added to the MP integer thus obtained to generate the next candidate solution, again as an MP integer, that is saved in (used to initialize) the newly defined variable v. Line 41 will then check to see if the next candidate solution is valid in the sense that it is a solution that falls in the range $[0, 2^n - 1]$, where $n = $ s.size(). If this is not the case, line 42 will simply return a null string to signal that all the candidate solutions have been generated. Otherwise, line 43 will convert the next candidate solution saved in v as an MP integer to a string named t by calling the member function of mpz_class named get_str(), and line 44 will simply pad the string just obtained t with a high enough number of leading 0's so that it is of the same size as the input string s and thus the size of a solution will remain intact. Finally, the next candidate solution in t is returned to the es_search() function as the new solution v, as shown in lines 45 and 26.

The evaluation operator is responsible for calculating the objective value of a solution. In this case, as shown in line 28, the evaluate() function is called to count how many 1's are in the new solution v by calling the C++ standard library function named count() in line 51. The determination operator is responsible for determining which solution will survive. In this case, as shown in line 30, the determination operator will be called to check the objective values of the current candidate solution s and the next candidate solution v (i.e., fs and fv) to see which one is better, as shown in lines 54–61. Finally, once all the candidate solutions are generated and checked, ES will also output the best solution in line 33.

3.1.3 Discussion of ES

It may seem that the ES method should solve the one-max problem easily and perfectly in the sense that it can always find the optimal solution by repeating the process of generating a new candidate solution, evaluating its objective value, and then comparing it with the objective value of the current solution. This is simply not the case because as the size of the one-max problem increases, the number of solutions ES has to check grows exponentially. Assuming that the size of the one-max problem is n, the number of solutions ES has to check is 2^n, rendering it impossible to check all the solutions in a reasonable time. This implies that it is certainly not a

---

[6] You are referred to the manual *GNU MP: The GNU Multiple Precision Arithmetic Library, Edition 6.2.0*, by Torbjörn Granlund and the GMP development team, which is available online at https://gmplib.org/gmp-man-6.2.0.pdf, for the details.

good idea to try to solve a complex optimization problem of a large size, e.g., a 1000-bit one-max problem, by ES.

3.2. Hill climbing (HC)

3.2.1 The basic idea of HC

Since ES may not be able to solve a large and complex optimization problem in a reasonable time, a possible way to find an "approximate solution" for this kind of problem in a reasonable time is the so-called HC. Of course, by an approximate solution, we mean that the solution is close to, but not necessarily, the optimal solution. Unlike ES, HC is a well-known greedy search algorithm. The search strategy of HC is to accept only a better solution as the next solution. This implies that if there are local optima located between the initial solution and the optimal solution, HC may get stuck at a local optimum.

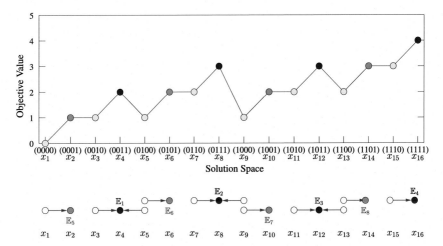

Figure 3.3 The search strategy of HC that accepts only a better solution as the next solution.

For the examples shown in Figs. 3.3 and 3.4, suppose HC has a 50/50 chance to move either left or right in solving the one-max problem; then, in addition to the global optimum (1111), HC may end up in one of the seven or three local optima. These two different situations are caused by the determination rule for comparing the new candidate solution and the best solution found so far. As shown in Fig. 3.3, if HC accepts only a better solution (i.e., $f(v) > f(s)$) as the next solution in solving the one-max

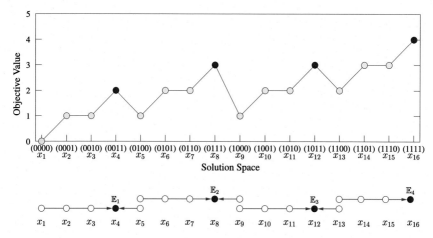

Figure 3.4 The search strategy of HC that accepts both better solutions and solutions that are equally good as the next solution.

problem, then the following seven solutions are all local optima: x_4 (0011), x_8 (0111), x_{12} (1011), x_2 (0001), x_6 (0101), x_{10} (1001), and x_{14} (1101). Unlike the end points $\mathbb{E}_1$, $\mathbb{E}_2$, $\mathbb{E}_3$, and $\mathbb{E}_4$, each of which is the best solution compared with its two nearest neighbors, the two nearest neighbors of the end points $\mathbb{E}_5$, $\mathbb{E}_6$, $\mathbb{E}_7$, and $\mathbb{E}_8$ are either equally good or even worse; the search process will thus also get stuck at these local optima. A possible solution to this problem is to violate the principle of HC by changing the determination rule so that it will accept not only a better solution but also a solution that is equally good as the next solution. Fig. 3.4 shows the possible result if HC accepts not only a better solution but also a solution that is equally good (i.e., $f(v) \geq f(s)$) as the next solution; then only the solutions x_4 (0011), x_8 (0111), and x_{12} (1011) will remain as the local optima, while x_{16} (1111) is the global optimum.

Fig. 3.5 shows the details of the examples described in Figs. 3.3 and 3.4, provided that the initial solution is x_9 (1000). In this case, it is apparent that HC will end up in one of the following three local optima: x_8 (0111), x_{10} (1001), and x_{12} (1011). This means that if the initial solution is x_9 (1000), it has a 50/50 chance to move to either x_8 (0111) or x_{10} (1001). If HC accepts only better solutions, it will get stuck at either x_8 or x_{10}. If HC accepts not only better solutions but also solutions that are equally good, then it has a 50/50 chance to move to either x_8 (0111) or x_{12} (1011). All of these examples show that HC has a pretty good chance to end up in a local optimum.

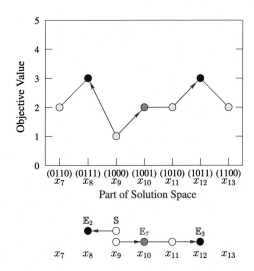

Figure 3.5 The details between x_7 and x_{13} of Figs. 3.3 and 3.4.

As shown in Algorithm 2, instead of checking all the candidate solutions, HC will check only a portion of the candidate solutions. The basic idea of HC can be summarized as two steps: (1) create a new solution v that is a neighbor of the current solution s and (2) consider only a better solution as the next solution.

Algorithm 2 Hill climbing.

1 Randomly create the initial solution s	☐I
2 $f_s = \text{Evaluate}(s)$	☐E
3 **While** the termination criterion is not met	
4 $\quad v = \text{NeighborSelection}(s)$	☐T
5 $\quad f_v = \text{Evaluate}(v)$	☐E
6 $\quad$ If f_v is better than f_s	☐D
7 $\qquad s = v$	
8 $\qquad f_s = f_v$	
9 $\quad$ EndIf	
10 **End**	
11 **Output** s	☐O

The first step can typically be made by creating a new solution v from the current solution s by randomly adjusting a small portion of the subsolutions of the current solution s, as shown in line 4 of Algorithm 2. In this way, v will be similar to s but not exactly the same. We thus call it a neighbor of s. Because HC will check only a portion of the candidate solutions in the solution space during the convergence process, it is pretty much like a mountain climber who is attempting to find a way to reach the top of a mountain around the place he or she is currently located. The second step is then made to determine which of the solutions will survive as the next solution, thus somehow determining the search direction of the search process. It will typically take the new solution v as the next solution only if it is better than the current solution s. Once the new solution v is created and the objective value is calculated (lines 4–5), HC will then compare the new solution (also called the new candidate solution) with the current solution to determine which one is better, thus to be kept as the next solution. Although Algorithm 2 is very similar to Algorithm 1, there are two major differences. The first one is that HC will randomly create the initial solution for different simulations of the same problem while ES will start with the same initial solution for different simulations of the same problem. The second one is that HC will "randomly" create a new solution that is nearby the current solution while ES will "systematically" generate a new solution with a value that is one larger than that of the current solution.

3.2.2 Implementation of HC for the one-max problem

To simplify the discussion that follows and to make the comparison between all search algorithms more consistent, we will use the "template" described in Fig. 3.6 to implement all the search algorithms described in this book. Since the unified framework for metaheuristics to be described in the next section will be based on this concept, we will tentatively call this kind of design concept a template to explain how HC is implemented. As shown in Fig. 3.6, a typical way to develop a metaheuristic algorithm for solving optimization problems is to take into account how the implementation is done and how the simulation is performed. Based on this template, a search algorithm will perform the function Initialization() once and then the operators OperatorA(), OperatorB(), ..., OperatorZ() repeatedly for a certain number of iterations specified for each run of the algorithm. Unlike ES, the final results of different runs of HC and other metaheuristics will not always be the same because they are stochastic search algorithms. To evaluate the performance of this kind of search algorithm, several runs are typically needed because a single run of such an algorithm will not be able to reflect its actual performance in terms of both the result and the computation time. A good simulation for any metaheuristic algorithm will usually be carried out for at least 30 runs based on the law of large numbers in probability theory to measure its performance. Note that the number of runs of most simulations is set to 100 in this book to measure the performance of each metaheuristic algorithm accurately. It is a very simple template for the implementation of metaheuristics, which could help us compare one metaheuristic algorithm with another.

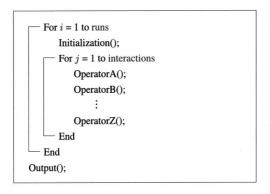

Figure 3.6 Template for developing a search algorithm.

Now, let us turn our attention to the implementation of HC for solving the one-max problem. The code resides in the directory `src/c++/1-onemax`, and the commands to compile this program are exactly the same as those for ES, i.e., `make clean dep all` after the command `cd src/c++/1-onemax` is performed. There are two ways to run this program: by invoking the program directly `./search` or by invoking the given script `./search.sh`, which will then invoke the program. The program takes five parameters. The first one is either an acronym or an abbreviated name of the algorithm, which is used to determine which search algorithm will be run. The following parameters are, respectively, the number of runs, the number of evaluations, the size of the one-max problem, and the filename of the initial seeds. In other words, the command to run this program is as follows:

```
$ ./search <algorithm> <#runs> <#evaluations> <#bits> <file-
name>
```

For example, it can be invoked by the following command:

```
$ ./search hc 100 1000 10 ""
```

This means that the above command will use HC as the search algorithm and carry it out for 100 runs, with 1000 evaluations per run. The size of the one-max problem is set equal to 10, and there are no initial seeds. The last parameter (`<filename>`) is there to allow the search algorithm to read the initial solutions from a file so that different (meta)heuristic algorithms may start with the same initial solutions to make the comparisons as fair as possible.

Before we proceed to discuss the implementation of HC, we should first have a look at Fig. 3.7, which shows the relationship between the implementation of HC and the script. In addition to the script `search.sh`, HC for the one-max program is composed of the following modules: `main.cpp`, `lib.h`, and `hc.h`. Of course, the relationship between the implementation of HC differs a bit from that of ES, as shown in Fig. 3.2. The driver module (program) is named "main.cpp," but it now resides in the directory named `src/c++/1-onemax/main`. The HC module named "hc.h" provides not only the interface but also the implementation of HC. The tool module named "lib.h" is a library of tools for all the search algorithms. The class `hc_r` defined in `hc_r.h` is derived from the class `hc` defined in `hc.h`. This is so that the transit() method of `hc` can be overridden to give it a new definition while reusing the rest of the code. This is a technique to change the behavior of a class without changing its underlying structure. This is one way to guarantee that both `hc_r` and `hc` behave exactly the same except the transition operator and is basically a design pattern called "template method."[7]

3.2.2.1 Main function

The driver module `main.cpp`, shown in Listing 3.4, can be considered as a driver module of driver modules, each of which is associated with an implemented search algorithm. This design is mainly aimed to reduce the number of files (one for each driver module) needed and the amount of code duplicated while at the same time making it work exactly the same as if each driver was implemented as a separate module (file). Here is how the driver of drivers, namely the main() function shown in lines 33–71, works exactly. In brief, it will first check if the name of the algorithm to be

---

[7] You are referred to the book *Design Patterns: Elements of Reusable Object-Oriented Software* (New York: Addison-Wesley, 1995) by Gamma et al. for more details.

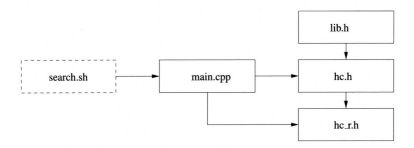

Figure 3.7 Relationship between the implementation of HC and the script file.

performed is given. If not, it will return immediately after printing an error message, as shown in lines 35–38. Otherwise, it will move on to checking if the name of the algorithm is valid, that is, if it is a name registered in the map defined in lines 20–31. If not, it will again return immediately after printing an error message, as shown in lines 40–44. Otherwise, it will move on to checking if the number of parameters given is as specified in the map defined in lines 20–31. If not, it will again return immediately after printing an error message telling the user of the program how the program should be invoked, as shown in lines 48–54. Otherwise, since all are set now, the driver of drivers will simply print out the common part of the parameters supplied by the user of this program before passing the control to the driver of the search algorithm, which will first print the rest of the parameters, if any, and then call the search algorithm specified by the user of the program. For instance, for HC, it is the driver in lines 73–84 that will be called by the driver of drivers to perform the search. One more thing done by the driver of drivers is keeping track of the computation time taken by each search algorithm, as shown in lines 63–65, and print it before it returns to the caller, as shown in line 68.

Listing 3.4: src/c++/1-onemax/main/main.cpp: The main function of main functions of algorithms whose implementations are given in this directory.

```
1 #include <iostream>
2 #include <iomanip>
3 #include <string>
4 #include <map>
5 #include <tuple>
6
7 using namespace std;
8
9 extern int main_hc(int argc, char** argv);
```

```
10 extern int main_sa(int argc, char** argv);
11 extern int main_sa_refinit(int argc, char** argv);
12 extern int main_ts(int argc, char** argv);
13 extern int main_hc_r(int argc, char** argv);
14 extern int main_ga(int argc, char** argv);
15 extern int main_gar(int argc, char** argv);
16 extern int main_gax2(int argc, char** argv);
17 extern int main_gaxu(int argc, char** argv);
18 extern int main_se(int argc, char** argv);
19
20 static map<const string, tuple<int(*)(int, char**), int, const char*>> dispatcher {
21     {"hc",          {main_hc,         6, ""}},
22     {"sa",          {main_sa,         8, "<min_temp> <max_temp>"}},
23     {"sa_refinit", {main_sa_refinit, 10, "<min_temp> <max_temp> <#samples> <#same bits"}},
24     {"ts",          {main_ts,         8, "<#neighbors> <tabulist_size>"}},
25     {"hc_r",        {main_hc_r,       6, ""}},
26     {"ga",          {main_ga,        10, "<popsize> <cr> <mr> <#players>"}},
27     {"gar",         {main_gar,       10, "<popsize> <cr> <mr> <#players>"}},
28     {"gax2",        {main_gax2,      10, "<popsize> <cr> <mr> <#players>"}},
29     {"gaxu",        {main_gaxu,      10, "<popsize> <cr> <mr> <#players>"}},
30     {"se",          {main_se,        11, "<#searchers> <#regions> <#samples> <#players> <
       scatter_plot?>"}},
31 };
32
33 int main(int argc, char** argv)
34 {
35     if (argc < 2) {
36         cerr << "# The name of algorithm is missing!" << endl;
37         return 1;
38     }
39
40     const auto& iter = dispatcher.find(argv[1]);
41     if (iter == dispatcher.end()) {
42         cerr << "# The specified algorithm '" << argv[1] << "' is not found!" << endl;
43         return 1;
44     }
45
46     const auto& t = iter->second;
47
48     if (argc != get<1>(t)) {
49         cerr << "Usage: ./search"
50             << " <algname> <#runs> <#evals> <#patterns> <filename_ini>"
51             << " " << get<2>(t)
52             << endl;
53         return 1;
54     }
55
56     cerr << "# name of the search algorithm: '" << argv[1] << "'" << endl
57         << "# number of runs: " << argv[2] << endl
58         << "# number of evaluations: " << argv[3] << endl
59         << "# number of patterns: " << argv[4] << endl
60         << "# filename of the initial seeds: '" << argv[5] << "'" << endl;
61
62     // now do and time the search
63     const clock_t begin = clock();
64     const int rc = get<0>(t)(argc, argv);
65     const clock_t end = clock();
66
67     const double cpu_time = static_cast<double>(end - begin) / CLOCKS_PER_SEC;
68     cerr << "# CPU time used: " << cpu_time << " seconds." << endl;
69
70     return rc;
71 }
72
73 #include "hc.h"
```

```
74 int main_hc(int argc, char** argv)
75 {
76     hc hc_search(atoi(argv[2]),
77                  atoi(argv[3]),
78                  atoi(argv[4]),
79                  argv[5]
80                  );
81     hc::solution s = hc_search.run();
82     cerr << s << endl;
83     return 0;
84 }
85
86 #include "sa.h"
87 int main_sa(int argc, char** argv)
88 {
89     cerr << "# minimum temperature: " << argv[6] << endl
90          << "# maximum temperature: " << argv[7] << endl;
91     sa sa_search(atoi(argv[2]),
92                  atoi(argv[3]),
93                  atoi(argv[4]),
94                  argv[5],
95                  atof(argv[6]),
96                  atof(argv[7])
97                  );
98     sa::solution s = sa_search.run();
99     cerr << s << endl;
100    return 0;
101 }
102
103 #include "sa_refinit.h"
104 int main_sa_refinit(int argc, char** argv)
105 {
106    cerr << "# minimum temperature: " << argv[6] << endl
107         << "# maximum temperature: " << argv[7] << endl
108         << "# number of samplings: " << argv[8] << endl
109         << "# number of same bits: " << argv[9] << endl;
110    sa_refinit sa_refinit_search(atoi(argv[2]),
111                                 atoi(argv[3]),
112                                 atoi(argv[4]),
113                                 argv[5],
114                                 atof(argv[6]),
115                                 atof(argv[7]),
116                                 atoi(argv[8]),
117                                 atoi(argv[9])
118                                 );
119    sa_refinit::solution s = sa_refinit_search.run();
120    cerr << s << endl;
121    return 0;
122 }
123
124 #include "ts.h"
125 int main_ts(int argc, char** argv)
126 {
127    cerr << "# number of neighbors: " << argv[6] << endl
128         << "# size of the tabu list: " << argv[7] << endl;
129    ts ts_search(atoi(argv[2]),
130                 atoi(argv[3]),
131                 atoi(argv[4]),
132                 argv[5],
133                 atoi(argv[6]),
134                 atoi(argv[7])
135                 );
136    ts::solution s = ts_search.run();
137    cerr << s << endl;
138    return 0;
```

```
139 }
140
141 #include "hc_r.h"
142 int main_hc_r(int argc, char** argv)
143 {
144     hc_r hc_search(atoi(argv[2]),
145                    atoi(argv[3]),
146                    atoi(argv[4]),
147                    argv[5]
148                    );
149     hc_r::solution s = hc_search.run();
150     cerr << s << endl;
151     return 0;
152 }
153
154 #include "ga.h"
155 int main_ga(int argc, char** argv)
156 {
157     cerr << "# population size: " << argv[6] << endl
158          << "# crossover rate: " << argv[7] << endl
159          << "# mutation rate: " << argv[8] << endl
160          << "# number of players: " << argv[9] << endl;
161     ga ga_search(atoi(argv[2]),
162                  atoi(argv[3]),
163                  atoi(argv[4]),
164                  argv[5],
165                  atoi(argv[6]),
166                  atof(argv[7]),
167                  atof(argv[8]),
168                  atoi(argv[9])
169                  );
170     ga::population p = ga_search.run();
171     cerr << p << endl;
172     return 0;
173 }
174
175 #include "gar.h"
176 int main_gar(int argc, char** argv)
177 {
178     cerr << "# population size: " << argv[6] << endl
179          << "# crossover rate: " << argv[7] << endl
180          << "# mutation rate: " << argv[8] << endl
181          << "# number of players: " << argv[9] << endl;
182     gar gar_search(atoi(argv[2]),
183                    atoi(argv[3]),
184                    atoi(argv[4]),
185                    argv[5],
186                    atoi(argv[6]),
187                    atof(argv[7]),
188                    atof(argv[8]),
189                    atoi(argv[9])
190                    );
191     gar::population p = gar_search.run();
192     cerr << p << endl;
193     return 0;
194 }
195
196 // two-point crossover
197 #include "gax2.h"
198 int main_gax2(int argc, char** argv)
199 {
200     cerr << "# population size: " << argv[6] << endl
201          << "# crossover rate: " << argv[7] << endl
202          << "# mutation rate: " << argv[8] << endl
203          << "# number of players: " << argv[9] << endl;
```

```
204    gax2 gax2_search(atoi(argv[2]),
205                     atoi(argv[3]),
206                     atoi(argv[4]),
207                     argv[5],
208                     atoi(argv[6]),
209                     atof(argv[7]),
210                     atof(argv[8]),
211                     atoi(argv[9])
212                     );
213    gax2::population p = gax2_search.run();
214    cerr << p << endl;
215    return 0;
216 }
217
218 // uniform crossover
219 #include "gaxu.h"
220 int main_gaxu(int argc, char** argv)
221 {
222    cerr << "# population size: " << argv[6] << endl
223         << "# crossover rate: " << argv[7] << endl
224         << "# mutation rate: " << argv[8] << endl
225         << "# number of players: " << argv[9] << endl;
226    gaxu gaxu_search(atoi(argv[2]),
227                     atoi(argv[3]),
228                     atoi(argv[4]),
229                     argv[5],
230                     atoi(argv[6]),
231                     atof(argv[7]),
232                     atof(argv[8]),
233                     atoi(argv[9])
234                     );
235    gaxu::population p = gaxu_search.run();
236    cerr << p << endl;
237    return 0;
238 }
239
240 #include "se.h"
241 int main_se(int argc, char** argv)
242 {
243    cerr << "# number of searchers: " << argv[6] << endl
244         << "# number of regions: " << argv[7] << endl
245         << "# number of samples: " << argv[8] << endl
246         << "# number of players: " << argv[9] << endl
247         << "# scatter plot?: " << argv[10] << endl;
248    se se_search(atoi(argv[2]),
249                 atoi(argv[3]),
250                 atoi(argv[4]),
251                 argv[5],
252                 atoi(argv[6]),
253                 atoi(argv[7]),
254                 atoi(argv[8]),
255                 atoi(argv[9]),
256                 atoi(argv[10])
257                 );
258    se_search.run();
259    return 0;
260 }
```

To add a new algorithm, and thus a new driver for the new algorithm, to the driver module of driver modules, here is a step-by-step description of what you have to do: (1) declare a main function (driver) for the new algorithm, say, main_*x*, where *x* is a unique name of the new algorithm, as shown in lines 9–18; (2) add the new algorithm to the dispatcher by adding a new entry, which includes in order the name of the algorithm (x given in the first step) as a string "x", the main function just declared (main_x declared in the first step), the number of parameters required, and a string specifying the parameters specific to this particular algorithm, as shown in lines 20–31; and (3) add the implementation of main_x (see lines 73–84 for an example of adding the search algorithm HC). Note that we are assuming here the number of parameters required by all the algorithms is no less than six. We can use the following aforementioned command

```
$ ./search hc 100 1000 10 ""
```

again to run HC for solving the one-max problem, and here is what happens. First, the main() function of the driver module will check to see if all the required parameters are given and valid. If they are, the main() function will move on to printing the parameters before it passes the control to the driver for the HC search algorithm because the value of argv[1] is "hc" in this case, meaning that HC will be used to solve the optimization problem in question. The driver for HC will then use the given parameters to create an instance of hc named hc_search, as shown in lines 73–84. After that, the member function run() of the object hc_search will be called to start the search process, that is, to perform the HC search algorithm, and then the search result will be assigned to a variable of type hc::solution named s, as shown in line 81. The internal of the instance hc_search of the class hc, as shown in Listing 3.5, will be discussed in great detail shortly. Finally, once the search is done, the final results of HC are printed in line 82.

3.2.2.2 Search function of HC

Following the coding style of this book, main.cpp is the driver of the program, but here it is hc.h that consists of both the interface and the implementation of HC, as shown in Listing 3.5. In what follows, we will describe in detail the implementation of HC.

3.2.2.2.1 Declaration of parameters and functions

As always, the definition of a class can be divided into two parts: public and private—i.e., the interface part and the implementation part. There is no exception to the class hc for HC, as shown in lines 11–39. Needless to say, the public or interface part specifies how the outside world communicates with an object of the class hc. For example, the type solution, the constructor hc(), and the member function run() are used by main.cpp to create an object of hc to start the search. They can be used by other search algorithms to ease the implementation of hybrid heuristic algorithms.[8] The private or implementation part specifies all the helping functions and data members needed for the implementation of the class hc. The operators init(), evaluate(), transit(), and determine() are all helping functions for creating the initial solution, measuring the fitness of a solution, adjusting the solution, and determining which solution is better. The data members num_runs, num_evals, num_patterns_sol, and filename_ini are the placeholders for the parameters passing in from main.cpp, namely, the number of runs, the number of evaluations, the number of subsolutions, and the filename of the initial solution for HC. The other two data members, d_sol and d_obj_val, are the placeholders for the so far best solution and the objective value of the so far best solution. The constructor, defined in lines 41–51, takes as input four parameters and returns as output an anonymous object of type hc. It will be called by other modules to create an instance of hc to perform an HC search.

3.2.2.2.2 The main loop

The most important part of this program is the function run() defined in lines 53–90. Typically, it implements the search strategy of a search algorithm. First of all, it will declare and initialize the variables d_avg_obj_val and d_avg_obj_val_eval for, respectively, the average objective value of the final solutions found in all runs and the average objective value of the solutions found in each evaluation. This is followed by the main search process in lines 58–79. In addition to the initialization operator init() to be performed at the start of each run, it consists of the operators transit(), evaluate(), and determine(), which will be performed repeatedly for the number of runs as specified in the variable d_num_runs. For each run, the number of evaluations as specified in the variable d_num_evals will be performed to search for the possible solution. To be more specific, each iteration of HC is made up of the following three operators: transit(), evaluate(), and determine().

---

[8] In this book, by a hybrid heuristic algorithm we mean a search algorithm that is a combination of two or more metaheuristics.

3.2.2.2.3 Additional functions

The initialization operator init() is defined in lines 92–105. In addition to initializing the vector d_sol, this operator will either read the initial solution in from the specified input file (lines 96–100) or randomly create the initial solution (lines 101–104). The transition operator transit() is responsible for adjusting the solution. As shown in line 109, this operator will simply call the transit() function defined in lib.h and returns whatever it returns.[9] The evaluation operator evaluate() is responsible for assessing the fitness of a solution. As defined in lines 112–115, this operator will simply count the number of 1's in a solution for the one-max problem. The determination operator determine() compares the objective values of the new candidate solution and the current solution. HC will then use this operator to determine whether to replace the current solution by the new candidate solution or not, as shown in lines 117–123. The variables v and s represent, respectively, the new candidate solution and the current solution, while the variables fv and fs represent, respectively, the objective values of these two solutions. If fv is better than fs, then the values of s and fs will be replaced by those of v and fv.

Listing 3.5: src/c++/1-onemax/main/hc.h: The implementation of HC.

```
1 #ifndef __HC_H_INCLUDED__
2 #define __HC_H_INCLUDED__
3
4 #include <string>
5 #include <fstream>
6 #include <numeric>
7 #include "lib.h"
8
9 using namespace std;
10
11 class hc
12 {
13 public:
14     typedef vector<int> solution;
15
16     hc(int num_runs,
17        int num_evals,
18        int num_patterns_sol,
19        string filename_ini);
20
21     virtual ~hc() {}
22
23     solution run();
24
25 private:
26     void init();
```

[9] Note the scope resolution operator "::" in front of the transit() function. Without this operator, the transit() function will simply call itself recursively, thus ending up being an infinite loop.

```cpp
27      virtual int evaluate(const solution& s);
28      virtual solution transit(const solution& s);
29      virtual void determine(const solution& v, int fv, solution& s, int &fs);
30
31  private:
32      int d_num_runs;
33      int d_num_evals;
34      int d_num_patterns_sol;
35      string d_filename_ini;
36
37      solution d_sol;
38      int d_obj_val;
39  };
40
41  inline hc::hc(int num_runs,
42               int num_evals,
43               int num_patterns_sol,
44               string filename_ini)
45      : d_num_runs(num_runs),
46        d_num_evals(num_evals),
47        d_num_patterns_sol(num_patterns_sol),
48        d_filename_ini(filename_ini)
49  {
50      srand();
51  }
52
53  inline hc::solution hc::run()
54  {
55      double d_avg_obj_val = 0;
56      vector<double> d_avg_obj_val_eval(d_num_evals, 0.0);
57
58      for (int r = 0; r < d_num_runs; r++) {
59          int eval_count = 0;
60
61          // 0. Initialization
62          init();
63          d_obj_val = evaluate(d_sol);
64          d_avg_obj_val_eval[eval_count++] += d_obj_val;
65
66          while (eval_count < d_num_evals) {
67              // 1. Transition
68              solution tmp_sol = transit(d_sol);
69
70              // 2. Evaluation
71              int tmp_obj_val = evaluate(tmp_sol);
72
73              // 3. Determination
74              determine(tmp_sol, tmp_obj_val, d_sol, d_obj_val);
75
76              d_avg_obj_val_eval[eval_count++] += d_obj_val;
77          }
78          d_avg_obj_val += d_obj_val;
79      }
80
81      // 4. Output
82      d_avg_obj_val /= d_num_runs;
83
84      for (int i = 0; i < d_num_evals; i++) {
85          d_avg_obj_val_eval[i] /= d_num_runs;
86          cout << fixed << setprecision(3) << d_avg_obj_val_eval[i] << endl;
87      }
88
89      return d_sol;
90  }
91
```

```
 92 inline void hc::init()
 93 {
 94     d_sol = solution(d_num_patterns_sol);
 95
 96     if (!d_filename_ini.empty()) {
 97         ifstream ifs(d_filename_ini.c_str());
 98         for (int i = 0; i < d_num_patterns_sol; i++)
 99             ifs >> d_sol[i];
100     }
101     else {
102         for (int i = 0; i < d_num_patterns_sol; i++)
103             d_sol[i] = rand() % 2;
104     }
105 }
106
107 inline hc::solution hc::transit(const solution& s)
108 {
109     return ::transit(s);
110 }
111
112 inline int hc::evaluate(const solution& s)
113 {
114     return accumulate(s.begin(), s.end(), 0);
115 }
116
117 inline void hc::determine(const solution& v, int fv, solution& s, int &fs)
118 {
119     if (fv > fs) {
120         fs = fv;
121         s = v;
122     }
123 }
124
125 #endif
```

3.2.2.3 Library function

The tool library for HC and all other search algorithms is given in Listing 3.6. We use it to emphasize that tools to be shared by all search algorithms should be centralized somewhere as a library or something else to reduce redundancy. Lines 4–11 show that the tool library will first include all the header files for functions that it uses. Lines 13–27 will then declare the functions in the namespace std that are eventually used in this library so that they can be called directly instead of being fully qualified by prefixing each of them with the name of the namespace ("std::" in this case) to inform the compiler where to find it before the types solution, population, pheromone_1d, and pheromone_2d are defined, as either a vector (a one-dimensional array) of T or a vector of vectors (a two-dimensional array) of T, where T is either int or double. As shown in lines 31–65, it will then overload the output operator "<<" for these types to make it possible to print the values saved in variables of these types in the same way as for the built-in types. Next comes a function to calculate the distance between two points, as shown in lines 67–74.

Lines 76–99 provide two transition operators as an example of defining operators as global functions so that they can be shared by as many metaheuristic algorithms as possible. The first transition operator (lines 76–90) is similar to the transit() operator of ES defined in the file es_search.cpp. The first difference between these two versions of the transition operator is that they are used to transit two different encodings of the solution—one as a string and the other as a vector of integers. To be more specific, the one defined in es_search.cpp is used to transit a solution encoded as a bit string while the one defined in lib.h is used to transit a solution encoded as a vector of binary numbers. More precisely, the version of transit() defined in lib.h will do the transition in three steps: (1) convert the vector of binary numbers to a bit string, (2) adjust the solution encoded as a bit string using the GNU MP library, and (3) convert it back to a vector of binary numbers. Steps 1 and 3 are required because the GNU MP library can only convert a string to an MP integer and vice versa. The second difference, as shown in lines 81–82, is that the transit() operator defined in lib.h will create a new candidate solution v by adding one to or subtracting one from the current solution s, that is, $v = s \pm 1$. This means that the next solution v will move the search either to the left (-1) or to the right $(+1)$ of the current solution s in the solution space shown in Fig. 3.3 with a 50/50 chance. The second transition operator is shown in lines 92–99 and is named transit_r(). The basic idea of this operator is to first randomly choose a subsolution s_i of the current solution s and then invert its value (i.e., from "1" to "0" or from "0" to "1") so that the Hamming distance between the current solution and the new solution will always be 1. This operator makes it possible to jump from one solution to another that is not limited to the solution either to the left or to the right of the current solution.

Listing 3.6: src/c++/1-onemax/main/lib.h: The tool library.

```
 1 #ifndef __LIB_H_INCLUDED__
 2 #define __LIB_H_INCLUDED__
 3
 4 #include <iostream>
 5 #include <iomanip>
 6 #include <vector>
 7 #include <string>
 8 #include <cmath>
 9 #include <chrono>
10 #include <tuple>
11 #include <gmpxx.h>
12
13 using std::ostream;
14 using std::cerr;
15 using std::endl;
16 using std::fixed;
```

```
17  using std::setprecision;
18  using std::string;
19  using std::vector;
20  using std::tuple;
21  using std::make_tuple;
22  using std::tie;
23
24  using solution = vector<int>;
25  using population = vector<solution>;
26  using pheromone_1d = vector<double>;
27  using pheromone_2d = vector<pheromone_1d>;
28
29
30
31  inline ostream&
32  operator<<(ostream& os, const solution& s)
33  {
34      os << "# " << s[0];
35      for (size_t i = 1; i < s.size(); i++)
36          os << ", " << s[i];
37      return os;
38  }
39
40  inline ostream&
41  operator<<(ostream& os, const population& p)
42  {
43      for (size_t i = 0; i < p.size()-1; i++)
44          os << p[i] << endl;
45      os << p[p.size()-1];
46      return os;
47  }
48
49  inline ostream&
50  operator<<(ostream& os, const pheromone_1d& p)
51  {
52      os << "# " << p[0];
53      for (size_t i = 1; i < p.size(); i++)
54          os << ", " << p[i];
55      return os;
56  }
57
58  inline ostream&
59  operator<<(ostream& os, const pheromone_2d& p)
60  {
61      for (size_t i = 0; i < p.size()-1; i++)
62          os << p[i] << endl;
63      os << p[p.size()-1];
64      return os;
65  }
66
67  inline double
68  distance(const vector<double>& a, const vector<double>& b)
69  {
70      double d = 0;
71      for (size_t i = 0; i < a.size(); i++)
72          d += (a[i]-b[i]) * (a[i]-b[i]);
73      return sqrt(d);
74  }
75
76  inline solution
77  transit(const solution& s)
78  {
79      string is(s.size(), '0');
80      transform(s.begin(), s.end(), is.begin(), [] (int i) { return i + '0'; });
81      const double r = static_cast<double>(rand()) / RAND_MAX;
```

```
 82    const mpz_class v = mpz_class(is, 2) + (r < 0.5 ? 1 : -1);
 83    if (v < 0 || v > (mpz_class(1) << s.size()) - 1)
 84        return s;
 85    string t = v.get_str(2);
 86    t.insert(0, s.size() - t.size(), '0');;
 87    vector<int> os(s.size(), 0);
 88    transform(t.begin(), t.end(), os.begin(), [] (int i) { return i - '0'; });
 89    return os;
 90 }
 91
 92 inline solution
 93 transit_r(const solution& s)
 94 {
 95    solution t(s);
 96    const int i = rand() % t.size();
 97    t[i] = !t[i];
 98    return t;
 99 }
100
101 inline void
102 srand()
103 {
104    unsigned int seed = std::chrono::system_clock::now().time_since_epoch().count();
105    cerr << "# seed: " << seed << endl;
106    srand(seed);
107 }
108
109 #endif
```

3.2.3 Discussion of HC

In Sections 3.2.1 and 3.2.2, we discussed the basic idea of HC and how the implementation is done. In addition to the fact that the basic idea of HC is very simple and it is thus very easy to implement, the computation time can also be controlled. This means that we do not have to check all the possible solutions; rather, we can check only a certain number of possible solutions. In other words, the user can determine how much computation resource he or she wants to invest. HC provides a way to find an approximate solution in a reasonable time. This is quite different from ES, which is aimed at finding the optimal solution; it may thus take forever to get its job done. It can be easily seen that the final result of HC depends to a certain degree on the initial solution and the transition operator, because HC considers only neighbor solutions that are better than the current solution. Also, because HC will check only part of all the possible solutions, so it may fall into a local optimum for complex optimization problems. Although the main disadvantage of HC is that it easily falls into a local optimum, there exist ways to mitigate this issue, which will be discussed in later chapters. Since the focus of this chapter is on understanding the performance of HC and ES, some comparisons between these two search algorithms will be made in the next section.

3.3. Comparisons between ES and HC

The empirical analysis described in this book is conducted on an HP Z6 desktop workstation with two 2.10 GHz Intel Xeon Silver 4110 CPUs and 16 GB of memory running Fedora 34 with Linux 5.12.15-300.fc34.x86_64, and the programs are written in C++ and compiled using g++. All results described herein are the averages of 100 runs, and for each run, the search algorithm was performed with 1000 evaluations[10] for both the one-max and deceptive problems.

3.3.1 Simulation results of ES and HC for the one-max problem

As mentioned before, we implemented two transition operators for HC; the one with transit() is called HC-LR, while the one with transit_r() is called HC-Rand. In this simulation, we will keep using HC-LR and HC-Rand to denote the same thing. The next solution v of HC-LR will be the one that is one smaller or one larger than the current solution s (i.e., $v = s - 1$ or $v = s + 1$), while the next solution v of HC-Rand will be created by inverting a randomly chosen subsolution of s (i.e., "1" becomes "0" and "0" becomes "1"). Again, only a better solution will be accepted as the next solution v.

As shown in Fig. 3.8, HC for the one-max problem of size $n = 10$ might or might not be able to find the optimal solution, where the objective value of the optimal solution is 10 in this case. This example shows that the transition operator of HC has a strong impact on whether or not it is able to solve this optimization problem (find the optimal solution). It can be easily seen that the results of HC-LR are far from optimum, while HC-Rand can find the optimal solution even at an early stage of the convergence process. The results further show that HC-Rand is able to find the optimal solution faster than ES in terms of the convergence speed.

Fig. 3.9 gives a simple example to explain why HC-Rand-based methods are better than HC-LR-based methods for solving the one-max problem. First, let us recall the search strategy of HC-LR shown in Fig. 3.3. If HC-LR starts the search at x_9 (1000), it will have a 50/50 chance to get stuck at either x_8 (0111) or x_{10} (1001) because the transition operator of HC-LR limits each move of the search to either the previous one or the next one of the current solution. That is, HC-LR will then move to either

---

[10] Since ES and HC check one and only one solution per iteration, the number of iterations is essentially the same as the number of evaluations as far as ES and HC are concerned.

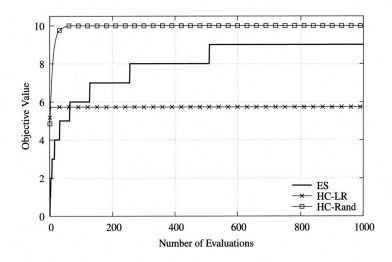

Figure 3.8 Comparison of ES and HC for the one-max problem of size $n = 10$.

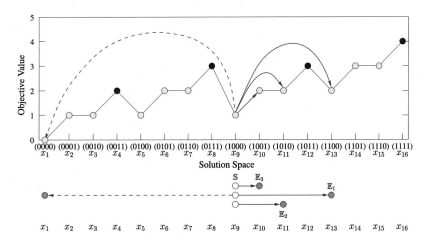

Figure 3.9 The search strategy of HC-Rand when starting from the candidate solution x_9.

x_8 (0111) or x_{10} (1001) in this case and eventually get stuck there. Such a transition operator reveals that a search strategy that accepts only better solutions as the next solution is very likely to fall into a local optimum.

As shown in Fig. 3.9, since the transition operator transit_r() of HC-Rand will randomly select a subsolution of the current solution to invert its value, if HC-Rand also starts the search at x_9, the next candidate solution will be x_1, x_{13}, x_{11}, or x_{10} because for a one-max problem of size $n = 4$, four possible solutions will be created as the next solution, in this case x_1, x_{13}, x_{11}, and x_{10}. Because the objective value of x_1 is 0, which is not better than the objective value of x_9, the next possible solution will be x_{13}, x_{11}, or x_{10}. This example shows that transit_r() makes it possible for HC to not get stuck at a local optimum for a one-max problem during the convergence process, e.g., x_4, x_8, and x_{12}. For instance, if it happens to be the case that the second bit of the current solution x_9 (1000) is selected and its value is inverted, the next solution will be $x_{13} = (1100)$. This makes it possible for the search of HC-Rand to not get stuck at the local optimum x_{12}. In other words, HC-Rand has a chance to find the global optimum x_{16} because the next possible solutions of x_{13} (1100) will be x_5 (0100), x_9 (1000), x_{15} (1110), and x_{14} (1101).

In summary, although both ES and HC-Rand find the optimal solution of a one-max problem of size $n = 10$ within 1024 evaluations in this case, the number of evaluations, and thus the computation time, of ES grows exponentially with problem size. This can be easily justified by having a look at the one-max problem of size $n = 100$. In this case, ES has to check all 2^{100} solutions, meaning that it has to perform 2^{100} evaluations, to find the optimal solution; however, our observation shows that HC-Rand takes only about 965 evaluations on average to find the optimal solution of the same problem. These results confirm that ES cannot solve large-scale optimization problems in a reasonable time, but HC can.

3.3.2 Simulation results of ES and HC for the deceptive problem

Since ES is not able to solve large-scale optimization problems in a reasonable time, we will now apply HC to the BSD-2 problem $\mathbb{P}_4$ to better understand its performance in solving complex optimization problems. As mentioned in Section 2.2.2, the BSD-2 problem $\mathbb{P}_4$ can be regarded as a deceptive problem that is defined as $f(s) = |B2D(s) - 2^{n-2}|$, where B2D is a function that returns the decimal value of the solution s that encodes a bit string $s = \{s_1, s_2, \ldots, s_n\}$ and $s_i \in \{0, 1\}$. As shown in lines 16–22 of Listing 3.7, all we have to do is to modify the evaluation operator evaluate() of hc to make it suitable for the BSD-2 problem.

Listing 3.7: src/c++/2-deception/main/hcdp.h: The evaluate() operator of HC for the BSD-2 problem.

```
16 inline int hcdp::evaluate(const solution& s)
17 {
```

```
18    int n = 0;
19    for (size_t i = 0; i < s.size(); i++)
20        n += s[i] << ((s.size() - 1) - i);
21    return abs(n - (1 << (s.size()-2)));
22 } //
```

Fig. 3.10 gives the simulation results of ES, HC-LR, and HC-Rand for the BSD-2 problem $\mathbb{P}_4$ of size $n = 4$. The results show that ES is able to

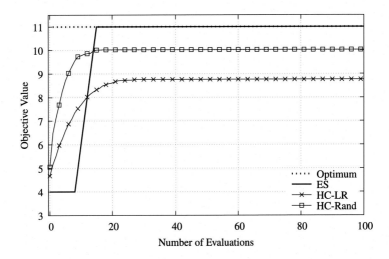

Figure 3.10 Simulation results of HC for a deceptive problem of size $n = 4$.

find the optimal solution quickly, because only 16 checks (evaluations) are needed, for there are in total 16 candidate solutions in this case. The results also show that HC-LR is unable to find the optimal solution sometimes when applying it to an optimization problem that has one or more local optima, even if the trap is not that complex, as shown in Fig. 2.7. In addition, for the one-max problem, the results further show that once it gets stuck in a local optimum, HC-LR is not able to escape from the local optimum no matter how many iterations are performed. It can be easily seen that the transition operator may have a strong impact on the performance of HC. For this simulation, HC-Rand is able to find a better result than HC-LR. The search strategy of HC-Rand is similar to that of HC-LR, that is, the only move that is allowed for each step is to a solution that is better than the current solution. As such, the transit_r() operator provides no guarantee that it can always escape local optima. In solving the BSD-2 problem $\mathbb{P}_4$, HC-Rand may sometimes get stuck at a local optimum at the early stage of the convergence process; as such, there is no guarantee that it will always find the optimal solution.

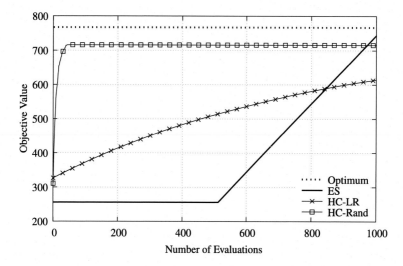

Figure 3.11 Simulation results of HC for a deceptive problem of size $n = 10$.

Fig. 3.11 shows that the number of evaluations ES requires to find the optimal solution increases significantly. In this case, ES is unable to find the optimal solution within 1000 evaluations. Based on the above discussion, it is reasonable to expect that ES will take 1024 evaluations to find the optimal solution because it has to check all possible candidate solutions, and with $n = 10$ there are exactly $2^n = 2^{10} = 1024$ possible candidate solutions. The best result of ES of the first 513 checks (evaluations of the solutions $(0000000000)_2$–$(1000000000)_2$ or 0–512) is 256. This explains why the convergence curve is flat from the first evaluation to the 513th evaluation. These results further show the performance of HC because both HC-based algorithms are able to find a better result than ES at the early stage of the convergence process. This implies that HC provides an alternative way to find an approximate solution for large and complex optimization problems.

Fig. 3.12 further explains why ES can be used for small complex optimization problems, such as the BSD-2 problem of size $n = 4$, because the number of possible candidate solutions is small. This implies that ES is able to find the optimal solution in a short time. As far as this example is concerned, ES only needs to search 16 possible solutions to find the optimal solution. Fig. 3.13 shows that the search strategy of HC-LR makes it possible to either get stuck at the local optimum x_1 or find the global optimum

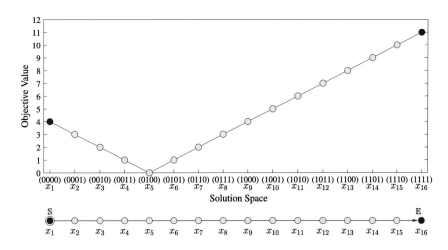

Figure 3.12 The search strategy of ES for the BSD-2 problem of size $n = 4$.

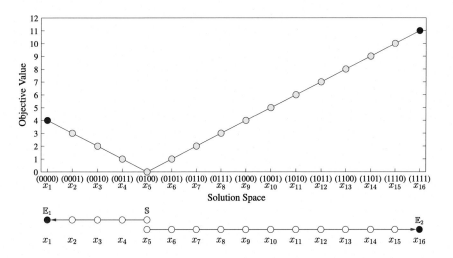

Figure 3.13 The search strategy of HC-LR for the BSD-2 problem of size $n = 4$.

x_{16}, depending on the starting point. If the starting point is in the range $x_1–x_4$, HC-LR will certainly end up getting stuck at the local optimum x_1. If the starting point is in the range $x_6–x_{16}$, HC-LR will surely end up finding the optimal solution. If HC-LR starts with the solution x_5, it has a 50/50 chance to get stuck at the local optimum or to find the global optimum. More precisely, assuming that it is equally likely to start the search at a particular solution, the probability for HC-LR to get stuck at the local optimum x_1 is $4/16 + 0.5/16 = 4.5/16$, while the probability for it to find the global optimum x_{16} is $11/16 + 0.5/16 = 11.5/16$.

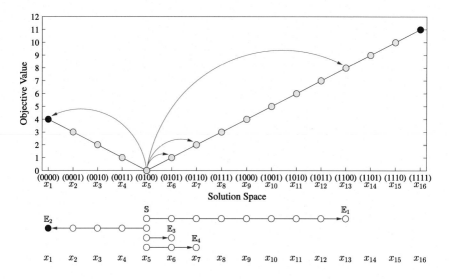

Figure 3.14 The search strategy of HC-Rand for a deceptive problem of size $n = 4$.

Fig. 3.14 further shows the performance of HC-Rand. In this example, we are again assuming that the starting point is the solution x_5 (0100). The next solution will then be x_{13}, x_1, x_7, or x_6. This implies that HC-Rand has a good chance to move to a solution in the region x_6–x_{16}. The probability of moving to the left region x_1–x_4 or to the right region x_6–x_{16} depends somehow on the location of the current solution. Obviously, a move to the right region is a guarantee to find the optimal solution, and a move to the left region is destined to a local optimum in this case. A detailed discussion on this will be given in the next section.

3.4. Summary of ES and HC

In this chapter, we show the possible search situations of ES and HC in solving the one-max and BSD-2 problems by presenting examples of the search strategies in the landscape of solution space and their simulation results. Note that the landscape of the "*solution space*" of an optimization problem seen by a search can be different when (1) different ways are used to represent the solutions or (2) different ways are used to generate new candidate solutions from current solutions. The solution space that a search algorithm sees can also be called the "*search space*." The main difference between the solution and search spaces is that the so-called solution space represents the set of all possible candidate solutions of an optimization problem $\mathbb{P}$, while the so-called search space represents the set of all possible candidate solutions that a search algorithm can see in solving an optimization problem $\mathbb{P}$. For this reason, the search and solution spaces that can be the same or different depend on how a search algorithm searches for an optimal solution. If a search algorithm cannot search all the possible candidate solutions in a solution space during the convergence process because of the way it is designed or because of the way solutions are encoded, then the search space of a search algorithm will be smaller than the solution space of the optimization problem in question. When designing a search algorithm, we would normally want to make these two spaces the same so that the search algorithm will not miss any possible good candidate solutions in the solution space. So we assume that all the search algorithms discussed in this book have the chance to search any possible candidate solutions in the solution space. This is the reason why we will use the term "solution space" to represent both the solution space of an optimization problem and the search space of a search algorithm.

Another interesting thing described in this chapter is the design of HC-LR and HC-Rand. The simulation results show that the transition operator of HC-Rand is capable of increasing the diversity of searches of HC. To better understand the search performance of HC-Rand, Table 3.1 shows all the possible next solutions of any current solution in solving the one-max and BSD-2 problems of size $n = 4$. In this table, s represents the current solution, v represents the possible next solution, $f(s)$ represents the objective value of s, and $f(v)$ represents the objective value of v. The greater than or equal to sign "$\geq$" indicates that $f(s) \geq f(v)$, so that v cannot be the next solution; that is, the move from s to v is not possible in this case, so no move is made. If the column labeled "Move" is assigned a Y, it means that $f(s) < f(v)$, so that the move from s to v is possible and v will replace s as the current solution. Otherwise, HC-Rand will simply randomly create another possible solution and then try again to see if it is possible to move. Moreover, R_L, R_R, and R_M denote, respectively, the range x_1–x_4, the range x_6–x_{16}, and the singleton x_5. Since there are 16 solutions in the solution space, there are 16 possible starting points. The table lists all the possible next solutions for each starting point to justify the characteristic of HC-Rand for the two optimization problems in question.

Table 3.1 shows that if the current solution or starting point is in the range R_R (x_6–x_{16}), then it has a very good chance to stay in that range and is thus able to move to a better solution to solve the BSD-2 problem. Table 3.2 shows the probabilities of moving to different regions. If the current solution or starting point is in the range x_7–x_{16}, the probability of finding a better next solution in R_R is 1. The expected value for the next solution to move to or stay in the range R_R is 78.125%, while that for the next solution to move to or stay in the range R_L is only 21.875%. This information explains why HC-Rand has a higher chance to find a good solution in solving the BSD-2 problem.

Tables 3.1 and 3.2 further show that if the current solution or starting point is x_1 (0000), it will get stuck at this location because the objective values of the possible solutions x_9 (1000), x_5 (0100), x_3 (0010), and x_2 (0001) are 4, 0, 2, and 3, none of which are larger than the objective value of x_1 (0000). This explains why there is no guarantee that HC-Rand is able to find the global optimum all the time. This problem can be solved by using ">=" instead of ">" in the comparison of current solution and possible next solution. This means replacing the relational expression "fv > fs" in line 119 of Listing 3.5 by the relational expression "fv >= fs." In this way, the next solution of x_1 will be x_9, a move from one region to another, that is, from the left region R_L to the right region R_R in this case. This makes it possible to find the optimal solution instead of getting stuck at x_1.

Figs. 3.15 and 3.16 further show what happens if we use ">=" instead of ">" in the comparison of the current solution and the possible next solution for solving the BSD-2 problem of sizes $n = 4$ and $n = 10$. The results of HC-Rand-M show that with this modification, HC-Rand will be able to find the optimal solution in most cases. This implies that the transition and determination mechanisms have a strong impact on the performance of a search algorithm.

In summary, in contrast to ES, HC is a greedy algorithm, which provides an alternative way for solving optimization problems because it can find an approximate solution in a reasonable time. HC has the following characteristics: (1) it is simple to implement and (2) it is capable of finding an approximate solution quickly. As a result, this method is used in many modern information systems for solving many engineering problems. However, without a good transition and determination mechanism, the end result of HC will be far from optimal for most optimization problems. Although HC easily falls into a local optimum, it takes a very reasonable

Table 3.1 Probability of HC-Rand to solve the one-max and BSD-2 problems.

$\mathbb{P}_1$ & $\mathbb{P}_4$		$\mathbb{P}_1$		$\mathbb{P}_4$			
s	v	$f(s):f(v)$	Move	$f(s):f(v)$	Range of s^t	Range of s^{t+1}	Move
x_1 (0000)	x_9 (1000)	0 < 1	Y	4 ≥ 4	R_L	R_L	N
	x_5 (0100)	0 < 1	Y	4 ≥ 0		R_L	N
	x_3 (0010)	0 < 1	Y	4 ≥ 2		R_L	N
	x_2 (0001)	0 < 1	Y	4 ≥ 3		R_L	N
x_2 (0001)	x_{10} (1001)	1 < 2	Y	3 < 5	R_L	R_R	Y
	x_6 (0101)	1 < 2	Y	3 ≥ 1		R_L	N
	x_4 (0011)	1 < 2	Y	3 ≥ 1		R_L	N
	x_1 (0000)	1 ≥ 0	N	3 < 4		R_L	Y
x_3 (0010)	x_{11} (1010)	1 < 2	Y	2 < 6	R_L	R_R	Y
	x_7 (0110)	1 < 2	Y	2 ≥ 1		R_L	N
	x_1 (0000)	1 ≥ 0	N	2 < 4		R_L	Y
	x_4 (0011)	1 < 2	Y	2 ≥ 1		R_L	N
x_4 (0011)	x_{12} (1011)	2 < 3	Y	1 < 7	R_L	R_R	Y
	x_8 (0111)	2 < 3	Y	1 < 3		R_R	Y
	x_2 (0001)	2 ≥ 1	N	1 < 3		R_L	Y
	x_3 (0010)	2 ≥ 1	N	1 < 2		R_L	Y
x_5 (0100)	x_{13} (1100)	1 < 2	Y	0 < 8	R_M	R_R	Y
	x_1 (0000)	1 ≥ 0	N	0 < 4		R_L	Y
	x_7 (0110)	1 < 2	Y	0 < 2		R_R	Y
	x_6 (0101)	1 < 2	Y	0 < 1		R_R	Y
x_6 (0101)	x_{14} (1101)	2 < 3	Y	1 < 9	R_R	R_R	Y
	x_2 (0001)	2 ≥ 1	N	1 < 3		R_L	Y
	x_8 (0111)	2 < 3	Y	1 < 3		R_R	Y
	x_5 (0100)	2 ≥ 1	N	1 ≥ 0		R_R	N
x_7 (0110)	x_{15} (1110)	2 < 3	Y	2 < 10	R_R	R_R	Y
	x_3 (0010)	2 ≥ 1	N	2 ≥ 2		R_R	N
	x_5 (0100)	2 ≥ 1	N	2 ≥ 0		R_R	N
	x_8 (0111)	2 < 3	Y	2 < 3		R_R	Y
x_8 (0111)	x_{16} (1111)	3 < 4	Y	3 < 11	R_R	R_R	Y
	x_4 (0011)	3 ≥ 2	N	3 ≥ 1		R_R	N
	x_6 (0101)	3 ≥ 2	N	3 ≥ 1		R_R	N
	x_7 (0110)	3 ≥ 2	N	3 ≥ 2		R_R	N
x_9 (1000)	x_1 (0000)	1 ≥ 0	N	4 ≥ 4	R_R	R_R	N
	x_{13} (1100)	1 < 2	Y	4 < 8		R_R	Y
	x_{11} (1010)	1 < 2	Y	4 < 6		R_R	Y
	x_{10} (1001)	1 < 2	Y	4 < 5		R_R	Y

continued on next page

Table 3.1 (*continued*)

		$\mathbb{P}_1$		$\mathbb{P}_4$			
s	v	$f(s):f(v)$	Move	$f(s):f(v)$	Range of s^t	Range of s^{t+1}	Move
x_{10} (1001)	x_2 (0001)	$2 \geq 1$	N	$5 \geq 3$		R_R	N
	x_{14} (1101)	$2 < 3$	Y	$5 < 9$	R_R	R_R	Y
	x_{12} (1011)	$2 < 3$	Y	$5 < 7$		R_R	Y
	x_9 (1000)	$2 \geq 1$	N	$5 \geq 4$		R_R	Y
x_{11} (1010)	x_3 (0010)	$2 \geq 1$	N	$6 \geq 2$		R_R	N
	x_{15} (1110)	$2 < 3$	Y	$6 < 10$	R_R	R_R	Y
	x_9 (1000)	$2 \geq 1$	N	$6 \geq 4$		R_R	N
	x_{12} (1011)	$2 < 3$	Y	$6 < 7$		R_R	Y
x_{12} (1011)	x_4 (0011)	$3 \geq 2$	N	$7 \geq 1$		R_R	N
	x_{16} (1111)	$3 < 4$	Y	$7 < 11$	R_R	R_R	Y
	x_{10} (1001)	$3 \geq 2$	N	$7 \geq 5$		R_R	N
	x_{11} (1010)	$3 \geq 2$	N	$7 \geq 6$		R_R	N
x_{13} (1100)	x_5 (0100)	$2 \geq 1$	N	$8 \geq 0$		R_R	N
	x_9 (1000)	$2 \geq 1$	N	$8 \geq 4$	R_R	R_R	N
	x_{15} (1110)	$2 < 3$	Y	$8 < 10$		R_R	Y
	x_{14} (1101)	$2 < 3$	Y	$8 < 9$		R_R	Y
x_{14} (1101)	x_6 (0101)	$3 \geq 2$	N	$9 \geq 1$		R_R	N
	x_{10} (1001)	$3 \geq 2$	N	$9 \geq 5$	R_R	R_R	N
	x_{16} (1111)	$3 < 4$	Y	$9 < 11$		R_R	Y
	x_{13} (1100)	$3 \geq 2$	N	$9 \geq 8$		R_R	N
x_{15} (1110)	x_7 (0110)	$3 \geq 2$	N	$10 \geq 2$		R_R	N
	x_{11} (1010)	$3 \geq 2$	N	$10 \geq 6$	R_R	R_R	N
	x_{13} (1100)	$3 \geq 2$	N	$10 \geq 8$		R_R	N
	x_{16} (1111)	$3 < 4$	Y	$10 < 11$		R_R	Y
x_{16} (1111)	x_8 (0111)	$4 \geq 3$	N	$11 \geq 3$		R_R	N
	x_{12} (1011)	$4 \geq 3$	N	$11 \geq 7$	R_R	R_R	N
	x_{14} (1101)	$4 \geq 3$	N	$11 \geq 9$		R_R	N
	x_{15} (1110)	$4 \geq 3$	N	$11 \geq 10$		R_R	N

amount of time to solve an optimization problem; it is thus still a feasible way for solving many complex optimization problems.

The above observations do not mean that we should throw ES and HC away. In fact, these two algorithms are still very useful for solving optimization problems; in particular, they are used to assist other search algorithms in different ways nowadays. Some local search methods are basically variants of ES, which limit the search regions and are very useful in helping other search algorithms to fine-tune the final search result. On the other hand, HC can also be used to fine-tune the final search result when combined

Table 3.2 Probability of moving to R_L and R_R in solving the BSD-2 problem.

s	Move to R_L	Move to R_R
x_1 (0000)	$\frac{1}{16} \times \frac{4}{4} = \frac{4}{64}$	$\frac{1}{16} \times \frac{0}{4} = \frac{0}{64}$
x_2 (0001)	$\frac{1}{16} \times \frac{3}{4} = \frac{3}{64}$	$\frac{1}{16} \times \frac{1}{4} = \frac{1}{64}$
x_3 (0010)	$\frac{1}{16} \times \frac{3}{4} = \frac{3}{64}$	$\frac{1}{16} \times \frac{1}{4} = \frac{1}{64}$
x_4 (0011)	$\frac{1}{16} \times \frac{2}{4} = \frac{2}{64}$	$\frac{1}{16} \times \frac{2}{4} = \frac{2}{64}$
x_5 (0100)	$\frac{1}{16} \times \frac{1}{4} = \frac{1}{64}$	$\frac{1}{16} \times \frac{3}{4} = \frac{3}{64}$
x_6 (0101)	$\frac{1}{16} \times \frac{1}{4} = \frac{1}{64}$	$\frac{1}{16} \times \frac{3}{4} = \frac{3}{64}$
x_7 (0110)	$\frac{1}{16} \times \frac{0}{4} = \frac{0}{64}$	$\frac{1}{16} \times \frac{4}{4} = \frac{4}{64}$
x_8 (0111)	$\frac{1}{16} \times \frac{0}{4} = \frac{0}{64}$	$\frac{1}{16} \times \frac{4}{4} = \frac{4}{64}$
x_9 (1000)	$\frac{1}{16} \times \frac{0}{4} = \frac{0}{64}$	$\frac{1}{16} \times \frac{4}{4} = \frac{4}{64}$
x_{10} (1001)	$\frac{1}{16} \times \frac{0}{4} = \frac{0}{64}$	$\frac{1}{16} \times \frac{4}{4} = \frac{4}{64}$
x_{11} (1010)	$\frac{1}{16} \times \frac{0}{4} = \frac{0}{64}$	$\frac{1}{16} \times \frac{4}{4} = \frac{4}{64}$
x_{12} (1011)	$\frac{1}{16} \times \frac{0}{4} = \frac{0}{64}$	$\frac{1}{16} \times \frac{4}{4} = \frac{4}{64}$
x_{13} (1100)	$\frac{1}{16} \times \frac{0}{4} = \frac{0}{64}$	$\frac{1}{16} \times \frac{4}{4} = \frac{4}{64}$
x_{14} (1101)	$\frac{1}{16} \times \frac{0}{4} = \frac{0}{64}$	$\frac{1}{16} \times \frac{4}{4} = \frac{4}{64}$
x_{15} (1110)	$\frac{1}{16} \times \frac{0}{4} = \frac{0}{64}$	$\frac{1}{16} \times \frac{4}{4} = \frac{4}{64}$
x_{16} (1111)	$\frac{1}{16} \times \frac{0}{4} = \frac{0}{64}$	$\frac{1}{16} \times \frac{4}{4} = \frac{4}{64}$
Expected value	0.21875	0.78125

with other global search algorithms. In addition to ES and HC, in recent years, several successful applications of metaheuristics showed that they can provide a much more efficient way to solve many optimization problems than ES and HC do in terms of both computation time and the end result. We will therefore turn our attention to metaheuristics in the next chapter.

It can be easily seen from the simulation results described in Section 3.3.1 that ES is not suitable for solving large-scale optimization problems, for it has to check every possible solution in the solution space. Although ES is guaranteed to find the optimal solution, it may take an unreasonable amount of time, say, days, years, or even centuries for large optimization problems. An important indicator that would help us decide if ES is suitable for the problem we are facing is whether we can use it to solve the problem in a "reasonable amount of time." This means that if ES cannot give us the answer in a reasonable amount of time, then it is not suitable for the problem we are facing even though it is guaranteed to find the optimal solution and get its objective value. Otherwise, ES should be used to solve

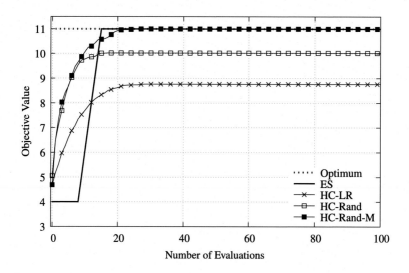

Figure 3.15 Simulation results of HC for a deceptive problem of size $n = 4$.

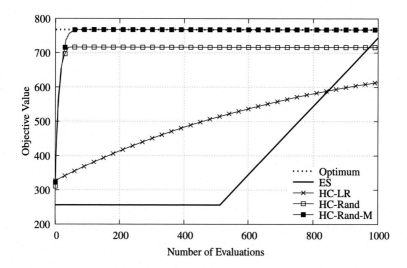

Figure 3.16 Simulation results of HC for a deceptive problem of size $n = 10$.

the problem. Another indicator is whether we need the "exact result." In practice, we do not need the exact result for many problems; rather, an approximate solution is generally acceptable or even good enough, for example in situations such as real-time traffic control. Although it is nice to find optimal solutions for optimization problems, the reality is that we often simply do not have enough computing power to find optimal solutions in a reasonable time. In practice, an alternative way to solve optimization problems is to find approximate solutions instead.

Supplementary source code

Here are the hyperlinks to the directories where the source code described in this chapter resides.

https://github.com/cwtsaiai/metaheuristics_2023/src/c++/0-es/1-onemax/

https://github.com/cwtsaiai/metaheuristics_2023/src/c++/1-onemax/

https://github.com/cwtsaiai/metaheuristics_2023/src/c++/2-deception/

CHAPTER FOUR

Metaheuristic algorithms

Before we move on to the details of each metaheuristic algorithm, this chapter will start with a brief introduction to the metaheuristic algorithms in Section 4.1, followed by a detailed description of a unified framework, which is designed to provide a "shortcut" to the world of metaheuristics, in Section 4.2. This framework provides a central idea of metaheuristics, thus making it easier for the novice to understand the basic ideas of metaheuristics. After that, some further comparisons between the exhaustive search (ES), greedy algorithm, and metaheuristic algorithms will be made in Section 4.3.

4.1. What is a metaheuristic algorithm?

It is generally believed that the term "metaheuristic," which is formed from two Greek words—*meta* and *heuristic*—was coined by Glover (Glover, 1986). One of the main attributes of a metaheuristic algorithm is that it performs a certain set of operators for a certain number of iterations to search for the optimal solution with great shrewdness. Unlike generating and checking all the candidate solutions systematically, the basic idea of metaheuristics is to find an approximate solution out of a very large solution space for a complex problem in a "reasonable time" via a "strategic guess." Compared to the ES and hill climbing (HC) algorithms described in Chapter 3, a metaheuristic algorithm will neither check all the candidate solutions of a complex optimization problem like ES nor fall into a local optimum at early iterations as easily as HC. Since metaheuristics (Blum and Roli, 2003; Glover and Laguna, 1997; Reeves, 1993) have been widely used and discussed for years, they can be regarded as a distinguished school of contemporary search methods.

Although the term "metaheuristic" was introduced in the 1980s, several metaheuristic algorithms (Blum and Roli, 2003) were actually presented in the 1960s or even earlier. For example, evolutionary algorithms (EAs) were developed in three sources (Beyer and Schwefel, 2002) that can be dated back to the 1960s. They are evolutionary programming (EP) (Fogel et al., 1966), developed in San Diego, the genetic algorithm (GA) (Holland, 1962), developed in Ann Arbor, and evolution strategies (ES)

Handbook of Metaheuristic Algorithms
https://doi.org/10.1016/B978-0-44-319108-4.00017-4

(Rechenberg, 1965), developed in Berlin. In addition to the EAs, many other kinds of metaheuristic algorithms (Blum and Roli, 2003) have also been presented for solving large-scale or complex optimization problems, by using a well-defined function to evaluate the searched solutions and a strategic guess to smartly determine the search directions during the convergence process. Among them, the tabu search (TS) (Glover, 1989, 1990a; Glover and Laguna, 1997), simulated annealing (SA) (Černý, 1982; Kirkpatrick et al., 1983), GA (Holland, 1962, 1975), differential evolution (DE) (Storn and Price, 1995, 1997), ant colony optimization (ACO) (Colorni et al., 1992; Dorigo and Gambardella, 1997; Dorigo et al., 1996; Dorigo and Stützle, 2004), and particle swarm optimization (PSO) (Kennedy and Eberhart, 1995; Poli et al., 2007) are the representative metaheuristic algorithms presented before 2000. Because of the successful use cases of these kinds of search algorithms, metaheuristic algorithms spring up like mushrooms. Several engineering and commercial applications show that metaheuristic algorithms are able to find good results in a reasonable time. This is one of the factors that boost the development of metaheuristic algorithms.

The year 1990 can be regarded as the first watershed in the development of metaheuristic algorithms. However, the available computing power may not satisfy the requirements of some complicated metaheuristic algorithms. Hence, some of them are difficult to realize, and for those that can be realized, the performance is not optimal. Because of this, in the early stages, TS and SA were very popular and widely used in several studies and intelligent computer systems. Computer technology has made rapid progress in the 1990s. The Internet and the world wide web (WWW) make it possible to interconnect millions of computers and users together on a global network; thus, communication and information sharing between researchers have been much faster. With the development of computer systems and distributed computing environments, several successful use cases of EAs have shown their possibilities. As a consequence, the number of studies on EA has increased significantly in that period. Another paradigm of metaheuristic algorithms is swarm intelligence, which was proposed in the middle of the 1990s. ACO and PSO are two representative search algorithms of this paradigm which both use search experience of the entire population and social behavior to determine the search strategies and improve the search performance of metaheuristic algorithms. Based on our experience, ACO-based algorithms are very useful for solving combinatorial optimization problems, whereas PSO-based algorithms are very efficient and effective for solving continuous optimization problems.

Since the late 1990s, the number of metaheuristic algorithms has exploded. A set of new metaheuristic algorithms have also been presented to provide "better ways" to solve optimization problems. They include the estimation of distribution (EDA) algorithm (Baluja, 1994; Larraanaga and Lozano, 2001; Mühlenbein and Paaß, 1996; Pelikan et al., 2000), harmony search (HS) (Geem et al., 2001), artificial bee colony (ABC) (Karaboga, 2005), the imperialist competitive algorithm (ICA) (Atashpaz-Gargari and Lucas, 2007), gravitational search algorithm (GSA) (Rashedi et al., 2009), cuckoo search (CS) (Yang and Deb, 2009), the bat algorithm (BA) (Yang, 2010), and spiral optimization (SO) (Kenichi and Keiichiro, 2010). In addition to these new metaheuristic algorithms, many other kinds of new metaheuristic algorithms have also been presented after the year 2000. They can be found in several articles and technical reports, and even Wikipedia has a page to list new metaheuristic algorithms (https:// en.wikipedia.org/wiki/List_of_metaphor-based_metaheuristics or https:// en.wikipedia.org/wiki/Talk:Metaheuristic/List_of_Metaheuristics). It can be easily seen that most of them have a more complicated design to improve their search performance. Also, some of them pay more attention to balancing the intensification and diversification than those presented before 2000. It has been shown that making every search meaningful is a good way to enhance the search performance of a metaheuristic algorithm. This may be one of the reasons why more and more studies use "the number of evaluations" to replace "the number of iterations" for evaluating the performance of a metaheuristic algorithm because each search can be regarded as an investment of computation resource (Tsai, 2015), just like drawing a card (checking a candidate solution) from a deck of cards (all possible candidate solutions). As such, using the number of evaluations provides a way that is more precise than using the number of iterations to evaluate the effect (improvement or outcome) of adding an additional unit of computation resource for the search. Since the 2000s or even earlier, another promising research direction has been to dynamically adjust the parameter settings of a metaheuristic algorithm during the convergence process. For example, Ratnaweera et al. (Ratnaweera et al., 2004) presented a modified version of PSO that dynamically adjusts the acceleration coefficients during the convergence process in order to find a better result than PSO with fixed acceleration coefficients.

From the year 2010 or even earlier, some groups have attempted to apply metaheuristic algorithms in high-performance computing environments. Using distributed or parallel computing systems to accelerate the response time of metaheuristic algorithms is an intuitive approach adopted in some early studies (Alba, 2005; Alba et al., 2013). Some of the parallel metaheuristics are not only able to provide the end results to the user more quickly; they also can find better results than metaheuristic algorithms on a single machine because the parallel computing mechanism leads them to increase the search diversity during the convergence process. When we look at these distributed and parallel computing environments, the cloud computing platform (e.g., Hadoop, Spark, Microsoft Azure, Amazon EC2, or Google Compute Engine) can now provide an easy way to use a distributed computing system to further reduce the response time of metaheuristics. In such a distributed computing system, the communication cost is a critical issue for reducing the overall response time of metaheuristics because it may spend more time to find the results than when using a single machine if the computation cost is not high enough to offset the communication cost between the master and its workers. Another noteworthy research trend is the continuous optimization problem, such as single-objective or multi-objective real-parameter optimization problems. The contribution of PSO (Kennedy and Eberhart, 1995) and DE (Storn and Price, 1997) cannot go unnoticed because they both encode each solution as a vector and transit the current solution to another candidate solution via velocity updates, for vectors provide a useful way for metaheuristic algorithms to find better results than the transition operator using perturbation or recombination.

4.2. A unified framework for metaheuristic algorithms

The appearance of metaheuristic algorithms has come with methods to classify them. These classification methods (Blum and Roli, 2003; Sultan et al., 2008) include: (1) nature-inspired vs. non-nature inspired, (2) dynamic vs. static objective function, (3) one vs. various neighborhood structures, (4) memory usage vs. memoryless methods, (5) with vs. without local search method, and (6) population-based vs. single-solution-based search. Very many metaheuristic algorithms are out there today and most of them are described in different ways, which is a barrier for beginners in this research domain. In this section, a unified framework for metaheuristics (UFM) will be used to describe well-known metaheuristic algorithms and their variants to overcome this barrier. We do believe that this will make it easier for the audience of this book to understand how metaheuristic algorithms discussed in this book are realized.

Algorithm 3 Unified framework for metaheuristics.

1 $s = \text{Initialization}(\mathbb{I})$	$\boxed{\text{I}}$
2 $f_s = \text{Evaluation}(s)$	$\boxed{\text{E}}$
3 **While** the termination criterion is not met	
4 $v = \text{Transition}(s)$	$\boxed{\text{T}}$
5 $f_v = \text{Evaluation}(v)$	$\boxed{\text{E}}$
6 $s, f_s = \text{Determination}(v, s, f_s, f_v)$	$\boxed{\text{D}}$
7 **End**	
8 Output s	$\boxed{\text{O}}$

As shown in Algorithm 3, an outline of a UFM is used to explain the basic idea of metaheuristics, which contain five main operators: *initialization, transition, evaluation, determination*, and *output*. In this framework, $\mathbb{I}$ denotes the input dataset, s denotes the current solution, v denotes the candidate solution, f_s denotes the objective value of s, and f_v denotes the objective value of v. Also, s and v can denote either a single solution or a set of solutions, where each solution has n elements or is an n-tuple. To simplify the discussion that follows, the five main operators will be denoted by **I**, **T**, **E**, **D**, and **O**. In fact, we have used this notation in the previous discussion of ES (Algorithm 1 in Section 3.1.1) and HC (Algorithm 2 in Section 3.2.1) because this notation also fits these two search algorithms. The role of each operator is discussed below.

1. **Initialization:** The initialization operator normally plays the roles of reading the input file (e.g., dataset), initializing all the parameters of a metaheuristic algorithm, and determining the initial solutions, which is normally based on a random process.

2. **Transition:** The transition operator usually plays the role of varying the search directions, such as perturbing a portion of the subsolutions of the current solution to generate a new candidate solution or generating a set of new candidate solutions each based on two or more of the current solutions.

3. **Evaluation:** The evaluation operator is responsible for measuring the quality of solutions, such as calculating the objective value of each solution to be used by the determination operator to distinguish the quality of all the solutions. An intuitive way is to use an objective function to measure the quality of a solution for the problem in question. However, some metaheuristics do not use the "objective value" directly to measure their solutions; rather, the objective value of a solution has to undergo some sort of transformation to obtain the so-called "fitness value."

4. **Determination:** The determination operator plays the role of deciding the search directions by using information the evaluation operator provides during the convergence process. The performance of a metaheuristic algorithm depends to a large extent on the performance of this operator. A "good" search strategy for this operator will make it possible for the metaheuristic algorithm to find a better solution faster or to avoid falling into a local optimum at early iterations.

5. **Output:** In spite of the fact that this operator seems to be trivial, the reality is that it can be either simple or complex depending on how much information we want to display for the metaheuristic algorithm. It can be as simple as displaying only the final result of the metaheuristic algorithm, or it can be as complex as displaying the trajectory of convergence of the metaheuristic algorithm to better understand the performance of a metaheuristic algorithm.

As shown in Algorithm 3, the initialization and output operators will be performed only once; the transition, evaluation, and determination operators, however, will be performed repeatedly for a certain number of times. To understand a new metaheuristic algorithm, the very first thing to do is certainly to understand the basic ideas of the transition, evaluation, and determination operators, which will in turn help us understand the search attributes of the metaheuristic algorithm. For this reason, we will use this framework to explain the other metaheuristics in the following chapters.

4.3. Comparisons of metaheuristics with exhaustive and greedy search

From the perspective of algorithm design, the transition, evaluation, and determination operators can be regarded as the "core" of the algorithm, which is aimed at searching for the solution; the initialization operator can be regarded as the "interface" between the input dataset and the metaheuristic algorithm; and the output operator can be regarded as the "interface" between the solution and the human being or the other algorithms, including metaheuristics. A metaheuristic algorithm is, of course, different from ES and greedy algorithms. To better understand the main differences between the ES, greedy, and metaheuristic algorithms, Table 4.1 shows a comparison between the main operators of these algorithms that are employed in most cases.

For ES, the initial solution is typically given; that is, not created randomly. Its transition and determination operators will follow some pre-

Table 4.1 Comparison between exhaustive search, greedy, and metaheuristic algorithms.

Operator	Exhaustive search	Greedy algorithm	Metaheuristic algorithm
Initialization	Non-random	Random	Random
Transition	Systematical	Partially random	Partially random
Evaluation	Objective value	Objective value	Objective/fitness value
Determination	Systematical	Deterministic	Strategically
Output	Optimum	Local optimum	Local optimum

defined rules to systematically check all candidate solutions based on the objective value. This kind of search method is guaranteed to find the optimal solution; however, it may take an unreasonable amount of time, especially when the solution space is large.

Unlike ES, the initial solution of the greedy algorithm will be created randomly. Since this kind of search method will not check all candidate solutions during the convergence process, there is no guarantee that it will find the optimal solution; it is, however, guaranteed to find an approximate solution in a reasonable time. A typical operation of the transition operator is to randomly change a portion of the subsolutions of the current solution to generate a new candidate solution. The determination operator will follow some predefined rules to find the approximate solution "deterministically," for example determining the next search direction greedily. Similar to ES, most greedy algorithms use the objective value of a solution to measure the quality of the solution obtained from the transition operator.

Compared to the ES and greedy algorithms, a metaheuristic algorithm is like an evolutionary version of a greedy algorithm in the sense that much more randomness kicks in during the convergence process. Like the greedy algorithm, the transition operator of a metaheuristic algorithm will also use a random procedure to generate the new candidate solution. Unlike the greedy algorithm, the new candidate solution will be generated from one or more searched solutions and relevant information the metaheuristic algorithm has. Although more randomness kicks in, the determination operator of metaheuristics still relies on a strategic guess to find the approximate solution instead of counting on a random walk in the solution space. In addition to information provided by the objective value, some metaheuristic algorithms use additional information to distinguish between solutions or to search for directions that have a better chance to find potential solutions. For example, the information provided by the accumulated search experience of a search algorithm—such as the trajectory of a solution, the personal best solution from the very beginning to the current iteration, and the (so far) global best solution, all of which affect the search direction of the next iteration—is used by PSO (Kennedy and Eberhart, 1995) to facilitate its search for a solution.

In summary, the computation cost of ES is higher than that of greedy and metaheuristic algorithms for solving optimization problems in most cases; however, it is guaranteed to find the global optimum, for it will check every possible solution in the solution space. This makes it impractical to use ES to solve optimization problems. An alternative solution is the so-called greedy algorithm. Besides the addition of randomness to the search process, it also reduces the computation time of the search process because it is greedy; thus, it will not check every candidate solution in the solution space. One disadvantage of some greedy algorithms is that they are more likely to fall into a local optimum at early iterations (also called premature[1]); as a result, the final solution is usually far from optimal. To overcome this issue, metaheuristic algorithms not only maintain the randomness but also employ a strategic guess to search possible good solutions and prevent the search from falling into a local optimum at early iterations. Hence, mechanisms to keep the search from falling into a local optimum are embedded in most metaheuristic algorithms. For instance, embedded in SA (Černý, 1982; Kirkpatrick et al., 1983) is a probability mechanism to allow the search process to accept a non-improving solution as the next solution while embedded in GA (Holland, 1975) is a mutation operator to randomly perturb a portion of the subsolutions of the current solution to create a new candidate solution. The basic idea of such mechanisms is typically to prevent the search process from staying in the same region in the solution space for a long time to avoid the search from falling into a local optimum. From these perspectives, it is easily seen that the key factor is the design of the transition, evaluation, and determination operators of metaheuristics, which are different from those of the ES and greedy algorithms. In the next chapter, we will turn our attention to the details of metaheuristics, beginning with the basic ideas underlying these algorithms, followed by how they are implemented and how they work.

---

[1] By the premature problem of a search algorithm, we mean that the search process of a search algorithm is likely to fall into a local optimum at early iterations, thus getting stuck in a particular region, making it very difficult or impossible to improve its result.

CHAPTER FIVE

Simulated annealing

5.1. The basic idea of simulated annealing (SA)

The simulated annealing (SA) algorithm was independently presented by Kirkpatrick et al. (Kirkpatrick et al., 1983) and Černý et al. (Černý, 1982, 1985). Some similar methods can also be found in other studies (Khachaturyan et al., 1981; Pincus, 1970). The basic idea of SA is to occasionally accept non-improving solutions, which means that SA will not always move to a better solution.

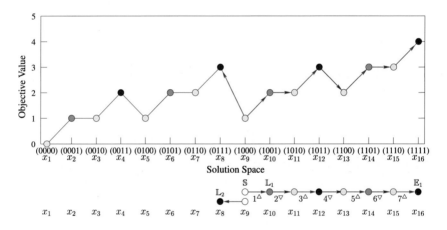

Figure 5.1 The search strategy of SA.

As shown in Fig. 5.1, the search strategy of SA is to start with a random possible solution in the solution space and then use the Metropolis acceptance criterion (Metropolis et al., 1953) P_a to determine whether a worse solution is to be accepted or not. In this example, $t^\triangle$ and $t^\triangledown$ indicate that the solution at the $(t + 1)$-th iteration is either better or not better than the solution at the t-th iteration, respectively. This example shows that if the starting point is x_9 and SA accepts only a new candidate solution that is better than the current solution, i.e., it has no escape mechanism, the search will get stuck at one of the two local optima x_8 and x_{10} (denoted $\mathbb{L}_1$ and $\mathbb{L}_2$). Fortunately, the search strategy of SA is to have a small chance to accept a non-improving solution as the next solution; as a consequence, if SA is currently at x_{10}, it still has a chance to move to x_{11} even though x_{11} is not better than x_{10} and then go on to an even better solution at a later stage of the convergence process. In this way, it has a certain chance to "escape" a local optimum during the convergence process. Different from HC, which might get stuck at one of the two local optima $\mathbb{L}_1$ and $\mathbb{L}_2$ if the search starts at x_9, SA may end up at the optimal solution $\mathbb{E}_1$ or at other local optima.

As shown in Algorithm 4, the main differences between SA and HC are as described in lines 1, 7, and 11, namely, the temperature at the t-th iteration Ψ_t, the probabilistic acceptance criterion P_a, and the annealing schedule.

Algorithm 4 Simulated annealing.

1 Set the initial temperature Ψ_0	I
2 Randomly create the initial solution s	I
3 $f_s = $ Evaluate(s)	E
4 **While** the termination criterion is not met	
5 $v = $ NeighborSelection(s)	T↦E
6 $f_v = $ Evaluate(v)	E
7 If $r < P_a$	D
8 $s = v$	
9 $f_s = f_v$	
10 Endif	
11 Update Ψ_t according to the annealing schedule	D
12 **End**	
13 **Output** s	O

To emulate the annealing process for a minimization optimization problem, SA will first calculate the difference between the objective values of the new candidate solution v and the current solution s to see whether it will accept the new candidate solution or not, as follows:

$$\Delta_f^{\min} = f(v) - f(s). \tag{5.1}$$

In case the difference between the objective values $\Delta_f^{\min}$ is less than 0, SA will accept the new candidate solution as the current solution, which means that v will replace s; otherwise, SA will calculate a probability $P_a^{\min}$ to decide whether or not to accept a non-improving candidate solution, as follows:

$$P_a^{\min} = \exp\left(\frac{-(f(v) - f(s))}{\Psi_t}\right) = \exp\left(\frac{f(s) - f(v)}{\Psi_t}\right), \tag{5.2}$$

where $f()$ denotes the evaluation function, s denotes the current solution, v denotes the new solution, and Ψ_t denotes the temperature at the t-th iteration.

To apply SA to a maximization optimization problem, all we have to do is to negate the difference in Eq. (5.1) so that it looks as follows:

$$\Delta_f^{max} = -\Delta_f^{min} = -(f(v) - f(s)) = f(s) - f(v). \tag{5.3}$$

Such a modification makes it possible to check Δ_f^{max} in a similar way; that is, if $\Delta_f^{max} < 0$, SA will accept the new candidate solution as the current solution; otherwise, SA will again calculate a probability P_a^{max} to decide whether or not to accept a non-improving candidate solution, as follows:

$$P_a^{max} = \exp\left(\frac{-(f(s) - f(v))}{\Psi_t}\right) = \exp\left(\frac{f(v) - f(s)}{\Psi_t}\right). \tag{5.4}$$

In summary, for minimization optimization problems, SA uses Δ_f^{min} and P_a^{min} to determine whether or not to accept a new candidate solution, while for maximization optimization problems, Δ_f^{max} and P_a^{max} are used. Table 5.1 gives some examples to show how the values of Δ_f^{min}, P_a^{min}, Δ_f^{max}, and P_a^{max} change when Ψ_t, $f(v)$, and $f(s)$ assume different values.

Table 5.1 The values of P_a^{min} and P_a^{max} of SA.

Ψ_t	$f(v)$	$f(s)$	Δ_f^{min}	P_a^{min}	Δ_f^{max}	P_a^{max}
0.9	1.0	2.0	−1.0	3.038	1.0	0.329
0.9	1.0	1.1	−0.1	1.118	0.1	0.895
0.2	1.0	2.0	−1.0	148.413	1.0	0.007
0.2	1.0	1.1	−0.1	1.649	0.1	0.607
0.9	2.0	1.0	1.0	0.329	−1.0	3.038
0.9	1.1	1.0	0.1	0.895	−0.1	1.118
0.2	2.0	1.0	1.0	0.007	−1.0	148.413
0.2	1.1	1.0	0.1	0.607	−0.1	1.649

In this table, Ψ_t assumes two different values (i.e., 0.9 and 0.2) to show how it affects the value of P_a. For example, for a maximization optimization problem, assume $f(v) = 1.0$ and $f(s) = 2.0$. This means that the new candidate solution v is worse than the current solution s because $\Delta_f^{max} = f(s) - f(v) > 0$; thus, SA will not always accept the candidate solution as the current solution. In this case, $P_a^{max} = 0.329$ and 0.007 for $\Psi_t = 0.9$ and 0.2, respectively. This indicates that a lower temperature Ψ_t

implies a smaller probability P_a to accept a non-improving candidate solution as the current solution. Another interesting thing that can be observed from this table is that the difference between $f(v)$ and $f(s)$ can also affect the value of P_a. For example, for the maximization optimization problem again, assume $\Psi_t = 0.9$ and $f(v) = 1.0$. It can be easily seen that when the value of $f(s)$ goes from 2.0 to 1.1, the value of $P_a^{\max}$ goes from 0.329 up to 0.895. This means that in case $f(v)$ is worse than $f(s)$, a smaller $\Delta_f^{\max}$ implies a higher probability $P_a^{\max}$ to accept a non-improving candidate solution as the current solution. The same goes for $P_a^{\min}$ for minimization optimization problems.

Now, let us move back to the discussion of Algorithm 4. As shown in line 7, if the new solution v satisfies the probabilistic acceptance criterion because $r < P_a$, where r is a uniformly distributed random number in the range $[0, 1]$, the current solution s and its objective value f_s will be replaced by the new solution v and its objective value f_v, respectively. The probabilistic acceptance criterion P_a, as defined in Eqs. (5.2) and (5.4), can be divided into three cases:

- The new solution v is better than the current solution s. In this case, SA will always accept the new solution owing to the fact that P_a will always be greater than 1.0 and $r \in [0, 1]$ will always be smaller than P_a.
- The new solution v is worse than the current solution s. In this case, SA will accept the new solution only if $r < P_a$, where r is as defined above.
- The new solution v is worse than the current solution s. In this case, SA will not accept the new solution because $r \geq P_a$.

A simple way to emulate the annealing schedule is to define the annealing schedule as follows:

$$\Psi_{t+1} = \Psi_t \times \alpha, \tag{5.5}$$

where α is typically somewhere in the range $(0, 1)$. It is easily seen that a larger value of α will slow down the annealing rate, whereas a smaller value of α will speed up the annealing rate.

5.2. Implementation of SA for the one-max and deceptive problems

As shown in Listing 5.1, in addition to the parameters for HC, which are the number of runs, the number of iterations, the length of the solution, and the filename for the initial solution, SA takes three additional parameters, namely, the number of candidates, the maximum temperature, and the minimum temperature for the determine() and annealing() operators to emulate the annealing schedule. A simple example to run the SA program provided in this book is as given below.

```
$ ./search sa 100 1000 10 "" 0.00001 1.0
```

In this example, the parameters say that "SA" is used as the search algorithm, and it will be carried out for "100" runs, with "1000" evaluations per run, with the length of the solution being "10." The empty string "" indicates that without a filename specified, the initial solutions will be created by a random process. The minimum temperature is set equal to "0.00001," and the maximum temperature is set equal to "1.0."

Listing 5.1: src/c++/1-onemax/main/sa.h: The implementation of SA.

```
 1 #ifndef __SA_H_INCLUDED__
 2 #define __SA_H_INCLUDED__
 3
 4 #include <iostream>
 5 #include <string>
 6 #include <fstream>
 7 #include <numeric>
 8 #include "lib.h"
 9
10 using namespace std;
11
12 class sa
13 {
14 public:
15     typedef vector<int> solution;
16
17     sa(int num_runs,
18        int num_evals,
19        int num_patterns_sol,
20        string filename_ini,
21        double min_temp,    // sa-specific parameter
22        double max_temp     // sa-specific parameter
23        );
24
25     virtual ~sa() {}
26
27     solution run();
28
29 private:
30     void init();
31     virtual int evaluate(const solution& s);
32     virtual solution transit(const solution& s);
33
34     // begin the sa functions
35     bool determine(int tmp_obj_val, int obj_val, double temperature);
36     double annealing(double temperature);
```

```
37      // end the sa functions
38
39 private:
40      int num_runs;
41      int num_evals;
42      int num_patterns_sol;
43      string filename_ini;
44
45      solution sol;
46      int obj_val;
47
48      solution best_sol;
49      int best_obj_val;
50
51      // begin the sa parameters
52      double min_temp;
53      double max_temp;
54      double curr_temp;
55      // end the sa parameters
56 };
57
58 inline sa::sa(int num_runs,
59              int num_evals,
60              int num_patterns_sol,
61              string filename_ini,
62              double min_temp,  // sa-specific parameter
63              double max_temp   // sa-specific parameter
64              )
65     : num_runs(num_runs),
66       num_evals(num_evals),
67       num_patterns_sol(num_patterns_sol),
68       filename_ini(filename_ini),
69       min_temp(min_temp),
70       max_temp(max_temp)
71 {
72      srand();
73 }
74
75 inline sa::solution sa::run()
76 {
77      double avg_obj_val = 0;
78      vector<double> avg_obj_val_eval(num_evals, 0);
79
80      for (int r = 0; r < num_runs; r++) {
81          int eval_count = 0;
82
83          // 0. Initialization
84          init();
85          obj_val = evaluate(sol);
86          avg_obj_val_eval[eval_count++] += obj_val;
87
88          while (eval_count < num_evals) {
89              // 1. Transition
90              solution tmp_sol = transit(sol);
91
92              // 2. Evaluation
93              int tmp_obj_val = evaluate(tmp_sol);
94
95              // 3. Determination
96              if (determine(tmp_obj_val, obj_val, curr_temp)) {
97                  obj_val = tmp_obj_val;
98                  sol = tmp_sol;
99              }
100
101             if (obj_val > best_obj_val) {
```

```
102                    best_obj_val = obj_val;
103                    best_sol = sol;
104                }
105
106            curr_temp = annealing(curr_temp);
107
108                avg_obj_val_eval[eval_count++] += best_obj_val;
109            }
110          avg_obj_val += best_obj_val;
111      }
112
113      // 4. Output
114      avg_obj_val /= num_runs;
115
116      for (int i = 0; i < num_evals; i++) {
117          avg_obj_val_eval[i] /= num_runs;
118          cout << fixed << setprecision(3) << avg_obj_val_eval[i] << endl;
119      }
120
121      return sol;
122 }
123
124 inline void sa::init()
125 {
126      curr_temp = max_temp;
127      best_obj_val = 0;
128
129      sol = solution(num_patterns_sol);
130
131      if (!filename_ini.empty()) {
132          ifstream ifs(filename_ini.c_str());
133          for (size_t i = 0; i < sol.size(); i++)
134              ifs >> sol[i];
135      }
136      else {
137          for (size_t i = 0; i < sol.size(); i++)
138              sol[i] = rand() % 2;
139      }
140 }
141
142 inline sa::solution sa::transit(const solution& s)
143 {
144      return ::transit_r(s);
145 }
146
147 inline int sa::evaluate(const solution& s)
148 {
149      return accumulate(s.begin(), s.end(), 0);
150 }
151
152 inline bool sa::determine(int tmp_obj_val, int obj_val, double temperature)
153 {
154      double r = static_cast<double>(rand()) / RAND_MAX;
155      double p = exp((tmp_obj_val - obj_val)/temperature);
156      return r < p;
157 }
158
159 inline double sa::annealing(double temperature)
160 {
161      return 0.9 * temperature;
162 }
163
164 #endif
```

5.2.1 Declaration of functions and parameters

Although the structure of `sa.h` is very similar to that of `hc.h` for solving the one-max problem, there are some differences. Lines 51–55 show the additional parameters required by SA. The variables (or data members to be more specific) `max_temp`, `min_temp`, and `curr_temp` will be used to represent the maximum, minimum, and current temperatures for the annealing schedule. The variables `best_obj_val` and `best_sol` will be used for keeping the best-so-far solution of SA. Lines 58–70 show that the variables `num_runs`, `num_evals`, `num_patterns_sol`, `filename_ini`, `max_temp`, and `min_temp` will receive their values from `main.cpp`, which are to be provided by the user of the program.

5.2.2 The main loop

The main search function of SA is given in lines 75–122, following the template of Fig. 3.6 and Algorithm 4. Like HC, SA also performs the initialization operator init() before entering the convergence process. The transit operator of SA randomly creates a candidate solution v from the current solution s like HC. This candidate solution v will be assigned to the variable `tmp_sol`, as shown in line 90. Since SA uses the acceptance probability to determine whether to accept a new solution or not, as shown in line 96, it will call the function determine() to make the decision. In lines 101–104, SA will check to see if the objective value of the new solution is better than the best-so-far solution. If the answer is positive, it will update the best-so-far solution `best_sol` and its objective value `best_sol_val` accordingly. In addition, SA will call the annealing() operator at the end of each iteration to decrease the temperature of the annealing schedule iteration by iteration, as shown in line 106.

5.2.3 Additional functions

The init() operator of SA will also initialize the current temperature in the very beginning, as shown in line 126. The transit() operator of SA creates a new candidate solution from the current solution, as shown in lines 142–145. This operator, just like the transit() operator of HC-Rand, will simply invert a randomly chosen subsolution of the current solution to create a candidate solution. This means that we can use the same transition operator of HC defined in `lib.h` as the transition operator of SA. As shown in lines 152–157, the determine() operator will first randomly generate a random number uniformly distributed in the range [0, 1] and assign it to the variable r; it will then compute the acceptance probability p in line 155 using Eq. (5.4) because the one-max problem is a maximization optimization problem. Given the values in r and p, SA can now decide whether to accept the new solution or not. The very last operator of SA is the operator to decrease the temperature. As shown in lines 159–162, this operator is named annealing(), which will decrease the temperature for the annealing schedule by 10% each iteration.

Like HC, only the evaluation operator of SA, as shown in Listing 5.2, needs to be changed to apply the SA code, as shown in Listing 5.1, to the deceptive problem $\mathbb{P}_4$ described in Section 2.2.2. In fact, only lines 18–21 of the evaluation operator given in Listing 5.2 need to be changed so that the objective value is computed as $f(s) = |B2D(s) - 2^{n-2}|$, where n is the number of bits in a solution.

Listing 5.2: src/c++/2-deception/main/sadp.h: The evaluate() operator of SA for the deceptive problem.

```
16 inline int sadp::evaluate(const solution& s)
17 {
18     int n = 0;
19     for (size_t i = 0; i < s.size(); i++)
20         n += s[i] << ((s.size() - 1) - i);
21     return abs(n - (1 << (s.size()-2)));
22 } //
```

5.3. Simulation results of SA

In this section, the one-max and deceptive problems are used as examples to help us better understand the performance of SA. For the simulation results described herein, the minimum temperature is set equal to 0.00001 and the maximum temperature is set equal to 1.0. Each simulation is carried out for 100 runs and 1000 evaluations are performed each run.

5.3.1 Simulation results of SA for the one-max problem

As shown in Table 5.2, the number of iterations required to find the optimal solution by HC-Rand and SA will be increased as the problem size increases, and SA does not always beat HC for solving different one-max problems. This might be due to the fact that the one-max problem is not a complex optimization problem; thus, HC-Rand is able to escape local optima in the solution space.

Table 5.2 Number of iterations needed by HC and SA to find the optimal solution.

Problem size	HC-Rand	SA
$n = 10$	59	79
$n = 20$	145	148
$n = 40$	277	306
$n = 60$	486	460
$n = 80$	786	576
$n = 100$	965	826

Fig. 5.2 gives the trajectories of convergence of these two algorithms, which are very similar to each other for the one-max problem of size $n = 100$. Similar situations can also be found for these two algorithms for solving the one-max problem of different sizes. However, the story is not over yet because this kind of optimization problem cannot show the superiority of SA.

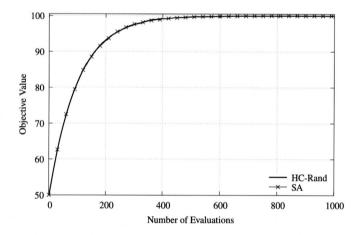

Figure 5.2 Convergence of HC and SA for the one-max problem of size $n = 100$.

5.3.2 Simulation results of SA for the deceptive problem

Fig. 5.3 compares the performance of HC and SA for a deceptive problem $\mathbb{P}_4$ of size $n = 4$. The result of HC-Rand shown here is exactly the same as that given in Fig. 3.10. For this deceptive problem, the objective value of the optimal solution is 11.0, but the end results of HC-Rand are far away from it on average. The results show that HC might fall into a local optimum and is not able to escape from it during the convergence process. This implies that HC is unable to find the optimal solution no matter how many iterations are performed when the convergence process falls into a local optimum. The figure also shows that SA is able to find the optimal solution for the deceptive problem $\mathbb{P}_4$ of size $n = 4$. This reveals that SA can find a better result than HC because it is capable of escaping local optima by accepting worse solutions during the convergence process.

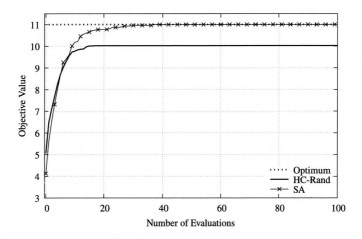

Figure 5.3 Simulation results of HC for a deceptive problem of size $n = 4$.

Similar results can be found in Fig. 5.4 for the deceptive problem $\mathbb{P}_4$ of size $n = 10$. SA, of course, can also find a better result than HC-Rand. An interesting thing that can be observed from this figure is that the convergence speed of SA might be lower than that of HC-Rand in a few periods at the early convergence process, but SA might be able to improve the best-so-far solution continuously at the later stages of the convergence process.

5.4. Discussion

A brief introduction to SA is given in this chapter, which shows that it is very simple and thus very easy to implement. The main difference between HC and SA is the determination operator; therefore, "a slight modification" of HC would make it an SA, as shown in Section 5.2. Based on the comparisons given in Section 5.3, we can now further infer that SA is able to find a better solution than HC because its determination operator

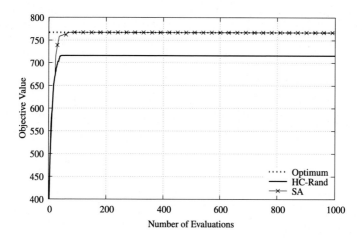

Figure 5.4 Simulation results of HC for a deceptive problem of size $n = 10$.

allows it to move to a worse solution during the convergence process. This means that compared to HC, which only accepts better solutions for later iterations, SA accepts not only better solutions but also worse solutions with a certain probability. This kind of search strategy makes it "a good strategy to escape local optima" and find better final solutions compared with HC when the problem in question is complex. The simulation results described in Section 5.3 also show that SA is able to escape local optima if a sufficient number of iterations are performed. These simulation results make it seem that SA is a perfect search algorithm for solving complex optimization problems; however, it has some open issues:

• The first open issue is how to determine the optimal parameter setting of SA because this will affect the convergence process. For example, if the maximum temperature (i.e., the initial temperature) is decreased, it will not only converge faster, but this will also reduce the probability of accepting a worse solution faster. There exists no perfect parameter setting of SA for all optimization problems because data are different, and so are the solution spaces, even for the same type of optimization problems. Different parameter settings may be more suitable for different data; therefore, how to determine the optimal parameter setting of SA has been an open issue.

• Two other open issues of SA are how to compute the acceptance probability and how to lower the temperature, because this will also affect the search strategy and the convergence speed of SA. For this reason, different studies may present different ways to compute the acceptance probability and to lower the temperature. Just like for the parameter setting, there exists no perfect solution for these issues.

The discussion in Sections 5.1, 5.2, and 5.3 is on a simple version, but certainly not the only version, of SA. In fact, the emergence of SA has inspired studies on enhancing its search performance (Greening, 1990; Ingber, 1989; Mahfoud and Goldberg, 1995; Szu and Hartley, 1987), applying it to different optimization problems (Ingber, 1993; Rutenbar, 1989), and even proving the convergence of the annealing schedule (Ferreira and de Queiroz, 2018). Different from the study by (Kirkpatrick et al., 1983) in which the Boltzmann distribution was used in the design of the generating function (for the generation of a new candidate solution), acceptance criteria (Eqs. (5.2) and (5.4)), and annealing schedule (Eq. (5.5)), in an early study (Szu and Hartley, 1987), the Cauchy distribution was used as the generating function $g(s)$ to generate a new candidate solution and $\Psi_t = \Psi_0/t$, where Ψ_0 is the initial temperature and Ψ_t is the temperature at the t-th iteration, was used as the annealing schedule to accelerate the convergence speed. In a later study (Ingber, 1989), each dimension of the generating function and annealing schedule was considered separately to further speed up the convergence speed of SA.

Different from changing the generating function and annealing schedule, some early studies (Greening, 1990; Mahfoud and Goldberg, 1995) attempted to parallelize the computations of SA. In (Greening, 1990), Ram et al. presented two parallel SAs. The first one first lets the host node randomly generate n solutions and then sends these solutions to all the computer nodes, each of which will run the SA to find a good solution. Each computer node will then return the solution it finds to the host node, which will then output the best of them as the solution. The second one is very similar to the first one, except that after receiving the results from all computer nodes, the host node will use them as the initial solutions of a genetic algorithm (GA) to further improve the end results. In (Mahfoud and Goldberg, 1995), Mahfoud and Goldberg first discussed some different parallel versions of SA, such as distributing the objective evaluation to several processors or simultaneously generating and evaluating the new candidate solution of the current solution, and then presented a parallel SA combined with GA, called parallel recombinative simulated annealing (PRSA). The basic idea of this algorithm is to divide the population (a set of candidate solutions) into a certain number of subpopulations, each of which will be run independently on a different computer node for a certain number of iterations. For each subpopulation, the crossover and mutation operators of

GA are used to transit the current solutions, and then the parent (current solutions) and offspring (new candidate solutions) are compared by using the acceptance criterion of SA to determine which solutions are to be kept for the next iteration.

It can be easily seen from the design of PRSA that combining SA with another metaheuristic algorithm which has been investigated in some early studies (Katsigiannis et al., 2012; Lin et al., 1993) is another critical research direction. For instance, Lin et al. (Lin et al., 1993) used SA to generate a set of candidate solutions as the initial population of GA. A later study (Katsigiannis et al., 2012) combined SA with tabu search (TS) by using the output solution of SA as the initial solution of TS. Since the combination of SA with another metaheuristic algorithm is perhaps able to find a better result than SA alone, a recent study (Assad and Deep, 2018) combined SA with harmony search (HS) (Geem, 2009; Geem et al., 2001), producing an algorithm called HS-SA. The main difference between HS-SA and hybrid algorithms of (Katsigiannis et al., 2012; Lin et al., 1993) is that HS-SA uses only the acceptance criterion of SA as an operator to determine whether to accept a new candidate solution or not, but SA in hybrid algorithms of (Katsigiannis et al., 2012; Lin et al., 1993) is used to search the solution. Another study (Zhang et al., 2018) combined HS, SA, and chaotic search (Chen and Aihara, 1995; Nozawa, 1992) to solve the renewable energy optimization problem. In this hybrid algorithm, SA and chaotic search are used to randomly generate the new candidate solution for HS. In (Abdel-Basset et al., 2021), SA was combined with Harris hawks optimization (HHO) (Heidari et al., 2019) for feature selection of classification problems. In this hybrid algorithm, HHO is first used to find a set of good candidate solutions, the best of which will then be used as a solution of SA for further improvement. In (Bi et al., 2021), SA, GA, and PSO were combined into a hybrid algorithm for optimizing the energy of offloading in a mobile-edge system. These three metaheuristic algorithms can be regarded as the search operators of this hybrid algorithm, which are performed repeatedly each iteration. In this hybrid algorithm, first GA is used to generate the new candidate solutions s from the current solutions, and then PSO is used to adjust these new candidate solutions s by using the acceptance criterion of SA.

Many different versions of SA have been presented. A summary of these is provided in a set of survey papers (Ingber, 1993; Nourani and Andresen, 1998; Romeo and Sangiovanni-Vincentelli, 1991), which also provide a comprehensive discussion. A framework of SA (shown in Algorithm 5) was presented and discussed in (Romeo and Sangiovanni-Vincentelli, 1991), which provides an alternative way to realize SA. This framework is a little

Algorithm 5 Framework of simulated annealing.

1 Set the initial temperature Ψ_0	I
2 Randomly create the initial solution s	I
3 f_s = Evaluate(s)	E
4 **While** the outer-loop termination criterion is not met	
5 **While** the inner-loop termination criterion is not met	
6 v = NeighborSelection(s)	T ↦ E
7 f_v = Evaluate(v)	E
8 If $r < P_a$	D
9 $s = v$	
10 $f_s = f_v$	
11 Endif	
12 **End**	
13 Update Ψ_t according to the annealing schedule	D
14 **End**	
15 **Output** s	O

bit different from Algorithm 4 because there are two loops (outer and inner loops). This implies that a certain number of new candidate solutions will be generated and evaluated before the temperature is updated.

In a later survey paper (Ingber, 1993), several applications of SA are discussed. Among them are circuit design, data analysis, imaging, biology, physics, geophysics, finance, and military applications. Parallel SAs are discussed in the surveys (Ingber, 1993; Romeo and Sangiovanni-Vincentelli, 1991). The exponential, logarithmic, linear, cooling, and constant thermodynamic speed annealing strategies were discussed and compared in (Nourani and Andresen, 1998) to show the convergence speed and end results using these strategies. Moreover, a more detailed and complete discussion of SA can be found in the studies (Delahaye et al., 2019; Henderson et al., 2003; Nikolaev and Jacobson, 2010).

Supplementary source code

Here are the hyperlinks to the directories where the source code described in this chapter resides.

https://github.com/cwtsaiai/metaheuristics_2023/src/c++/1-onemax/
https://github.com/cwtsaiai/metaheuristics_2023/src/c++/2-deception/

CHAPTER SIX

Tabu search

6.1. The basic idea of tabu search (TS)

It is more or less implied in the design of most metaheuristics that operators are able to avoid falling into local optima at early iterations during the convergence process. In addition to simulated annealing (SA), tabu search (TS) (Glover, 1986) is another single-solution-based metaheuristic algorithm. The idea of TS is very much like that of the fairy tale of Hansel and Gretel (also called the candy house),[1] who laid a trail of white pebbles to mark the path that they have traveled so that they can follow it back home. In TS, however, we can avoid traveling the same path that has been traveled recently. This story can help us understand how TS was designed

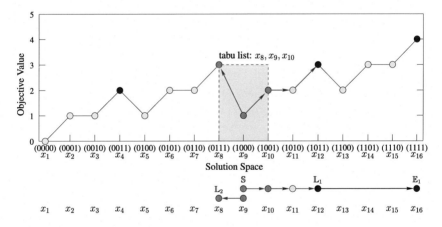

Figure 6.1 The search strategy of TS.

by Glover (Glover, 1989, 1990a; Glover and Laguna, 1997) to avoid searching the same solutions repeatedly too frequently to prevent the search from cycling. The search process of TS keeps track of recently visited solutions by storing them in a queue (called tabu list) to avoid the later search process from searching for the same candidate solutions. As shown in Fig. 6.1, if the starting point is x_9, after checking its two neighbors, x_8 and x_{10}, these three searched solutions will be kept in the tabu list. Now, if the current

---

[1] This fairy tale is available online at https://en.wikipedia.org/wiki/Hansel_and_Gretel.

solution is x_{10}, the search of TS will avoid searching the searched solutions in the tabu list again for a while. As a result, x_8, x_9, and x_{10} will not be the next solution of x_{10}. Instead, the next solution will be either x_7 or x_{11} if the TS search can only move one step to the right or to the left. In case TS chooses x_{11} as the next solution, the search of TS might trend to x_{12} later.

Algorithm 6 Tabu search.

---

1 Empty the tabu list $\mathbb{T}$		$\boxed{\text{I}}$
2 Randomly create the initial solution s		$\boxed{\text{I}}$
3 f_s = Evaluate(s)		$\boxed{\text{E}}$
4 **While** the termination criterion is not met		
5 v, f_v = NonTabuNeighborSelection(s, $\mathbb{T}$)		$\boxed{\text{T}} \mapsto \boxed{\text{E}} \mapsto \boxed{\text{D}}$
6 If f_v is better than f_s		$\boxed{\text{D}}$
7 $s = v$		
8 $f_s = f_v$		
9 Endif		
10 **End**		
11 **Output** s		$\boxed{\text{O}}$

Algorithm 6 shows a simple version of TS. It can be easily seen that the structure of TS is very similar to that of HC and SA, except for the tabu list and the method to create the solution from the neighbors of the current solution. The tabu list is emptied and the initial solution is randomly created just like for SA in lines 1–2. As shown in line 5, the NonTabuNeighborSelection() operator provides a unique method to create a best candidate solution v from the neighbor solutions of the current solution s. We mark this operator by "$\mathbf{T} \mapsto \mathbf{E} \mapsto \mathbf{D}$" because it will perform the following operations: first transit the current solution s to generate a set of new candidate solutions, then evaluate those not in the tabu list, and finally determine the best one and use it as the new candidate solution v. Besides, the NonTabuNeighborSelection() operator will take care of updating the tabu list by appending the new candidate solution v to the tabu list and, if the capacity of the tabu list is exceeded, removing the oldest entry. The determination operator of TS is composed of a tabu list and the so-called aspiration criterion (or criteria). As shown in lines 6–9, a simple way to use the determination operator is to accept the new candidate solution v as the current solution s if v is better than s. Note that this kind of procedure can be regarded as the aspiration criterion (or criteria) of TS that plays (or play) the role of checking the new candidate solution v to see if it is "good enough" in terms of the predefined condition(s) to override the current solution. Bear in mind that we can simply accept the best new candidate solution v as the current solution s disregarding if v is better than s.

This is eventually a typical approach that can also be found in most studies of TS.

6.2. Implementation of TS for the one-max and deceptive problems

A simple command to run TS is as follows:

```
$ ./search ts 100 1000 10 "" 3 7
```

The first three parameters of this command state that it will use TS as the search algorithm, it will carry it out for 100 runs, and 1000 evaluations will be performed for each run. The remaining four parameters state that the length of the solution is 10; there is no initial solution file; the new candidate solution v is generated by the select_neighbors_not_in_tabu() operator from three neighbor solutions of the current solution; and the size of the tabu list is set to seven. As shown in Listing 6.1, the implementation of TS is very similar to that of SA, except the tabu list and how a new candidate solution is created.

Listing 6.1: src/c++/1-onemax/main/ts.h: The implementation of TS.

```
 1 #ifndef __TS_H_INCLUDED__
 2 #define __TS_H_INCLUDED__
 3
 4 #include <iostream>
 5 #include <string>
 6 #include <fstream>
 7 #include <numeric>
 8 #include "lib.h"
 9
10 using namespace std;
11
12 class ts
13 {
14 public:
15     typedef vector<int> solution;      // ts-specific vector
16     typedef vector<solution> tabu_list; // ts-specific vector
17
18     ts(int num_runs,
19        int num_evals,
20        int num_patterns_sol,
21        string filename_ini,
22        int num_neighbors, // ts-specific parameter
23        int siz_tabulist   // ts-specific parameter
24        );
25
26     virtual ~ts() {};
27
```

```
28     solution run();
29
30 protected:
31     int eval_count;
32
33 private:
34     void init();
35     virtual solution transit(const solution& s);
36
37 protected:
38     virtual int evaluate(const solution& s);
39
40     // begin the ts functions
41     virtual tuple<solution, int, int> select_neighbor_not_in_tabu(const solution& s);
42     bool in_tabu(const solution& s);
43     void append_to_tabu_list(const solution& s);
44     // end the ts functions
45
46 private:
47     int num_runs;
48     int num_evals;
49     int num_patterns_sol;
50     string filename_ini;
51
52     solution sol;
53     int obj_val;
54
55     solution best_sol;
56     int best_obj_val;
57
58     // begin the ts parameters
59     int num_neighbors;
60     tabu_list tabulist;
61     int siz_tabulist;
62     // end the ts parameters
63 };
64
65 // begin the auxiliary function
66 inline bool operator==(const ts::solution& s1, const ts::solution& s2)
67 {
68     for (size_t i = 0; i < s1.size(); i++)
69         if (s1[i] != s2[i])
70             return false;
71     return true;
72 }
73 // end the auxiliary function
74
75 inline ts::ts(int num_runs,
76              int num_evals,
77              int num_patterns_sol,
78              string filename_ini,
79              int num_neighbors, // ts-specific parameter
80              int siz_tabulist   // ts-specific parameter
81              )
82     : num_runs(num_runs),
83       num_evals(num_evals),
84       num_patterns_sol(num_patterns_sol),
85       filename_ini(filename_ini),
86       num_neighbors(num_neighbors),
87       siz_tabulist(siz_tabulist)
88 {
89     srand();
90 }
91
92 inline ts::solution ts::run()
```

```
93  {
94      double avg_obj_val = 0;
95      vector<double> avg_obj_val_eval(num_evals, 0.0);
96
97      for (int r = 0; r < num_runs; r++) {
98          // 0. initialization
99          init();
100         obj_val = evaluate(sol);
101         eval_count = 0;
102         avg_obj_val_eval[0] += obj_val;
103
104         while (eval_count < num_evals-1) {
105             // 1. transition and evaluation
106             auto[tmp_sol, tmp_obj_val, n_evals] = select_neighbor_not_in_tabu(sol);
107             if (n_evals > 0) {
108                 // 2. determination
109                 if (tmp_obj_val > obj_val) {
110                     sol = tmp_sol;
111                     obj_val = tmp_obj_val;
112                 }
113
114                 if (obj_val > best_obj_val) {
115                     best_obj_val = obj_val;
116                     best_sol = sol;
117                 }
118
119                 avg_obj_val_eval[eval_count] += best_obj_val;
120                 for (int i = 1; i < n_evals; i++)
121                     avg_obj_val_eval[eval_count-i] = avg_obj_val_eval[eval_count];
122             }
123         }
124         avg_obj_val += best_obj_val;
125     }
126
127     avg_obj_val /= num_runs;
128
129     for (int i = 0; i < num_evals; i++) {
130         avg_obj_val_eval[i] /= num_runs;
131         cout << fixed << setprecision(3) << avg_obj_val_eval[i] << endl;
132     }
133
134     return sol;
135 }
136
137 inline void ts::init()
138 {
139     tabulist = tabu_list(siz_tabulist, solution(num_patterns_sol));
140     best_obj_val = 0;
141     eval_count = 0;
142
143     sol = solution(num_patterns_sol);
144     if (!filename_ini.empty()) {
145         ifstream ifs(filename_ini.c_str());
146         for (int i = 0; i < num_patterns_sol; i++)
147             ifs >> sol[i];
148     }
149     else {
150         for (int i = 0; i < num_patterns_sol; i++)
151             sol[i] = rand() % 2;
152     }
153 }
154
155 inline ts::solution ts::transit(const solution& s)
156 {
157     return ::transit_r(s);
```

```
158 }
159
160 inline tuple<ts::solution, int, int> ts::select_neighbor_not_in_tabu(const solution& s)
161 {
162     int n_evals = 0;
163     int f_t = 0;
164     solution t(num_patterns_sol);
165     for (int i = 0; i < num_neighbors; i++) {
166         solution v = transit(s);
167         if (!in_tabu(v)) {
168             n_evals++;
169             int f_v = evaluate(v);
170             if (f_v > f_t) {
171                 f_t = f_v;
172                 t = v;
173             }
174             if (eval_count >= num_evals-1)
175                 break;
176         }
177     }
178     if (n_evals > 0)
179         append_to_tabu_list(t);
180     return make_tuple(t, f_t, n_evals);
181 }
182
183 inline bool ts::in_tabu(const solution& s)
184 {
185     for (size_t i = 0; i < tabulist.size(); i++)
186         if (tabulist[i] == s)
187             return true;
188     return false;
189 }
190
191 inline int ts::evaluate(const solution& s)
192 {
193     eval_count++;
194     return accumulate(s.begin(), s.end(), 0);
195 }
196
197 inline void ts::append_to_tabu_list(const solution& s)
198 {
199     tabulist.push_back(s);
200     if (static_cast<int>(tabulist.size()) >= siz_tabulist)
201         tabulist.erase(tabulist.begin());
202 }
203
204 #endif
```

6.2.1 Declaration of functions and parameters

The implementation of TS is defined as a class named ts. There is no exception to the class ts for TS. In what follows, we will describe the interface part and the implementation part in turn. For the class ts, the interface part begins with two typedefs. First, solution is defined as an alias of vector<int> in line 15. Then, tabu_list is defined as an alias of vector<solution> in line 16. This means that the tabu list is implemented as a vector of solutions that have been found during the convergence process of TS. The constructor is then declared in lines 18–24. It will take its parameters from the user of the program or from the other programs, which can be divided into two parts: parameters required by all the search methods and parameters required by TS. The parameters required by all the programs are the number of runs, the number of iterations, the number of subsolutions of a solution, and the filename of the file of the initial solution. The two parameters required by TS are the number of neighbor solutions of the current solution that will be checked to generate the new candidate solution each iteration (i.e., num_neighbors) and the size of the tabu list (i.e., siz_tabulist). As shown in line 26, the virtual destructor—even when defined as an empty function as in this case—is required for a class that has any virtual function defined so that an instance of a derived class of such a class can be correctly deleted through a pointer to the base class. The prototype of the key function, namely, run(), is then given, the definition of which will be given shortly. This ends the interface part of the class ts. In addition to the init(), transit(), and evaluate() functions, lines 41–43 define three more helping functions for TS to create the neighbor solutions (select_neighbor_not_in_tabu()), to check whether a solution is in the tabu list or not (in_tabu()), and to append a solution to the tabu list (append_to_tabu_list()), respectively. Lines 59–61 then define three variables that are specific to TS, that is, num_neighbors for the number of neighbors to be checked, tabulist for the tabu list, and siz_tabulist for the size of the tabu list. The operator "==" is overloaded in lines 66–72 to check if two solutions are the same. It is used by the function in_tabu() to check if a solution is already in the tabu list.

6.2.2 The main loop

The most important function of TS is the run() function, as shown in lines 92–135. As mentioned before, the structure of this function is very similar to that of SA; thus, we will illustrate only the parts that are unique to TS. The first unique aspect is that TS uses an alternative way to create the neighbor solution, by calling the select_neighbors_not_in_tabu() function. Because the new candidate solutions are evaluated in the select_neighbors_not_in_tabu() function, the evaluations will not be repeated in the main loop. This is a little bit different from the implementations of HC and SA. The second unique thing is that TS will update the tabu list if the new candidate solution is not in the tabu list, by calling the append_to_tabu_list() function, and at the time new candidate solutions are generated by the select_neighbors_not_in_tabu() function. In addition to resetting the best objective value best_obj_val to 0, the init() function is mainly responsible for creating the initial solution and the tabu list, as shown in lines 137–153.

6.2.3 Additional functions

As mentioned before, the way the neighbor solutions are created by the select_neighbors_not_in_tabu() function for TS is different from the way they are created for other metaheuristics. Lines 160–181 show how this works exactly. Like SA, this function will create a candidate solution. Unlike SA and other metaheuristics, as far as this implementation is concerned, the new candidate solution of TS will be selected from a set of neighbor solutions of the current solution, which is essentially "the best of those not in the tabu list," as shown in lines 167–176, where the in_tabu() function is called to check whether the new candidate solution is in the tabu list. The in_tabu() function checks whether a candidate solution is in the tabu list or not, as shown in lines 183–189. As mentioned before, since the operator "==" has been overloaded for the implementation of this function, the equality of two solutions can be easily determined by comparing them using this operator. It will return true if the input solution is already in the tabu list, and false otherwise. Another function for TS is the append_to_tabu_list() function, as shown in lines 197–202. This function is responsible for appending the new solution to the end of the tabu list while at the same time removing the first one, which is typically the oldest one, from the tabu list. Because each new candidate solution generated by the select_neighbors_not_in_tabu() function that is not in the tabu list will be evaluated by the evaluate() function, it has to be counted as an evaluation of TS. For this reason, a few changes must be made, as shown in lines 31, 101–102, 120–121, 141, and 193.

Listing 6.2: src/c++/2-deception/main/tsdp.h: Using polymorphism to customize the implementation of TS for a deceptive problem.

```
17  inline int tsdp::evaluate(const solution& s)
18  {
19      eval_count++;
20      int n = 0;
21      for (size_t i = 0; i < s.size(); i++)
22          n += s[i] << ((s.size() - 1) - i);
23      return abs(n - (1 << (s.size()-2)));
24  }
25
26  inline tuple<tsdp::solution, int, int> tsdp::select_neighbor_not_in_tabu(const solution& s)
27  {
28      auto[t, f_t, n_evals] = ts::select_neighbor_not_in_tabu(s);
29      if (n_evals == 0) {
```

```
30      do {
31          for (size_t i = 0; i < s.size(); i++)
32              t[i] = rand() % 2;
33      } while (in_tabu(t));
34      append_to_tabu_list(t);
35      f_t = evaluate(t);
36      n_evals = 1;
37   }
38   return make_tuple(t, f_t, n_evals);
39 } //
```

To apply the implementation of TS (Listing 6.1) to the deceptive problem $\mathbb{P}_4$, the situation is very similar to those of HC and SA. That is, all we have to do is to change the evaluation operator so that it now looks like as shown in lines 17–24 of Listing 6.2. Since the new candidate solutions generated by the select_neighbor_not_in_tabu() function in Listing 6.1 to solve a small deceptive problem are frequently in the tabu list, the search process of TS may get stuck at generating new candidate solutions because all the generated new candidate solutions already happen to be in the tabu list. To deal with this issue, the select_neighbor_not_in_tabu() function was modified to ensure that every time it is called, it will generate a solution, as shown in lines 26–39 of Listing 6.2. To be more specific, it first calls the select_neighbor_not_in_tabu() function in Listing 6.1 to generate a new candidate solution from the current solution, as shown in line 28. If all neighbor solutions of the current solution are in the tabu list, that is, n_evals is equal to 0, as shown in line 29, it will randomly generate a new candidate solution that is not in the tabu list to prevent the search process from getting stuck at a particular solution or region.

6.3. Simulation results of TS

In this section, the one-max and deceptive problems will be used to illustrate the performance of TS, just like they were used to evaluate the performance of SA in Section 5.3. For the simulation results described here, a solution is generated each iteration, and the maximum size of the tabu list is set to seven. All simulations are carried out for 100 runs, and 1000 evaluations are performed in each run for both the one-max and deceptive problems.

6.3.1 Simulation results of TS for the one-max problem

Compared with Table 5.2 in Section 5.3.1, one more method is analyzed in Table 6.1 for the one-max problem. The results, all of which are the average of 100 runs, show that TS is able to find the optimal solution much faster than the other two search algorithms for problems of sizes $n = 10$, $n = 20$, and $n = 60$. Note that there is no guarantee that TS will definitely beat the other search algorithms; rather, the results just show that TS is able to find the optimal solution much faster than the other two algorithms for the one-max problem in some cases.

Table 6.1 Number of iterations needed by HC, SA, and TS to find the optimal solution.

Problem size	HC-Rand	SA	TS
$n = 10$	59	79	33
$n = 20$	145	148	96
$n = 40$	277	306	325
$n = 60$	486	460	374
$n = 80$	786	576	600
$n = 100$	965	826	901

In fact, although the number of runs for each problem size is set to 100, the average result of these different runs for each search algorithm still fluctuates. This means that the results described in Table 6.1 for the performance of HC-Rand, SA, and TS in solving the one-max problem of the same size may be different from simulation to simulation. Fig. 6.2 shows that the trajectory of the convergence of TS is very similar to that of SA for solving the one-max problem of size $n = 100$.

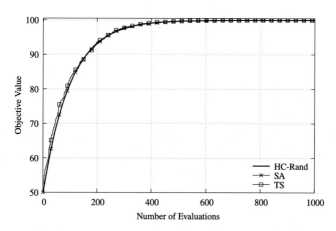

Figure 6.2 Convergence of HC, SA, and TS for the one-max problem of size $n = 100$.

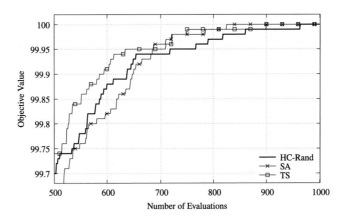

Figure 6.3 Enlargement of a portion of Fig. 6.2.

Although all three search algorithms take less than 1000 evaluations to find the optimal solution, a close look at the results in Fig. 6.2 reveals that TS may be able to find a better solution much faster than HC and SA at the early stage of the convergence process, as shown in Fig. 6.3.

6.3.2 Simulation results of TS for the deceptive problem

To better understand the performance of TS, we also apply it to the deceptive problem $\mathbb{P}_4$. As shown in Figs. 6.4 and 6.5, the convergence characteristic of TS for solving the deceptive problem of different sizes is very similar to that of SA. From these comparisons, it can be easily seen that both SA and TS are equipped with some mechanisms to prevent the search process from falling into a local optimum; therefore, both of them are able to find the optimal solution for $\mathbb{P}_4$ of sizes $n = 4$ and $n = 10$, as shown in Figs. 6.4 and 6.5. The results further show that metaheuristic algorithms equipped with mechanisms to avoid falling into local optima at early iterations are able to find better results than greedy and heuristic algorithms.

6.4. Discussion

TS is capable of avoiding the issue of falling into local optima. In contrast to SA, which escapes local optima by accepting worse solutions from time to time, TS prevents the search process from checking the same solutions again and again for a while. It can be imagined that the search process of TS is like using soil to fill the holes that can be regarded as the

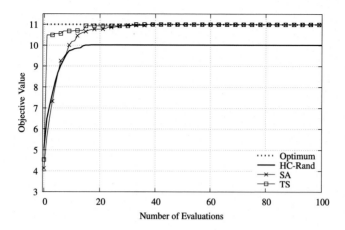

Figure 6.4 Simulation results of TS for the deceptive problem of size $n = 4$.

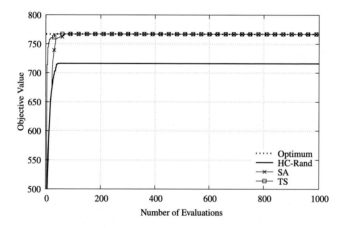

Figure 6.5 Simulation results of TS for the deceptive problem of size $n = 10$.

local optima in the solution space that have been traveled so that we can disregard these regions for a while. Glover coined the name "short-term memory" for the items (tabus) stored in the tabu list (Glover and Kochenberger, 2003). These tabus are very useful because they keep track of the solutions that TS has checked recently to avoid the search process from searching the same regions repeatedly. The simulation results of Section 6.3 show that TS is a very useful search algorithm for solving optimization problems.

Similar to SA, some open issues regarding the parameter settings need to be addressed for TS. One of the important things to determine for TS is the size of the tabu list, for one simple reason. Although a larger tabu list means a larger short-term memory to keep track of the solutions that have been checked recently, thus avoiding the issue of searching the same regions again and again, it will take much more memory to keep track of the solutions and much more time to check if a new candidate solution is in the tabu list. Although the size of a tabu list may have a strong impact on the performance of TS, TS is still a high-performance metaheuristic algorithm, and it is very simple and thus very easy to implement. Glover (Glover et al., 1993), the inventor of TS, paid special attention to the mechanism to determine the size of the tabu list. In this early study, he discussed the impact that the size of the tabu list and the dynamic approach may have on the performance of TS. This research issue was also addressed in another study (Battiti and Tecchiolli, 1994), in which Battiti and Techiolli presented a self-learning mechanism to automatically determine an appropriate tabu list size based on the occurrence of search cycles. That is, the size of the tabu list will be dynamically increased when search cycles are detected by this self-learning mechanism or dynamically decreased when all new candidate solutions are in the tabu list, so that none of them satisfies the aspiration criterion. Another example can be found in (Salhi, 2002), in which the size of the tabu list was determined by many factors, such as the neighborhood, the problem size, the neighborhood size of the i-attribute, its frequency of occurrence in the previous solutions, and the current number of iterations. In addition to determining or varying the size of the tabu list during the convergence process, the aspiration criteria and parallelization of TS were also discussed in some early studies (Glover, 1990b; Glover et al., 1993; Talbi et al., 1999; Tsubakitani and Evans, 1998), in which all the operators of TS since their public appearance were extensively investigated.

Furthermore, studies of TS were not only focused on enhancing the performance of its operators; some early studies applied TS in different scenarios, such as graph coloring (Hertz and de Werra, 1988), the job-shop scheduling problem (Dell'Amico and Trubian, 1993), the vehicle routing problem (Gendreau et al., 1994), data clustering (Al-Sultan, 1995), feature selection (Zhang and Sun, 2002), and even water network optimization (da Conceição and Ribeiro, 2004). These applications show that many researchers were very interested in applying TS in real-world situations. Since TS is very simple and easy to implement, the first thing researchers have to do is to formulate the problem in question. The second objective is to find better results than with existing methods. For example, Nara et al. (Nara et al., 2001) used TS to find a better placement plan of distributed generators (DGs) to reduce the distribution loss. In this study, Nara and his colleagues first defined the placement problem as an optimization problem with the goal of minimizing the distribution loss, that is, to obtain a good placement plan to determine the allocation and size of DGs. A simple TS can then be used to find an applicable placement plan of DGs. In a later study (Lin et al., 2002), Lin et al. further defined an economic dispatch problem in power systems and then attempted to combine TS with a genetic algorithm (GA) to solve this optimization problem and minimize the generation cost of all power units. Another critical research direction is how to generate appropriate solutions from the current solution, when we try to apply a metaheuristic algorithm like TS to a real optimization problem. In (Alvarez-Valdes et al., 2002), simple move, swap, and multi-swap neighbor generations are developed to generate appropriate solutions to the course scheduling problem. In (Toth and Vigo, 2003), Toth and Vigo pointed out that the generation of appropriate solutions for TS is useful in solving the vehicle routing problem. In this study, they observed that long arcs have a small chance to be part of a high-quality vehicle routing path; thus, a neighbor generation mechanism was developed to avoid involving only "long" arcs in the routing path.

How to combine TS with other (meta)heuristic algorithms is another research direction that has been investigated and discussed for years. For example, in (Mantawy et al., 1999), SA, TS, and GA were integrated to solve the unit commitment scheduling problem. In this hybrid metaheuristic algorithm, SA and TS can be regarded as the search operators of GA, TS is used to generate new candidate solutions from the current solutions, and SA is used to determine whether the new candidate solutions generated by the mutation operator can be kept for the next iteration (also called the generation in GA) or not. A similar concept can also be found in (Li and Gao, 2016), where TS can be regarded as a search or local operator of GA. Another combination of TS with a metaheuristic algorithm can be found in (Shen et al., 2008), where TS is combined with particle swarm optimization (PSO). The same combination was used in (Mantawy et al., 1999), where TS can be regarded as a search operator of PSO. However, the main difference between the hybrid algorithms of (Shen et al., 2008) and (Mantawy et al., 1999) is that instead of passing searched solutions from one metaheuristic algorithm to another sequentially each iteration, 90% of the solutions are generated by PSO, while the remaining 10% of the solutions are generated by TS in each iteration. Of course, there are other ways to combine TS with search algorithms, for example with variable neighborhood search (Schermer et al., 2019) and with artificial neural networks (Karamichailidou et al., 2021).

From the studies of (Glover, 1989, 1990a,b; Glover and Kochenberger, 2003; Glover and Laguna, 1997; Glover et al., 2007, 1993), it can be easily seen that Glover was devoted to the study of TS for many years. Several excellent books and survey papers were written to explain the basic ideas of TS, possible ways to enhance the performance of TS, and the kinds of applications for which TS can be used. Of course, we can find surveys of TS by other researchers (Crainic, 2005; Gendreau and Potvin, 2005; Talbi et al., 1999). Among them, the interesting research topic of developing parallel TS methods was discussed in (Crainic, 2005; Talbi et al., 1999). These studies revealed different ways to parallelize TS, such as search control cardinality by one or more processes, search from the same or different initial points or strategies for different processes, and even information exchange between different processes. The studies we discussed in this section are just the tip of the iceberg of abundant studies on TS. In fact, there are tens of thousands of studies on TS. According to the argument of Glover (Glover, 1990b) and our observations, because TS is simple and easy to implement, it can be used either alone or as a high-level algorithm to guide other search algorithms to solve optimization problems. The strategy to keep a set of solutions to guide the search is still very useful for metaheuristic algorithms today. Hence, we can still find recent studies (Alinaghian et al., 2021; Karamichailidou et al., 2021; Li et al., 2022; Schermer et al., 2019) that used TS to solve optimization problems.

Supplementary source code

Here are the hyperlinks to the directories where the source code described in this chapter resides.

https://github.com/cwtsaiai/metaheuristics_2023/src/c++/1-onemax/
https://github.com/cwtsaiai/metaheuristics_2023/src/c++/2-deception/

CHAPTER SEVEN

Genetic algorithm

7.1. The basic idea of genetic algorithm (GA)

As the name suggests, the genetic algorithm (GA) (Holland, 1962, 1975) is inspired by Darwin's theory of evolution. Holland presented a technique for emulating the evolution theory on a computer, which, importantly, can be used to solve many optimization problems. The technique for emulating the evolution theory presented by Holland is composed of the following operators: the "selection," "crossover," and "mutation" operators in addition to the initialization and output operators. To understand GA, we need to know some terminology of GA. First, each iteration during the convergence process is called a generation. Different from other single-solution-based metaheuristic algorithms, such as simulated annealing (SA) and tabu search (TS), which typically search one and only one candidate solution per iteration, as a population-based metaheuristic algorithm, GA will search more than one candidate solution per generation. All the solutions of each generation are called a *population*. Each solution is called a *chromosome* or an *individual*. Each subsolution of a solution is called a *gene*.

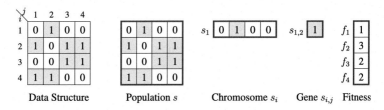

Figure 7.1 The relationship between population s, chromosome s_i, and gene $s_{i,j}$.

As shown in Fig. 7.1, this means that GA will create a set of solutions $s = \{s_1, s_2, \ldots, s_m\}$, each of which contains n subsolutions, denoted by $s_i = \{s_{i,1}, s_{i,2}, \ldots, s_{i,n}\}$, in the very beginning. In other words, s represents the population (a set of solutions), s_i represents a chromosome (a solution), and $s_{i,j}$ represents a gene (a subsolution). Besides, m represents the number of chromosomes in the population (also called population size) and n represents the number of subsolutions in a chromosome. The objective value of each chromosome will also be transformed to a value called the *fitness value* by a so-called fitness function.

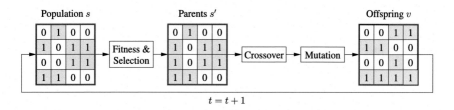

Figure 7.2 The relationship between parents and children.

Each current solution selected by the selection operator is called a *parent*, while each new candidate solution is called an *offspring* during the convergence process. As shown in Fig. 7.2, this means that the parents s' will be selected by the selection operator from the current population s at iteration t. The fitness values of all the chromosomes will also be calculated by the selection operator. The fitness value of a chromosome can be either its objective value or a rank (of all the chromosomes). Once the parents are selected by the selection operator, the crossover and mutation operators are used to generate a new population v, which is called the offspring of the parents s'. The offspring at iteration t will become the current population s of iteration $t + 1$. This is referred to as the reproduction process.

Figs. 7.3 and 7.4 show how the selection, crossover, and mutation operators of GA work, to help us understand the convergence process of GA. As shown in Fig. 7.3, GA will first randomly select some of the chromosomes based on the fitness value of each chromosome. Fig. 7.3(a) shows the situation in which only the selection operator is used, i.e., no other transition operators are used. In this case, the distribution of all the chromosomes will be shifted and changed from left to right on the x-axis from generation $t = 1$ to generation $t = 3$. This means that the average objective value of the population will be increased while the variance is decreased. This example explains that the selection operator can be used to pick up better candidate solutions and remove worse candidate solutions, that is, it acts like "natural selection." From the perspective of statistics, the selection operator will narrow the distribution of all the chromosomes during the convergence process. If GA uses only the selection operator but none of the transition operators (e.g., crossover or mutation), it will not generate any new candidate solutions even though the average objective value of all the chromosomes is raised from 1.5 to 2.5. In this case, all the chromosomes will still be similar to each other, and the best objective value of the candidate solutions will not be changed from $t = 1$ to $t = 3$. Fig. 7.3(b) shows that if GA uses not only the selection operator to select better chromosomes for the next generation but also the crossover and mutation operators as the transition operators to generate new chromosomes for the population, this makes it possible for GA to find better candidate solutions. In this example, the average objective value of all the chromosomes will be increased from 1.5 to 3, while the best objective value will exceed 3.

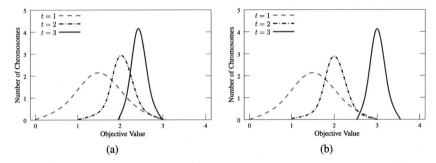

(a)   (b)

Figure 7.3 The strategies of the selection, crossover, and mutation operators of GA: (a) using only the selection operator and (b) using all the three operators.

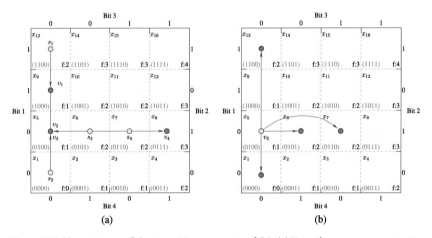

(a)   (b)

Figure 7.4 The strategy of the transition operator of GA. (a) How the crossover operator works. (b) How the mutation operator works.

Fig. 7.4 further shows that GA will typically apply the crossover and mutation operators to the chromosomes selected by the selection operator. Unlike hill climbing (HC), SA, and TS, which search only one candidate solution per iteration, GA searches multiple candidate solutions per generation. As such, a two-dimensional landscape will be used to describe

the solution space of GA in this chapter instead of the one-dimensional
landscape as used in Chapters 3, 5, and 6 to facilitate the description and
understanding of GA. In this figure, f:w in each grid represents the objective value of candidate solution x_i for the one-max problem. For instance,
f:0 in the grid at the bottom-left corner represents that the objective value
of candidate solution x_1 is zero. Bit j represents the value of the j-th bit.
The bit string of each candidate solution is given in the bottom-left corner
of the grid, which is composed of the values of Bit 1 (left), Bit 2 (right),
Bit 3 (top), and Bit 4 (bottom). Fig. 7.4(a) gives a simple example with
four chromosomes s_1, s_2, s_3, and s_4 at x_{13}, x_1, x_6, and x_7, respectively. After a one-point crossover, the new chromosomes v_1 and v_2 at x_9 and x_5
might be created from s_1 and s_2 if the crossover point is between the first
and second genes while the chromosomes v_3 and v_4 at x_5 and x_8 might
be created from s_3 and s_4 if the crossover point is between the third and
fourth genes. This means that the same number of offspring v_1, v_2, v_3, and
v_4 will be created from the parents s_1, s_2, s_3, and s_4. Since the crossover
point for parents s_1 and s_2 is between the first and second genes, the parents will first be divided into two parts, as follows: (1) and (100) for s_1 and
(0) and (000) for s_2, the second parts of which will then be exchanged to
generate the offspring v_1 (1000) and v_2 (0100). In this example, the offspring of s_3 and s_4 will also be generated similarly except that the crossover
point may be located at a different position, i.e., between the third and
fourth genes. As such, the solutions in the population will then be changed
from $s_1 = (1100)$, $s_2 = (0000)$, $s_3 = (0101)$, and $s_4 = (0110)$ to $v_1 = (1000)$,
$v_2 = (0100)$, $v_3 = (0100)$, and $v_4 = (0111)$. The crossover and mutation operators will be discussed in detail later.

Fig. 7.4(b) further shows that the mutation operator will randomly
change the value of some genes of each chromosome. This example shows
that v_3 is created by the crossover operator and it is possible for GA to invert
the value of a bit of v_3. So, v_3 at x_5 may be moved to x_{13}, x_1, x_6, or x_7. One
of the important things to bear in mind is that not all the chromosomes will
be involved in crossover and mutation. This means that only a certain percentage of the chromosomes will be selected by the selection operator to be
involved in crossover and mutation. The percentage for crossover is referred
to as the crossover rate, while the percentage for mutation is referred to as
the mutation rate. The rates for these two operators are generally quite different. Typically, the crossover rate is set equal to 60%, while the mutation
rate is set to 1%. This scenario somehow explains that the search directions
of GA are unlike those of SA and TS, which are basically moved along the

solution space; GA is, however, kind of jumping among all the solutions in the solution space.

Algorithm 7 Genetic algorithm.

1 Randomly create the initial population s	$\boxed{\text{I}}$
2 **While** the termination criterion is not met	
3 $\quad f = \text{Fitness}(s)$	$\boxed{\text{E}}$
4 $\quad s' = \text{Selection}(s, f)$	$\boxed{\text{D}}$
5 $\quad v_c = \text{Crossover}(s')$	$\boxed{\text{T}}$
6 $\quad v_m = \text{Mutation}(v_c)$	$\boxed{\text{T}}$
7 $\quad s = v_m$	$\boxed{\text{D}}$
8 **End**	
9 **Output** s	$\boxed{\text{O}}$

As shown in Algorithm 7, a set of initial solutions for GA called the population will be randomly created in line 1. The Fitness(), Selection(), Crossover(), and Mutation() operators will then be performed repeatedly until a predefined termination condition is met, say, for a certain number of generations. As shown in line 3, GA will compute the fitness values of all the chromosomes during the convergence process. A very intuitive implementation of the Fitness() function is to use the objective value of each chromosome as the fitness value. Once having the fitness value of each chromosome, the Selection() operator can then pick up some suitable chromosomes v from the current chromosomes s. A very simple but well-known selection operator is the roulette wheel selection operator (Holland, 1962, 1975). It will first compute the survival probability of each chromosome based on its fitness value and then use the probabilities thus obtained to randomly pick a certain number of chromosomes as the parents s'. The probability of each chromosome being elected as a parent can be computed as

$$p_i = \frac{f_i}{\sum_{j=1}^{m} f_j}, \qquad (7.1)$$

where p_i is the survival probability of the i-th chromosome, f_j is the fitness value of the j-th chromosome, m is the number of chromosomes in the population, and $\sum_{j=1}^{m} f_j$ is the fitness value of all the chromosomes in the population. It is expected that the number of s_i's in s' will be proportional to the fitness value of s_i. This implies that if a chromosome has a fitness value that is higher than those of the others, its survival probability will be higher than those of the others too. As a consequence, it will have a higher chance to be selected as a parent for producing the offspring for the next generation. Another well-known selection operator, called the tournament selection operator (Brindle, 1980), is much easier to implement than the wheel selection operator. Tournament selection will be performed as many times as necessary, and each time it is performed, the operator will randomly select a certain number of solutions (called players), say, three, from the population and then pick the one that has the best fitness value as a parent. As shown in line 5, after the Selection() operator is performed, GA will use the Crossover() operator as the transition operator to change the structure of each chromosome. This operator typically plays the role of exchanging the information between two chromosomes.

For example, suppose that two chromosomes (0010) and (0101) are picked by the selection operator and are named s'_1 and s'_2, respectively. The crossover operator will then randomly generate a number r uniformly distributed in the range $[0, 1]$. If it satisfies the crossover rate c_r (i.e., $r \leq c_r$), the crossover operator will randomly choose a point and exchange the genes of these two chromosomes, as follows:

If the chosen crossover point is 2, GA will first split these two chromosomes at the position between the second and third genes, as follows:

```
Parent 1 s₁′: (00 | 10)
Parent 2 s₂′: (01 | 01)
```

GA will then exchange the last two genes of these two chromosomes to create the new offspring, named v_{c_1} and v_{c_2}, respectively, as follows:

```
Offspring 1 vc₁: (00 | 01)
Offspring 2 vc₂: (01 | 10)
```

Another transition operator of GA is the Mutation() operator, which is usually responsible for fine-tuning the solution so that it is capable of escaping from a local optimum. Generally, the mutation operator works as follows: First, for each gene, a number r uniformly distributed in the range $[0, 1]$ is randomly generated. If it satisfies the mutation rate m_r (i.e., $r \leq m_r$), the mutation operator will then invert the value of this gene. To make the description as consistent as possible, we will use the above example to explain how the mutation operator works, as follows:

If the value randomly generated for the third gene of offspring 1 v_{c_1} satisfies the mutation rate, it would look like

Offspring 1 v_{c_1} : (00 0 1)

GA will then change it from 0 to 1 or from 1 to 0, so it would look like

Offspring 1 v_{m_1}: (00 1 1)

Compared to SA and TS, GA is much more complicated not only in the design of the transition operator but also in the design of the determination operator. The main differences between GA, SA, and TS are fourfold:

- **Multiple search directions:** Compared to the single-solution-based metaheuristic algorithms (e.g., SA and TS) that search only one solution at a time, GA searches for more than one solution at a time during the convergence process. Since GA will search for multiple directions or regions at a time, its search diversity will normally be much higher than single-solution-based metaheuristics that search for only one direction or region at a time during the convergence process.
- **Selection operator:** Another characteristic of GA is that it uses the selection and fitness function operators to determine solutions to be searched, not just based on the objective value of each solution. This kind of mechanism keeps the search process of GA from looking for the best solution in the population all the time so that it will not always choose the solution with the best objective or fitness value to search its neighbors again and again. Consequently, GA will not easily get stuck in local optima at early iterations.

- **Crossover operator:** This operator is one of the transition operators of GA, which plays the role of exchanging information between parent chromosomes, such as moving portions of the genes of a chromosome to another. This kind of mechanism allows GA to restructure its solutions to form new solutions in such a way that the structure of the new solutions is not confined to the structure of the initial solutions and may even inherit partial structures from their parents; as a consequence, the search process of GA will quickly jump from one region to another in the solution space during the convergence process.
- **Mutation operator:** This operator is another transition operator of GA, which ensures that the search process of GA is capable of escaping from a local optimum by changing the value of some genes randomly. Of course, this kind of mechanism will also play the role of fine-tuning the chromosomes of GA because only a few genes will be changed at a time.

7.2. Implementation of GA for the one-max and deceptive problems

Before we turn our discussion to how GA is implemented, here is a simple example to show how it is run.

```
$ ./search ga 100 1000 10 "" 10 0.6 0.01 3
```

This example shows that GA is used as the underlying search algorithm, which is carried out for 100 runs, and 1000 evaluations are performed each run. Note that this is so that the comparison between GA and other metaheuristic algorithms is made as fair as possible. Eventually, the same number of evaluations (candidate solutions checked) during the convergence process is used for all the metaheuristic algorithms compared in this book. In this case, $1000 = 100 \times 10$, that is, 1000 evaluations is equivalent to 100 generations performed times 10 chromosomes checked per generation. Besides, the length of the solution is 10, and there is no file of initial solutions. The population size is set to 10, the crossover rate is set to 60%, the mutation rate is set to 1%, and the number of players for tournament selection is set to three.

Listing 7.1 shows how the main functions and operators of GA are implemented.

Listing 7.1: src/c++/1-onemax/main/ga.h: The implementation of GA.

```
1 #ifndef __GA_H_INCLUDED__
2 #define __GA_H_INCLUDED__
3
4 #include <string>
5 #include <fstream>
6 #include <numeric>
7 #include "lib.h"
8
9 using namespace std;
10
11 class ga
12 {
13 public:
14     typedef vector<int> solution;
15     typedef vector<solution> population;
16     typedef vector<int> obj_val_type;
17
18     ga(int num_runs,
19         int num_evals,
20         int num_patterns_sol,
21         string filename_ini,
22         int pop_size,
23         double crossover_rate,
24         double mutation_rate,
25         int num_players);
26
27     virtual ~ga() {}
28
29     population run();
30
31 private:
32     // begin the ga functions
33     virtual population select(const population& curr_pop, const obj_val_type& curr_obj_vals, int
        num_players);
34     virtual population crossover(population& curr_pop, double cr);
35     population mutate(population& curr_pop, double mr);
36     // end the ga functions
37
38 private:
39     void init();
40     virtual obj_val_type evaluate(const population& curr_pop);
41     void update_best_sol(const population& curr_pop, const obj_val_type& curr_obj_vals);
42
43 private:
44     int num_runs;
45     int num_evals;
46     int num_patterns_sol;
47     string filename_ini;
48
49     solution best_sol;
50     int best_obj_val;
51
52     // data members for population-based algorithms
53     population curr_pop;
54     obj_val_type curr_obj_vals;
55
56     // data members for ga-specific parameters
57     int pop_size;
58     double crossover_rate;
59     double mutation_rate;
60     int num_players;
61 };
62
```

```
63  inline ga::ga(int num_runs,
64                int num_evals,
65                int num_patterns_sol,
66                string filename_ini,
67                int pop_size,
68                double crossover_rate,
69                double mutation_rate,
70                int num_players
71                )
72      : num_runs(num_runs),
73        num_evals(num_evals),
74        num_patterns_sol(num_patterns_sol),
75        filename_ini(filename_ini),
76        pop_size(pop_size),
77        crossover_rate(crossover_rate),
78        mutation_rate(mutation_rate),
79        num_players(num_players)
80  {
81      srand();
82  }
83
84  inline ga::population ga::run()
85  {
86      double avg_obj_val = 0;
87      vector<double> avg_obj_val_eval(num_evals, 0);
88
89      for (int r = 0; r < num_runs; r++) {
90          int eval_count = 0;
91          double best_so_far = 0;
92
93          // 0. initialization
94          init();
95
96          while (eval_count < num_evals) {
97              // 1. evaluation
98              curr_obj_vals = evaluate(curr_pop);
99              update_best_sol(curr_pop, curr_obj_vals);
100
101             for (int i = 0; i < pop_size; i++) {
102                 if (best_so_far < curr_obj_vals[i])
103                     best_so_far = curr_obj_vals[i];
104                 if (eval_count < num_evals)
105                     avg_obj_val_eval[eval_count++] += best_so_far;
106             }
107
108             // 2. determination
109             curr_pop = select(curr_pop, curr_obj_vals, num_players);
110
111             // 3. transition
112             curr_pop = crossover(curr_pop, crossover_rate);
113             curr_pop = mutate(curr_pop, mutation_rate);
114         }
115         avg_obj_val += best_obj_val;
116     }
117     avg_obj_val /= num_runs;
118
119     for (int i = 0; i < num_evals; i++) {
120         avg_obj_val_eval[i] /= num_runs;
121         cout << fixed << setprecision(3) << avg_obj_val_eval[i] << endl;
122     }
123
124     return curr_pop;
125 }
126
127 inline void ga::init()
```

```
128 {
129     curr_pop = population(pop_size, solution(num_patterns_sol));
130     best_obj_val = 0;
131
132     if (!filename_ini.empty()) {
133         ifstream ifs(filename_ini.c_str());
134         for (int p = 0; p < pop_size; p++)
135             for (int i = 0; i < num_patterns_sol; i++)
136                 ifs >> curr_pop[p][i];
137     }
138     else {
139         for (int p = 0; p < pop_size; p++)
140             for (int i = 0; i < num_patterns_sol; i++)
141                 curr_pop[p][i] = rand() % 2;
142     }
143 }
144
145 inline ga::obj_val_type ga::evaluate(const population& curr_pop)
146 {
147     obj_val_type obj_vals(pop_size, 0);
148     for (int p = 0; p < pop_size; p++)
149         obj_vals[p] = accumulate(curr_pop[p].begin(), curr_pop[p].end(), 0);
150     return obj_vals;
151 }
152
153 // tournament selection
154 inline ga::population ga::select(const population& curr_pop, const obj_val_type& curr_obj_vals,
    int num_players)
155 {
156     population tmp_pop(curr_pop.size());
157     for (size_t i = 0; i < curr_pop.size(); i++) {
158         int k = rand() % curr_pop.size();
159         double f = curr_obj_vals[k];
160         for (int j = 1; j < num_players; j++) {
161             int n = rand() % curr_pop.size();
162             if (curr_obj_vals[n] > f) {
163                 k = n;
164                 f = curr_obj_vals[k];
165             }
166         }
167         tmp_pop[i] = curr_pop[k];
168     }
169     return tmp_pop;
170 }
171
172 inline void ga::update_best_sol(const population& curr_pop, const obj_val_type& curr_obj_vals)
173 {
174     for (size_t i = 0; i < curr_pop.size(); i++) {
175         if (curr_obj_vals[i] > best_obj_val) {
176             best_obj_val = curr_obj_vals[i];
177             best_sol = curr_pop[i];
178         }
179     }
180 }
181
182 // one-point crossover
183 inline ga::population ga::crossover(population& curr_pop, double cr)
184 {
185     const size_t mid = curr_pop.size()/2;
186     for (size_t i = 0; i < mid; i++) {
187         const double f = static_cast<double>(rand()) / RAND_MAX;
188         if (f <= cr) {
189             int xp = rand() % curr_pop[0].size();        // crossover point
190             for (size_t j = xp; j < curr_pop[0].size(); j++)
191                 swap(curr_pop[i][j], curr_pop[mid+i][j]);
```

```
192          }
193      }
194      return curr_pop;
195 }
196
197 inline ga::population ga::mutate(population& curr_pop, double mr)
198 {
199      for (size_t i = 0; i < curr_pop.size(); i++){
200          for (size_t j = 0; j < curr_pop[0].size(); j++){
201              const double f = static_cast<double>(rand()) / RAND_MAX;
202              if (f <= mr)
203                  curr_pop[i][j] = !curr_pop[i][j];
204          }
205      }
206      return curr_pop;
207 }
208
209 #endif
```

Listing 7.2 shows how easily an operator can be replaced by another operator by defining it as a virtual function. Here, it shows how the tournament selection defined in lines 154–170 of Listing 7.1 is replaced by roulette wheel selection.

Listing 7.2: src/c++/1-onemax/main/gar.h: The implementation of GA using the roulette wheel selection operator.

```
1 #ifndef __GAR_H_INCLUDED__
2 #define __GAR_H_INCLUDED__
3
4 #include "ga.h"
5
6 class gar: public ga
7 {
8 public:
9      using ga::ga;
10
11 private:
12      population select(const population& curr_pop, const obj_val_type& curr_obj_vals, int
         num_players) override;
13 };
14
15 // roulette wheel selection
16 inline gar::population gar::select(const population& curr_pop, const obj_val_type& curr_obj_vals
    , int num_players)
17 {
18      // 1. compute the probabilities of the roulette wheel
19      const double total = accumulate(curr_obj_vals.begin(), curr_obj_vals.end(), 0);
20      population tmp_pop(curr_pop.size());
21      for (size_t i = 0; i < curr_pop.size(); i++) {
22          // 2. select the individuals for the next generation
23          double f = total * static_cast<double>(rand()) / RAND_MAX;
24          for (size_t j = 0; j < curr_pop.size(); j++) {
25              if ((f -= curr_obj_vals[j]) <= 0) {
26                  tmp_pop[i] = curr_pop[j];
27                  break;
28              }
29          }
30      }
```

```
31      return tmp_pop;
32 }
33
34 #endif
```

7.2.1 Declaration of functions and parameters

As shown in Listing 7.1, since GA relies on the so-called population to search for the optimal solution, it can be implemented as an instance of a vector of vectors, the type of which is named population and is typedefed as shown in lines 14–15. The constructor is declared in lines 18–25. It takes eight parameters from either the user or other programs. Again, the parameters can be divided into two parts: those that are required by all the metaheuristics and those that are required by GA. The parameters required by all the metaheuristics are the number of runs, the number of generations, the number of subsolutions in a solution, and the name of the file of the initial solutions. The parameters required for GA only are the population size (pop_size), the crossover rate (crossover_rate), and the mutation rate (mutation_rate). The evaluation and transition operators have to be redefined because they differ from those of the other metaheuristics. Some additional operators have to be defined also to achieve the goal of GA. As shown in lines 33–35 and lines 40–41, the operators for GA are select(), crossover(), mutate(), evaluate(), and update_best_sol(). Note that the tournament selection operator is a replacement of the roulette wheel selection operator to select the suitable chromosomes for the next generation. We will discuss and compare these two selection operators later. As we discussed earlier, the chromosomes of GA are represented by a vector of vectors named curr_pop, as shown in line 53. To keep track of the fitness value of each chromosome of GA, a vector named curr_obj_vals is enough, as shown in line 54. Moreover, the number of chromosomes, the crossover rate, and the mutation rate are as defined in the variables pop_size, crossover_rate, mutation_rate, and num_players, as shown in lines 57–59.

7.2.2 The main loop

The main loop of GA is shown in lines 84–125. The initialization operator of GA will first create a set of chromosomes and then evaluate their fitness values at the very first generation. Also, because the offspring will be generated by the selected parents using the fitness function and selection operators, the order of the search operators is a little bit different from those of SA and TS. During the convergence process, GA will first evaluate the fitness value of each chromosome, as shown in line 98, and invoke the update_best_sol() operator, as shown in line 99, to keep track of the best-so-far solution. GA will then perform the selection operator to pick up the suitable chromosomes, as shown in line 109. Finally, it will perform the crossover() and mutation() operators to transit the structure of the chromosomes, as shown in lines 112–113. The evaluate(), update_best_sol(), select(), crossover(), and mutation() operators will be performed repeatedly for each generation until the termination criterion is met.

7.2.3 Additional functions

As shown in lines 127–143 of Listing 7.1, the very first thing the initialization function init() does is create a vector of vectors to keep track of all the chromosomes in line 129, followed by a reset of the best-so-far objective value in line 130. It will then create a set of chromosomes as the initial candidate solutions, either randomly or from a file, as shown in lines 132–142. As shown in lines 145–151 of Listing 7.1, the evaluation operator is similar to those in Listings 5.1 and 6.1 for the one-max problem; the main difference is that GA has to compute the objective values of all the chromosomes, implying that GA has to evaluate more than one solution at each generation. Since GA has to deal with more than one chromosome at each iteration, it requires an additional procedure to check which chromosome is the best of each iteration and also if it is the best-so-far solution so that the best-so-far solution is updated accordingly at each iteration, as shown in lines 172–180.

As shown in lines 154–170 of Listing 7.1, the selection operator used here performs tournament selection. As the name suggests, this operator will randomly select a certain number of players, and it will then pick up the best one as one of the parents for the next generation (lines 158–167). In addition to tournament selection, another well-known selection operator performs roulette wheel selection. The implementation of this operator is shown in Listing 7.2. It can be divided into two parts: (1) compute the probability of each chromosome using Eq. (7.1), as shown in lines 19 and 23 of Listing 7.2, and (2) use it to select chromosomes as the parents for the next generation, by generating a number uniformly distributed in the range [0, 1] and checking the region into which it falls so that the chromosome corresponding to that region will be selected, as shown in lines 24–29 of Listing 7.2.

How the crossover operator of GA works is shown in lines 183–195, which say that $m/2$ crossovers will be performed, where m is the number of chromosomes, as follows: A number uniformly distributed in the range [0, 1] is first generated and then checked to see if it satisfies (i.e., is less than or equal to) the crossover rate. If it does, crossover is performed. This means that a certain percentage of the chromosomes will be crossovered, while the others will remain intact, depending on the crossover rate. This program uses so-called one-point crossover to restructure the chromosomes, which works by first randomly generating a crossover point and then swapping genes on and after the crossover point. As shown in lines 197–207, this program employs a simple mutation. First, for each chromosome, a random number is generated for each gene; if the generated number for a gene satisfies the mutation rate, then the value of this gene is inverted (line 203). Note that to make it easier for the audience of the book to understand how a GA is implemented, the implementation described here is a little bit different from the implementation described in (Goldberg, 1989) and some other GA studies. This means that the implementation described in this book can be modified to improve its performance.

As shown in Listing 7.3, like HC, SA, and TS, to use this program to solve the deceptive problem $\mathbb{P}_4$, all we have to do is change the evaluation operator. Unlike the other single-solution-based metaheuristic algorithms compared in this chapter, the fitness value of each chromosome has to be computed by invoking the objective function of $\mathbb{P}_4$.

Listing 7.3: src/c++/2-deception/main/gadp.h: The evaluate() operator of GA for a deceptive problem.

```
16  inline vector<int> gadp::evaluate(const population& curr_pop)
17  {
18      const size_t pop_size = curr_pop.size();
19      const size_t num_patterns_sol = curr_pop[0].size();
20      vector<int> n(pop_size, 0);
21      for (size_t p = 0; p < pop_size; p++){
22          for (size_t i = 0; i < num_patterns_sol; i++)
23              n[p] += curr_pop[p][i] << ((num_patterns_sol - 1) - i);
24          n[p] = abs(n[p] - (1 << (num_patterns_sol-2)));
25      }
26      return n;
27  } //
```

7.3. Simulation results of GA

In this section, we will again use the one-max and deceptive problems to evaluate the performance of GA. The population size of GA is set to 10, the crossover rate is set to 60% (i.e., 0.6 of the seventh input parameter), and the mutation rate is set to 1% (i.e., 0.01 of the eighth input parameter). All the simulations were carried out for 100 runs, and 1000 evaluations, which is equivalent to 100 generations, were performed each run for the one-max and deceptive problems, respectively.

7.3.1 Simulation results of GA for the one-max problem

In Table 7.1, "GA-r" and "GA-t" denote, respectively, GA using roulette wheel selection and GA using tournament selection. The results show that GA does not perform well for the one-max problem compared to SA and TS. Although it seems that GA-t is able to find the optimal solution faster than HC in terms of the number of iterations performed, the reality is that GA is much slower than HC if we take into consideration the number of evaluations performed by these two algorithms. In some cases, GA-r cannot find the optimal solution in 1000 evaluations. In the table, a value in brackets (i.e., [*value*]) is used to show the end results of this algorithm and also to indicate that the optimal solution cannot be found in 1000 evaluations. From these results, it can be easily seen that GA is not particularly good at local search; therefore, there is no guarantee that it will always find the optimal solution for the one-max problem, just like many single-solution-based metaheuristic algorithms. The results also show that the performance of GA-r and that of GA-t are different.

Table 7.1   Number of evaluations needed by HC, SA, TS, and GA to find the optimal solution.

Case	HC	SA	TS	GA-r	GA-t
$n = 10$	59	79	33	[9.77]	692
$n = 20$	145	148	96	[17.29]	873
$n = 40$	277	306	325	[30.20]	[39.95]
$n = 60$	486	460	374	[42.50]	[59.37]
$n = 80$	786	576	600	[53.94]	[77.41]
$n = 100$	965	826	901	[64.56]	[93.82]

Although GA does not beat the other algorithms for the one-max problem in these comparisons, it does not mean that the search capability of GA is not good. In fact, GA is a "global search algorithm," and its local search ability is not good, as observed by most researchers. The results do not imply that the roulette wheel selection operator is worse than the tournament selection operator, even if the results show that GA-t outperforms GA-r. Our experience shows that the roulette wheel selection operator is useful for problems where the objective values of their candidate solutions are significantly different from each other, whereas tournament selection is useful for problems where the objective values of their candidate solutions are very close to each other.

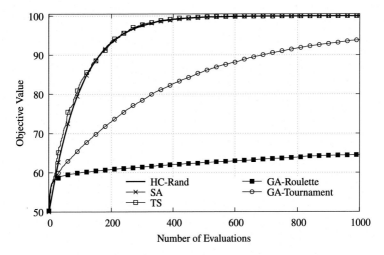

Figure 7.5 Simulation results of GA using different selection operators for the one-max problem of size $n = 100$.

The results for the one-max problem of size $n = 100$ described in Fig. 7.5 show that the convergence speed of GA with the tournament selection operator is much higher than that with the roulette wheel selection operator. Two interesting things can be observed from the comparisons in Table 7.1 and Figs. 7.5 and 7.6. First, GA gives the best result at generation 1, which consists of the first 10 evaluations. This can be easily explained because GA checks 10 chromosomes per iteration, but the other search algorithms check only one solution per iteration. Although GA and the other search algorithms check the same number of candidate solutions in the first 10 evaluations, GA has a larger chance to find better results than the other search algorithms at early evaluations, as shown in Fig. 7.6. Second, the convergence speed of GA-r is lower than that of all the other search algorithms, not only SA, TS, and GA-t, but also HC, in terms of the number of evaluations.

To better understand the characteristics of the convergence trend of GA, Fig. 7.7 shows the results of GA with roulette wheel and tournament selection operators for 50,000 evaluations to see if it is possible for GA to find the optimal solution in solving the one-max problem of size $n = 100$.

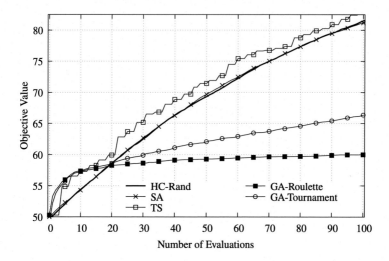

Figure 7.6 The details of Fig. 7.5.

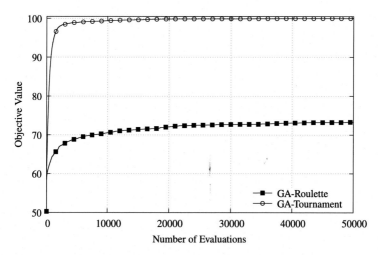

Figure 7.7 Simulation results of GA using different selection operators for the one-max problem of size $n = 100$.

The results show that given enough time, GA may be able to find the optimal solution in most cases for the one-max optimization problem. In this example, GA with tournament selection is able to find the optimal solution for the one-max problem of size $n = 100$ at around 50,000 evaluations on average.

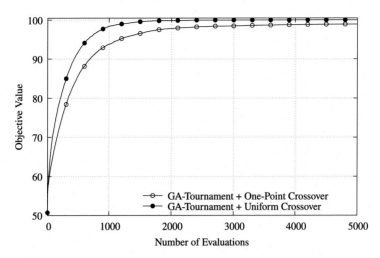

Figure 7.8 Comparison of GA using one-point and uniform crossover operators for the one-max problem of size $n = 100$.

Fig. 7.8 further compares GA with tournament selection and different crossover operators, namely, one-point and uniform crossover operators. Unlike the one-point crossover operator, which exchanges segments between the parents, the uniform crossover operator gives each gene of a chromosome a 50% chance to be exchanged with the gene at the same locus in the other parent. The results show that GA with uniform crossover operator can find the optimal solution for the one-max problem of size $n = 100$ at around 3000 evaluations on average. Although GA with uniform crossover operator still cannot find the optimal solution for the one-max problem of size $n = 100$ within 1000 evaluations on average, the improvement is significant compared to GA with one-point crossover operator for solving this problem. The results also show that the performance of GA for solving the one-max problem can be improved by using a more suitable operator to replace the operator originally used in GA, e.g., using a uniform crossover operator to replace the one-point crossover operator.

In addition to using suitable operators for GA, how to determine the values of the parameters of GA is another crucial issue. As shown in Fig. 7.9, we also use different numbers of players in tournament selection for solving the one-max problem to better understand whether the parameter settings will affect the end results of GA. This figure shows that GA is capable of finding the optimal solution for the one-max problem within 1500 eval-

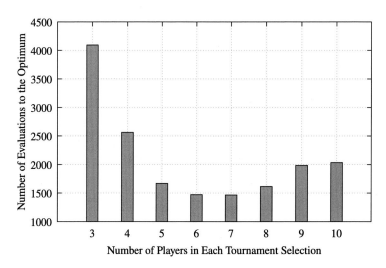

Figure 7.9 Comparison of GA using uniform crossover with different numbers of players in the tournament selection for the one-max problem of size $n = 100$.

uations when the number of players of the tournament selection is set to six or seven. The results further show that the convergence speed can be accelerated if the parameters of GA are set to suitable values. Although this change still cannot make GA capable of finding the optimal solution for the one-max problem within 1000 evaluations, it implies that the performance of GA can be enhanced one way or another. The issue of setting the parameters to suitable values (hyperparameter optimization problem) is an advanced research topic of metaheuristic algorithms, which we will discuss in Section 19.4.

7.3.2 Simulation results of GA for the deceptive problem

As shown in Figs. 7.10 and 7.11, even though the convergence speed of GA is lower than that of other search algorithms for the one-max problem, it is close to that of other search algorithms for the deceptive problem of sizes 4 and 10. According to our observations, unlike SA and TS, which can find the optimal solution very quickly for simple optimization problems, GA is very useful for complex optimization problems. This means that GA is typically worse than SA and TS for simple optimization problems. On the other hand, for complex optimization problems, GA and its variants (e.g., GA with some useful local search methods) may be able to find a better result than SA and TS.

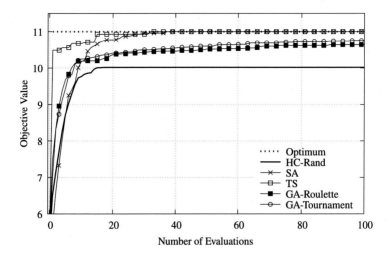

Figure 7.10 Simulation results of GA for the deceptive problem of size $n = 4$.

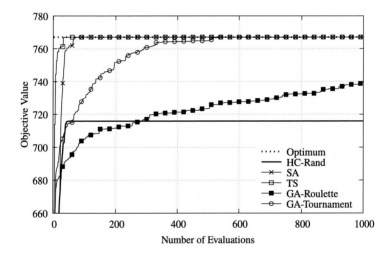

Figure 7.11 Simulation results of GA for the deceptive problem of size $n = 10$.

As mentioned in Section 7.3.1, using suitable operators or appropriate parameters may enhance the convergence speed or improve the end results of GA; therefore, we also use different numbers of players in the tournament selection for solving the deceptive problem to understand the impact the number of players may have on the performance of GA. Fig. 7.12 shows that although the time required to find the optimal solution can be signif-

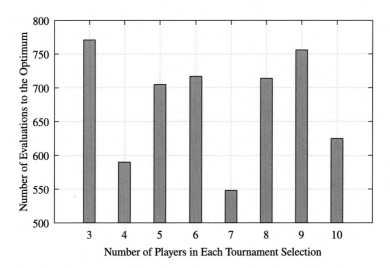

Figure 7.12 Comparison of GA using uniform crossover with different numbers of players in the tournament selection for the deceptive problem of size $n = 10$.

icantly shortened with different numbers of players (e.g., four or seven) compared to three players, the trend of this parameter setting is not clear. The results imply that how to determine good values for the parameters of GA can be regarded as a complex optimization problem, and a good setting of the parameters will be very useful for enhancing the search performance of GA and improving its end result.

7.4. Discussion

GA was pioneered by Holland (Holland, 1962, 1975), his colleagues, and his students. This algorithm is definitely one of the most important population-based metaheuristic algorithms. Although GA was invented earlier than SA and TS, it was not widely used in the 1970s and 1980s for a very simple reason. The design of GA is much more complex than that of SA and TS; as a consequence, it is not only more difficult to implement than SA and TS, it also takes much more memory space. Even Holland himself was surprised that the GA field was quiet for decades, as described in the preface of the 1992 edition of his book (Holland, 1992).

> When this book was originally published I was very optimistic, envisioning extensive reviews and a kind of "best seller" in the realm of monographs. Alas! That did not happen. After five years I did regain some optimism because the book did not "die," as is usual with monographs, but kept on selling at 100–200

copies a year. Still, research in the area was confined almost entirely to my students and their colleagues, and it did not fit into anyone's categories...

In fact, GA was presented almost 20 years earlier than SA and TS, and it took about 18 years to get the second edition of this book published. This can be easily justified by the fact that the concept GA presents was too new to be understood and it is hard to implement on the computers of the 1970s and 1980s. However, GA has been applied to many research domains since 1990, and it still has a strong impact on almost every research domain (Holland, 1975) today. This success can be partly ascribed to the book by Goldberg (Goldberg, 1989), in which the source code of GA is attached. This makes it much easier for the audience to understand how exactly GA works and how it is implemented.

Generally speaking, the main advantage of GA is that it is capable of solving complex optimization problems. To make it easier for the audience to understand, in this chapter, we apply it to two simple optimization problems, namely, the one-max and deceptive problems. Although the simulation results show that its performance is worse than that of SA and TS, the performance of GA can be enhanced by appropriate modifications, thus making it better than SA and TS. Also, since GA consists of several operators and parameters, it is more flexible than other metaheuristics in the sense that users of GA can modify these operators and parameters to make them suit the optimization problem in question. For example, in an early study (Syswerda, 1989), Syswerda presented uniform crossover for GA, which provides better results than the one-point and two-point crossovers do in solving the one-max problem, the traveling salesman problem, and some other optimization problems. This implies that suitable operators are able to enhance the performance of GA. In a later study (De Jong and Spears, 1992), De Jong and Spears discussed and analyzed the one-point, two-point, n-point, and uniform crossovers. Among them, n-point and uniform crossover can be regarded as multi-point crossover.

In addition to the crossover, the mutation was also discussed in some early studies (Bäck, 1993; Srinivas and Patnaik, 1994). In (Bäck, 1993), Bäck presented a dynamic method for adjusting the mutation rate during the convergence process of GA. It works by setting the mutation rate equal to a higher value initially, say, 0.9, and then decreasing it toward $1/n$, where n is the number of subsolutions of a solution. In a later study (Srinivas and Patnaik, 1994), Srinivas and Patnaik presented the so-called adaptive adjustment methods for crossover and mutation, which are defined by

$$p_c = \begin{cases} k_1 \left(\frac{f_{\max} - f'}{f_{\max} - \bar{f}} \right), & \text{if } f' \geq \bar{f}, \\ k_3, & \text{otherwise,} \end{cases} \qquad (7.2)$$

$$p_m = \begin{cases} k_2 \left(\frac{f_{\max} - f}{f_{\max} - \bar{f}} \right), & \text{if } f \geq \bar{f}, \\ k_4, & \text{otherwise,} \end{cases} \qquad (7.3)$$

where p_c denotes the crossover rate, p_m denotes the mutation rate, $f_{\max}$ denotes the maximum fitness value of the population, $\bar{f}$ denotes the average fitness value of the population, f denotes the fitness value of the solution, and f' denotes the larger of the fitness values of the solutions to be crossed. Besides, k_1 and k_3 are set equal to 1.0, and k_2 and k_4 are set equal to 0.5. These two adaptive adjustment methods will lower or raise the crossover and mutation rates for chromosomes with a fitness value that is better or worse than the average fitness value of the population, respectively.

The selection operator, of course, was also discussed in some early studies (Kuo and Hwang, 1996; Thierens and Goldberg, 1994). Thierns and Goldberg (Thierens and Goldberg, 1994) analyzed the convergence characteristics of the proportionate (roulette wheel), tournament, and truncation selection operators. Kuo and Hwang (Kuo and Hwang, 1996) pointed out that increasing the sampling rate of schemata (particular solutions) that are above average provides no guarantee to find the global optimum because the global optimum might be a relatively isolated peak of the solution space. For this reason, they presented a disruptive selection with a non–monotonic fitness function for GA. The basic idea is that even worse solutions have a chance to contain useful information for the search. The fitness value of a chromosome is defined as the difference between the objective value of the chromosome and the average objective value of all chromosomes in the population. Their experimental results show that disruptive selection is capable of accelerating the convergence speed of GA in finding the optima of a deceptive problem compared to directional selection, e.g., proportional selection.

Different from developing better search operators for GA, the focus of some early studies (Harik and Lobo, 1999; Potts et al., 1994) was on the division of the population. Potts et al. (Potts et al., 1994) used multiple populations to replace the single population of canonical GA. In their design, GA has four populations—one with a higher mutation rate in which all genes have a lower probability of being "1," one with a medium mutation rate in which all genes have a 50/50 chance to be "1" or "0," one with a lower mutation rate in which all genes have a higher probability of being

"1," and the last one playing the role of keeping highly fit chromosomes from the other three populations. Harik and Lobo in a later study (Harik and Lobo, 1999) provided a discussion on either doubling the population size or using multiple populations to find a way for determining the population size in solving an optimization problem. The strategy of doubling the population size involves starting with a small population (four chromosomes) and then generating a new population twice as large as the previous one once the previous one converges. The strategy of multiple populations involves starting with a set of populations of various sizes. Smaller populations might be catched up by a larger population if their average fitness values are worse than that of the larger population. In this case, smaller populations will simply be eliminated.

To better understand the fundamentals on the design of canonical GA, the reader is referred to excellent books (Davis, 1991; Goldberg, 1989; Holland, 1975; Michalewicz, 1996; Mitchell, 1996) and tutorial and review papers (Beasley et al., 1993a,b; Katoch et al., 2021; Whitley, 1994). Among them, the schema theorem and building-block hypothesis will help the reader understand the characteristics of GA. The solution space is typically represented by a hypercube to explain the relationship between schema,[1] schemata, hyperplane, and hypercube and how GA works in the solution space. Different encoding schemata and operators for selection, crossover, and mutation were also discussed in these books and tutorial and review papers. We can also find many branches of GA. Examples are steady-state GA (De Jong, 1975; Syswerda, 1991), GENITOR (Whitley, 1989), GENITOR II (Whitley and Starkweather, 1990), the non-dominated sorting GA (NSGA) (Srinivas and Deb, 1994), NSGA-II (Deb et al., 2002), and the estimation of distribution algorithm (EDA) (Larraanaga and Lozano, 2001; Mühlenbein and Paaß, 1996). Unlike canonical GA, which uses offspring to replace all the parents in the population, for GENITORs and steady-state GA, only a few offspring will be generated each generation to replace the worst individuals of the population instead of their parents. The idea is to let the parents and offspring compete for survival to the next generation; therefore, it has also been called overlapping generations (De Jong, 1975). These reproduction plans, of course, are similar to evolution strategies (ES) (Beyer and Schwefel, 2002; Rechenberg, 1965). Another interesting branch is to apply GA to the multi-objective problem

---

[1] A set of chromosomes with genes at the same loci matching a set of special genes, e.g., a set of chromosomes with genes at loci 6–8 matching 1∗0, where "∗" is a wildcard that can assume a value of 0 or 1.

(Deb and Kalyanmoy, 2001). Since NSGA-II (Deb et al., 2002) uses the non-dominated sort and crowding distance mechanisms, it provides good results in solving multi-objective problems. As a consequence, it has become one of the most popular multi-objective optimization algorithms. EDA (Larraanaga and Lozano, 2001; Mühlenbein and Paaß, 1996) can be regarded as an improved version of GA, the aim of which is to construct a probabilistic model to describe promising candidate chromosomes during the convergence process and then use the probabilistic model to generate the offspring.

Unlike modifying the selection, crossover, and mutation operators and the fitness function, how to combine GA with other (meta)heuristic algorithms is also an important research direction of GA. Such kinds of successful results of GA can be found in many studies. Krishna and Murty (Krishna and Narasimha Murty, 1999) combined GA with k-means and produced the genetic k-means algorithm (GKA) to solve the data clustering problem. No need to say, hybrid GAs will give better results than k-means and other metaheuristics alone. Another study (Kelly and Davis, 1991) was aimed at the classification problem, in which GA was used to find the suitable weights associated with the attributes of data for the k-nearest neighbor classification algorithm to calculate the inter-datum distances. In (Liaw, 2000), Liaw used TS as the local search operator of GA for solving the open shop scheduling problem. A later study (Gonçalves et al., 2005) presented a local search operator for GA in solving the job shop scheduling problem in which the local search operator is responsible for identifying the critical path and swapping genes on it to improve the makespan. In (Kao and Zahara, 2008), Kao and Zahara presented a hybrid algorithm combining GA and particle swarm optimization (PSO) to solve the multi-modal function optimization problem. The basic idea of this hybrid algorithm is to first generate $4N$ solutions, where N is the number of dimensions of the problem, each iteration. The top-ranked $2N$ solutions will be used as the population of GA, while the bottom-ranked $2N$ solutions will be used as the population of PSO, and all the solutions found by GA and PSO will be ranked again at the end of each iteration.

From the literature on the development of GA, it can be easily seen that GA has been applied to a variety of domains, such as computer science, engineering, and even management. Eventually, almost everything we can imagine has a trace of GA. It can be used in engineering applications, such as electromagnetics (Weile and Michielssen, 1997); it can also be used in our daily lives, such as fashion (Kim and Cho, 2000); it is in fact almost everywhere. Its application to neural networks (NNs) (Ding et al., 2011; Leung et al., 2003; Miller et al., 1989; Montana and Davis, 1989; Schaffer et al., 1992; Tong and Mintram, 2010; Whitley et al., 1990) and deep NNs (DNNs) (Ali and Ahmed, 2019; Chung and Shin, 2020; Li et al., 2021; Loussaief and Abdelkrim, 2018; Real et al., 2017; Rostami et al., 2021; Sun et al., 2019, 2020; Xie and Yuille, 2017) is a very interesting research domain. In the early stage, researchers aimed to apply GA to NNs to improve their performance, which can be regarded as applying an unsupervised learning algorithm (GA) to a supervised learning algorithm (NN). In (Schaffer et al., 1992), Schaffer et al. discussed several combinations of GA and NN, which can be divided into two categories: supportive (those that are used sequentially) and collaborative (those that are used simultaneously). In the supportive category, GA is used to select suitable features of data or hyperparameters for an NN, while in the collaborative category, GA is used to determine the weights or topologies of an NN. In another study (Whitley et al., 1990), Whitley et al. discussed how to apply GA to NN; they used GA to determine the weights of connections and neurons and pruned the connections of NN to further reduce its size. A simple example of a combination of GA and NN can be found in an early study (Miller et al., 1989) in which Miller and Todd attempted to encode the network architecture as a bit string of the connections between different neurons. Another combination of GA and NN can be found in (Montana and Davis, 1989), where each chromosome is composed of a set of real numbers to represent the weights of connections and neurons in an NN. Leung et al. (Leung et al., 2003) provided a new crossover operator to generate one offspring v from two parents based on the minimum and maximum values of the gene at each locus and a new mutation operator to generate three offspring from v. In a few later studies (Ding et al., 2011; Tong and Mintram, 2010), the use of GA to select suitable features for NN (Tong and Mintram, 2010) and the use of GA to learn connection weights (Ding et al., 2011) were investigated.

Since DNN has become a new paradigm of NN in recent years, GA has once again been used to enhance its performance. In (Loussaief and Abdelkrim, 2018), GA is used to find suitable hyperparameters for a convolutional NN (CNN), such as the number and size of filters in convolutional layers. In (Chung and Shin, 2020), GA is used to determine the number and size of filters in convolutional layers as well as the size of filters in pooling layers. Of course, that GA can be used to select a set of suitable features for NN implies that it can also be used to select features for DNN (Ali and Ahmed, 2019; Rostami et al., 2021). As one of the most well-known DNNs, the structure of CNN has become more and more complex; thus, how to automatically generate a suitable CNN architecture for different optimization problems has become a promising research topic, called neural architecture search (NAS). Some recent studies (Real et al., 2017; Sun et al., 2019, 2020; Xie and Yuille, 2017) attempted to use GA to find a good neural architecture. In addition to using simple GA (Real et al., 2017; Xie and Yuille, 2017) for NAS, in (Sun et al., 2020), Sun et al. presented a new crossover operator to deal with problems the solutions of which are of different lengths and a new mutation operator to add, remove, and modify genes of a solution to further enhance the performance of GA for NAS.

In summary, our observation shows that like other metaheuristics, GA still has some open issues that need to be addressed. An important open issue of GA is that it has *too many parameters*. As a result, how to determine the values of these parameters, which may have a strong impact on the end results, has been a problem. Another open issue of GA is its *local search ability*. The results of Section 7.3 show that without local search ability, simple GA cannot provide a better result. For this reason, several studies attempted to combine GA with local search methods or single-solution-based metaheuristics to provide a better result. Compared to other single-solution-based metaheuristics, another open issue of GA is the *high computation cost*. The computation cost of each generation of GA is much higher than that of each iteration of single-solution-based metaheuristics; therefore, comparison based on the same number of iterations is unfair. More recently, several studies suggested comparing metaheuristics using the number of evaluations. Although this is still not very accurate, it is much fairer than using the same number of iterations. Also, several recent studies have presented solutions to reduce the computation time of GA, such as dynamically adjusting the number of chromosomes or keeping the elitist chromosomes to accelerate the convergence speed. This chapter is just a starting point of GA; there exist many interesting methods to be discovered.

Supplementary source code

Here are the hyperlinks to the directories where the source code described in this chapter resides.

https://github.com/cwtsaiai/metaheuristics_2023/src/c++/1-onemax/
https://github.com/cwtsaiai/metaheuristics_2023/src/c++/2-deception/

CHAPTER EIGHT

Ant colony optimization

8.1. The basic idea of ant colony optimization (ACO)

Ant colony optimization (ACO) was pioneered by Dorigo and his colleagues (Colorni et al., 1992; Dorigo and Gambardella, 1997; Dorigo et al., 1996; Dorigo and Stützle, 2004). It is not a single algorithm, but a collection of ant-based search algorithms. The ant system (AS) (Dorigo et al., 1996) and ant colony system (ACS) (Dorigo and Gambardella, 1997) are two well-known versions of ACO algorithms. As the name suggests, ACO is inspired by the behavior of ants. For this kind of metaheuristic algorithm, the routing path (tour) of an ant represents a solution to the problem in question. Each ant constructs a solution step by step by using the distance information and the searched information saved in the pheromone table.

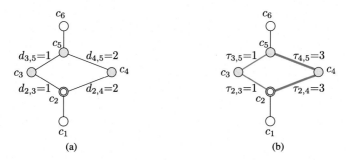

Figure 8.1 Strategies of ACO. (a) Distance information. (b) Pheromone information.

ACO and its variants are typically very useful in solving optimization problems the solution structures of which are a sequence of subsolutions, such as the traveling salesman problem (TSP). Since ACO is also a population-based metaheuristic algorithm, it will create a set of solutions $s = \{s_1, s_2, \ldots, s_m\}$ in the very beginning, where s denotes a set of routing paths of ants, $s_i = \{s_{i,1}, s_{i,2}, \ldots, s_{i,n}\}$ denotes a routing path of an ant (a solution), and $s_{i,j}$ denotes the j-th subsolution of the i-th ant. Note that m and n denote the number of ants in the population and the number of subsolutions in an ant, respectively. As shown in Fig. 8.1, where each node represents a city of the TSP, for an ant starting at node c_1 and now standing on node c_2, there are two candidate nodes (i.e., c_3 and c_4) that can be se-

139

lected as the next node. Fig. 8.1(a and b) shows that ACO-based algorithms usually use both distances between nodes and the pheromone table together to determine the node for the next move. In this example, $d_{\cdot,\cdot}$ denotes the distance between two nodes; for example, $d_{2,4}$ denotes the distance between nodes c_2 and c_4. The pheromone values on all the edges eventually represent the search experience of all the ants; for example, $\tau_{2,4}$ represents the pheromone value of the edge between node c_2 and node c_4. The higher the pheromone value of an edge, the higher the number of ants traveling through it, and the lower the pheromone value of an edge, the lower the number of ants traveling through it. In this example, the edge connecting nodes c_2 and c_3 is the best choice if the ant considers only the distance information; however, the edge connecting nodes c_2 and c_4 is the best choice if the ant takes into account the pheromone information. An edge that has a larger pheromone value means that a larger number of ants choose this edge as a subsolution of their solutions. The distances between nodes can be regarded as the local information seen by an ant that guides the ant to choose an edge that is as short as possible, while the pheromone on the edges can be regarded as the global information contributed by all the ants since the very beginning that guides an ant to choose the edge that has been traveled by a larger number of ants. From these observations, it can be easily seen that ACO takes into account the distance and pheromone information at the same time in the determination of search directions, that is, in the determination of new candidate solutions, thus avoiding searching particular solutions repeatedly and increasing the chance of finding better solutions.

Figure 8.2 How ants choose the next node. (a) Without distance and pheromone information. (b) With distance and pheromone information.

Fig. 8.2 is based on Fig. 8.1 to further illustrate how an ACO-based algorithm works. In Fig. 8.2(a), suppose that six ants start at node c_1 and now stand on node c_2, at which they can choose to go to either node c_3 or node c_4. Without the distance and pheromone information, the ants will choose either node c_3 or node c_4 with a 50/50 chance; as a consequence, three of the six ants go to node c_3 while the other three go to node c_4. As shown in Fig. 8.2(b), with the integrated distance and pheromone information, the ants will then choose the next node based on this information. That is, two of the six ants go to node c_3, while the other four go to node c_4.

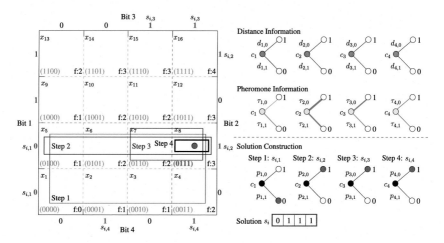

Figure 8.3 Strategies of ACO for solving the one-max problem.

As we mentioned before, a distinguishing feature of ACO is that a solution is constructed step by step. Fig. 8.3 is based on the example described in Fig. 7.4 to show how ACO creates a solution for the one-max problem based on the distance and pheromone information. In this figure, the distance and pheromone information is used in calculating the probabilities $p_{i,0}$ for c_i to 0 and $p_{i,1}$ for c_i to 1. For example, for an ant standing on node c_1 during the solution construction process, it has two choices, 0 and 1, and it chooses 0 in this case. Eventually, an ant, when standing on nodes c_2, c_3, and c_4, will use exactly this probability model to determine the remaining subsolutions step by step based on the distance and pheromone information to construct a complete solution s_i. The solution (0111) means that it will choose "0" as the first subsolution (i.e., Bit 1) and "1," "1," and "1" as the second, third, and fourth subsolutions later to create a complete solution.

As shown in Algorithm 8, ACO will first initialize the pheromone values of all the edges in line 1. The operators SolutionConstruction(), Evaluation(), PheromoneUpdate(), and LocalSearch() will then be performed for a certain number of iterations. Note that the evaluation operator

Evaluation() described here is responsible for calculating the objective values of all ants, e.g., tour distances in solving the TSP. Also note that the local search operator LocalSearch() is optional to ACO, but it can be used to improve the quality of a solution.

Algorithm 8 Ant colony optimization.

1 Initialize the pheromone values τ for all the edges $\boxed{\text{I}}$
2 **While** the termination criterion is not met
3        $s = \text{SolutionConstruction}(\tau, \text{dist})$ using Eq. (8.1) $\boxed{\text{T}}\!\mapsto\!\boxed{\text{E}}\!\mapsto\!\boxed{\text{D}}$
4        $f = \text{Evaluation}(s)$ $\boxed{\text{E}}$
5        $\tau = \text{PheromoneUpdate}(s, f)$ using Eq. (8.4) $\boxed{\text{T}}$
6        $s = \text{LocalSearch}(s, f)$ [Optional] $\boxed{\text{T}}\!\mapsto\!\boxed{\text{E}}\!\mapsto\!\boxed{\text{D}}$
7 **End**
8 **Output** s $\boxed{\text{O}}$

8.1.1 The ant system (AS)

The operator SolutionConstruction() shown in line 3 will first randomly put each ant a^k on a node c_i as the starting point of each ant, and it will then compute the transition probability from the node c_i to the other nodes that have not been visited. A well-known model presented in (Dorigo et al., 1996) for computing the transition probability of AS is as follows:

$$s_{k,i} = p_{i,j}^k = \frac{[\tau_{i,j}]^\alpha [\eta_{i,j}]^\beta}{\sum_{g \in \mathbb{N}_i^k} [\tau_{i,g}]^\alpha [\eta_{i,g}]^\beta}, \tag{8.1}$$

where $s_{k,i}$ denotes the i-th subsolution of the k-th ant, $p_{i,j}^k$ denotes the transition probability of the k-th ant at node c_i choosing node c_j as the next node, $\tau_{i,j}$ and $\tau_{i,g}$ denote the pheromone values, $\eta_{i,j}$ and $\eta_{i,g}$ denote the heuristic values, α and β denote the importance of $\tau_{i,j}$ and $\eta_{i,j}$, and $\mathbb{N}_i^k$ denotes the set of candidate subsolutions that can be chosen by ant k at node i, that is, the subsolutions that have not been visited yet by ant k and thus can be selected by ant k at node i. In (Dorigo et al., 1996), $\eta_{i,j}$ is called the visibility, which is defined as $1/d_{i,j}$, where $d_{i,j}$ is the distance between node c_i and node c_j. AS chooses the nodes step by step using Eq. (8.1). At each step, the number of nodes that can be chosen will be decreased, and the transition probability will also be recomputed for choosing the next node. A complete solution will be constructed once all the nodes have been chosen and visited.

As shown in line 5, after all the solutions are constructed by all the ants, AS will perform the PheromoneUpdate() operator to update the pheromone values of all the edges. This operator will first compute the pheromone value of each ant as follows:

$$\Delta \tau_{i,j}^{k} = \begin{cases} \frac{Q}{L_k}, & \text{if the } k\text{-th ant chooses edge } e_{i,j} \text{ as a part of the tour,} \\ 0, & \text{otherwise,} \end{cases} \quad (8.2)$$

where Q is a constant and L_k is the tour length of the k-th ant. That is, the pheromone value of each ant $\Delta \tau_{i,j}^{k}$ is inversely proportional to the quality of the solution. Next, given the values $\Delta \tau_{i,j}^{k}$ of each ant, this operator will compute the accumulated pheromone value for each edge $e_{i,j}$ as follows:

$$\Delta \tau_{i,j} = \sum_{k=1}^{m} \Delta \tau_{i,j}^{k}, \quad (8.3)$$

where m is the number of ants. Finally, this operator will update the pheromone value of each edge as follows:

$$\tau_{i,j} = \rho \times \tau_{i,j} + \Delta \tau_{i,j}, \quad (8.4)$$

where ρ is a weight in the range $[0, 1)$.

8.1.2 The ant colony system (ACS)

It can be easily seen that the way the transition probability of AS (as shown in Eq. (8.1)) is computed is similar to the roulette wheel selection of the genetic algorithm (GA) (as shown in Eq. (7.1)); therefore, it takes a lot of computations when using AS. Dorigo et al. (Dorigo and Gambardella, 1997) presented a more useful way to compute the transition probability $\mathcal{S}_{i,j}^{k}$, which is defined as follows:

$$s_{k,i} = \mathcal{S}_{i,j}^{k} = \begin{cases} \arg\max_{j \in \mathbb{N}_i^k} [\tau_{i,j}] \times [\eta_{i,j}]^{\beta}, & \text{if } q \leq q_0, \\ p_{i,j}^{k}, & \text{otherwise,} \end{cases} \quad (8.5)$$

where $s_{k,i}$ represents the i-th subsolution of the k-th ant, q is a uniformly distributed random number in the range $[0, 1]$, and q_0 is a parameter in the range $[0, 1]$ to determine *exploitation* or *biased exploration*. When $q > q_0$, ACS will use Eq. (8.1) to compute its transition probability except that α will be set equal to 1 in this case. When $q \leq q_0$, ACS will choose the edge $e_{i,j}$ that has the maximum value of $[\tau_{i,j}] \times [\eta_{i,j}]^{\beta}$. This means that ACS will not always use $p_{i,j}^{k}$ (Eq. (8.1)) to compute its transition probability but has a certain chance to use it to choose the best possible subsolution. This change may eventually reduce the computation time of this operator and accelerate the convergence speed of ACO. This can be easily explained by observing that if $q \leq q_0$, this transition operator does not have to calculate the probability of each feasible subsolution; rather, it will simply use the maximum value of $[\tau_{i,j}] \times [\eta_{i,j}]^\beta$ to choose the next subsolution, thus making it a greedy search. Compared to AS, two kinds of update methods are used by ACS: local update and global update.

For the local update, ACS will update the pheromone value of the edge right after it was traveled by an ant. This means that the local update of ACS will be done once all the ants finished their tours. The local rule is defined as follows:

$$\tau_{i,j} = (1 - \rho) \times \tau_{i,j} + \rho \times \Delta\tau_{i,j}, \tag{8.6}$$

where ρ is a parameter in the range $(0, 1)$ for determining the evaporation rate of the pheromone of each edge. As mentioned by Dorigo, there are several ways to compute $\tau_{i,j}$. One possible way is to set it to $\tau_0 = (n \times L_{nn})^{-1}$, where n is the number of subsolutions (cities) and L_{nn} is the tour length produced by the nearest neighbor heuristic method.

For the global update, ACS will update the pheromone values of all the edges of the global best tour after all the ants complete their tours at each iteration by using the following global update rule:

$$\tau_{i,j} = (1 - \sigma) \times \tau_{i,j} + \sigma \times \Lambda\tau_{i,j}, \tag{8.7}$$

where σ is the pheromone decay parameter in the range $(0, 1)$ and

$$\Lambda\tau_{i,j} = \begin{cases} L_{gb}^{-1}, & \text{if } e_{i,j} \in \text{global best tour}, \\ 0, & \text{otherwise}, \end{cases} \tag{8.8}$$

where L_{gb} is the length of the global best tour, which is defined as the shortest tour from the very beginning to the current iteration.

8.2. Implementation of ACO for the traveling salesman problem

Since ACO was designed for the TSP, this section will give a brief introduction to the basic idea of ACO by using it to solve the TSP. For more details of the TSP, you are referred to Section 2.2.3. Also, because

ACS is one of the classic ACO-based algorithms, this section will show how it is implemented to explain the details of ACO. Here is an example command line for invoking ACS for the TSP.

```
$ ./search aco 100 25000 51 "" "eil51.tsp" 20 0.1 2.0 0.1 0.9
```

The first term "aco" means that the search program will use ACS as the underlying search algorithm. The second term states that it will be carried out for 100 runs, while the third term says that 25,000 evaluations (1250 iterations) will be performed per run. Since this implementation is for the TSP, the fourth term "51" represents the number of cities. The fifth term is the name of the file where the initial solutions reside, and since the filename is an empty string, it means that no initial solutions are given in this case. The sixth term is the name of the file where the benchmark for TSP is stored. In this case, it is named "eil51.tsp," in which the number "51" indicates the number of cities. The following terms—20, 0.1, 2.0, 0.1, and 0.9—denote, in order, the number of ants and the values of α, β, ρ, and q_0 of ACS.

Listing 8.1 shows an implementation of the main functions and operators of ACS.

Listing 8.1: src/c++/3-tsp/main/aco.h: The implementation of ACO.

```
 1 #ifndef __ACO_H_INCLUDED__
 2 #define __ACO_H_INCLUDED__
 3
 4 #include <iostream>
 5 #include <string>
 6 #include <vector>
 7 #include <fstream>
 8 #include <list>
 9 #include <limits>
10
11 using namespace std;
12
13 class aco
14 {
15 public:
16     typedef vector<int> solution;
17     typedef vector<solution> population;
18     typedef vector<double> pattern;
19     typedef vector<pattern> instance;
20     typedef vector<double> pheromone_1d;
21     typedef vector<pheromone_1d> pheromone_2d;
22     typedef vector<double> dist_1d;
23     typedef vector<dist_1d> dist_2d;
24     typedef vector<double> obj_val_t;
```

```
25
26    aco(int num_runs,
27          int num_evals,
28          int num_patterns_ins,
29          string filename_ini,
30          string filename_ins,
31          int pop_size,
32          double alpha,
33          double beta,
34          double rho,
35          double q0);
36
37    population run();
38
39 private:
40    void init();
41    void construct_solutions(population& curr_pop, const dist_2d& dist, pheromone_2d& ph, double
         beta, double q0);
42    obj_val_t evaluate(const population& curr_pop);
43    void update_global_ph(pheromone_2d& ph, const solution& best_sol, double best_obj_val);
44    void update_local_ph(pheromone_2d& ph, int curr_city, int next_city);
45    void update_best_sol(population& curr_pop, obj_val_t& curr_obj_vals);
46    void show_optimum();
47
48 private:
49    int num_runs;
50    int num_evals;
51    int num_patterns_sol;
52    string filename_ini;
53    string filename_ins;
54
55    double avg_obj_val;
56    vector<double> avg_obj_val_eval;
57    double best_obj_val;
58    solution best_sol;
59    instance ins_tsp;
60    solution opt_sol;
61
62    // data members for population-based algorithms
63    int pop_size;
64    population curr_pop;
65    population temp_pop;
66    obj_val_t curr_obj_vals;
67
68    // data member for aco-specific parameters
69    double alpha;
70    double beta;
71    double rho;
72    double q0;
73    double tau0;
74    pheromone_2d ph;
75    pheromone_2d tau_eta;
76    dist_2d dist;
77 };
78
79 inline aco::aco(int num_runs,
80                 int num_evals,
81                 int num_patterns_sol,
82                 string filename_ini,
83                 string filename_ins,
84                 int pop_size,
85                 double alpha,
86                 double beta,
87                 double rho,
88                 double q0
```

```
 89                    )
 90       : num_runs(num_runs),
 91         num_evals(num_evals),
 92         num_patterns_sol(num_patterns_sol),
 93         filename_ini(filename_ini),
 94         filename_ins(filename_ins),
 95         pop_size(pop_size),
 96         alpha(alpha),
 97         beta(beta),
 98         rho(rho),
 99         q0(q0)
100 {
101       srand();
102 }
103
104 inline aco::population aco::run()
105 {
106       avg_obj_val = 0.0;
107       obj_val_t avg_obj_val_eval(num_evals, 0.0);
108
109       for (int r = 0; r < num_runs; r++) {
110           int eval_count = 0;
111           double best_so_far = numeric_limits<double>::max();
112
113           // 0. initialization
114           init();
115
116           while (eval_count < num_evals) {
117               // 1. transition
118               construct_solutions(curr_pop, dist, ph, beta, q0);
119
120               // 2. evaluation
121               curr_obj_vals = evaluate(curr_pop);
122
123               for (size_t i = 0; i < curr_obj_vals.size(); i++) {
124                   if (best_so_far > curr_obj_vals[i])
125                       best_so_far = curr_obj_vals[i];
126                   if (eval_count < num_evals)
127                       avg_obj_val_eval[eval_count++] += best_so_far;
128               }
129
130               // 3. determination
131               update_best_sol(curr_pop, curr_obj_vals);
132               update_global_ph(ph, best_sol, best_obj_val);
133           }
134           avg_obj_val += best_obj_val;
135       }
136       avg_obj_val /= num_runs;
137
138       for (int i = 0; i < num_evals; i++) {
139           avg_obj_val_eval[i] /= num_runs;
140           cout << fixed << setprecision(3) << avg_obj_val_eval[i] << endl;
141       }
142
143       show_optimum();
144
145       return curr_pop;
146 }
147
148 inline void aco::init()
149 {
150       // 1. parameters initialization
151       //
152       // Well, some of them should be moved to the constructor because
153       // they need to be initialized once and only once; some of them
```

```
154    // should be declared and/or initialized where it is used.
155    ins_tsp = instance(num_patterns_sol, pattern(2, 0.0));
156    curr_pop = temp_pop = population(pop_size, solution(num_patterns_sol, 0));
157    opt_sol = solution(num_patterns_sol, 0);
158    ph = pheromone_2d(num_patterns_sol, pheromone_1d(num_patterns_sol, tau0));
159    dist = dist_2d(num_patterns_sol, dist_1d(num_patterns_sol, 0.0));
160    obj_val_t curr_obj_vals(pop_size, 0.0);
161    solution best_sol(num_patterns_sol, 0);
162    best_obj_val = numeric_limits<double>::max();
163
164    // 2. input the TSP benchmark
165    if (!filename_ins.empty()) {
166        ifstream ifs(filename_ins.c_str());
167        for (int i = 0; i < num_patterns_sol ; i++)
168            ifs >> ins_tsp[i][0] >> ins_tsp[i][1];
169    }
170
171    // 3. input the optimum solution
172    string filename_ins_opt(filename_ins + ".opt");
173    if (!filename_ins_opt.empty()) {
174        ifstream ifs(filename_ins_opt.c_str());
175        for (int i = 0; i < num_patterns_sol ; i++) {
176            ifs >> opt_sol[i];
177            --opt_sol[i];
178        }
179    }
180
181    // 4. initial solutions
182    if (!filename_ini.empty()) {
183        ifstream ifs(filename_ini.c_str());
184        for (int k = 0; k < pop_size; k++)
185            for (int i = 0; i < num_patterns_sol; i++)
186                ifs >> curr_pop[k][i];
187    }
188    else {
189        for (int k = 0; k < pop_size; k++) {
190            for (int i = 0; i < num_patterns_sol; i++)
191                curr_pop[k][i] = i;
192            for (int i = num_patterns_sol-1; i > 0; --i)        // shuffle the solutions
193                swap(curr_pop[k][i], curr_pop[k][rand() % (i+1)]);
194        }
195    }
196
197    // 5. construct the distance matrix
198    for (int i = 0; i < num_patterns_sol; i++) {
199        ph[i][i] = 0.0;
200        for (int j = 0; j < num_patterns_sol; j++)
201            dist[i][j] = i == j ? 0.0 : distance(ins_tsp[i], ins_tsp[j]);
202    }
203
204    // 6. create the pheromone table, by first constructing a solution using the nearest
       neighbor method
205    const int n = rand() % curr_pop.size();
206    list<int> cities(curr_pop[n].begin(), curr_pop[n].end());
207    best_sol[0] = *cities.begin();
208    cities.erase(cities.begin());
209    double Lnn = 0.0;
210    for (int i = 1; i < num_patterns_sol; i++) {
211        const int r = best_sol[i-1];
212        auto min_s = cities.begin(); // simply choose a city
213        double min_dist = numeric_limits<double>::max();
214        for (auto it = cities.begin(); it != cities.end(); it++) {
215            const int s = *it;
216            if (dist[r][s] < min_dist) {
217                min_dist = dist[r][s];
```

```
218                 min_s = it;
219             }
220         }
221         const int s = *min_s;
222         best_sol[i] = s;
223         cities.erase(min_s);
224         Lnn += dist[r][s];
225     }
226     const int r = best_sol[num_patterns_sol-1];
227     const int s = best_sol[0];
228     Lnn += dist[r][s];
229
230     tau0 = 1 / (num_patterns_sol * Lnn);
231     for (int r = 0; r < num_patterns_sol; r++)
232         for (int s = 0; s < num_patterns_sol; s++)
233             ph[r][s] = r == s ? 0.0 : tau0;
234 }
235
236 inline void aco::construct_solutions(population& curr_pop, const dist_2d& dist, pheromone_2d& ph
    , double beta, double q0)
237 {
238     int num_patterns_sol = curr_pop[0].size();
239
240     // 1. tau eta
241     pheromone_2d tau_eta(num_patterns_sol, pheromone_1d(num_patterns_sol));
242     for (int r = 0; r < num_patterns_sol; r++)
243         for (int s = 0; s < num_patterns_sol; s++)
244             tau_eta[r][s] = r == s ? 0.0 : ph[r][s] * pow(1.0 / dist[r][s], beta);
245
246     // 2. construct the solution: first, the first city of the tour
247     population temp_pop = curr_pop;
248     for (size_t j = 0; j < curr_pop.size(); j++) {
249         solution& asol = temp_pop[j];
250         const int n = rand() % num_patterns_sol;
251         curr_pop[j][0] = asol[n];
252         asol.erase(asol.begin() + n);
253     }
254
255     // 3. then, the remaining cities of the tour
256     for (int i = 1; i < num_patterns_sol; i++) {
257         // for each step
258         for (size_t j = 0; j < curr_pop.size(); j++) {
259             // for each ant
260             solution& asol = temp_pop[j];
261             int r = curr_pop[j][i-1];
262             int s = asol[0];
263             int x = 0;
264             const double q = static_cast<double>(rand()) / RAND_MAX;
265             if (q <= q0) {
266                 // 3.1. exploitation
267                 double max_tau_eta = tau_eta[r][s];
268                 for (size_t k = 1; k < asol.size(); k++) {
269                     if (tau_eta[r][asol[k]] > max_tau_eta) {
270                         s = asol[k];
271                         x = k;
272                         max_tau_eta = tau_eta[r][s];
273                     }
274                 }
275             }
276             else {
277                 // 3.2. biased exploration
278                 double total = 0.0;
279                 for (size_t k = 0; k < asol.size(); k++)
280                     total += tau_eta[r][asol[k]];
281
```

```
282                    // 3.3. choose the next city based on the probability
283                    double f = total * static_cast<double>(rand()) / RAND_MAX;
284                    for (size_t k = 0; k < asol.size(); k++) {
285                        if ((f -= tau_eta[r][asol[k]]) <= 0) {
286                            s = asol[k];
287                            x = k;
288                            break;
289                        }
290                    }
291                }
292                curr_pop[j][i] = s;
293                asol.erase(asol.begin() + x);
294
295                // 4.1. update local pheromones, 0 to n-1
296                update_local_ph(ph, r, s);
297            }
298        }
299
300        // 4.2. update local pheromones, n-1 to 0
301        for (size_t k = 0; k < curr_pop.size(); k++) {
302            // for each ant
303            const int r = curr_pop[k][curr_pop[0].size()-1];
304            const int s = curr_pop[k][0];
305            update_local_ph(ph, r, s);
306        }
307 }
308
309 inline aco::obj_val_t aco::evaluate(const population& curr_pop)
310 {
311     obj_val_t tour_dist(curr_pop.size(), 0.0);
312     for (size_t k = 0; k < curr_pop.size(); k++) {
313         for (size_t i = 0; i < curr_pop[0].size(); i++) {
314             const int r = curr_pop[k][i];
315             const int s = curr_pop[k][(i+1) % curr_pop[0].size()];
316             tour_dist[k] += dist[r][s];
317         }
318     }
319     return tour_dist;
320 }
321
322 inline void aco::update_best_sol(population& curr_pop, obj_val_t& curr_obj_vals)
323 {
324     for (size_t i = 0; i < curr_pop.size(); i++) {
325         if (curr_obj_vals[i] < best_obj_val) {
326             best_obj_val = curr_obj_vals[i];
327             best_sol = curr_pop[i];
328         }
329     }
330 }
331
332 inline void aco::update_global_ph(pheromone_2d& ph, const solution& best_sol, double
        best_obj_val)
333 {
334     for (size_t i = 0; i < best_sol.size(); i++) {
335         const int r = best_sol[i];
336         const int s = best_sol[(i+1) % best_sol.size()];
337         ph[s][r] = ph[r][s] = (1 - alpha) * ph[r][s] + alpha * (1 / best_obj_val);
338     }
339 }
340
341 inline void aco::update_local_ph(pheromone_2d& ph, int r, int s)
342 {
343     ph[s][r] = ph[r][s] = (1 - rho) * ph[r][s] + rho * tau0;
344 }
345
```

```
346 inline void aco::show_optimum()
347 {
348     double opt_dist = 0;
349     for (int i = 0; i < num_patterns_sol; i++) {
350         const int r = opt_sol[i];
351         const int s = opt_sol[(i+1) % num_patterns_sol];
352         opt_dist += dist[r][s];
353     }
354     cerr << "# Current route: " << best_sol << endl;
355     cerr << "# Optimum distance: " << fixed << setprecision(3) << opt_dist << endl;
356     cerr << "# Optimum route: " << opt_sol << endl;
357 }
358
359 #endif
```

8.2.1 Declaration of functions and parameters

As shown in lines 18–23, several typedefs (aliases of types) are declared for solving the TSP. The typedefs `pattern` and `instance` are declared for creating suitable variables for storing the coordinates of the cities of the TSP benchmark. The typedefs `pheromone_1d` and `pheromone_2d` are declared for creating suitable matrices to keep track of the pheromone information, while the typedefs `dist_1d` and `dist_2d` are declared for creating suitable matrices to record all the distances between cities.

As shown in lines 26–35, in addition to the parameters for all the meta-heuristic algorithms, the aco() function requires the additional parameters `alpha`, `beta`, `rho`, and `q0`, which correspond to the parameters α, β, ρ, and q_0 of ACS. In addition to the abovementioned typedefs, this program also declares the functions for initialization (line 40), solution construction (line 41), objective evaluation (line 42), global pheromone update (line 43), local pheromone update (line 44), best-so-far solution update (line 45), and showing the best and optimum results of the TSP benchmark (line 46) of ACS for the TSP.

In addition to the typedefs and functions described above, also defined are variables for the filenames of the initial solution and benchmark (lines 52–53) as well as variables for the input TSP dataset (line 59), the optimal solution (line 60), α (line 69), β (line 70), ρ (line 71), q_0 (line 72), τ (line 73), the pheromone vector (line 74), the η vector (line 75), and the distance vector (line 76).

8.2.2 The main loop

The main function of ACS is shown in lines 104–146. Like the implementation of GA, simulated annealing (SA), tabu search (TS), and hill climbing (HC) described in the previous chapters, ACS will also be carried out for a certain number of runs, to perform the operators of ACS, as shown in lines 118, 121, 131, and 132, repeatedly for a certain number of iterations each run. The design of this program follows the unified framework for metaheuristics (i.e., Algorithm 3) described in Section 4.1. Unlike the implementation of GA, SA, TS, and HC for the one-max problem, the implementation of ACS is for the TSP; therefore, several functions need to be redesigned.

8.2.3 Additional functions

Since the implementation for most of the additional functions of this program is different from the other implementations given in the previous chapters, this section will discuss in detail the following operators: init(), construct_solutions(), evaluate(), update_best_sol(), update_global_ph(), update_local_ph(), and show_optimum(). As shown in lines 148–234, the implementation of the init() operator can be divided into six parts: (1) initialize the parameters, (2) input the TSP benchmark, (3) input the optimal solution, (4) initialize the initial solution, (5) construct a two-dimensional vector for the pheromone matrix, and (6) construct the solution using the nearest neighbor method. These procedures will be discussed in detail below.

1. Initialize the parameters: As shown in lines 155–162, the parameters initialized are `ins_tsp` for the solutions of ACS, `curr_pop` and `temp_pop` for the cities, `opt_sol` for the optimum tour, `ph` for the pheromone matrix, `dist` for the distances between all the cities, and `best_obj_val` for the best objective value. Besides, the temporary variables for the current objective values `curr_obj_vals` and the best solution `best_sol` are created.

2. Input the TSP benchmark: As shown in lines 165–169, the coordinates of all the cities for a TSP benchmark residing in a text file will be read one by one to the variable `ins_tsp`.

3. Input the optimal solution: As shown in lines 172–179, the optimal solution (i.e., tour) will be read from a text file and stored in the vector `opt_sol`.

4. Initialize the initial solution: As shown in lines 182–195, this operator will either read the initial solution from a text file or randomly create *p* solutions (i.e., tours) for the TSP.

5. Construct the distance matrix: As shown in lines 198–202, the distances between all the cities will be computed and stored in the distance matrix `dist`. Although it takes a lot of memory space to keep this infor-

mation, it will avoid the problem of recomputing the distances between nodes.

6. Construct the solution using the nearest neighbor method: Since $1/(n \times L_{nn})$ is used in (Dorigo and Gambardella, 1997) as the initial value of the pheromone of each edge, the nearest neighbor method is required to construct the solution L_{nn}. For this reason, it will use the first city as the starting point and then choose the nearest neighbor city as the next city for the tour. This process will be repeated until all the cities are visited, and then it will come back to the first city. The edges passing through will form a complete tour for the TSP, the distance of which will be taken as L_{nn}, as shown in lines 205–228. This operator will compute $1/(n \times L_{nn})$, where n is the number of cities for the TSP, and then it will use this value as the initial pheromone (i.e., τ_0) of each edge, as shown in lines 230–233.

As shown in lines 236–307, the construct_solutions() operator can be divided into four parts: (1) compute the value $[\tau_{i,j}] \times [\eta_{i,j}]^\beta$ of each edge, (2) pick the first city for each ant, (3) select the remaining cities of the TSP tour for each ant, and (4) locally update the pheromone. These procedures will be discussed in detail below.

1. Compute $[\tau_{i,j}] \times [\eta_{i,j}]^\beta$: As shown in lines 242–244, a two-dimensional array, which acts as a matrix, is declared to keep the values $[\tau_{i,j}] \times [\eta_{i,j}]^\beta$, and then Eq. (8.5) is used to compute the value of each edge. For this reason, for $i \neq j$, `tau_eta[i][j]` will be set to $[\tau_{i,j}] \times [\eta_{i,j}]^\beta$; otherwise, it will be set to 0.0.

2. Pick the first city: As shown in lines 248–253, each ant will randomly select a city as the first city of the tour and then remove it from the candidate cities.

3. Select the remaining cities: As shown in lines 256–298, each ant will then select the next cities one by one to create a complete TSP tour, which can be divided into three parts: (1) compute the exploitation probability, (2) compute the biased exploration probability, and (3) select the city using Eq. (8.5). In this part, the variables r and s represent, respectively, the city on which the k-th ant currently stands and the city to which it will then move, and the variable x represents the city that has been selected as the next city, thus to be removed from the candidate cities.

For $q \leq q0$, the exploitation of ACS using Eq. (8.5) is shown in lines 265–275. Its purpose is to find the city that has the maximum value of `tau_eta[i][j]` (i.e., $[\tau_{i,j}] \times [\eta_{i,j}]^\beta$) as the next city while the

k-th ant is at the i-th city. If the j-th city is selected as the next city of the tour, then it will be removed from the candidate cities.

In addition to choosing the city with the maximum value of `tau_eta[i][j]` as the next city, for $q > q_0$ (i.e., for the biased exploration), ACS will then use Eq. (8.1) to compute the probability of the candidate cities (i.e., the transition probability). After that, ACS will then randomly select the next city by using Eq. (8.1) as roulette wheel selection in GA, as shown in lines 276–293.

4. Locally update the pheromone table: ACS will update the pheromone value of the edge traversed by an ant right after a subsolution is constructed (i.e., right after a city is picked as the next city of the tour) by calling the update_local_ph() function, which will update the pheromone value using Eq. (8.6), as shown in lines 341–344.

The distance of a TSP tour is computed as shown in lines 309–320. The evaluate() operator has to compute the distance from the first city to the second city, the distance from the second city to the third city, and so on, up to the distance from the second-last city to the last city, and then the distance from the last city back to the first city. The objective value of the k-th ant is defined as the distance of the TSP tour created by the k-th ant, i.e., the summation of all the distances between the cities of the TSP tour. As shown in lines 322–330, the update_best_sol() operator will first find the ant that traverses the shortest distance among all ants in the current iteration and then use it to update the best-so-far solution if ACS finds that this distance is shorter than the best-so-far solution at this iteration. For ACS, global and local updates are two different methods to increase the pheromone value of the edges between cities, which will affect the search trends of ACS. As shown in lines 332–339, the global update operator update_global_ph() will update the pheromone value of edges of the best-so-far solution using Eq. (8.7). Unlike the global update operator, the local update operator update_local_ph() will only update the pheromone value of edges that have been visited by any ant at each iteration using Eq. (8.6), as shown in lines 341–344. For both global update and local update, the edge from the i-th city to the j-th city $e_{i,j}$ and the edge from the j-th city to the i-th city $e_{j,i}$ have to be updated together no matter which edge, be it $e_{i,j}$ or $e_{j,i}$, belongs to the best-so-far solution or was visited by an ant. The show_optimum() function is an optional operator, which can be used to show the optimal solution to further understand the difference between the solution found by ACS and the optimal solution from TSPLIB, as shown in lines 346–357.

8.3. Simulation results of ACO for the traveling salesman problem

In this section, we will use eil51, a benchmark for the TSP from TSPLIB,[1] (Reinelt, 1991; TSPLIB, 1995), to evaluate the performance of ACS. The benchmark eil51 has 51 cities. The TSPLIB also provides the optimal solution for this benchmark. Since the implementation of ACS is based on (Dorigo and Gambardella, 1997), most parameters will just follow its settings; that is, β is set to 2.0, q_0 to 0.9, α to 0.1, ρ to 0.1, and τ_0 to $1/(n \times L_{nn})$. The number of ants is set to 20. As in the previous chapters, all simulations were carried out for 100 runs, and 1250 iterations (25,000 evaluations) were performed each run.

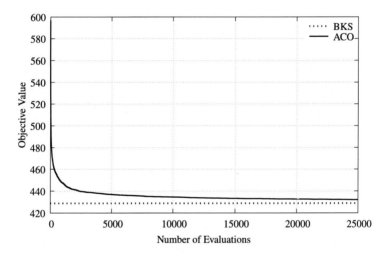

Figure 8.4 Simulation results of ACS for eil51.

As shown in Fig. 8.4, ACS is able to find an approximate solution that is very close to the optimal solution from the TSPLIB. Since there are 20 ants and 1250 iterations were performed, this is essentially equivalent to 25,000 evaluations each run. Note that BKS in Fig. 8.4 represents the best known solution of eil51. The solution of ACS is about 432.094 on average after 1250 iterations. It is easily seen from its convergence trend that a better solution can be found by increasing the numbers of evaluations and iterations. In some runs, ACS is able to find a "good solution" or even a solution that is better than the optimal solution provided by

[1] It can also be downloaded from http://elib.zib.de/pub/mp-testdata/tsp/tsplib/tsp/.

(Reinelt, 2007). Fig. 8.5(a and b) shows that the total distances of these two solutions not rounded up or down to the nearest integer by the nint() function of (Reinelt, 1995, 1991) are 429.983 and 428.872, respectively. According to our observation, the difference comes up because the way the distance between each pair of cities is represented is different. Fig. 8.5(a) shows the result of rounding the distance between each pair of cities up or down to the nearest integer by the nint() function of (Reinelt, 1995, 1991), while Fig. 8.5(b) gives the result of our implementation where the distance between each pair of cities is represented by a real number.

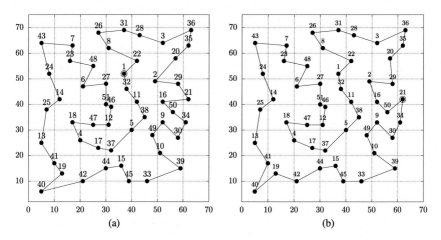

Figure 8.5 Comparison of (a) the optimal solution provided by TSPLIB and (b) a good solution found by ACS for eil51.

8.4. Discussion

It can be easily seen from the results described in the previous section that ACO-based algorithms are very powerful for the TSP. The study described in (Dorigo and Stützle, 2004) shows that ACS-based algorithms can also find better results for scheduling or routing problems. According to our observation, ACO-based algorithms can find good results for these problems because its search strategy implies swarm consciousness. ACO-based algorithms use a pheromone table to keep track of all the search experiences of all the ants, and each ant will refer to this information to construct its solution. This kind of information can be regarded as the so-called *global search information*. In addition, ACO-based algorithms also use the distance information between each pair of cities, thus allowing them to pick good subsolutions during the convergence process. This kind of information can be regarded as the so-called *local search information* or domain knowledge. The solution of ACO-based algorithms is a combination of the results of global search and the results of local search. As a result, ACO-based algorithms can find *good* solutions for the TSP.

In addition to AS (Dorigo et al., 1991, 1996) and ACS (Dorigo and Gambardella, 1997), a variety of ACOs can be found in the literature, which include ant-density (Dorigo et al., 1991), ant-quantity (Dorigo et al., 1991), ant-cycle (Dorigo et al., 1991), elitist AS (Dorigo et al., 1991), Ant-Q (Gambardella and Dorigo, 1995), the $\mathcal{MAX}$-$\mathcal{MIN}$ ant system ($\mathcal{MMAS}$) (Stützle and Hoos, 1996), rank-based AS (Bullnheimer et al., 1997), evolving ACO (Botee and Bonabeau, 1998), the hypercube framework for ACO (HCF) (Blum and Dorigo, 2004), beam–ACO (Blum, 2005b), and so forth (Dorigo and Stützle, 2010). Ant-density, ant-quantity, ant-cycle, and elitist AS were presented in (Dorigo et al., 1991). The main difference of these four pheromone accumulation models lies in how the quantity of pheromone for the edge between node c_i and node c_j is determined if an ant goes from c_i to c_j. Recall Eq. (8.2), in which the parameters Q and L_k are used to control the quantity of pheromone accumulation for an edge traversed by an ant in AS. This means that the quantity of pheromone $\Delta \tau_{i,j}^k = \frac{Q}{L_k}$, where Q is a constant and L_k is the tour length of the k-th ant, will be left on the edge $e_{i,j}$ if the k-th ant chooses to go from node c_i to node c_j. Different from Eq. (8.2), the pheromone accumulation models for ant-density and ant-quantity are, respectively, $\Delta \tau_{i,j}^k = Q$ and $\Delta \tau_{i,j}^k = \frac{Q}{d_{i,j}}$, where $d_{i,j}$ is the distance between c_i and c_j, if edge $e_{i,j}$ was visited by the k-th ant. For ant-cycle, $\Delta \tau_{i,j}^k = \frac{Q}{L_k}$, which is essentially Eq. (8.2), if edge $e_{i,j}$ was visited by the k-th ant. The basic idea of elitist AS is to reinforce the best-so-far tour by a quantity $e \times \frac{Q}{L^*}$, where e is the number of elitist ants and L^* is the length of the best-so-far tour, for all the edges of this tour in each iteration. A similar idea employed by Ant-Q (Gambardella and Dorigo, 1995) is to reinforce the best edge as follows:

$$\tau_{i,j} = (1 - \alpha) \times \tau_{i,j} + \alpha \times \left(\Delta \tau_{i,j} + \gamma \max_{z \in \mathbb{N}_i}(\tau_{i,z}) \right), \tag{8.9}$$

where α is the learning step, γ is the discount factor, and $\mathbb{N}_i$ is the set of candidate nodes that can be chosen by an ant standing at node c_i. The transition probability for an ant of Ant-Q to select node c_j as the next node of node c_i is almost the same as Eq. (8.5) of ACS. As shown in Eq. (8.10), the only difference between Ant-Q and ACS is that the pheromone value $\tau_{i,j}$ is raised to the power of δ in Ant-Q, but not in ACS:

$$S_{i,j}^k = \begin{cases} \arg\max_{j \in \mathbb{N}_i^k} [\tau_{i,j}]^\delta \times [\eta_{i,j}]^\beta, & \text{if } q \leq q_0, \\ p_{i,j}^k, & \text{otherwise.} \end{cases} \qquad (8.10)$$

The basic idea of $\mathcal{MMAS}$ (Stützle and Hoos, 1996) is to bound all pheromone values $\tau_{i,j}$ so that they fall in a range $[\tau_{\min}, \tau_{\max}]$. Moreover, for $\mathcal{MMAS}$, only the "best one ant" will be used to update the pheromone values in each iteration, where the best one ant can be either the best one of all ants in an iteration (called the iteration-best tour) or the best-so-far ant since the beginning of the run (called the global best tour). Thus,

$$\tau_{i,j} = \rho \times \tau_{i,j} + \frac{1}{L_{\text{best}}}, \qquad (8.11)$$

where L_{best} represents the total routing distance of the best ant. In (Bullnheimer et al., 1997), Bullhheimer et al. presented rank-based AS, which considers only the ω best ants to update the pheromone values for the edges they visited. The μ-th best ant will be associated with the weight $\sigma - \mu$, while the weight of the best ant is set equal to σ. The pheromone accumulation model of the $\omega = \sigma - 1$ best ants will then be

$$\Delta\tau_{i,j}^\mu = \begin{cases} (\sigma - \mu)\frac{Q}{L_\mu}, & \text{if the } k\text{-th ant chooses edge } e_{i,j} \text{ as part of the tour,} \\ 0, & \text{otherwise,} \end{cases}$$
$$(8.12)$$

and the pheromone accumulation model of the best ant will be

$$\Delta\tau_{i,j}^* = \begin{cases} \sigma\frac{Q}{L^*}, & \text{if the } k\text{-th ant chooses edge } e_{i,j} \text{ as part of the tour} \\ 0, & \text{otherwise,} \end{cases} \qquad (8.13)$$

where μ is the ranking index of the ant, σ is the number of elitist ants, L_μ is the tour length of the μ-th best ant, and L^* is the tour length of the best ant. The pheromone accumulation model of rank-based AS can therefore be defined as follows:

$$\tau_{i,j} = \rho \times \tau_{i,j} + \Delta\tau_{i,j} + \Delta\tau_{i,j}^*, \qquad (8.14)$$

where $\Delta\tau_{i,j} = \sum_{\mu=1}^{\sigma-1} \Delta\tau_{i,j}^\mu$. Similar to other metaheuristics, ACO-based algorithms also face the issue of parameters. Compared to other metaheuristics, ACO-based algorithms typically have too many parameters. How to determine the values of these parameters is another open issue of all ACO-based algorithms. In (Botee and Bonabeau, 1998), Botee and Bonabeau presented evolving ACO by using GA to find the approximate values for the parameters of ACO. In (Blum, 2005b), Blum presented a hybrid method for solving the open shop scheduling problem, called beam-ACO, by modifying the solution construction mechanism of ACO and beam search (Ow and Morton, 1988). Two methods are used in beam-ACO, one using the weight function to weigh the different possibilities of extending a partial solution and the other using a lower bound to eliminate a set of partial solutions at each step of solution construction.

Since most ACO-based algorithms use Eq. (8.1), which is computationally very expensive, to compute the transition probability, there is no doubt that they usually take a long time to converge. Compared to other metaheuristics, this mechanism will certainly increase the computation time of ACO-based algorithms; therefore, they can be easily accelerated by reducing the computation time of the transition probability method. In fact, Dorigo and his colleagues modified the transition probability method from Eq. (8.1) to Eq. (8.5), as discussed in (Dorigo and Gambardella, 1997), to reduce the computation costs of ACO-based algorithms. On the other hand, parallel computing methods provide an efficient way to accelerate the response time of ACO-based algorithms (Pedemonte et al., 2011; Randall and Lewis, 2002; Stützle, 1998). In (Stützle, 1998), Stützle presented two master–slave parallelization methods for ACO-based algorithms, one of which has the master node play the roles of trail update, solution construction, and solution delivery to the slave nodes for local search, while the other has the master node play only the role of trail update and the slave nodes play the roles of solution construction and local search. Many different kinds of parallel ACO-based algorithms, such as parallel independent ant colonies, parallel interacting, parallel ants, and parallel evaluation of solution elements, can be found in (Pedemonte et al., 2011; Randall and Lewis, 2002). Different parallel computing architectures, such as multi-core processors, graphics processing units (GPUs), and grid environments, were also discussed in (Pedemonte et al., 2011). To classify parallel ACO-based algorithms, Martín et al. further used the number of colonies, cooperation, and granularity to provide a comprehensive hierarchical taxonomy for parallel ACOs, which reveals that parallel computing is a critical research topic of ACO.

It can be found from the history of the development of ACO that some researchers attempted to apply ACO to different optimization problems, that is, not only TSPs, but also scheduling, classification, vehicle routing, multi-objective problems, etc. (Bell and McMullen, 2004; Doerner et al., 2004; Liang and Smith, 2004; Merkle et al., 2002; Parpinelli et al., 2002). In (Merkle et al., 2002), ACO is used for solving the resource-constrained project scheduling problem. The pheromone accumulation model is modified to reinforce the best found schedule; so, it now becomes

$$\tau_{i,j} = \tau_{i,j} + \rho \frac{1}{2T^*}, \qquad (8.15)$$

where T^* is the makespan of the best found schedule. Another example can be found in (Parpinelli et al., 2002), in which ACO is used to extract the classification rules from the input data, called ant-miner. An interesting aspect of this algorithm is that it has only a single ant in each iteration and it uses a pruning method to reduce the number, while at the same time improving the quality, of the classification rules. Since canonical ACO is inspired by ant behavior, a complete solution is constructed step by step (subsolution by subsolution) by an ant. ACO can be used to solve combinatorial optimization problems; however, this kind of search strategy is not suitable for continuous optimization problems, such as function and multi-objective optimization problems. Some studies (Socha and Dorigo, 2008; Yang et al., 2017) aimed to develop ACO for continuous domains. In (Socha and Dorigo, 2008), the transition probability and solution construction operators are modified by selecting a Gaussian function from a set of Gaussian functions to generate (sample) a random number as the i-th subsolution of the k-th ant.

ACO was combined with other (meta)heuristic algorithms to further enhance its performance in several studies (Duan and Yu, 2007; Engin and Güçlü, 2018; Hoseini and Shayesteh, 2010; Huang and Liao, 2008; Mahi et al., 2015). In (Engin and Güçlü, 2018), the mutation operator of GA is used for the candidate solution of ACO when it falls into a local optimum; moreover, the crossover operator of GA is used for the two best candidate solutions of ACO. Of course, there are many ways to combine ACO with other (meta)heuristic algorithms. Among them are ACO with the memetic algorithm (Duan and Yu, 2007), ACO with TS (Huang and Liao, 2008), ACO with SA and GA (Hoseini and Shayesteh, 2010), and ACO with PSO and 3-opt (Mahi et al., 2015). In summary, as mentioned before, since most ACO-based algorithms use the solution construction procedure to build a solution, they provide an effective way for combinatorial optimization problems. As a result, most ACO-based algorithms are typically used to solve combinatorial optimization problems. A number of survey papers (Blum, 2005a; Dorigo et al., 2006; Dorigo and Blum, 2005; Dorigo and Stützle, 2010) provide simple examples to explain how ACO works, discussions on different versions of ACO to show possible ways to enhance the performance of ACO, and applications that use ACO.

Supplementary source code

Here are the hyperlinks to the directories where the source code described in this chapter resides.

https://github.com/cwtsaiai/metaheuristics_2023/src/c++/3-tsp/

CHAPTER NINE

Particle swarm optimization

9.1. The basic idea of particle swarm optimization (PSO)

Particle swarm optimization (PSO) was presented in (Kennedy and Eberhart, 1995). Even though Kennedy and Eberhart showed just the basic idea of PSO in this paper, many researchers soon realized that it is an effective and efficient search algorithm for optimization problems; therefore, PSO has been popular among researchers using metaheuristics to solve complex problems from the 1990s. Like ACO, PSO is based on the concept of using the search experiences of the swarm to affect the search directions of individuals. These two search algorithms have become the iconic algorithms of swarm intelligence.

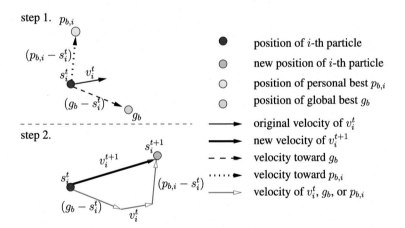

Figure 9.1 The impact of different information.

As shown in Fig. 9.1, each solution of PSO is called a particle, and its next search direction is affected by the currently available information. In this example, s_i^t represents the position of the i-th particle at iteration t, s_i^{t+1} represents the position of the i-th particle at iteration $t+1$, v_i^t represents the velocity of the i-th particle at iteration t, v_i^{t+1} represents the velocity of the i-th particle at iteration $t+1$, $p_{b,i}$ represents the best-so-far solution of the i-th particle, and g_b represents the best-so-far solution of all the particles. Note that v_i^t and v_i^{t+1} can also be regarded as the original and new trajectories of the i-th particle. The first step of this example states that v_i^{t+1} will be affected by three velocities: the original trajectory v_i^t, the velocity toward g_b that can be calculated by $g_b - s_i^t$, and the velocity toward $p_{b,i}$ that can be calculated by $p_{b,i} - s_i^t$. To simplify the discussion that follows, as far as this example is concerned, we assume that the three velocities v_i^t, $g_b - s_i^t$, and $p_{b,i} - s_i^t$ are of equal importance in the sense of how they affect the momentum of each particle. The second step of this example presents how the forces of these three velocities are combined to generate the new velocity v_i^{t+1} for moving the i-th particle to the new position s_i^{t+1}. Hence, for PSO, the positions of particles can be regarded as the solutions of the optimization problem in question. The search strategy of PSO is typically affected by the personal best $p_{b,i}$, the global best g_b, and the velocity v_i^t with different weights, meaning that the new position s_i^{t+1} of the i-th particle is composed of these three pieces of information.

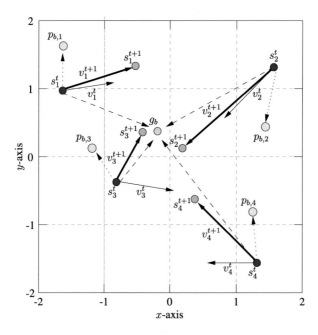

Figure 9.2 How particles are used to search the solution space.

As shown in Fig. 9.2, because the search trajectory of each particle is typically affected by the original trajectory v_i^t, the global best g_b, and the personal best $p_{b,i}$ of *different weights* for most PSO algorithms, every velocity of a particle will then have a different impact on the new position of the particle. The global best is an accumulation of the search experiences of all the particles. This explains that PSO uses not only local search information (i.e., the personal best $p_{b,i}$ and v_i^t), but also global search information (i.e., the global best g_b). Compared to using the pheromone values for all possible subsolutions of ACO, the accumulated searched information of PSO has been simplified from a pheromone table to a global best position g_b. This implies that the positions of all the particles of a PSO will refer to the position g_b; therefore, PSO can find an approximate solution in a large continuous space very quickly.

Algorithm 9 Particle swarm optimization.

1 Initialize the positions s^0 and velocities v^0 of all particles	$\boxed{\text{I}}$
2 Initialize the personal best of particles p_b^0 and global best g_b^0	$\boxed{\text{I}}$
3 **While** the termination criterion is not met	
4 $\quad v^{t+1} = \text{NewVelocity}(s_i^t, v_i^t, g_b^t, p_b^t)$ using Eq. (9.1)	$\boxed{\text{T}}$
5 $\quad s^{t+1} = \text{NewPosition}(s_i^t, v^{t+1})$ using Eq. (9.2)	$\boxed{\text{T}}$
6 $\quad f^{t+1} = \text{Evaluation}(s^{t+1})$	$\boxed{\text{E}}$
7 $\quad p_b^{t+1} = \text{LocalBestUpdate}(s^t, s^{t+1}, f^t, f^{t+1})$	$\boxed{\text{D}}$
8 $\quad g_b^{t+1} = \text{GlobalBestUpdate}(s^t, s^{t+1}, f^t, f^{t+1})$	$\boxed{\text{D}}$
9 $\quad t = t + 1$	
10 **End**	
11 **Output** s	$\boxed{\text{O}}$

Algorithm 9 gives an outline of PSO. Although PSO uses many kinds of information to guide the search directions, it still matches the unified framework described in Section 4.1. As shown in lines 1–2, PSO will first initialize the positions $s^0 = \{s_1^0, s_2^0, \ldots, s_m^0\}$ and velocities $v^0 = \{v_1^0, v_2^0, \ldots, v_m^0\}$ of all the particles and then it will initialize the positions of the personal best solution $p_b^0 = \{p_{b,1}^0, p_{b,2}^0, \ldots, p_{b,m}^0\}$ of all the particles and the position of the global best solution g_b^0.[1] In the main loop, line 4 shows that the new velocity of each particle (which can be regarded as the search direction or trajectory) v_i^{t+1} is a function of its position s_i^t, individual trajectory v_i^t, global best g_b, and local best $p_{b,i}^t$ at each iteration. The new position of each particle s_i^{t+1} will then be guided by the new velocity v_i^{t+1}, implying that PSO will use this information to adjust the search directions of particles, as shown in line 5. The fitness value of each particle will be calculated by the evaluation operator, as shown in line 6. Given the new velocities,

---

[1] Here, m is the number of particles.

positions, and fitnesses of all the particles, the local and global bests can then be computed, as shown in lines 7–8.

The following equations are used by the original PSO (Kennedy and Eberhart, 1995) to compute the new velocity v_i^{t+1} and the new position s_i^{t+1} of each particle:

$$v_i^{t+1} = v_i^t + c_1\varphi_1(p_{b,i}^t - s_i^t) + c_2\varphi_2(g_b^t - s_i^t) \qquad (9.1)$$

and

$$s_i^{t+1} = s_i^t + v_i^{t+1}, \qquad (9.2)$$

where φ_1 and φ_2 are two uniformly distributed random numbers used to determine the influence of $p_{b,i}$ and g_b and c_1 and c_2 are two constant values to affect the cognitive and social learning weights of PSO during the convergence process. In (Shi and Eberhart, 1998), Shi and Eberhart added a new parameter into Eq. (9.1) called the inertia weight ω to balance the global and local search abilities of PSO. The new velocity v_i^{t+1} of each particle is now defined as

$$v_i^{t+1} = \omega v_i^t + c_1\varphi_1(p_{b,i}^t - s_i^t) + c_2\varphi_2(g_b^t - s_i^t). \qquad (9.3)$$

Based on experiments on the Schaffer function, Shi and Eberhart (Shi and Eberhart, 1998) further pointed out that PSO is more like a "local search algorithm" when ω is small (e.g., $\omega < 0.8$) but more like a "global search algorithm" when ω is large (e.g., $\omega > 1.2$) and has the best chance to find the global optimum when ω is medium (e.g., $0.8 < \omega < 1.2$).

9.2. Implementation of PSO for the function optimization problem

As mentioned previously, PSO is useful for solving continuous optimization problems. A simple example for solving the single-objective function optimization problem $\mathbb{P}_6$ (e.g., the Ackley function) will be given to provide a brief introduction to the basic idea of PSO. For more details of the problem $\mathbb{P}_6$, please see also Section 2.3.1. Note that the implementation of PSO described in this section and some of the parameter settings are based, respectively, on the original idea of (Kennedy and Eberhart, 1995; Shi and Eberhart, 1998) and on the technical report (Hassan, 2004).

Here is an example command line for invoking PSO for the Ackley function.

```
$ ./search pso 100 1000 2 "" "mvfAckley" 10 0.5 1.5 1.5 -32.0
32.0
```

The first term "pso" means that the `search` program will use PSO as the underlying search algorithm. The second term states that it will be carried out for 100 runs, while the third term says that 1000 evaluations will be performed per run. The fourth term is the number of subsolutions. For a function optimization problem, it is the number of dimensions of the function. This means that the function in this case is two-dimensional. The fifth term (i.e., the first string) is the name of the file where the initial solutions reside, and since the filename is an empty string, it means that no initial solutions are given in this case. The sixth term is the name of the function to be benchmarked. In this case, it is named "mvfAckley." As far as this program is concerned, the measurement is based on the implementation described in (Adorio and Diliman, 2005); therefore, more test functions can be added to this program if we want. The terms 10, 0.5, 1.5, 1.5, -32.0, and 32.0 denote, respectively, the number of particles, ω, c_1, c_2, and the lower and upper bounds of all dimensions of the solution space. Listing 9.1 shows an implementation of the main functions and operators of PSO.

Listing 9.1: src/c++/4-function/main/pso.h: The implementation of PSO.

```
 1 #ifndef __PSO_H_INCLUDED__
 2 #define __PSO_H_INCLUDED__
 3
 4 #include <iostream>
 5 #include <fstream>
 6 #include <limits>
 7 #include "pso.h"
 8 #include "lib.h"
 9 #include "functions.h"
10
11 using namespace std;
12
13 class pso
14 {
15 public:
16     using solution = vector<double>;
17     using population = vector<solution>;
18
19     pso(int num_runs,
```

```
20          int num_evals,
21          int num_dims,
22          string filename_ini,
23          string filename_ins,
24          int pop_size,
25          double omega,
26          double c1,
27          double c2,
28          double vmin,
29          double vmax);
30
31      population run();
32
33      vector<double> evaluate(string function_name, int dims, const population& curr_pop);
34      population new_velocity(const population& curr_pop, population& velocity, population& pbest,
            solution& gbest, double omega, double c1, double c2, double vmin, double vmax);
35      population new_position(population& curr_pop, population& velocity, double vmin, double vmax
            );
36
37 private:
38      void init();
39      void update_pb(vector<double>& pbest_obj_vals, vector<double>& curr_obj_vals);
40      void update_gb(double& gbest_obj_val, vector<double>& curr_obj_vals);
41      void update_best_sol(const double& gbest_obj_val, const solution& gbest);
42
43 private:
44      int num_runs;
45      int num_evals;
46      int num_dims;
47      string filename_ini;
48      string filename_ins;
49
50      double avg_obj_val;
51      vector<double> avg_obj_val_eval;
52      double best_obj_val;
53      solution best_sol;
54
55      // data members for population-based algorithms
56      int pop_size;
57      population curr_pop;
58      population velocity;
59      vector<double> curr_obj_vals;
60
61      // data members for pso-specific parameters
62      population pbest;
63      solution gbest;
64      vector<double> pbest_obj_vals;
65      double gbest_obj_val;
66
67      double omega;
68      double c1;
69      double c2;
70      double vmin;
71      double vmax;
72 };
73
74 inline pso::pso(int num_runs,
75                 int num_evals,
76                 int num_dims,
77                 string filename_ini,
78                 string filename_ins,
79                 int pop_size,
80                 double omega,
81                 double c1,
82                 double c2,
```

```
83                   double vmin,
84                   double vmax
85                   )
86      : num_runs(num_runs),
87        num_evals(num_evals),
88        num_dims(num_dims),
89        filename_ini(filename_ini),
90        filename_ins(filename_ins),
91        pop_size(pop_size),
92        omega(omega),
93        c1(c1),
94        c2(c2),
95        vmin(vmin),
96        vmax(vmax)
97  {
98      srand();
99  }
100
101 inline pso::population pso::run()
102 {
103     avg_obj_val = 0.0;
104     vector<double> avg_obj_val_eval(num_evals, 0.0);
105
106     for (int r = 0; r < num_runs; r++) {
107         int eval_count = 0;
108         double best_so_far = numeric_limits<double>::max();
109
110         // 0. initialization
111         init();
112
113         while (eval_count < num_evals) {
114             // 1. compute new velocity of each particle
115             velocity = new_velocity(curr_pop, velocity, pbest, gbest, omega, c1, c2, vmin, vmax)
116     ;
117             // 2. adjust all the particles to their new positions
118             curr_pop = new_position(curr_pop, velocity, vmin, vmax);
119
120             // 3. evaluation
121             curr_obj_vals = evaluate(filename_ins, num_dims, curr_pop);
122
123             for (size_t i = 0; i < curr_obj_vals.size(); i++) {
124                 if (best_so_far > curr_obj_vals[i])
125                     best_so_far = curr_obj_vals[i];
126                 if (eval_count < num_evals)
127                     avg_obj_val_eval[eval_count++] += best_so_far;
128             }
129
130             // 4. update
131             update_pb(pbest_obj_vals, curr_obj_vals);
132             update_gb(gbest_obj_val, curr_obj_vals);
133             update_best_sol(gbest_obj_val, gbest);
134         }
135         avg_obj_val += best_obj_val;
136     }
137
138     avg_obj_val /= num_runs;
139
140     for (int i = 0; i < num_evals; i++) {
141         avg_obj_val_eval[i] /= num_runs;
142         cout << fixed << setprecision(3) << avg_obj_val_eval[i] << endl;
143     }
144
145     cout << best_sol << endl;
146
```

```
147      return curr_pop;
148 }
149
150 inline void pso::init()
151 {
152      // 1. parameter initialization
153      curr_pop = population(pop_size, solution(num_dims, vmax));
154      velocity = population(pop_size, solution(num_dims, vmax));
155      pbest = population(pop_size, solution(num_dims, vmax));
156      pbest_obj_vals = vector<double>(pop_size, numeric_limits<double>::max());
157      gbest = solution(num_dims, vmax);
158      gbest_obj_val = numeric_limits<double>::max();
159      curr_obj_vals = vector<double>(pop_size, numeric_limits<double>::max());
160      best_sol = vector<double>(num_dims, vmax);
161      best_obj_val = numeric_limits<double>::max();
162
163      // 2. initialize the positions and velocities of particles
164      if (!filename_ini.empty()) {
165          ifstream ifs(filename_ini.c_str());
166          for (int p = 0; p < pop_size; p++) {
167              for (int i = 0; i < num_dims; i++) {
168                  ifs >> curr_pop[p][i] >> velocity[p][i];
169              }
170          }
171      }
172      else {
173          for (int p = 0; p < pop_size; p++)
174              for (int i = 0; i < num_dims; i++) {
175                  curr_pop[p][i] = vmin + (vmax - vmin) * rand() / (RAND_MAX + 1.0);
176                  velocity[p][i] = curr_pop[p][i] / num_evals;
177              }
178      }
179
180      // 3. evaluation
181      curr_obj_vals = evaluate(filename_ins, num_dims, curr_pop);
182
183      // 4. update
184      update_pb(pbest_obj_vals, curr_obj_vals);
185      update_gb(gbest_obj_val, curr_obj_vals);
186      update_best_sol(gbest_obj_val, gbest);
187 }
188
189 inline vector<double> pso::evaluate(string function_name, int dims, const population& curr_pop)
190 {
191      vector<double> obj_vals(pop_size, 0.0);
192      auto it = function_table.find(function_name);
193      if (it != function_table.end()) {
194          const auto& f = it->second;
195          for (size_t i = 0; i < curr_pop.size(); i++)
196              obj_vals[i] = f(curr_pop[i].size(), curr_pop[i]);
197      }
198      return obj_vals;
199 }
200
201 inline void pso::update_pb(vector<double>& pbest_obj_vals, vector<double>& curr_obj_vals)
202 {
203      for (size_t i = 0; i < curr_obj_vals.size(); i++)    {
204          if (curr_obj_vals[i] < pbest_obj_vals[i]) {
205              pbest_obj_vals[i] = curr_obj_vals[i];
206              pbest[i] = curr_pop[i];
207          }
208      }
209 }
210
211 inline void pso::update_gb(double& gbest_obj_val, vector<double>& curr_obj_vals)
```

```
212 {
213     for (size_t i = 0; i < curr_obj_vals.size(); i++) {
214         if (curr_obj_vals[i] < gbest_obj_val) {
215             gbest_obj_val = curr_obj_vals[i];
216             gbest = curr_pop[i];
217         }
218     }
219 }
220
221 inline void pso::update_best_sol(const double& gbest_obj_val, const solution& gbest)
222 {
223     if (gbest_obj_val < best_obj_val) {
224         best_obj_val = gbest_obj_val;
225         for (size_t i = 0; i < gbest.size(); i++)
226             best_sol[i] = gbest[i];
227     }
228 }
229
230 inline pso::population pso::new_velocity(const population& curr_pop, population& velocity,
        population& pbest, solution& gbest, double omega, double c1, double c2, double vmin, double
        vmax)
231 {
232     population new_v(curr_pop.size(), solution(curr_pop[0].size()));
233     for (size_t i = 0; i < curr_pop.size(); i++) {
234         for (size_t j = 0; j < curr_pop[i].size(); j++) {
235             double r1 = static_cast<double>(rand()) / RAND_MAX;
236             double r2 = static_cast<double>(rand()) / RAND_MAX;
237             new_v[i][j] = omega * velocity[i][j] + c1 * r1 * (pbest[i][j] - curr_pop[i][j]) + c2
        * r2 * (gbest[j] - curr_pop[i][j]);
238             if (new_v[i][j] < vmin)
239                 new_v[i][j] = vmin;
240             else if (new_v[i][j] > vmax)
241                 new_v[i][j] = vmax;
242         }
243     }
244     return new_v;
245 }
246
247 inline pso::population pso::new_position(population& curr_pop, population& velocity, double vmin
        , double vmax)
248 {
249     population new_pos(curr_pop.size(), solution(curr_pop[0].size()));
250     for (size_t i = 0; i < curr_pop.size(); i++) {
251         for (size_t j = 0; j < curr_pop[i].size(); j++) {
252             new_pos[i][j] = curr_pop[i][j] + velocity[i][j];
253             if (new_pos[i][j] < vmin)
254                 new_pos[i][j] = vmin;
255             else if (new_pos[i][j] > vmax)
256                 new_pos[i][j] = vmax;
257         }
258     }
259     return new_pos;
260 }
261
262 #endif
```

9.2.1 Declaration of functions and parameters

As shown in lines 19–29, in addition to the parameters that most population-based metaheuristics have in common, that is, num_runs, num_evals, num_dims, filename_ini, and pop_size, this program requires three additional parameters for PSO, namely, omega, c1, and c2, and three additional parameters for function optimization problems, namely, filename_ins, vmin, and vmax. As shown in lines 33–41, this program declares six functions (operators) to achieve the goal of PSO, namely, evaluate(), new_velocity(), new_position(), update_pb(), update_gb(), and update_best_sol(). Because PSO was designed based on vector computation and because this program is used for function optimization, not only the PSO itself but also the implementation of most operators are different from the other metaheuristics described in the previous chapters. The variable filename_ins, as shown in line 48, is for the name of the benchmark for function optimization. Its value will be used by the program to determine which evaluation function is used to measure the search results. As shown in lines 62–71, this program also declares several new variables for PSO. To make it easier to compare and compute the personal best, pbest and pbest_obj_vals are declared for the solutions p_b and their corresponding objective values. Besides, gbest and gbest_obj_val are declared for the global best solution g_b and its corresponding objective value. The variables omega, c1, and c2 are for the parameters ω, c_1, and c_2. The other variables vmin and vmax are, respectively, the lower and upper bounds of all dimensions of the solution space.

9.2.2 The main loop

As shown in lines 101–148, the main loop of PSO, which follows the unified framework described in Section 4.1, will be carried out for a certain number of runs, in each of which the operators of PSO will be performed for a certain number of evaluations until the stop condition is met. For PSO, the implementation includes the following operators: init(), evaluate(), new_velocity(), new_position(), update_pb(), update_gb(), and update_best_sol().

9.2.3 Additional functions

The init() operator is responsible for generating the positions (initial solutions) and velocities of all the particles and initializing all the parameters of PSO, as shown in lines 150–187. The parameters of PSO to be initialized by this operator are the position of each particle (curr_pop), the velocity of each particle (velocity), the personal best solution of each particle (pbest), the global best solution (gbest), the objective value of each particle (curr_obj_vals), the objective value of each personal best solution (pbest_obj_vals), and the objective value of the global best solution (best_obj_val). Since some of the PSO-based algorithms do not use the global best solution to guide the search directions of the particles, we also declare another variable to keep track of the best-so-far solution (i.e., best_sol) for the other types of PSO-based algorithms. In addition to the option of reading the initial solutions in from a text file, the initial solutions and velocities can be randomly created on the condition that they can only assume values within the lower and upper bounds of each dimension of the solution space, as shown in lines 172–178. Note that the initialization of velocity is based on (Hassan, 2004), which decreases the value of velocities of all the particles from $[s_{min}, s_{max}]$ to $[s_{min}/t_{max}, s_{max}/t_{max}]$, to keep the particles from moving too fast, that is,

$$v_{ij}^0 = \frac{s_{min} + \mathcal{R}(s_{max} - s_{min})}{t_{max}}, \tag{9.4}$$

where v_{ij}^0 is the initial velocity of the j-th dimension of the i-th particle, $\mathcal{R}$ is a random number uniformly distributed in the range [0, 1], s_{min} and s_{max} are the lower and upper bounds of the solution space, and t_{max} is the maximum number of evaluations.[2] In this implementation, in addition to initializing parameters and vectors of PSO, the init() operator will also evaluate the particles it generates to get the personal and global bests and the best-so-far solution via evaluate(), update_pb(), update_gb(), and update_best_sol(), as shown in lines 181–186. The evaluate() operator, as shown in lines 189–199, is responsible for computing the objective value of each particle. The operator is designed in such a way that it is able to be extended for other function optimization problems. For this reason, the measurement of the Ackley function (please refer to Definition 10 in Section 2.3.1) resides in the file "functions.h," as shown in Listing 9.2.

Listing 9.2: src/c++/4-function/main/functions.h: The implementation of the Ackley function.

```
1 #ifndef __FUNCTIONS_H_INCLUDED__
2 #define __FUNCTIONS_H_INCLUDED__
3
4 #include <map>
5 #include <string>
6 #include <vector>
7 #include <cmath>
```

---

[2] In this implementation, the maximum number of evaluations instead of the maximum number of iterations is used as t_{max}.

```
 8
 9 using namespace std;
10
11 static constexpr auto pi = acos(-1);
12
13 inline double mvfAckley(int n, const vector<double>& x)
14 {
15     double s1 = 0.0;
16     double s2 = 0.0;
17     for (int i = 0; i < n; i++) {
18         s1 += x[i] * x[i];
19         s2 += cos(2.0 * pi * x[i]);
20     }
21     return -20.0 * exp(-0.2 * sqrt(s1/n)) + 20.0 - exp(s2/n) + exp(1.0);
22 }
23
24 // here is the table for all the functions
25 map<const string, double(*)(int, const vector<double>&)> function_table {
26     {"mvfAckley", mvfAckley}
27 };
28
29 #endif
```

As shown in lines 201–209 of Listing 9.1, this program will compare the solution of each particle with its personal best solution. If the solution is better than its personal best solution, then this program will update its personal best solution and the corresponding objective value. A similar comparison can also be found in lines 211–219, except that this time, the program will check to see if any solution is better than the global best solution. If it is, then the program will update the global best solution and the corresponding objective value. As far as the implementation of this program is concerned, the global best solution can be regarded as the best-so-far solution. This program will also update the best-so-far solution at each iteration, as shown in lines 221–228. The new_velocity() operator, as shown in lines 230–245, computes the new velocity of each particle (i.e., v_i^{t+1}) using Eq. (9.3). This operator will also check the velocity of each particle to prevent it from moving out of the bound (e.g., vmin or vmax) of the predefined solution space. Another key operator of PSO, as shown in lines 247–260, computes the new position of each particle. This new_position() operator uses Eq. (9.2) to adjust the new position of each particle s_i^{t+1} dimension by dimension, that is, $s_{ij}^{t+1} = s_{ij}^t + v_{ij}^{t+1}$. Like the new_velocity() operator, this operator will also check the position of each particle to prevent it from moving out of the bound of the given solution space. This operator will move it back to the bound when the particle is moved outside the given solution space.

9.3. Simulation results of PSO for the function optimization problem

In this section, the Ackley function—a function optimization problem (also called the global optimization problem)—is used to evaluate the performance of PSO. This benchmark is a well-known test function for the function optimization problem. The number of dimensions of this function is given, meaning that the user has to predefine it. The optimal solution is always $(0, 0, \ldots, 0)$, where the number of zeros is equivalent to the number of dimensions. In other words, the optimal solution has nothing to do with the number of dimensions. The implementation of PSO described herein is based on (Kennedy and Eberhart, 1995), and some of the parameter settings are based on those described in (Hassan, 2004). The parameters ω, c_1, and c_2 are set equal to 0.5, 1.5, and 1.5, respectively. The lower and upper bounds, v_{min} and v_{max}, of the solution space are set equal to -32.0 and 32.0, respectively. The number of particles is set equal to 10 for solving the Ackley function optimization problem. All tests were carried out for 100 runs, and the number of iterations performed in each run is 100, i.e., 1000 evaluations.

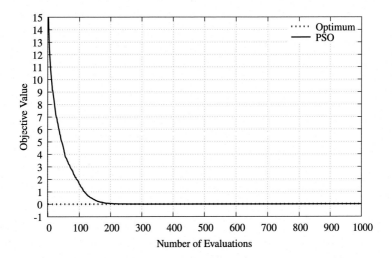

Figure 9.3 Simulation results of PSO for the Ackley function optimization problem.

As the results given in Fig. 9.3 show, PSO takes about 39 iterations (390 evaluations) on average to find the optimal solution[3] of the two-

---

[3] Here, we assume that the objective value of a candidate solution s_i is the same as that of the optimal solution if $f(s_i) < 0.0005$, that is, $f(s_i) = 0.000$ after rounding to the third decimal.

dimensional Ackley function optimization problem. Fig. 9.4 uses the results of a run as an example to show the positions of all the particles and the global best solution of PSO to help us better understand the search behavior of the particles of PSO. These results show that PSO converges very fast. As shown in Fig. 9.4(a), most of the particles are scattered in the solution space before entering the convergence process of PSO. Some of the particles are located on or near the bounds of the given solution space, such as the fifth particle and the ninth particle. After five iterations, the global best solution is very close to the optimal solution $(0.0, 0.0)$. All the particles are attracted close to the global best solution, as shown in Fig. 9.4(b). Fig. 9.4(c) shows that PSO has found the optimal solution in this run. All the particles are, of course, moved even closer to the global best solution in this case. From the positions of all the particles, it is easy to understand the search behavior, which explains that all the particles will, as expected, trend to the global best and their personal best very quickly.

9.4. Discussion

From the results of this chapter, it can be easily seen that PSO is a high-performance metaheuristic algorithm, especially for the function optimization problem. In the early stages of the development of PSO, some tutorial papers (Eberhart and Shi, 1998, 2001) focusing on how it works, possible research directions, and its applications were published. Eberhart and Shi (Eberhart and Shi, 1998) pointed out the distinguishing features of PSO by comparing it to the genetic algorithm (GA). They emphasized that PSO uses the previous best position to influence the position of a particle, while GA uses the crossover operator to change a chromosome. In a later study (Eberhart and Shi, 2001), Eberhart and Shi further discussed the research issues on inertia weights, constriction trajectories of PSO, and possible solutions of PSO for dynamic systems. To improve the performance of PSO, many extensions, modifications, and applications have been presented in recent years. Among them are parameter setting in velocity and position update, topology structure, multiple swarms, local optimum avoidance, and PSO for solving problems in discrete space. Many successful results for solving single-objective and multi-objective optimization problems using PSO and its variants can also be found today.

Since the early stages of the development of PSO, the parameter setting in velocity and position update has been addressed by several studies (Ratnaweera et al., 2004; van den Bergh and Engelbrecht, 2006; Zhan et

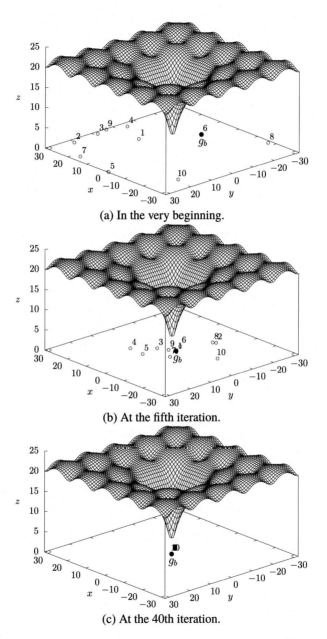

(a) In the very beginning.

(b) At the fifth iteration.

(c) At the 40th iteration.

Figure 9.4 An illustration of the positions of particles and the global best solution of PSO for the Ackley function optimization problem.

al., 2009). As mentioned in Section 9.1, Shi and Eberhart added the inertia weight ω to the velocity update equation of the original PSO to get Eq. (9.3) from Eq. (9.1) to improve its performance. This has become a critical research topic because the parameters of PSO in velocity update affect the search performance of PSO. It is not hard to imagine that the values of parameters have a strong impact on the trajectory of particles, the local best solutions, and the global best solution. They will also affect the search strategy of PSO. Hence, in (van den Bergh and Engelbrecht, 2006), van den Bergh and Engelbrecht gave a discussion on the influence the parameters of PSO may have on the particle trajectories. Among them are velocity clamping to control the increase in velocity, constricted and unconstricted trajectories, inertia weight, and particle trajectories. Whether or not to fix the values of these parameters during the entire convergence process and whether or not to dynamically control these parameters by linear or non-linear rules are two open issues discussed in several studies (Clerc and Kennedy, 2002; Ratnaweera et al., 2004; Shi and Eberhart, 1999; Zhan et al., 2009). In (Shi and Eberhart, 1999), Shi and Eberhart discussed several PSO methods with a linearly varying inertia weight ω, and in (Ratnaweera et al., 2004), Ratnaweera et al. used a mathematical representation to explain such a parameter adjustment method as follows:

$$\omega = (\omega_1 - \omega_2) \times \frac{t_{max} - t}{t_{max}} + \omega_2, \qquad (9.5)$$

where ω_1 and ω_2 are the initial and final values of the inertia weight ω, respectively, and t and t_{max} are the current iteration number and the maximum number of iterations, respectively. This parameter adjustment method is referred to as time-varying inertia weight (TVIW). Ratnaweera et al. in the same study (Ratnaweera et al., 2004) further presented the time-varying acceleration coefficients c_1 and c_2 for velocity update. They are defined as follows:

$$c_1^t = (c_{1f} - c_{1i}) \frac{t}{t_{max}} + c_{1i}, \qquad (9.6)$$

$$c_2^t = (c_{2f} - c_{2i}) \frac{t}{t_{max}} + c_{2i}, \qquad (9.7)$$

where c_{1f}, c_{1i}, c_{1f}, and c_{1i} are four constants. Clerc and Kennedy (Clerc and Kennedy, 2002) analyzed the convergence behavior of the original PSO (also called standard PSO in some studies) first and then presented some ways to implement the constriction coefficient. One of the well-known ways to update the position of a particle with constriction coefficient for the velocity update was defined as follows:

$$v_i^{t+1} = \chi \left(v_i^t + \varphi_1 (p_{b,i}^t - s_i^t) + \varphi_2 (g_b^t - s_i^t) \right) \tag{9.8}$$

and

$$s_i^{t+1} = s_i^t + v_i^{t+1}, \tag{9.9}$$

where $\varphi = \varphi_1 + \varphi_2 > 4$ and

$$\chi = \frac{2\kappa}{\varphi - 2 + \sqrt{\varphi^2 - 4\varphi}}, \tag{9.10}$$

where κ is a constraint value that is commonly smaller than one. By using these equations, the particles will then move on cyclic trajectories in the solution space. In (Zhan et al., 2009), Zhan et al. presented adaptive PSO (APSO), which uses a parameter adaptation scheme to adjust the parameters during the convergence process. The basic idea of this parameter adaptation scheme is to use a real-time evolutionary state estimation procedure to identify the current evolutionary state in each iteration. It first computes the evolutionary factor f as follows:

$$f = \frac{d_g - d_{\min}}{d_{\max} - d_{\min}} \in [0, 1], \tag{9.11}$$

where d_g is the mean distance between the global best particle g_b and all the particles, and $d_{\max}$ and $d_{\min}$ are the maximum and minimum mean distances between d_i and all the particles, respectively. The value of f will then be classified by a fuzzy membership function into one of four possible evolution states: exploration, exploitation, convergence, and jumping out. Then, c_1 and c_2 will be adjusted by using one of the following four equations:

$$\begin{cases} c_1 \triangle \text{ and } c_2 \triangle, & \text{if PSO is in the "convergence" state,} \\ c_1 \triangle \text{ and } c_2 \triangledown, & \text{if PSO is in the "exploitation" state,} \\ c_1 \triangle \text{ and } c_2 \triangledown, & \text{if PSO is in the "exploration" state,} \\ c_1 \triangledown \text{ and } c_2 \triangle, & \text{if PSO is in the "jumping out" state,} \end{cases} \tag{9.12}$$

where $\triangle$ and $\triangle$ indicate that the values of c_1 and c_2 will be raised and lowered, respectively. After that, the value of ω will be adjusted by

$$w = \frac{1}{1 + 1.5e^{-2.6f}}. \tag{9.13}$$

In addition to the strategies for updating velocity and position, the population topology (or neighborhood structure) (Kennedy, 1999; Poli et al., 2007) is another open issue because it affects the convergence of PSO. The topology can be fully connected, single-sighted, ring, or isolated, which will affect how the global best solution and the local best solutions are determined by PSO. In (Poli et al., 2007), Poli et al. divided the topologies of PSO into two types: static and dynamic. Different from the static topology, which usually uses a predefined topology as the communication structure between particles, the dynamic topology changes the communication structure between particles based on the given rule(s). A simple example of dynamic topology can be found in (van den Bergh and Engelbrecht, 2004), in which van den Bergh and Engelbrecht presented a variant of PSO, called cooperative PSO (CPSO), that uses multiple swarms to optimize different components of the solution. Each swarm will work on a set of subsolutions of the solution, and a set of particles in swarms will be replaced by the best-so-far particle. The communication structure between particles (topology) will also be changed. Liang and Suganthan (Liang and Suganthan, 2005) presented the dynamic multi-swarm particle swarm optimizer (DMS-PSO), in which a set of small-sized swarms is constructed first and then the members of these swarms are regrouped once for a certain number of iterations to change the communication structure between particles. Such kind of multi-swarm strategy can also be found in (Niu et al., 2007), in which a master–slave communication topology model for PSO was presented. All the slave swarms will send their best local particles to the master swarm, and the master swarm will select the best of all the received particles as the best particle of the master swarm. After that, the velocities of particles in the master swarm will be updated by

$$v_i^{m,t+1} = \omega v_i^{m,t} + c_1 \varphi_1 (p_{b,i}^{m,t} - s_i^{m,t}) + \Phi c_2 \varphi_2 (g_b^{m,t} - s_i^{m,t}) + (1 - \Phi) c_3 \varphi_2 (g_b^{s,t} - s_i^{m,t}),$$
(9.14)

where $v_i^{m,t}$, $s_i^{m,t}$, and $p_{b,i}^{m,t}$ represent the velocity, position, and personal best solution of the i-th particle in the master swarm at iteration t, $g_b^{s,t}$ and $g_b^{m,t}$ represent the best particle in the slave swarms and the best particle in the master swarm, respectively, and Φ is a migration factor that is defined as follows:

$$\Phi = \begin{cases} 0 & \text{if } f(g_b^{s,t}) < f(g_b^{m,t}), \\ 0.5 & \text{if } f(g_b^{s,t}) = f(g_b^{m,t}), \\ 1 & \text{if } f(g_b^{s,t}) > f(g_b^{m,t}), \end{cases}$$
(9.15)

where $f(g_b^{s,t})$ and $f(g_b^{m,t})$ are the fitness values of particles $g_b^{s,t}$ and $g_b^{m,t}$.

One of the advantages of PSO is its fast convergence; however, this is a double-edged sword, meaning that PSO may easily fall into local optima at early iterations. For this reason, how to prevent all the particles from moving toward the global best solution too quickly is an important open issue. Modifying the population topology is just one way to solve the problem. In fact, many recent studies have presented different methods to solve this problem. A simple approach is to reinitialize the position of a particle if it is too close to the center of a swarm (Clerc, 1993). Combining PSO with other (meta)heuristic algorithms provides an alternative way to enhance the search performance of PSO. Adding the mutation operator (Higashi and Iba, 2003; Stacey et al., 2003) of GA to PSO to disturb the solutions of PSO is also a useful way to avoid PSO from getting stuck at local optima at early iterations. Also, in (Angeline, 1998), a tournament selection operator of GA was added to PSO. All the particles will be sorted by their fitness values, and then the current positions and current velocities in the worst half of the population except their personal best positions will be replaced by the best half of the population. In a later study (Coello et al., 2004) on the multi-objective optimization problem, the historical record of best solutions found by a particle is used to keep a set of non-dominated solutions while the mutation operator of GA is used to increase the search diversity of PSO.

In (Gong et al., 2016), Gong et al. discussed two possible ways to combine PSO with GA, namely, the parallel and cascade frameworks. In the parallel framework, PSO and GA are loosely coupled, and the population is typically divided into two subpopulations, one of which is controlled by GA while the other is controlled by PSO. In the cascade framework, the population is generated by GA and then updated by PSO at each generation. Gong et al. then presented the genetic learning PSO (GL-PSO) by adding the mutation operator of GA to PSO to change the position of particles and the tournament operator of GA to replace the particles whose fitness values have not been improved for a certain number of iterations. In addition to the incorporation of GA into PSO, chaos search (Juang, 2004; Liu et al., 2005), simulated annealing (Wang and Li, 2004), and tabu search (Shen et al., 2008) were also combined with PSO to improve its search performance. A simple hybrid PSO can be found in (Wang and Li, 2004), in which SA is used as a search operator. In each iteration, each particle is searched independently by SA after its fitness value is evaluated. The velocity, position, personal best of each particle, and global best are then updated by the rules of the original PSO. Another hybrid PSO can be found in (Shen et al., 2008) in which 90% of the particles are searched by PSO, except for one that is picked from the explored neighborhood according to the aspiration criterion and tabu condition, and 10% of the particles are forced to fly randomly.

Also, how to apply PSO to different optimization problems or applications has been an important research topic since PSO was invented. Applying PSO to multi-objective optimization problems (Reyes-Sierra and Coello, 2006) is a popular research topic because PSO is capable of finding approximate solutions very quickly for continuous space problems. In the early stage of multi-objective PSO (MOPSO), the original PSO with a few modifications was used (Coello and Lechuga, 2002; Moore et al., 2000; Mostaghim and Teich, 2003). In (Moore et al., 2000), the particles of PSO are used to keep track of all non-dominated solutions (particles) of a multi-objective optimization problem. In a later study (Coello and Lechuga, 2002), a set of non-dominated solutions are stored in a repository, and then a set of hypercubes are generated as the coordinate systems of particles. The velocity update of each particle is then guided by its velocity, its personal best, and a solution in the repository. Mostaghim and Teich (Mostaghim and Teich, 2003) presented an interesting method to find the best local guide for each particle called Sigma that first computes a set of σ's with the coordinates of the objective space for particles and then uses the set of σ's to move particles directly toward the Pareto optimal front. To further improve the search ability of MOPSO, Tripathi et al. (Tripathi et al., 2007) presented a variant of PSO called time-varying MOPSO (TV-MOPSO), in which a time-varying control method is used to change the values of inertia (ω) and acceleration coefficients (c_1 and c_2) dynamically while the mutation operator is used to move the particles around randomly to boost the search diversity of TV-MOPSO. In a later study (Xue et al., 2013), Xue et al. applied MOPSO to the feature selection problem to take into account the number of features and classification accuracy at the same time.

Using PSO to solve the clustering problem (Omran et al., 2005; Tsai et al., 2015; van der Merwe and Engelbrecht, 2003) or enhance the performance of deep learning (Jiang et al., 2020; Qolomany et al., 2017) is another interesting research topic, which shows the possibilities of applying PSO to different kinds of problems. Some clues can be found from these studies about what we have to do and how to do it when we want to apply PSO to different problems or applications. An interesting example can be found in (van der Merwe and Engelbrecht, 2003), in which van der Merwe and Engelbrecht added the one-iteration k-means algorithm to PSO for solving the data clustering problem. Another example can be found in (Omran et al., 2005), in which the fitness function of PSO is aimed to minimize the intra-distance between pixels and their cluster means and maximize the inter-distance between any pair of clusters at the same time to solve the image clustering problem. In (Tsai et al., 2015), Tsai et al. further showed that the acceleration method for other metaheuristic algorithms, named pattern reduction, can also be used to speed up PSO. Since the deep learning paradigm was discussed everywhere, we can also find the footprint of PSO in studies on deep learning. A possible way to apply PSO to deep learning is to use it to solve the hyperparameter optimization problem (HPO), that is, to find a set of suitable hyperparameters. Using PSO to find the setting of hyperparameters, such as the number of hidden layers and the number of neurons in each layer, is a simple example of using PSO for the HPO (Qolomany et al., 2017). Another promising research direction is to use PSO or MOPSO to automatically find an approximate neural architecture (Jiang et al., 2020).

Because the original idea of PSO is to use vectors to represent its solutions, the performance of PSO for optimization problems in a discrete space is typically not good without a suitable modification. Changing the way the particles are encoded and modifying the velocity computation operator are some intuitive ways to solve this problem. A simple approach was presented in (Kennedy and Eberhart, 1997) in which the velocity $v_{i,d}$ and position $s_{i,d}$ in the d-th dimension of the i-th particle are constrained to the interval $[0.0, 1.0]$ and the velocity $v_{i,d}$ is used as a probability to determine if $s_{i,d}$ is 1 or 0. Just like using PSO to solve optimization problems in a discrete space, there still exist some open issues today although we can find a lot of successful applications using PSO. In summary, PSO and its variants have become a big family nowadays. Providing a comprehensive discussion that includes all interesting topics of PSO in a chapter of this book is simply impossible; therefore, we discuss only a few interesting and well-known topics here. There exist, however, some useful survey papers (Bonyadi and Michalewicz, 2017; Harrison et al., 2017; Lalwani et al., 2019; Poli et al., 2007; Shi, 2004; Wang et al., 2018a) that provide a more extensive and detailed discussion on PSO in a general or particular domain. In (Shi, 2004), Shi provides a good tutorial and some possible applications of PSO. Two interesting survey papers (Harrison et al., 2017; Lalwani et al., 2019) were focused on the self-adaptive parameter tuning methods and parallel PSO.

Finally, many kinds of PSO are discussed in a number of survey papers (Bonyadi and Michalewicz, 2017; Poli et al., 2007; Wang et al., 2018a), from the perspectives of inertia weight, constriction coefficients, population topology, population size, binary PSO, applying PSO to dynamic problems, hybrid and adaptive PSO, theoretical analyses (convergence and movement), and applications that provide us some good roadmaps to enter the research area of PSO.

Supplementary source code

Here are the hyperlinks to the directories where the source code described in this chapter resides.

https://github.com/cwtsaiai/metaheuristics_2023/src/c++/4-function/

CHAPTER TEN

Differential evolution

10.1. The basic idea of differential evolution (DE)

Differential evolution (DE), presented by Storn and Price (Storn and Price, 1995, 1997), can be regarded as a branch of evolutionary computation (EC). Like the genetic algorithm (GA), DE also contains crossover (also called recombination in DE), mutation, and selection operators. Unlike GA, which typically encodes the solution as a string of binary or integer numbers, DE usually encodes the solution as a vector of real numbers. Also, the mutation and crossover operators are designed for solutions so encoded. These are the reasons why DE can usually be used to optimize real parameters or real-valued functions, such as in continuous function optimization, effectively and efficiently.

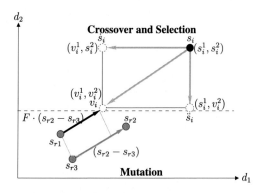

Figure 10.1 An illustration of the mutation and crossover operators of DE (Das and Suganthan, 2011).

Fig. 10.1 shows how DE works using a simple strategy called DE/rand/1/bin. Before entering the main loop, the current solutions s (also called parent vectors) are first randomly created by the initialization operator. The mutation, crossover, and selection operators are then performed repeatedly in the main loop. The mutation operator is responsible for creating new vectors v (called donor or mutant vectors), each of which is generated by adding the differences between a certain number of pairs of current vectors s randomly selected from the population. The crossover operator is responsible for creating new vectors u (called trial/offspring vectors), each of which is based on mapping current and donor vectors s and v. The selection operator is responsible for choosing the better one of the i-th offspring u_i and the i-th current vector s_i as the target vector s_i' for the next generation. We will discuss these operators in detail below.

First, the mutation operator creates a donor vector v_i by combining three current vectors s_{r1}, s_{r2}, and s_{r3} randomly selected from the population as follows:

$$v_i = s_{r1} + F(s_{r2} - s_{r3}), \tag{10.1}$$

where the scaling factor F is a predefined constant that generally falls in the range of 0 to 1 to adjust the impact of the distance between s_{r2} and s_{r3} on the selected current solution s_{r1}. In this way, the selected current solution s_{r1} will then move to the donor vector v_i.

The crossover operator will proceed to construct the offspring vector u_i, dimension by dimension, based on the donor vector v_i and the current vector s_i as follows:

$$u_i^d = \begin{cases} v_i^d & \text{if } \mathbf{r}_i^d \leq c_r \text{ or } d = r_d, \\ s_i^d & \text{otherwise}, \end{cases} \tag{10.2}$$

where u_i^d denotes the d-th dimension of the i-th offspring vector, v_i^d denotes the d-th dimension of the i-th donor vector, s_i^d denotes the d-th dimension of the i-th current vector, c_r denotes a predefined crossover rate, $\mathbf{r}_i^d$ denotes a random number uniformly distributed in the range [0, 1], and r_d denotes a random dimension index uniformly distributed in the range [1, D], where D is the number of dimensions of each vector. To be more specific, for each dimension, if $\mathbf{r}_i^d$ is no larger than the predefined crossover rate c_r or the dimension index d is equal to the randomly generated dimension index r_d, then u_i^d will be assigned v_i^d; otherwise, it will be assigned s_i^d. For the example given in Fig. 10.1, u_i can be one of the following: v_i, s_i, $\dot{s}_i$, and $\ddot{s}_i$, where $\dot{s}_i$ indicates that s_i^1 of s_i is replaced by v_i^1, while $\ddot{s}_i$ indicates that s_i^2 of s_i is replaced by v_i^2 because of the way the crossover operator works.

The last operator in the main loop is the selection operator, which is a little bit different from the tournament selection operators of GA because the comparison here is made between an offspring vector u_i and a parent vector s_i so as to find the better one s_i' to pass on to the next generation.

Mathematically, it is defined as follows:

$$s_i' = \begin{cases} u_i & \text{if } f(u_i) \leq f(s_i), \\ s_i & \text{otherwise,} \end{cases} \tag{10.3}$$

where $f()$ is the objective or fitness function used to evaluate the searched solution. Eq. (10.3) shows that the i-th new candidate solution at the next generation s_i' will be either u_i or s_i, depending on the fitness values of u_i and s_i.

Since the searched directions of DE are affected by mutation, we can easily find a lot of mutation strategies in the history of the development of DE (Das et al., 2016; Opara and Arabas, 2019). The five most representative mutation strategies are:

DE/rand/1	$v_i = s_{r1} + F(s_{r2} - s_{r3})$,	(10.4)
DE/best/1	$v_i = s_{\text{best}} + F(s_{r1} - s_{r2})$,	(10.5)
DE/current-to-best/1	$v_i = s_i + F(s_{\text{best}} - s_i) + F(s_{r1} - s_{r2})$,	(10.6)
DE/best/2	$v_i = s_{\text{best}} + F(s_{r1} - s_{r2}) + F(s_{r3} - s_{r4})$,	(10.7)
DE/rand/2	$v_i = s_{r1} + F(s_{r2} - s_{r3}) + F(s_{r4} - s_{r5})$,	(10.8)

where s_{rj} indicates the j-th randomly selected current vector and s_{best} indicates the best current vector of the current generation in terms of the objective or fitness value. It can be easily seen that the new candidate (target) vector v_i obtained by using Eq. (10.4) is based on three randomly selected vectors s_{r1}, s_{r2}, and s_{r3}. This strategy makes it possible for the searches of DE to not always trend to the best-so-far vector. As shown in Eq. (10.5), by substituting the best-so-far vector s_{best} for s_{r1}, the convergence speed of DE can be accelerated compared to the strategy of Eq. (10.4). If we want, the number of randomly selected current vectors can be increased and the best-so-far vector can be used in the mutation strategy. In this way, both diversification and intensification during the convergence process can be taken into account at the same time, as Eqs. (10.6), (10.7) and (10.8) show. From these equations, it can be easily seen that the naming rule of the mutation strategy in the study of DE is "DE/s_{rp}/n_{ps}," where s_{rp} is the reference point and n_{ps} is the number of pairs of randomly selected current vectors. Note that s_{rp} can refer to a randomly selected current vector, the i-th current vector, or the best-so-far vector. The strategies of crossover and control parameters of DE were also investigated in several studies (Das et al., 2016). A more thorough convention for naming the various transition strategies (i.e., mutation and crossover) is "DE/s_{rp}/n_{ps}/$\mathcal{D}$," where $\mathcal{D}$ represents the type of the crossover scheme, such as exponential (exp) or binomial (bin).

Algorithm 10 Differential evolution.

1 Initialize the population s of all parameter vectors	I
2 f_s = Evaluation(s)	E
3 **While** the termination criterion is not met	
4 v = Mutation(s) using Eq. (10.1)	T
5 u = Crossover(v) using Eq. (10.2)	T
6 f_u = Evaluation(u)	E
7 s', f_s' = Selection(s, u, f_s, f_u) using Eq. (10.3)	D
8 $s = s', f_s = f_s'$	D
9 **End**	
10 **Output** s	O

As shown in Algorithm 10, in addition to the initialization and evaluation operators in lines 1–2, DE contains the following operators: Mutation(), Crossover(), Evaluation(), and Selection(), as shown in lines 4–7. This means that these four operators will then be used to generate the target vector s' once the donor and trial vectors, v and u, are generated by using Eqs. (10.1), (10.2) and (10.3) in the main loop of DE every generation. To develop a new variant of DE, the mutation strategy can be replaced by a different movement method (i.e., transition operator), to move the position of the current vector s to the position of the trial vector v, such as Eq. (10.5), Eq. (10.6), Eq. (10.7), Eq. (10.8), or other mutation strategies (Das et al., 2016; Opara and Arabas, 2019).

10.2. Implementation of DE for the function optimization problem

Since DE is aimed at solving continuous optimization problems, the implementation described in this section is certainly also for a continuous optimization problem. We apply it to a single-objective function optimization problem $\mathbb{P}_6$, namely the Ackley function, so that we can measure its performance in solving such a problem. Note that the implementation of DE described in this section is based on the original idea of (Storn and Price, 1997), that is, the mutation strategy is DE/rand/1. Here is an example command line for invoking DE for the Ackley function.

```
$ ./search de 100 1000 2 "" "mvfAckley" 10 0.7 0.5 -32.0 32.0
```

The first term "de" means that the `search` program will use DE as the underlying search algorithm. The second term states that it will be carried out for 100 runs, while the third term says that 1000 evaluations (100 iterations) will be performed per run. The fourth term is the number of subsolutions. For a function optimization problem, it is the number of dimensions of the function. This means that the function in this case is two-dimensional. The fifth term is the name of the file in which the initial solutions reside, and since the filename is an empty string, no initial solutions are given in this case. The sixth term is the name of the function to be benchmarked. In this case, it is named "mvfAckley." As in the implementation of PSO, the measurement is also based on the implementation described in (Adorio and Diliman, 2005). The terms 10, 0.7, 0.5, -32.0, and 32.0 denote, respectively, the number of individuals, the scaling factor F, the crossover rate, and the lower and upper bounds of each dimension of the solution space. Listing 10.1 shows an implementation of the main functions and operators of DE.

Listing 10.1: src/c++/4-function/main/de.h: The implementation of DE.

```
1 #ifndef __DE_H_INCLUDED__
2 #define __DE_H_INCLUDED__
3
4 #include <iostream>
5 #include <fstream>
6 #include <limits>
7 #include "de.h"
8 #include "lib.h"
9 #include "functions.h"
10
11 using namespace std;
12
13 class de
14 {
15 public:
16     using solution = vector<double>;
17     using population = vector<solution>;
18
19     de(int num_runs,
20         int num_evals,
21         int num_dims,
22         string filename_ini,
23         string filename_ins,
24         int pop_size,
25         double F,
26         double cr,
27         double vmin,
```

```
28        double vmax);
29
30    virtual ~de() {}
31
32    population run();
33
34    vector<double> evaluate(string function_name, int dims, const population& curr_pop);
35    virtual population mutate(const population& curr_pop, double F, double vmin, double vmax,
      const solution& best_sol);
36    population crossover(const population& curr_pop, const population& curr_pop_v, double cr);
37    population select(const population& curr_pop, const population& curr_pop_u, solution&
      curr_obj_vals, solution& curr_obj_vals_u, string function_name);
38
39 private:
40    void init(population& curr_pop);
41
42 private:
43    int num_runs;
44    int num_evals;
45    int num_dims;
46    string filename_ini;
47    string filename_ins;
48
49    double best_obj_val;
50    solution best_sol;
51
52    // data members for population-based algorithms
53    int pop_size;
54    vector<double> curr_obj_vals;
55
56    // data members for de-specific parameters
57    double F;
58    double cr;
59    double vmin;
60    double vmax;
61 };
62
63 inline de::de(int num_runs,
64               int num_evals,
65               int num_dims,
66               string filename_ini,
67               string filename_ins,
68               int pop_size,
69               double F,
70               double cr,
71               double vmin,
72               double vmax
73               )
74    : num_runs(num_runs),
75      num_evals(num_evals),
76      num_dims(num_dims),
77      filename_ini(filename_ini),
78      filename_ins(filename_ins),
79      pop_size(pop_size),
80      F(F),
81      cr(cr),
82      vmin(vmin),
83      vmax(vmax)
84 {
85    srand();
86 }
87
88 inline de::population de::run()
89 {
90    population curr_pop(pop_size, solution(num_dims, vmax));
```

```
91    double avg_obj_val = 0.0;
92    vector<double> avg_obj_val_eval(num_evals, 0.0);
93
94    for (int r = 0; r < num_runs; r++) {
95        int eval_count = 0;
96        double best_so_far = numeric_limits<double>::max();
97
98        // 0. initialization (also called initialisation)
99        init(curr_pop);
100
101        while (eval_count < num_evals) {
102            // 1. mutation
103            population curr_pop_v = mutate(curr_pop, F, vmin, vmax, best_sol);
104
105            // 2. crossover
106            population curr_pop_u = crossover(curr_pop, curr_pop_v, cr);
107
108            // 3. evaluation
109            vector<double> curr_obj_vals_u = evaluate(filename_ins, num_dims, curr_pop_u);
110
111            // 4. selection
112            curr_pop = select(curr_pop, curr_pop_u, curr_obj_vals, curr_obj_vals_u, filename_ins
      );
113
114            for (size_t i = 0; i < curr_obj_vals.size(); i++) {
115                if (best_so_far > curr_obj_vals[i]) {
116                    best_so_far = curr_obj_vals[i];
117                    best_sol = curr_pop[i];
118                }
119                if (eval_count < num_evals)
120                    avg_obj_val_eval[eval_count++] += best_so_far;
121            }
122        }
123        avg_obj_val += best_so_far;
124    }
125
126    avg_obj_val /= num_runs;
127
128    for (int i = 0; i < num_evals; i++) {
129        avg_obj_val_eval[i] /= num_runs;
130        cout << fixed << setprecision(3) << avg_obj_val_eval[i] << endl;
131    }
132
133    cout << best_sol << endl;
134
135    return curr_pop;
136 }
137
138 inline void de::init(population& curr_pop)
139 {
140    // 1. parameter initialization
141    best_sol = vector<double>(num_dims, vmax);
142    best_obj_val = numeric_limits<double>::max();
143    curr_obj_vals = vector<double>(pop_size, numeric_limits<double>::max());
144
145    // 2. initialize the positions and velocities of particles
146    if (!filename_ini.empty()) {
147        ifstream ifs(filename_ini.c_str());
148        for (int p = 0; p < pop_size; p++) {
149            for (int i = 0; i < num_dims; i++)
150                ifs >> curr_pop[p][i];
151        }
152    }
153    else {
154        for (int p = 0; p < pop_size; p++)
```

```
155          for (int i = 0; i < num_dims; i++)
156              curr_pop[p][i] = vmin + (vmax - vmin) * rand() / (RAND_MAX + 1.0);
157      }
158
159      // 3. evaluate the initial population
160      curr_obj_vals = evaluate(filename_ins, num_dims, curr_pop);
161  }
162
163  inline vector<double> de::evaluate(string function_name, int dims, const population& curr_pop)
164  {
165      vector<double> obj_vals(curr_pop.size(), 0.0);
166      const auto it = function_table.find(function_name);
167      if (it != function_table.end()) {
168          const auto& f = it->second;
169          for (size_t i = 0; i < curr_pop.size(); i++)
170              obj_vals[i] = f(curr_pop[i].size(), curr_pop[i]);
171      }
172      return obj_vals;
173  }
174
175  inline de::population de::mutate(const population& curr_pop, double F, double vmin, double vmax,
          const solution& best_sol)
176  {
177      population new_pop(curr_pop.size(), solution(curr_pop[0].size()));
178      for (size_t i = 0; i < curr_pop.size(); i++) {
179          const solution& s1 = curr_pop[rand() % curr_pop.size()];
180          const solution& s2 = curr_pop[rand() % curr_pop.size()];
181          const solution& s3 = curr_pop[rand() % curr_pop.size()];
182          for (size_t j = 0; j < curr_pop[i].size(); j++) {
183              new_pop[i][j] = s1[j] + F * (s2[j]-s3[j]);
184              if (new_pop[i][j] < vmin)
185                  new_pop[i][j] = vmin;
186              if (new_pop[i][j] > vmax)
187                  new_pop[i][j] = vmax;
188          }
189      }
190      return new_pop;
191  }
192
193  inline de::population de::crossover(const population& curr_pop, const population& curr_pop_v,
          double cr)
194  {
195      population new_pop(curr_pop.size(), solution(curr_pop[0].size()));
196      for (size_t i = 0; i < curr_pop.size(); i++) {
197          const double s = rand() % curr_pop[0].size();
198          for (size_t j = 0; j < curr_pop[i].size(); j++) {
199              const double r = static_cast<double>(rand()) / RAND_MAX;
200              new_pop[i][j] = (r < cr || s == j) ? curr_pop_v[i][j] : curr_pop[i][j];
201          }
202      }
203      return new_pop;
204  }
205
206  inline de::population de::select(const population& curr_pop, const population& curr_pop_u,
          solution& curr_obj_vals, solution& curr_obj_vals_u, string function_name)
207  {
208      population new_pop(curr_pop.size(), solution(curr_pop[0].size()));
209      for (size_t i = 0; i < curr_pop.size(); i++) {
210          if (curr_obj_vals_u[i] < curr_obj_vals[i]) {
211              new_pop[i] = curr_pop_u[i];
212              curr_obj_vals[i] = curr_obj_vals_u[i];
213          }
214          else {
215              new_pop[i] = curr_pop[i];
216          }
```

```
217    }
218      return new_pop;
219  }
220
221 #endif
```

10.2.1 Declaration of functions and parameters

As shown in lines 19–28, in addition to the parameters that most population-based metaheuristics have in common, i.e., num_runs, num_evals, num_dims, filename_ini, and pop_size, this program requires two additional parameters for DE, namely, F and cr, and three additional parameters for function optimization problems, namely, filename_ins, vmin, and vmax. As shown in lines 34–40, this program declares five functions for DE, namely, evaluate(), mutate(), crossover(), select(), and init(). As shown in lines 57–60, this program also declares several new variables for DE. The variables F and cr are for the scaling factor F and the crossover rate c_r. The other variables vmin and vmax are, respectively, the lower and upper bounds of all dimensions of the solution space, which are used just like for PSO in solving function optimization problems.

10.2.2 The main loop

The main function of DE is shown in lines 88–136. It will perform the following operators of DE for a certain number of generations until the stop condition is met: init(), mutate(), crossover(), evaluate(), and select().

10.2.3 Additional functions

The init() operator of DE is responsible for generating the initial solutions and initializing all the parameters of DE, as shown in lines 138–161. The parameters of DE to be initialized by this operator are the positions of all parameter vectors (curr_pop), the best solution (best_sol), and the objective value of the global best solution (best_obj_val). How a searched vector for the Ackley function is evaluated by the evaluation operator evaluate() is shown in lines 163–173, which are similar to the implementation of PSO for this optimization problem described in Section 9.2. As shown in lines 175–191, the implementation of the mutation operator mutate() of DE described here is based on DE/rand/1. That is, for each parameter vector, it will first randomly select three parameter vectors (s1, s2, and s3), as shown in lines 179–181, and then calculate the i-th donor vector v_i by combining these three parameter vectors using the mutation strategy defined in Eq. (10.4), as shown in lines 183–187. In the case that the value of a dimension of the donor vector v_i exceeds the upper or lower bounds of that dimension, it will be clipped to either the upper bound vmax or the lower bound vmin of that dimension. As shown in lines 193–204, the implementation of the crossover operator crossover() will create the offspring vectors u_i one by one and dimension by dimension based on v_i and s_i, as shown in Eq. (10.2). In the case that the random value r ([0, 1]) is less than the crossover rate cr or the random number s ([1, D]) is equal to the dimension index j, the offspring vector u_i^j will be assigned v_i^j (curr_pop_v[i][j]); otherwise, it will be assigned s_i^j (curr_pop[i][j]). Note that D is the number of dimensions.

The implementation of the selection operator select() of DE is a little bit different from that of GA in the sense that GA will pick one of the individuals based on their fitness values and survival probabilities as the i-th parent to generate the offspring, while DE will compare the fitness values of the i-th offspring vector u_i (curr_obj_vals_u[i]) and the i-th current vector s_i (curr_obj_vals[i]) and choose the better one to pass it on to the next generation, as shown in lines 206–219. In brief, GA selects a new candidate solution from the current population, whereas DE selects a new candidate solution from a current solution and an offspring. If we take their selection and reproduction operators into account, GA can be regarded as a non-overlapping generation model, while DE can be regarded as an overlapping generation model.

10.2.4 Other mutation strategies

The mutation strategy used in the implementation described in Listing 10.1 is DE/rand/1. Here, we show how it can be easily replaced by another mutation strategy in such a way that all the code described in Listing 10.1 can be reused except the mutation function itself. As shown in Listing 10.2, Listing 10.3, Listing 10.4, and Listing 10.5, all we have to do is to make a derivation of the class described in Listing 10.1 while at the same time *overriding* the mutation function, that is, by making it an implementation of the other mutation strategies of DE, namely, DE/best/1, DE/current-to-best/1, DE/best/2, and DE/rand/2, as shown in Eq. (10.5), Eq. (10.6), Eq. (10.7), and Eq. (10.8), in this case. In this way, these new versions of DE can also be used to solve continuous optimization problems.

Listing 10.2: src/c++/4-function/main/deb1.h: The DE/best/1 mutation function of DE.

```
1 #ifndef __DEB1_H_INCLUDED__
2 #define __DEB1_H_INCLUDED__
3
4 #include "de.h"
5
6 class deb1: public de
7 {
8 public:
9     using de::de;
10
11 private:
12     population mutate(const population& curr_pop, double F, double vmin, double vmax, const
       solution& best_sol) override;
13 };
14
15 inline deb1::population deb1::mutate(const population& curr_pop, double F, double vmin, double
       vmax, const solution& best_sol)
16 {
17     population new_pop(curr_pop.size(), solution(curr_pop[0].size()));
18     for (size_t i = 0; i < curr_pop.size(); i++) {
19         const solution& s1 = curr_pop[rand() % curr_pop.size()];
20         const solution& s2 = curr_pop[rand() % curr_pop.size()];
21
22         for (size_t j = 0; j < curr_pop[i].size(); j++) {
23             new_pop[i][j] = best_sol[j] + F * (s1[j]-s2[j]);
24             if (new_pop[i][j] < vmin)
25                 new_pop[i][j] = vmin;
26             if (new_pop[i][j] > vmax)
27                 new_pop[i][j] = vmax;
28         }
29     }
30     return new_pop;
31 }
32
33 #endif
```

Listing 10.3: src/c++/4-function/main/decb1.h: The DE/current-to-best/1 mutation function of DE.

```
1 #ifndef __DECB1_H_INCLUDED__
2 #define __DECB1_H_INCLUDED__
3
4 #include "de.h"
5
6 class decb1: public de
7 {
8 public:
9     using de::de;
10
11 private:
12     population mutate(const population& curr_pop, double F, double vmin, double vmax, const
       solution& best_sol) override;
13 };
14
15 inline decb1::population decb1::mutate(const population& curr_pop, double F, double vmin, double
       vmax, const solution& best_sol)
16 {
17     population new_pop(curr_pop.size(), solution(curr_pop[0].size()));
```

```
18    for (size_t i = 0; i < curr_pop.size(); i++) {
19        const solution& s1 = curr_pop[rand() % curr_pop.size()];
20        const solution& s2 = curr_pop[rand() % curr_pop.size()];
21
22        for (size_t j = 0; j < curr_pop[i].size(); j++) {
23            new_pop[i][j] = curr_pop[i][j] + (F * (best_sol[j]-curr_pop[i][j])) + (F * (s1[j]-s2
      [j]));
24            if (new_pop[i][j] < vmin)
25                new_pop[i][j] = vmin;
26            if (new_pop[i][j] > vmax)
27                new_pop[i][j] = vmax;
28        }
29    }
30    return new_pop;
31 }
32
33 #endif
```

Listing 10.4: src/c++/4-function/main/deb2.h: The DE/best/2 mutation function of DE.

```
1  #ifndef __DEB2_H_INCLUDED__
2  #define __DEB2_H_INCLUDED__
3
4  #include "de.h"
5
6  class deb2: public de
7  {
8  public:
9      using de::de;
10
11 private:
12     population mutate(const population& curr_pop, double F, double vmin, double vmax, const
      solution& best_sol) override;
13 };
14
15 inline deb2::population deb2::mutate(const population& curr_pop, double F, double vmin, double
      vmax, const solution& best_sol)
16 {
17     population new_pop(curr_pop.size(), solution(curr_pop[0].size()));
18     for (size_t i = 0; i < curr_pop.size(); i++) {
19         const solution& s1 = curr_pop[rand() % curr_pop.size()];
20         const solution& s2 = curr_pop[rand() % curr_pop.size()];
21         const solution& s3 = curr_pop[rand() % curr_pop.size()];
22         const solution& s4 = curr_pop[rand() % curr_pop.size()];
23
24         for (size_t j = 0; j < curr_pop[i].size(); j++) {
25             new_pop[i][j] = best_sol[j] + (F * (s1[j]-s2[j])) + (F * (s3[j]-s4[j]));
26             if (new_pop[i][j] < vmin)
27                 new_pop[i][j] = vmin;
28             if (new_pop[i][j] > vmax)
29                 new_pop[i][j] = vmax;
30         }
31     }
32     return new_pop;
33 }
34
35 #endif
```

Listing 10.5: src/c++/4-function/main/der2.h: The DE/rand/2 mutation function of DE.

```
 1 #ifndef __DER2_H_INCLUDED__
 2 #define __DER2_H_INCLUDED__
 3
 4 #include "de.h"
 5
 6 class der2: public de
 7 {
 8 public:
 9     using de::de;
10
11 private:
12     population mutate(const population& curr_pop, double F, double vmin, double vmax, const
       solution& best_sol) override;
13 };
14
15 inline deb2::population der2::mutate(const population& curr_pop, double F, double vmin, double
       vmax, const solution& best_sol)
16 {
17     population new_pop(curr_pop.size(), solution(curr_pop[0].size()));
18     for (size_t i = 0; i < curr_pop.size(); i++) {
19         const solution& s1 = curr_pop[rand() % curr_pop.size()];
20         const solution& s2 = curr_pop[rand() % curr_pop.size()];
21         const solution& s3 = curr_pop[rand() % curr_pop.size()];
22         const solution& s4 = curr_pop[rand() % curr_pop.size()];
23         const solution& s5 = curr_pop[rand() % curr_pop.size()];
24
25         for (size_t j = 0; j < curr_pop[i].size(); j++) {
26             new_pop[i][j] = s1[j] + (F * (s2[j]-s3[j])) + (F * (s4[j]-s5[j]));
27             if (new_pop[i][j] < vmin)
28                 new_pop[i][j] = vmin;
29             if (new_pop[i][j] > vmax)
30                 new_pop[i][j] = vmax;
31         }
32     }
33     return new_pop;
34 }
35
36 #endif
```

10.3. Simulation results of DE for the function optimization problem

In this section, the Ackley function is used to evaluate the performance of DE in solving a single-objective function optimization problem. The implementation of DE described herein is based on (Storn and Price, 1997). The scaling factor F and the crossover rate c_r are set equal to 0.7 and 0.5, respectively, based on (Guo and Yang, 2015; Poikolainen et al., 2015). The lower and upper bounds of the solution space, v_{min} and v_{max}, are set equal to -32.0 and 32.0, respectively. The number of parameter vectors is set equal to 10 for solving the Ackley function optimization problem. All tests were carried out for 100 runs, and the number of iterations performed each run is 100, i.e., 1000 evaluations. As shown in Fig. 10.2, in addition to PSO, five DE-based algorithms (DE/rand/1/bin, DE/best/1/bin, DE/current-to-best/1/bin, DE/best/2/bin, and DE/rand/2/bin) are also used in solving the two-dimensional Ackley function optimization problem. It can be easily seen that DE/rand/2/bin converges slower than all the other search algorithms, while DE/current-to-best/1/bin converges faster than all the other DE-based algorithms. The results also show that PSO converges much faster than most DE-based algorithms. A detailed comparison between these search algorithms given in Fig. 10.3 shows that PSO converges faster than DE/current-to-best/1/bin during the first 300 evaluations; however, DE/current-to-best/1/bin is faster than PSO in finding the optimal solution[1] at evaluations 378 and 390, respectively.

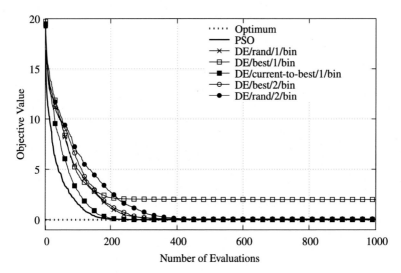

Figure 10.2 Simulation results of DE for the Ackley function optimization problem.

As shown in Fig. 10.4, although the results indicate that PSO with the parameter settings $\omega = 0.5$, $c_1 = 1.5$, and $c_2 = 1.5$ (called PSO) is slower than DE/current-to-best/1/bin in finding the optimal solution of the two-dimensional Ackley function optimization problem, this does not mean that PSO is ineffective or inefficient for solving other function optimization problems. Eventually, if we set the parameters of PSO to $\omega = 0.5$, $c_1 = 1.0$, and $c_2 = 1.5$ (called PSO+), the convergence speed of PSO will

---

[1] Here, we again assume that the objective value of a candidate solution s_i is the same as that of the optimal solution if $f(s_i) < 0.0005$, that is, $f(s_i) = 0.000$ after rounding to the third decimal.

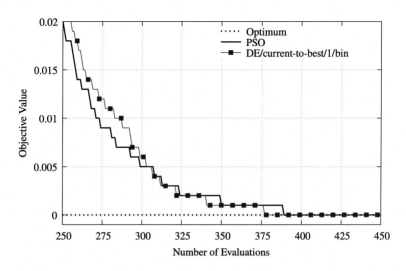

Figure 10.3 Details of the simulation results given in Fig. 10.2.

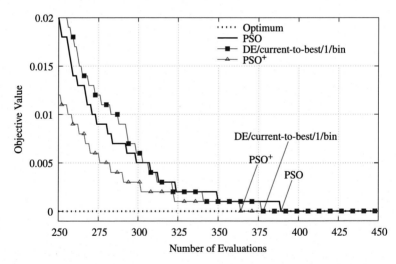

Figure 10.4 Comparison between DE/current-to-best/1/bin and PSO algorithms with different parameter settings.

be accelerated, and only 365 evaluations are needed for PSO to find the optimal solution in this case. The comparison given in Fig. 10.4 does not imply that PSO is better than DE or DE is better than PSO. Rather, it means that both PSO and DE are very effective and efficient in solving function optimization problems and their performance may be improved by using suitable parameter settings.

10.4. Discussion

Compared to other EC algorithms, GA (Holland, 1962), exhaustive search (Rechenberg, 1965), and evolutionary programming (Fogel et al., 1966), DE (Storn and Price, 1995, 1997) is young. In the early stage, studies were conducted to explain how DE works (Fleetwood, 2004; Storn, 1996) and what the main differences between DE and GA are (Tušar and Filipič, 2007). These studies provide us a simple guideline to understand the basic idea of DE. As one of the branches of EC, the functionality of the operators of DE is similar to that of GA. Different from most GAs that are useful in solving combinatorial optimization problems with a huge discrete solution space, most DEs are useful in solving global optimization problems with a continuous solution space. The main difference lies in the design of the transition operators (crossover and mutation), which are suitable for either a discrete solution space or a continuous solution space. Compared with GA, the design of crossover and mutation operators of DE is, in fact, quite similar to the design of the velocity update operator of PSO because both algorithms encode solutions as vectors and update the positions of solutions by the differences of solutions to determine the search directions and regions for the next generation. However, the main difference between DE and PSO is that PSO uses the current solution s_i, the personal best solution $p_{b,i}$, and the global best solution g_b to adjust the position of the current solution s_i, while DE uses a certain number of current solutions to find a candidate solution u_i and compares it with the current solution s_i to find the better one as the offspring solution s_i' for the next generation.

From the discussion and the simulation results in Section 10.3, it can be easily seen that DE is a high-performance metaheuristic algorithm for continuous optimization problems. Because DE contains selection, crossover, and mutation operators, just like GA, it can be regarded as a continuous version of GA. Some useful mechanisms developed in the previous studies of GA may be used to further enhance the performance of DE. Since many researchers focus on developing different mutation and crossover strategies, it has become a blossoming branch in EC. Because most studies of DE used a unified naming convention, i.e., "DE/s_{rp}/n_{ps}/$\mathcal{D}$" (Das et al., 2016; Opara and Arabas, 2019), researchers in this domain can easily repeat and reproduce DE and its variants. So, the different strategies of DE have become a set of powerful tools for solving continuous optimization problems. A set of variants for solving single-objective function optimization problems, which can be regarded as the epitome in the development of DE, has been presented in recent years. Since these improved DE algorithms add a number of new operators to DE, such as *a new parameter control strategy, a new mutation operator, a new crossover operator*, and *a new population reduction operator*, it is hard to categorize them. For this reason, we will discuss them in chronological order of presentation. Ali and Törn in an early study (Ali and Törn, 2000) attempted to change the scaling factor to diversify the searches of DE at early stages and intensify them at later stages during the convergence process. The scaling factor F is defined as

$$F = \begin{cases} \max\left(l_{\min}, 1 - \left|\frac{f_{\max}}{f_{\min}}\right|\right), & \text{if } \left|\frac{f_{\max}}{f_{\min}}\right| < 1, \\ \max\left(l_{\min}, 1 - \left|\frac{f_{\min}}{f_{\max}}\right|\right), & \text{otherwise}, \end{cases} \quad (10.9)$$

where $f_{\max}$ and $f_{\min}$ represent the maximum and minimum fitness values of the population, respectively, and $l_{\min}$ is the lower bound for the scaling factor F to ensure that it falls in the range $[l_{\min}, 1)$. In (Liu and Lampinen, 2002), Liu and Lampinen presented the so-called fuzzy adaptive DE (FADE), in which a fuzzy logic controller is used to adapt the parameters F and c_r for the mutation and crossover operators. Self-adaptive DE (SaDE), where two mutation operators (DE/rand/1/bin and DE/current-to-best/2/bin) are used in the candidate strategy pool, was first presented in (Qin and Suganthan, 2005). It was then extended in (Qin et al., 2009), where four mutation strategies (DE/rand/1/bin, DE/rand-to-best/2/bin, DE/rand/2/bin, and DE/current-to-best/1) are used in the candidate strategy pool. Each individual will then randomly select a mutation operator based on its success rate in generating an improved individual[2] in a certain number of generations. In SaDE, a set of crossover rates that were used to generate improved individuals are kept and used to adjust the crossover rate of the crossover operator. Teo in a later study (Teo, 2006) presented another variant of DE, called DE with self-adapting populations (DESAP), which considers not only the adaption mechanisms of F and c_r but also the population size. In (Brest et al., 2006), Brest et al. investigated the use of deterministic, adaptive, and self-adaptive strategies to control parameters in evolutionary algorithms (EAs) (Eiben et al., 1999; Hinterding et al., 1997)

---

[2] By an improved individual, we mean an individual that is better than its parent.

and noted that DE is nothing but an instance of EAs in the sense that a self-adaptive method is used to control its crossover rate and scaling factor. A variant of DE called jDE was presented in (Brest et al., 2006), the basic idea of which is to randomly change the values of the crossover rate and scaling factor for the i-th individual at each generation as follows:

$$c_{r,i}^{t+1} = \begin{cases} r_1, & \text{if } r_2 < \tau_1, \\ c_{r,i}^t & \text{otherwise}, \end{cases} \tag{10.10}$$

$$F_i^{t+1} = \begin{cases} F_l + r_3 \times F_u, & \text{if } r_4 < \tau_2, \\ F_i^t & \text{otherwise}, \end{cases} \tag{10.11}$$

where r_j, $j = 1, 2, 3, 4$, are random variables uniformly distributed in the range $[0, 1]$, τ_1 and τ_2 are set equal to 0.1, and F_l and F_u are set equal to 0.1 and 0.9, respectively.

Opposition-based DE (ODE) was presented in (Rahnamayan et al., 2008), in which a set of opposite individuals were added to improve the convergence rate of DE. In ODE, the j-th dimension of the i-th opposite individual o_i^j is generated by using the minimum and maximum values of the j-th dimension of the current population:

$$o_i^j = \min(s^j) + \max(s^j) - s_i^j, \tag{10.12}$$

where $\min(s^j)$ and $\max(s^j)$ represent the minimum and maximum values of the j-th dimension of the current population, respectively, and s_i^j, for $i = 1, 2, \ldots, m$ and $j = 1, 2, \ldots, n$, with m being the number of individuals in the current and opposite populations and n being the number of dimensions (variables) of each individual (solution), represents the j-th dimension of the i-th individual in the current population. At the end of each generation, ODE will select m individuals from the current and opposite populations to pass them on to the next generation. In (Zhang and Sanderson, 2009), Zhang and Sanderson presented an adaptive DE called JADE, which contains a new mutation operator (DE/current-to-$p$best), an optional external archive, and the parameter adaptation methods for c_r and F. The DE/current-to-$p$best mutation operator without the external archive is defined as

$$v_i = s_i + F_i(s_{\text{best}}^p - s_i) + F_i(s_{r1} - s_{r2}), \tag{10.13}$$

while the one with the external archive is defined as

$$v_i = s_i + F_i(s_{\text{best}}^p - s_i) + F_i(s_{r1} - \tilde{s}_{r2}), \tag{10.14}$$

where s_{best}^{p}, $p \in (0, 1]$, is an individual randomly chosen from the top $100p\%$ of individuals in the current population, F_i is the mutation rate that will be regenerated at each generation for the i-th individual s_i, $\tilde{s}_{r2}$ is randomly chosen from the union of the current population s and the archive $\mathbf{A}$ (i.e., $s \bigcup \mathbf{A}$), and the archive is used to keep a set of parent solutions that fail in the selection process. Some solutions in the archive will be randomly removed when the number of parent solutions in the archive exceeds a predefined threshold. The crossover rate for the i-th individual s_i of JADE, denoted $c_{r,i}$, is generated from a normal distribution (r^{n_i}) with mean μc_r and standard deviation 0.1 at each generation by

$$c_{r,i} = r^{n_i}(\mu c_r, 0.1), \qquad (10.15)$$

$$\mu c_r = (1 - c) \times \mu c_r + c \times \text{mean}_A(S_{c_r}), \qquad (10.16)$$

where c is a constant in the range $(0, 1]$, $\text{mean}_A(\cdot)$ is the arithmetic mean, and S_{c_r} is the set of all successful crossover rates, where a "successful crossover rate" $S_{c_{r,i}} = c_{r,i}$ means that v_i generated by crossover is better than s_i with crossover rate $c_{r,i}$. The mutation rate (scaling factor) F_i of JADE is generated by the Cauchy distribution (r^{c_i}) with the location parameter μF and scale parameter 0.1 at each generation:

$$F_i = r^{c_i}(\mu F, 0.1), \qquad (10.17)$$

$$\mu F = (1 - c) \times \mu F + c \times \text{mean}_L(S_F), \qquad (10.18)$$

$$\text{mean}_L(S_F) = \frac{\sum_{F \in S_F} F^2}{\sum_{F \in S_F} F}, \qquad (10.19)$$

where S_F represents the set of all successful mutation rates, μF represents the location parameter that is initialized to be 0.5 and will be updated by Eq. (10.18), and $\text{mean}_L(\cdot)$ represents the Lehmer mean defined by Eq. (10.19). Another improved version of DE, called non-linear simplex DE (NSDE), was presented in (Ali et al., 2009), in which the non-linear simplex method (NSM) is used to generate the new candidate solutions by using reflection, expansion, contraction, and reduction strategies. Inspired by the estimation of distribution algorithm (EDA) (Larraanaga and Lozano, 2001), the basic idea of compact DE (cDE) (Mininno et al., 2011) is to generate the individuals for the mutation operator by the probability model constructed with elite individuals and their fitness values.

Success-history-based parameter adaptation DE (SHADE), presented by Tanable and Fukunaga (Tanabe and Fukunaga, 2013), has been widely used in several recent studies. The basic idea of SHADE is to use a historical memory to keep a set of successful control parameter settings to automatically select the values of the control parameters, such as c_r and F. Similar to SaDE and JADE, a successful control parameter of SHADE means that an individual with c_r or F is better than its parent. As an improved version of JADE, SHADE uses $(\mu c_r, \mu F)$ as a single pair to guide the parameter adaption of DE, while JADE keeps and uses μc_r and μF independently. In a later study (Tanabe and Fukunaga, 2014), Tanable and Fukunaga presented an extended version of SHADE, named LSHADE, the basic idea of which is to linearly decrease the population size to improve the performance of SHADE. The linear function to reduce the population size can be defined as

$$m_{t+1} = \text{round}\left[\left(\frac{m^{\min} - m^{\text{init}}}{\text{NFE}^{\max}}\right) \times \text{NFE} + m^{\text{init}}\right], \qquad (10.20)$$

where m_{t+1} is the population size at the $(t+1)$-th generation, m^{init} is the population size at the first generation, $m^{\min}$ is the smallest population size, which is typically set equal to 4, NFE is the current number of fitness evaluations, and $\text{NFE}^{\max}$ is the maximum number of fitness evaluations. An improved version of SHADE, named jSO, was presented in (Brest et al., 2017), the underlying idea of which is to adjust the mutation rate periodically so that the mutation operator, called DE/current-to-$p$best-w/1, is defined by

$$v_i = s_i + F_w(s^p_{\text{best}} - s_i) + F(s_{r1} - s_{r2}), \qquad (10.21)$$

$$F_w = \begin{cases} 0.7 \times F & \text{if NFE} < 0.2\text{NFE}^{\max}, \\ 0.8 \times F & \text{if NFE} < 0.4\text{NFE}^{\max}, \\ 1.2 \times F & \text{otherwise,} \end{cases} \qquad (10.22)$$

where F_w is a scaling factor that will be changed periodically during the convergence process of DE and F is a predefined scaling factor. In (Stanovov et al., 2018), Stanovov et al. presented an improved version of LSHADE, named LSHADE-RSP, by adding the so-called rank-based selective pressure strategy to LSHADE. The idea is to apply rank selection of GA to the individuals of DE. This means that each individual of DE will be assigned a rank defined by

$$\text{Rank}_i = k \times (m - i) + 1, \qquad (10.23)$$

where k is a scaling factor for the greediness of the rank selection, m is the population size, and $i \in [1, m]$ is the index of a sorted fitness array. Besides, the probability of selecting the i-th individual is defined as

$$p_i = \frac{\text{Rank}_i}{\sum_{j=1}^{m} \text{Rank}_j}. \tag{10.24}$$

With Eqs. (10.23) and (10.24), Stanovov et al. then modified the current-to-pbest/1 mutation to make it a rank-based mutation scheme current-to-$p$best/r that is similar to Eq. (10.14) of JADE and Eq. (10.21) of jSO. So v_i of the mutation operator can be generated by

$$v_i = s_i + F_w(s_{\text{best}}^p - s_i) + F(s_{p_1} - s_{p_2}), \tag{10.25}$$

where F_w follows the setting of jSO defined in Eq. (10.22), p_1 and p_2 are the random indices chosen according to the probability p_i defined in Eq. (10.24), and F and c_r are computed using Cauchy and normal distributions to randomly choose from the successful parameter settings kept in the memory.

Another way to increase the search performance of DE is to use multiple populations. In (Wu et al., 2016), Wu et al. presented a new variant of DE, called multi-population ensemble DE (MPEDE), in which the population is first divided into four subpopulations. Three of them ($\mathbf{P}_1$, $\mathbf{P}_2$, and $\mathbf{P}_3$) are assigned, respectively, the DE/current-to-pbest/1, DE/current-to-rand/1, and DE/rand/1 mutation strategies. The remaining one ($\mathbf{P}_4$) is used as the reward subpopulation. In the very beginning, the subpopulation $\mathbf{P}_4$ is randomly assigned one of the three mutation strategies, and then it is assigned a possibly new mutation strategy that gets the best solution among the three mutation strategies once every t_g generations. A similar method using multiple populations can be found in (Wu et al., 2018), in which JADE, DE with composite trial vector generation strategies and control parameters (CoDE), and DE with ensemble of parameters and mutation strategies (EPSDE) are used for the subpopulations $\mathbf{P}_1$, $\mathbf{P}_2$, and $\mathbf{P}_3$, respectively, and the subpopulation $\mathbf{P}_4$ is used as the reward subpopulation. In a later study (Zhan et al., 2020), Zhan and his colleagues also presented a multi-population strategy called adaptive distributed DE (ADDE), which uses three subpopulations called exploration population using DE/current-to-rand/1, exploitation population using DE/current-to-pbest/1, and balance population using DE/current-to-rand/1 (in exploration state) and DE/current-to-pbest/1 (in exploitation state) in a master–slave framework for the single-objective function optimization problem. A new variant of DE called j2020 (Brest et al., 2020) is a derivation from jDE (Brest et al., 2006) and jDE100 (Brest et al., 2019). The unique characteristics of j2020 are that it has two populations, it uses the crowding mechanism, and it uses the restart mechanism. Like jDE100, j2020 also has a small population ($\mathbf{P}_s$) and a big population ($\mathbf{P}_b$), each of which will be searched by the mutation, crossover, and selection operators of jDE. The best individual of $\mathbf{P}_b$ will be passed to $\mathbf{P}_s$. Unlike jDE100, a crowding mechanism is used in j2020 after the crossover operation of the big population $\mathbf{P}_b$ to find the individual u_i' that is closest to u_i based on the Euclidean distance. Then, u_i' and u_i will compete using the selection operator to determine which one will survive for the next generation. Another difference between jDE100 and j2020 is that in jDE100, the randomly selected individuals s_{r1}, s_{r2}, and s_{r3} for the mutation operator all come from $\mathbf{P}_b$, but in j2020, s_{r1} is randomly selected from $\mathbf{P}_b$ while s_{r2} and s_{r3} are randomly selected from $\mathbf{P}_b \bigcup \mathbf{M}_s$, where $\mathbf{M}_s$ is a proper subset of $\mathbf{P}_s$, that is, $\mathbf{M}_s \subset \mathbf{P}_s$. This means that the new candidate individuals generated by the mutation operator will not only refer to the individuals of $\mathbf{P}_b$ but also to a small number of the individuals of $\mathbf{P}_s$. In (Hadi et al., 2021), Hadi et al. presented another improved DE algorithm that also uses multiple populations, called LSHADE-SPACMA. In each generation, this algorithm employs LSHADE and SPACMA to generate individuals for populations $\mathbf{P}_{\text{LSHADE}}$ and $\mathbf{P}_{\text{SPACMA}}$, respectively, and then concatenates them together as follows: $\mathbf{P} = \mathbf{P}_{\text{LSHADE}} \bigcup \mathbf{P}_{\text{SPACMA}}$. The adaptive guided DE (AGDE) mutation strategy (Mohamed and Mohamed, 2019) is then used to divide $\mathbf{P}$ into three clusters (best, better, and worst) to generate new candidate individuals. In a recent study (Bujok and Kolenovsky, 2021), Bujok and Kolenovsky presented a variant of DE called DE distance-based mutation selection, population size reduction, and optional external archive (DEDMNA), which integrates several strategies to improve the performance of DE. First, three mutation strategies are used to generate three mutant vectors u_1, u_2, and u_3. The one that is closest to the current individual s_i is then selected to compete with s_i as follows:

$$
d_k = \begin{cases} d(u_k, s_i) & \text{if } \frac{\text{NFE}}{\text{NFE}^{\max}} < r, \\ d(u_k, s_{\text{best}}) & \text{otherwise,} \end{cases} \qquad (10.26)
$$

where d_k is the distance between u_k and s_i or s_{best}, r is a random number uniformly distributed in the range $(0, 1)$, NFE is the current number of fitness evaluations, and $\text{NFE}^{\max}$ is the maximum number of fitness evaluations. The first case of Eq. (10.26) represents the exploration phase, while the second case represents the exploitation phase. The individual $u_{\min}$ (u_k that gives $\min(d_k)$) will be selected by the crossover operator to compete with the current individual s_i to generate a new trail vector by using the standard binomial crossover operator as shown in Eq. (10.2). In another recent study (Stanovov et al., 2021), a new research topic that uses the so-called nonlinear population size reduction to adjust the population size of DE during the convergence process was proposed.

Different from the aforementioned variants that either modify existing strategies of or add new strategies to the search operators (mutation, crossover, and selection) of DE, an alternative approach is to combine DE with other (meta)heuristic algorithms to improve their performance. For example, a hybrid metaheuristic algorithm was presented in (Zhang and Xie, 2003) that combines DE with PSO for solving the single-objective function optimization problem. In this hybrid metaheuristic algorithm, PSO is used to generate new candidate solutions at odd generations while DE is used to generate new candidate solutions at even generations. Another hybrid DE can be found in (Jadon et al., 2017), in which Jadon et al. combined artificial bee colony (ABC) (Karaboga, 2005) and DE. In this combination, DE is used as a search operator of ABC to generate the new candidate solution for the onlooker bee of ABC. Similar to other metaheuristic algorithms, DE can also be parallelized (Tasoulis et al., 2004; Teijeiro et al., 2016). A ring-network topology was used in (Tasoulis et al., 2004) to accelerate DE and improve its end results. The basic idea of parallel DE as described in (Tasoulis et al., 2004) is similar to the island model of GA, where first the population is divided into a set of subpopulations, then the evolution process of each subpopulation is performed independently for a certain number of generations, and finally the best individual of each subpopulation is moved to another subpopulation to share their search experience, which is called "migration" according to a predefined ring topology of subpopulations. In a later study (Teijeiro et al., 2016), Teijeiro et al. discussed and investigated a couple of parallel DE algorithms, including master–slave- and island model-based DE algorithms, and applied both of them to the Spark cloud computing environment. How to apply DE to the multi-objective optimization problem can be found in (Abbass et al., 2001; Liang et al., 2019; Robič and Filipič, 2005), in which it can be easily seen that the strategies used in GA for solving the multi-objective optimization problem, such as non-dominated sorting and the crowding distance metric, can also be used in DE. Mezura-Montes et al. (Mezura-Montes et al., 2008) showed some possible technologies that can be used by DE in solving the multi-objective optimization problem from the perspectives of promoting diversity (crowding distance and fitness sharing) and individual selection based on elitism. DE algorithms for the multi-objective optimization problem were then divided into three categories, namely, non-Pareto-based approaches, Pareto-based approaches (using dominance or ranking technologies), and combined approaches. Another interesting research topic can be found in (Deng et al., 2020; Ilonen et al., 2003), where DE is applied to neural networks. In (Ilonen et al., 2003), DE is used as an optimization method to train and find suitable weights of the neural network. An improved quantum-inspired DE (QDE) was presented in (Deng et al., 2020) to optimize the connection weights of a deep belief network (DBN), where each individual is encoded as a string of Q-bits and the Mexh wavelet function, standard normal distribution, adaptive quantum state update, and quantum non-gate mutation are used to avoid the searches from falling into local optima too quickly.

A large number of studies on DE have been presented in recent years, especially on solving the single-objective function optimization problem. Hence, a number of survey papers (Das et al., 2016; Das and Suganthan, 2011; Dragoi and Dafinescu, 2016; Neri and Tirronen, 2010; Opara and Arabas, 2019; Piotrowski, 2017; Price, 2013) have also been presented to discuss these variants of DE. In (Neri and Tirronen, 2010), Neri and Tirronen first divided these variants of DE into two classes, namely, DE integrating an extra component and modified structures of DE, and then discussed them from the perspectives of DE with trigonometric mutation, simplex crossover local search, population size reduction, scaling factor local search, self-adaptive control parameters, ODE, global-local search DE, and self-adaptive coordination of multiple mutation rules. All of these technologies have been widely used in recent studies on DE. A comprehensive survey can be found in (Das and Suganthan, 2011), where Das and Suganthan not only gave a simple tutorial to learn about DE but also discussed the search behaviors of search operators, such as the empirical distributions of candidate trial vectors with different values of c_r. Different variants of DE with different mutation, crossover, and selection operators were also discussed in this survey paper, along with hybrid DE algorithms, parallel DE algorithms, and DE algorithms for solving different optimization problems. Because several variants of DE were presented after 2011, Das and his colleagues extended their discussions on DE in a later survey paper (Das et al., 2016) to point out some important future avenues of DE. One of the important discussions given in this survey paper is on the adaptation strategies of F, c_r, and population size. Other important discussions are on the prominent variants of DE for solving the single-objective function optimization problem, population models, and DE for different optimization problems. How to determine the population size of DE and its impact was discussed in (Piotrowski, 2017) for solving the single-objective function optimization problems of the CEC2005 (Suganthan et al., 2005) and CEC2011 (Suganthan et al., 2011) datasets. Piotrowski observed that for DE, setting the population size to 100 provides good results for functions with less than 30 dimensions, and setting the population size to 200 may provide good results for higher-dimensional problems; as such, it can be used as the first guess for testing. However, for many DE algorithms, different population sizes may be needed because they will affect the convergence speed. Based on these observations, Piotrowski pointed out that the adaptive population size strategy is highly recommended when applying DE to the single-objective function optimization problem, especially when the number of dimensions is high. For the above discussions, the development of DE and its variants produced several useful technologies for solving the single-objective function optimization problem. Among them are: (1) control of parameters (c_r and F) by deterministic, adaptive, and self-adaptive strategies, (2) improved and multiple mutation or crossover operators, (3) a self-adapting population size (linear and non-linear population size reduction methods), (4) using the other (meta)heuristic algorithm as the search operator of DE or DE as the operator of another (meta)heuristic algorithm, (5) multiple populations, and (6) parallel computing. In summary, because the abovementioned technologies are useful for DE in solving the single-objective function optimization problem and other optimization problems, it is not difficult to imagine that all of them are useful for other metaheuristic algorithms in solving optimization problems in continuous space.

Supplementary source code

Here are the hyperlinks to the directories where the source code described in this chapter resides.

https://github.com/cwtsaiai/metaheuristics_2023/src/c++/4-function/

PART TWO

Advanced technologies

CHAPTER ELEVEN

Solution encoding and initialization operator

From the discussions in Chapters 1–10, it is easy to understand that the very first thing to do to apply a metaheuristic algorithm to an optimization problem is to decide how a solution for the problem in question is encoded or represented. Another thing related to the encoding of a solution that we need to take into account at the same time is the design of the initialization operator because these two things will strongly impact the search performance of a metaheuristic algorithm in solving an optimization problem. Hence, in this chapter, we will first discuss how to encode a solution of a metaheuristic algorithm for an optimization problem before we move on to the initialization operator of a metaheuristic algorithm because some designs use this operator to get a refined initial solution, which is somehow relevant to the encoding of a solution. Finally, we will provide a summary for the encoding of a solution and the initialization operator.

11.1. Encoding of solutions

The encoding of a solution is also referred to as the representation of a solution. As we mentioned in Chapter 2, optimization problems can typically be divided into two categories in terms of the space in which the solutions of the optimization problem are located, namely, discrete and continuous. For the discrete optimization problem, we typically use a search algorithm to find a good solution that contains a set of binary or integer numbers. For continuous optimization problems, a good solution we want to find is usually a set of real numbers. An intuitive way to encode the solution for a search algorithm $\mathcal{A}$ is to use a candidate solution of the problem $\mathbb{P}$ we want to solve to design the encoding of a solution. In Section 2.2.1, a binary bit string is used to represent a candidate solution for the one-max problem $\mathbb{P}_1$; therefore, we can use a binary bit string $s = \{s_1, s_2, \ldots, s_n\}$, where n is the problem size, to encode a solution of this problem. The solutions for an optimization problem can, of course, be encoded in many different ways. That is, in addition to the simplest one that encodes solutions of the one-max problem as a binary string, we can in fact use other ways to encode the solutions for a search algorithm to search for "good solutions" in the solution space of the optimization problem, if necessary. In this section, we will use a well-known combinatorial optimization problem, the clustering problem (Jain et al., 1999), as a simple example to explain how the solutions of an optimization problem can be encoded in different ways.

Definition 14 (The clustering problem $\mathbb{P}_9$). The clustering problem is defined as a problem of partitioning a given set of n objects $x = \{x_1, x_2, \ldots, x_n\}$ in d-dimensional space into k groups based on some predefined similarity metrics. An optimal clustering is typically a partitioning that minimizes the intra-cluster distance (e.g., distances between objects in the same group) and maximizes the inter-cluster distance (e.g., distances between objects in different groups). A simple way to encode a solution of the clustering problem is to use another set of the same size as x, denoted, say, $s = \{s_1, s_2, \ldots, s_n\}$, where each element, or subsolution, $s_i \in \{1, 2, \ldots, k\}$ represents a cluster to which the i-th object is assigned. The sum of squared errors (SSE) is a well-known measure to evaluate the clustering result. By using SSE, this optimization problem can be defined as

$$s^* := \underset{s \in \mathbf{A}}{\arg\min} f(s) = \sum_{j=1}^{k} \sum_{i=1}^{n_j} \| x_i^j - c_j \|^2,$$

where:

- **A** is the set of all possible solutions,
- k is the number of clusters,
- n_j is the number of objects in the j-th cluster,
- x_i^j is the i-th object in the j-th cluster,
- $c_j = (1/n_j) \sum_{i=1}^{n_j} x_i^j$ is the j-th centroid, and
- $n = \sum_{j=1}^{k} n_j$.

Note that objects are also called input data, data nodes, or patterns in different but relevant studies of data clustering. Also note that if SSE is used as the only measure to evaluate the searched solution of a clustering algorithm, the objective of such a clustering problem is likely to minimize only the total intra-cluster distance between the objects in the same group and their centroid. As shown in Fig. 11.1, given a set of eight objects (i.e., $n = 8$), denoted $x = \{x_1, x_2, x_3, x_4, x_5, x_6, x_7, x_8\}$, to be classified into two clusters (i.e., $k = 2$), denoted $\Pi = \{\Pi_1, \Pi_2\}$, one possibility is that $\Pi_1 = \{x_1, x_3, x_5, x_7\}$ and $\Pi_2 = \{x_2, x_4, x_6, x_8\}$, where Π_j represents the j-th cluster while c_j represents the centroid of the j-th cluster. The very first thing to do in applying a metaheuristic algorithm to a clustering problem is to decide how the solutions are encoded.

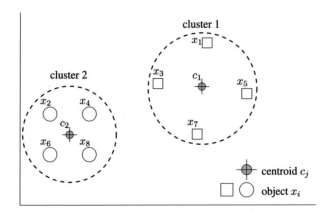

Figure 11.1 The relationship between objects and centroids.

To facilitate manual calculation, an even simpler example, which assumes that the objects are in one-dimensional space, is used. That is, given a set of six objects, denoted $x = \{x_1, x_2, x_3, x_4, x_5, x_6\} = \{1.0, 1.2, 2.1, 3.2, 3.4, 2.3\}$, to be classified into three clusters, denoted $\Pi = \{\Pi_1, \Pi_2, \Pi_3\}$, one possibility is that $\Pi_1 = \{x_1, x_2\}$, $\Pi_2 = \{x_3, x_6\}$, and $\Pi_3 = \{x_4, x_5\}$, meaning that $\Pi_1 = \{1.0, 1.2\}$, $\Pi_2 = \{2.1, 2.3\}$, and $\Pi_3 = \{3.2, 3.4\}$, where Π_j represents the j-th cluster and the centroid of the j-the cluster c_j can be easily calculated as the average of all objects in Π_j. For instance, $c_1 = (1.0 + 1.2)/2 = 1.1$, $c_2 = (2.1 + 2.3) = 2.2$, and $c_3 = (3.2 + 3.4)/2 = 3.3$ in this case.

Encoding in discrete space: As mentioned before, an intuitive way to encode the solution for $\mathbb{P}_9$ is $s = \{s_1, s_2, \ldots, s_n\}$, where $s_i \in \{1, 2, \ldots, k\}$, called cluster ID encoding. Suppose that there are six objects to be classified into three groups. Fig. 11.2 gives a simple example to illustrate how this encoding works. In this example, each subsolution $s_i \in \{1, 2, \ldots, k\}$ indicates to which group or cluster the i-th object x_i is assigned or classified.

1	1	2	3	3	2
s_1	s_2	s_3	s_4	s_5	s_6

Figure 11.2 Encoding of a solution by cluster ID $\mathbb{E}_1$.

Encoding in continuous space: Another encoding that is also widely used to represent the solution of a clustering problem is composed of a set of centroids. As shown in Fig. 11.3, for the same dataset, the solution becomes $s = \{s_1, s_2, \ldots, s_k\}$, where k is the number of clusters and s_i is the centroid of the i-th cluster. By using this encoding, the solution is of a length that is normally smaller than the encoding by cluster ID, but each subsolution will be a real number, i.e., a floating number instead of an integer. When the number of dimensions is larger than one, say, d, the solution will become $s = \{s_{1,1}, s_{1,2}, \ldots, s_{1,d}, s_{2,1}, s_{2,2}, \ldots, s_{2,d}, \ldots, s_{k,1}, s_{k,2}, \ldots, s_{k,d}\}$. Even though the length of a solution will increase as the number of dimensions increases, the length of this encoding is typically smaller than the encoding by cluster ID because k and d are usually much smaller than n, so $k \times d$ is typically smaller than n for a clustering problem. As a result, several studies (Bandyopadhyay and Maulik, 2002; Xu and Wunsch-II, 2005) use this encoding instead of the cluster ID encoding. In this example, the centroids used to represent the clustering results, $\Pi_1 = \{x_1, x_2\}$, $\Pi_2 = \{x_3, x_6\}$, and $\Pi_3 = \{x_4, x_5\}$, can be calculated as follows: $c_1 = (x_1 + x_2)/2 = (1.0 + 1.2)/2 = 1.1$, $c_2 = (x_3 + x_6)/2 = (2.1 + 2.3)/2 = 2.2$, and $c_3 = (x_4 + x_5)/2 = (3.2 + 3.4)/2 = 3.3$, respectively. Since this encoding method uses the centroids to represent a solution, i.e., each subsolution s_i is represented by its centroid c_i, we have $s = \{s_1, s_2, s_3\} = \{c_1, c_2, c_3\}$, as shown in Fig. 11.3.

1.1	2.2	3.3
s_1	s_2	s_3

Figure 11.3 Encoding of a solution by centroid $\mathbb{E}_2$.

Translation between $\mathbb{E}_1$ and $\mathbb{E}_2$: As shown in Fig. 11.4, although this encoding contains only the information of centroids, it can still be easily translated to the encoding of cluster ID by a two-step procedure. First, the distances between centroids and objects are computed so that the nearest centroid for each object can be found. After that, the encoding $s = \{s_1, s_2, s_3\}$ can be translated to the encoding $s' = \{s'_1, s'_2, \ldots, s'_6\}$, where each subsolution s'_i stands for the group to which the i-th object belongs. As shown in Fig. 11.5, we can also translate the encoding of cluster ID to the encoding of centroids. This translation is very simple. All we have to do is to find objects belonging to all the groups and then compute their centroids.

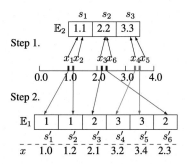

Figure 11.4 How to translate the solution from encoding of centroid to encoding of cluster ID.

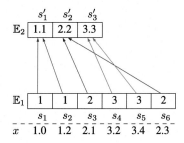

Figure 11.5 How to translate the solution from encoding of cluster ID to encoding of centroid.

Encoding by binary string: These are certainly not the only ways to encode the solution of a clustering problem. In fact, several encoding methods have been presented and used in studies on using metaheuristic algorithms to solve clustering problems (Tsai et al., 2015). Different from the encoding of centroids and cluster ID, the so-called binary encoding method is also used for encoding the solution of a metaheuristic algorithm in solving clustering problems. As shown in Fig. 11.6, this encoding contains an $n \times k$ binary string $s = \{s_{1,1}, s_{1,2}, \ldots, s_{1,n}, \ldots, s_{k,1}, s_{k,2}, \ldots, s_{k,n}\}$, $s_{ij} \in \{0, 1\}$, where n is the number of objects, k is the given number of groups for the clustering problem, and g_j is the j-th group.

	x_1	x_2	x_3	x_4	x_5	x_6
g_1	1	1	0	0	0	0
g_2	0	0	1	0	0	1
g_3	0	0	0	1	1	0

Figure 11.6 Encoding of a solution as a binary string $\mathbb{E}_3$.

Encoding by hybrid methods: Fig. 11.7 depicts another encoding for clustering, called hybrid encoding. This encoding can be divided into two parts. The first part keeps a set of centroids, while the second part is a binary string each bit of which is associated with a centroid in the first part. A value of 1 in the second part indicates that the clustering algorithm will use this centroid, while a value of 0 in the second part indicates that the clustering algorithm will not use this centroid. This encoding provides a flexible way for the clustering algorithm to dynamically determine the proper number of clusters during the convergence process.

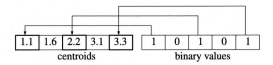

Figure 11.7 Encoding of a solution using a hybrid method $\mathbb{E}_4$.

Given these different encoding methods for the same optimization problem $\mathbb{P}_9$, the question that arises now is, why are these different encoding methods for the same optimization problem needed and developed by researchers? The reasons can be summarized as follows:

- Memory space: If the number of input objects of an optimization problem is very large, how to prevent the solution of these input objects from taking too much memory space will become a critical issue. The clustering problem, $\mathbb{P}_9$, is a typical example in the sense that if the encoding of the solution is cluster ID $\mathbb{E}_1$ (as shown in Fig. 11.2), the length of each solution will increase as the number of input objects increases. To reduce the usage of memory space, we can, of course, encode the solution as a set of centroids $\mathbb{E}_2$, as shown in Fig. 11.3. However, this new encoding method, $\mathbb{E}_2$, which is aimed at reducing the size of a solution, may bring up other issues, such as the computation costs to get the cluster ID of each object from the centroids to reconstruct a candidate solution for the clustering problem.

- Computation time: Another critical issue is how to avoid additional computation time for an encoding of the solution. If the main issue of a search algorithm is the computation time, encodings that can be accessed directly and quickly, that is, require no transformation and reconstruction, should be used. For the same example of solving $\mathbb{P}_9$, it is expected that the encoding of cluster ID $\mathbb{E}_1$ will take less time than the encoding of centroids $\mathbb{E}_2$ as far as getting the candidate solution of

$\mathbb{P}_9$ from the encoded solution (be it using encoding $\mathbb{E}_1$ or $\mathbb{E}_2$) is concerned because a clustering algorithm that uses encoding $\mathbb{E}_2$ may take additional time for the transformation and reconstruction of a candidate solution for the clustering problem during the clustering process.

- A particular algorithm: Different from the concerns of memory space and computation time, how to develop an appropriate encoding for a particular metaheuristic algorithm is a critical issue. For example, if we want to apply ant colony optimization (ACO) to the clustering problem, the encoding of a solution by centroids ($\mathbb{E}_2$) for particle swarm optimization (PSO) may not be appropriate because for ACO-based clustering algorithms it is required to take into account some specific operators and data structures such as the solution construction operator for each subsolution. Also, because ACO is suitable for combinatorial optimization problems, the encoding by a set of integer numbers will be much better than the encoding by a set of floating numbers, such as encoding method $\mathbb{E}_1$.

- Combinations of algorithms: A similar concern is the combination of two or more metaheuristic algorithms. In this case, the most important issue is how to use or design a general encoding method for all metaheuristic algorithms involved. Otherwise, if the involved metaheuristic algorithms need to use different encodings, an alternative way to deal with this issue is to develop transformation methods to transform the solutions between different encoding methods. For example, if we have a hybrid clustering algorithm that uses ACO and PSO for solving the clustering problem $\mathbb{P}_9$, encoding $\mathbb{E}_1$ can be used for ACO while encoding $\mathbb{E}_2$ can be used for PSO. All we need to do is to use transformation methods to transform solutions from one encoding to another for ACO and PSO, such as cluster ID to centroids, as shown in Fig. 11.5.

- A particular propose: Even though most encoding methods are developed to represent a candidate solution, a special case is where the encoding methods add additional information to all the solutions for a metaheuristic algorithm. For example, Das et al. (Das et al., 2008) presented a hybrid encoding method similar to $\mathbb{E}_4$ to allow the differential evolution algorithm (Storn and Price, 1997) to automatically find the appropriate number of clusters during its convergence process. As shown in Fig. 11.8, this kind of encoding can be divided into two parts: centroids and activation thresholds. The activation thresholds are used to determine whether the associated centroids will be used or not. By using this kind of encoding, the metaheuristic algorithm can then automatically find the possible number of clusters to classify the objects.

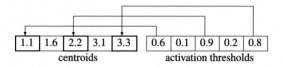

Figure 11.8 Encoding of a solution using a hybrid method $\mathbb{E}_5$.

11.2. Initialization operator

We do not know which candidate solutions will be good starting points for metaheuristic algorithms in most cases. In fact, we should assume that we know nothing at all about the entire solution space in the very beginning because different datasets may have different distributions, even for the same kind of optimization problem. For such uncertain situations, most metaheuristic algorithms typically use a random procedure to create initial solutions to avoid biasing to particular regions in the solution space. However, a random procedure is not a panacea for creating initial solutions of a metaheuristic algorithm, because when the initial solution is far away from the optimal solution the metaheuristic algorithm may spend a lot of time to move toward the near optimal solutions.

In several studies (Ahuja et al., 2000; Burke et al., 1999; Hill, 1999; Maaranen et al., 2006; Nowicki and Smutnicki, 1996; Todorovski and Rajicic, 2006) a refinement mechanism was used to replace the random procedure to create "good" initial solutions to enhance the performance of metaheuristic algorithms. Listing 11.1 provides a simple example to illustrate how to use a refinement mechanism to replace a random procedure in the initialization operator of a simulated annealing (SA) algorithm. As shown in lines 146–148 of Listing 11.1, unlike using a random procedure to create each subsolution, the refinement mechanism in the initialization operator creates the subsolutions block by block in the sense that it uses two parameters—num_samplings and num_same_bits—to determine how many refinement solutions are to be created before SA is invoked and how many subsolutions will assume the same value of a random procedure.

Listing 11.1: src/c++/1-onemax/main/sa_refinit.h: The refinement function of SA.

```
143    else {
144        solution tmp_sol(num_patterns_sol);
145
146        for (int j = 0; j < num_samplings; j++) {
147            for (int i = 0; i < num_patterns_sol; i += num_same_bits)
148                fill_n(tmp_sol.begin()+i, num_same_bits, rand() % 2);
149
150            int tmp_obj_val = evaluate(tmp_sol);
151
152            if (j == 0 || tmp_obj_val > obj_val) {
153                best_obj_val = obj_val = tmp_obj_val;
154                sol = tmp_sol;
155            }
156            avg_obj_val_eval[j] += obj_val;
157        }
158    } //
```

As shown in Fig. 11.9, each subsolution has a 50/50 chance to be either a "1" or a "0" in solving $\mathbb{P}_1$. Fig. 11.10 further shows that the refinement mechanism will randomly select a value as the value of each subsolution in a block of m subsolutions of a complete solution. In this case, $m = 2$.

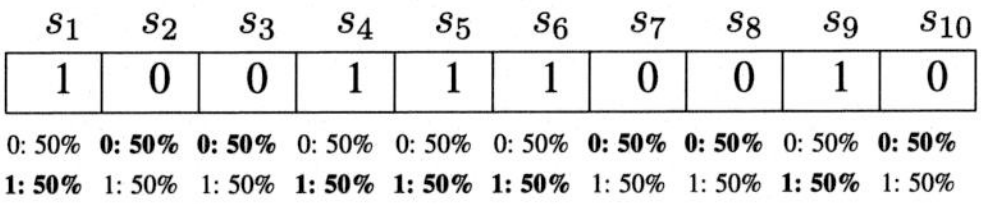

s_1	s_2	s_3	s_4	s_5	s_6	s_7	s_8	s_9	s_{10}
1	0	0	1	1	1	0	0	1	0

0: 50% **0: 50%** **0: 50%** 0: 50% 0: 50% 0: 50% **0: 50%** **0: 50%** 0: 50% **0: 50%**
1: 50% 1: 50% 1: 50% **1: 50%** **1: 50%** **1: 50%** 1: 50% 1: 50% **1: 50%** 1: 50%

Figure 11.9 How to create the solution by using a random procedure.

s_1	s_2	s_3	s_4	s_5	s_6	s_7	s_8	s_9	s_{10}
1	1	0	0	1	1	0	0	1	1

0: 50%   **0: 50%**   0: 50%   **0: 50%**   0: 50%
1: 50%   1: 50%   **1: 50%**   1: 50%   **1: 50%**

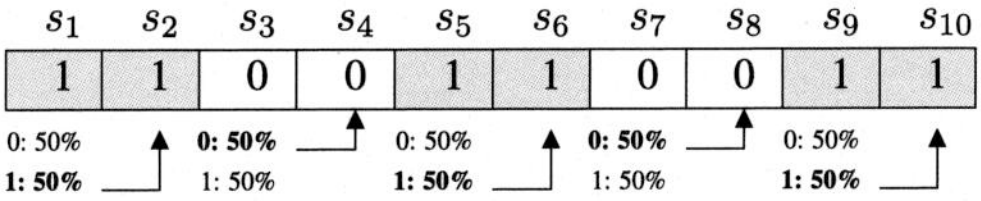

Figure 11.10 How to create the initial solution by using a refinement mechanism.

The performance of SA using a random procedure and SA using a refinement mechanism for creating the initial solutions is compared in Figs. 11.11 and 11.12. In these figures, init-sx denotes the refinement mechanism with setting x, which indicates how many subsolutions are to be in a block so that they will assume the same value, as shown in Fig. 11.10. For example, init-s2 indicates that each random procedure will randomly select a value for two subsolutions, while init-s4 indicates that each random procedure will randomly select a value for four subsolutions.

As shown in Fig. 11.12, the refinement mechanism may be able to create a good start point (i.e., a good initial solution) for metaheuristic

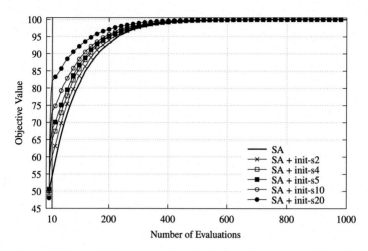

Figure 11.11 Difference between SA equipped with random initialization and SA equipped with refined initialization.

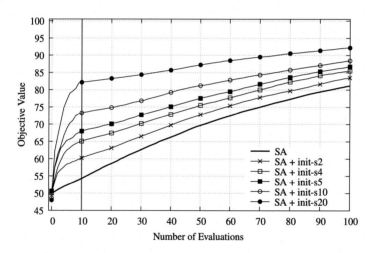

Figure 11.12 The first 100 evaluations of Fig. 11.11.

algorithms. The average objective value after the first 10 evaluations of SA with randomly created initial solutions is around 54, while the average objective values after the first 10 evaluations of SA with the best refined initial solutions of init-s2, init-s4, init-s5, init-s10, and init-s20 are all better. In this test, since the best refined initial solution is selected out of 10 sampled solutions, these evaluations need to be counted as the number of evaluations performed during the convergence process. This explains why despite being selected out of 10 candidate solutions, all of them outperform 10 evaluations of SA with a randomly created initial solution. A useful refinement mechanism will also speed up the convergence process of SA. As shown in Table 5.2, SA with a randomly created initial solution takes about 826 evaluations, whereas SA with the init-s20 refinement mechanism takes only 747 evaluations to find the optimal solution for the one-max problem of size $n = 100$. Fig. 11.12 further shows that the initial solution created using a refinement mechanism is much better than the initial solution created using a random procedure. As a result, metaheuristic algorithms may be able to reach the final state or find a good result more quickly if a good refinement mechanism is adopted. Of course, we need to take into account the additional costs required to create the candidate initial solutions if a refinement mechanism is used to replace the random procedure of an initialization operator for metaheuristic algorithms.

Fig. 11.13 further shows the distributions of different initial solution strategies. Fig. 11.13(a) shows that a general way to create the initial solution is to create each subsolution independently. The number of 1's in the initial solution can be regarded as a normal distribution. Fig. 11.13(b–f) shows the distributions of refinement mechanisms init-s2, init-s4, init-s5, init-s10, and init-s20. These distributions show that the refinement mechanism used will increase the probability of creating a solution that has much more "1's" than "0's," while it will also increase the probability of creating a solution that has much more "0's" than "1's." In brief, solutions created by this refinement mechanism will radically change the distribution so that the solution created will end up being either very "bad" or very "good." Fig. 11.13(f) gives a simple example to illustrate that the refinement mechanism used has an 18.75% chance to create a solution in which the number of "1's" is less than 40, while it has a 50.00% chance to create a solution in which the number of "1's" is more than 60. If some of the solutions created by this refinement mechanism are selected, SA can get a better initial solution in solving the one-max problem to further enhance its search performance.

The refinement mechanisms for metaheuristic algorithms can generally be divided into two types. The following descriptions detail how they can be used to enhance the performance of a metaheuristic algorithm.

- With domain knowledge: Although we have to assume that we do not know where the optimal solution is located in the solution space in the very beginning of the search, we sometimes understand the kinds of solutions that are near the optimum or are much better than so-

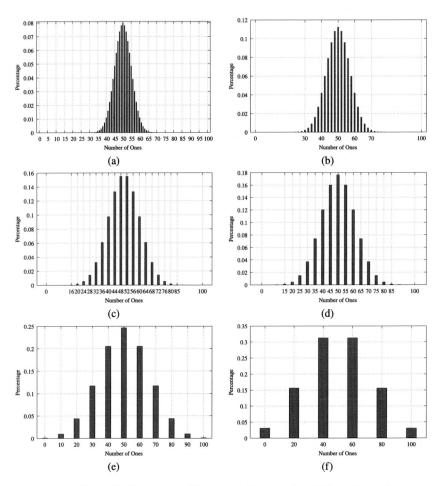

Figure 11.13 Examples illustrating different initial strategies. (a) An independent random procedure is used to create each subsolution. (b) Init-s2. (c) Init-s4. (d) Init-s5. (e) Init-s10. (f) Init-s20.

lutions created by random procedures. Using domain knowledge to create initial solutions is one of the most common ways to get better starting points for metaheuristic algorithms. It can be found in some studies (Burke et al., 1999; Todorovski and Rajicic, 2006) that use metaheuristic algorithms to solve scheduling optimization problems. Random procedures have a high probability of creating infeasible solutions, but when using infeasible initial solutions, metaheuristic algorithms generally require much more time to find feasible solutions than when a feasible solution is used as the starting point. This explains why some studies (Burke et al., 1999; Todorovski and Rajicic, 2006) using metaheuristic algorithms use domain knowledge to design the refinement initialization operator to create a set of feasible solutions to accelerate the convergence process. For example, in (Todorovski and Rajicic, 2006), the initialization operator of the genetic algorithm (GA) first checks the solutions (e.g., voltage angles) to see if they are feasible for solving the optimal power flow problem (OPFP). If not, it will adjust them to make them feasible. The simulation results of (Todorovski and Rajicic, 2006) show that GA with a refinement initialization operator based on domain knowledge gets better results than GA with random initial solutions for solving the OPFP.

- Without domain knowledge: In situations where we have no domain knowledge for the solution structure, there are still ways to create refined initial solutions for metaheuristic algorithms. As shown in Listing 11.1, a possible way to create initial solutions for a metaheuristic algorithm is to select the best solution or better solutions from a set of candidate solutions created by a random procedure. Another possible way is to create initial solutions using other heuristic or metaheuristic algorithms. For example, Ahuja et al. (Ahuja et al., 2000) employed the greedy randomized adaptive search procedure (GRASP) (Feo and Resende, 1995) to create better candidate solutions for GA in solving the quadratic assignment problem (QAP). The simulation results show that GA with GRASP to create the initial population may find a better final solution than GA with random initial solutions. In another study (Maaranen et al., 2006), Maaranen et al. analyzed the influence of the initial solution for GA. Their simulation results show that it affects the final results. However, initial solutions created by random procedures do not always yield worse results than initial solutions obtained with refinement methods having good diversity.

In summary, with good domain knowledge, we may be able to design a good refinement initialization operator to create "good" initial solutions to speed up the convergence process of GA while at the same time improving its final results. Without good domain knowledge, the use of a refinement initialization operator to get refined initial solutions may not always give better final results. Refinement initialization operators typically require more computation time than random procedures in the creation of initial solutions for metaheuristic algorithms; thus, we need to take into account this additional computation time when applying it to enhance the performance of a metaheuristic algorithm.

11.3. Discussion

This chapter first gives a brief introduction to the encoding of so-lutions for metaheuristic algorithms to highlight issues that need to be considered before applying them to new optimization problems. The first thing we need to do when we want to apply a metaheuristic algorithm to a new optimization problem is to choose a suitable encoding to represent the solution of the problem in question. A good encoding of solutions for a metaheuristic algorithm needs to take into account not only if the encoding can represent a complete solution of the problem in question, but also if the encoding is effective and efficient. Moreover, we need to consider if the initialization, transition, and evaluation operators need to be modified to solve the problem. For example, we can use the implementation of GA presented in Chapter 7 as the base and modify it to make it suitable for solving the traveling salesman problem (TSP). In this case, the encoding depicted in Chapter 7 can be inherited, meaning that the encoding does not need to be changed, as shown in Listing 11.2. According to Defini-tion 8 (Section 2.2.3), each solution for the TSP can assume a value in the set of sequences $c_\pi = \{\langle c_{\pi(1)}, c_{\pi(2)}, \ldots, c_{\pi(n)} \rangle\}$, that is, a permutation of the n cities. This implies that a solution of TSP can be encoded as an integer vec-tor or array, but the meaning of each subsolution is quite different. Hence, the initialization, transition, and evaluation operators have to be modified to make the algorithm suitable for the TSP.

Listing 11.2: src/c++/3-tsp/main/ga.h: Encoding of GA for the TSP.

```
15    typedef vector<int> solution;
16    typedef vector<solution> population; //
```

As shown in lines 171–176 of Listing 11.3, it will first assign the value of i to the i-th subsolution and then reorder the subsolutions randomly. With this kind of modification, GA for solving the one-max problem can now be used to solve the TSP, and it is guaranteed that it will not generate infeasible solutions during initialization, although this kind of modification by itself is not enough. The modification described here shows that the encoding and the design of the initialization operator are closely linked and insepa-rable. To make it capable of solving the TSP, of course, the other operators also need to be modified. Modification of the transition and evaluation operators of GA for solving the TSP will be discussed in the following chapters.

Listing 11.3: src/c++/3-tsp/main/ga.h: How the initial solutions of GA for solving the TSP are created.

```
163    // 4. initial solutions
164    if (!filename_ini.empty()) {
165        ifstream ifs(filename_ini.c_str());
166        for (int p = 0; p < pop_size; p++)
167            for (int i = 0; i < num_patterns_sol; i++)
168                ifs >> curr_pop[p][i];
169    }
170    else {
171        for (int p = 0; p < pop_size; p++) {
172            for (int i = 0; i < num_patterns_sol; i++)
173                curr_pop[p][i] = i;
174            for (int i = num_patterns_sol-1; i > 0; --i) // shuffle the solutions
175                swap(curr_pop[p][i], curr_pop[p][rand() % (i+1)]);
176        }
177    } //
```

From these discussions, it is easy to understand that solution encoding and initialization operators are not the most difficult part in the implementation of a metaheuristic algorithm, and the subtleties are usually ignored; however, both strongly impact the performance and final results of the metaheuristic algorithm. This explains why the very first thing that we need to do when we want to use a metaheuristic algorithm to solve a new complex optimization problem is to choose a good encoding and develop a good initialization operator.

Supplementary source code

Here are the hyperlinks to the directories where the source code described in this chapter resides.

https://github.com/cwtsaiai/metaheuristics_2023/src/c++/1-onemax/
https://github.com/cwtsaiai/metaheuristics_2023/src/c++/3-tsp/

CHAPTER TWELVE

Transition operator

12.1. Why use different transition operators

For a metaheuristic algorithm, the transition operator typically generates new candidate solutions for a later iteration during the convergence process. A new candidate solution can be generated either by changing a portion of the subsolutions of the current solution or as a brand-new feasible solution by a random process.

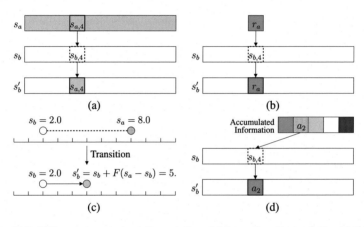

Figure 12.1 Different ways to transit a solution. (a) From the other solution(s). (b) By a random process. (c) Toward another solution. (d) From the accumulated information.

Fig. 12.1 shows four representatives of the transition operator: (1) change a portion of the subsolutions of the current solution based on the other solution(s) or (2) use a random process to generate a new solution, which can be regarded as crossover and mutation of GA, as shown in Fig. 12.1(a and b), respectively; (3) move the position of the current solution via the velocity of the other solution(s) toward a particular solution to create the new solution, which can be regarded as position update of particle swarm optimization (PSO) or differential evolution (DE), as shown in Fig. 12.1(c); and (4) create a new solution based on the accumulated community information, which can be regarded as solution construction of ant colony optimization (ACO) based on a pheromone table, as shown in Fig. 12.1(d). Note that in Fig. 12.1, s_a and s_b are two different solutions, s'_b is the new candidate solution of s_b, $s_{a,4}$ and $s_{b,4}$ are the fourth subsolutions of s_a and s_b, respectively, and F is a scaling factor set equal to 0.5. When we have insight into a metaheuristic algorithm, such as different designs in the encoding of a solution, it can be easily seen that the transition operator will also affect the search performance of the metaheuristic algorithm in different ways. In fact, in the discussions of simulated annealing (SA) (Section 5.4), tabu search (TS) (Section 6.4), the genetic algorithm (GA) (Section 7.4), ACO (Section 8.4), PSO (Section 9.4), and DE (in Section 10.4), several transition operators developed for different purposes have also been discussed. All of them illustrated that a solution can be transited or a new candidate solution can be generated in many ways to enhance the search performance of a metaheuristic algorithm. This is the reason why we turn our discussion to the design of the transition operator in this chapter. Below we give three major reasons why the transition operator has to be redesigned for the metaheuristic algorithm in solving an optimization problem.

- Different solution structures: Intuitively, a solution for the one-max problem can be encoded as a binary string, and the same is true for a solution for the knapsack problem. This implies that some simple transition operators for the one-max problem, such as the one-point and two-point crossover operators, can be used by GA in solving the knapsack problem and other optimization problems if the same solution encoding method is used. However, new suitable transition operators have to be designed for different kinds of optimization problems if they use different encoding methods. For example, one-point crossover of GA may not be useful for the floating-point solution encoding method. A simple way to design a crossover operator to transit the solution with floating-point encoding for GA is to move the value of the offspring gene to a random point between genes of parent solutions at the same locus. If we assume that the i-th genes of chromosomes s_a and s_b are 0.1 and 0.9, respectively, the i-th gene of the offspring s'_a of chromosome s_a can be generated by a random value uniformly distributed in the range $[0.1, 0.9]$. We can also use vectors to represent the solutions, and the transition operator will then change the current solution by changing the velocities of the vectors. Yes, this is exactly how the transition operator of PSO and DE for function optimization problems is designed. These examples explain that we have to design a suitable transition operator for a metaheuristic algorithm using a different encoding method to make its search more useful.

- To avoid creating infeasible solutions: There exists no transition operator that will work for all optimization problems. For example, using the so-called one-point crossover as the transition operator will typically not create any infeasible solution in solving the one-max optimization problem; however, it may create an infeasible solution for the traveling salesman problem (TSP), which will reduce the search performance of a metaheuristic algorithm, as shown in the example of Fig. 12.2. As such, we need to develop a new transition operator to avoid ineffective searches.

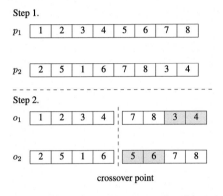

Figure 12.2 Why a transition operator may create infeasible solutions in solving the traveling salesman problem.

In this example, the one-point crossover operator for solving the TSP may create solutions that are infeasible. Suppose that $p_1 = (1, 2, 3, 4, 5, 6, 7, 8)$ and $p_2 = (2, 5, 1, 6, 7, 8, 3, 4)$ are the two parents of the current solutions. Further suppose that the crossover point is between the fourth and fifth genes of the chromosomes. After the one-point crossover is performed, the offspring o_1 and o_2 may contain some infeasible subsolutions, such as 3 and 4 in o_1 and 5 and 6 in o_2, that are duplicated, thus making it impossible for them to be a feasible solution of the TSP. Apparently, in this case, some recovery mechanisms are needed to make an infeasible solution feasible. It can be easily seen from this example that using a metaheuristic algorithm to solve different optimization problems, different transition operators may need to be developed to avoid the metaheuristic algorithm from creating infeasible solutions during the convergence process to reduce the additional recovery costs.

- To enhance the search performance: In addition to developing appropriate transition operators for metaheuristic algorithms to create feasible solutions in solving optimization problems, many studies (Das et al., 2016; Pavai and Geetha, 2016) have been conducted to develop better transition operators to enhance their search performance. For example, as mentioned in Section 10.4, the basic idea of SHADE (Tanabe and Fukunaga, 2013) is to keep a set of "good" control parameters in a historical memory to find the values of c_r and F to adjust the crossover rate and scaling factor (step size) of the transition operator. By using this mechanism, DE is able to automatically adjust the transition operator to suit different optimization problems or different stages of the convergence process. Another example is to use k-means as one of the transition operators for GA in solving the data clustering problem to improve the clustering results, as described in (Bandyopadhyay and Maulik, 2002; Krishna and Narasimha Murty, 1999). Because k-means was originally developed for data clustering, using it as the transition operator for a metaheuristic algorithm is just like using domain knowledge to enhance its search performance. This example also shows that using particular domain knowledge to develop the transition operator is a good way to enhance the search performance of a metaheuristic algorithm.

The above three reasons explain why we have to develop different transition operators for different encoding methods and different optimization problems. In addition to the development of an appropriate transition operator, how to choose appropriate parameters for the transition operator is another important issue that we have to take into consideration when developing a metaheuristic algorithm for solving an optimization problem. Studies have been conducted to develop adaptive and self-adaptive strategies to adjust the control parameters of the transition operator of DE (Brest et al., 2006) and to develop a time-varying parameter adjustment method for PSO (Ratnaweera et al., 2004). In fact, how to choose appropriate parameters for the transition operator is not a new topic. An example can be found in (Grefenstette, 1999), in which four different ways to change the mutation rate during the convergence process were discussed, namely, (1) fixed mutation, (2) genetic mutation, (3) fixed hypermutation, and (4) genetic hypermutation. As the name suggests, the fixed mutation method is the traditional way of fixing the mutation rate (i.e., the probability of mutating each subsolution of a solution) during the convergence process.

For the genetic mutation method, three additional bits (interpreted as an integer value k between 0 and 7) are added to each solution to control the mutation rate (m_r), which is defined as

$$m_r = \alpha_M + \frac{k}{8},\tag{12.1}$$

where α_M is the minimum mutation rate. As shown in Fig. 12.3, for the solution s_1, in addition to the subsolutions $s_{1,1}, s_{1,2}, \ldots, s_{1,5}$, the integer value represented by $s_{1,1}^k$, $s_{1,2}^k$, and $s_{1,3}^k$ will change the mutation rate of this solution. In this example, if α_M is set equal to 0.03 and $k = 2$ based on the bit values of $s_{1,1}^k$, $s_{1,2}^k$, and $s_{1,3}^k$, then the mutation rate will be $0.03 + 2/8 = 0.28$. This example shows that the mutation rate is embedded in the solution so that it is up to the determination operator to select the suitable mutation rate to transit this solution during the convergence process.

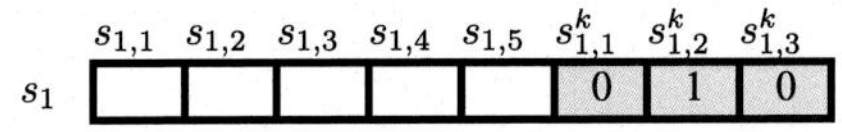

Figure 12.3 The genetic mutation method.

As shown in Fig. 12.4, for the fixed hypermutation method, the solutions are changed by first randomly resetting a fixed fraction h_r of solutions and then mutating the remaining solutions by a mutation operator with mutation rate m_r. For example, $h_r = 2/6$ implies that two of the six solutions (s_1 and s_2) will be replaced by a random solution while the remaining solutions (s_3, s_4, s_5, and s_6) will be mutated by a mutation operator with mutation rate m_r.

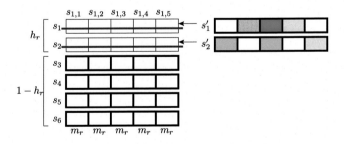

Figure 12.4 The fixed hypermutation method.

Fig. 12.5 gives a simple example to illustrate the genetic hypermutation, which also uses h_r and m_r to determine if a solution will be replaced by a

random solution or by mutation. This implies that three additional mutation control bits (s_1^k, s_2^k, and s_3^k) are added to each solution to determine the
rate of replacement by a random solution. In this example, the solutions
s_2 and s_4 will be replaced by new random solutions, while the remaining
solutions will be mutated by the mutation operator with mutation rate m_r.
The simulation results described in (Grefenstette, 1999) show that the hypermutation method outperforms the fixed mutation method for dynamic
landscapes (environment).

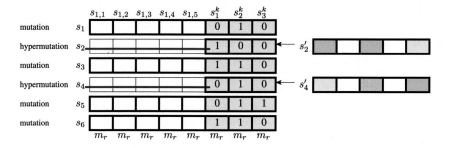

Figure 12.5 The genetic hypermutation method.

In summary, there are several ways to develop a good transition operator. Using domain knowledge, adopting another search algorithm, and
even using a local search operator are all possible ways to develop a good
transition operator. Since *adopting another search algorithm* as a transition operator can be regarded as a hybrid heuristic algorithm while *using a local
search mechanism* can be regarded as adding an additional transition operator,
we will discuss them in Chapters 15 and 16, respectively. Of course, the
strategies of *using a historical memory to save a set of good parameters or solutions*,
using an adaptive or self-adaptive mechanism to control parameters, and *using a predefined schedule to adjust the transition operator* are all promising for enhancing
the search performance of the transition operator. From the perspective of
the impact of a transition operator, the additional computation time and
the improvement of the end results caused by the transition operator are
two major issues that need to be taken into account when using a new or
different transition operator.

12.2. Different transition operators of GA for solving the TSP

Three different crossover operators (partially mapped crossover [PMX], cycle crossover [CX], and order crossover [OX]) are used in GA for solving the TSP as examples to show the impact different transition operators may have on the performance of metaheuristic algorithms. They are discussed in great detail in (Larrañaga et al., 1999).

As shown in Fig. 12.6, suppose that the current solutions are $p_1 = (1, 2, 3, 4, 5, 6, 7, 8)$ and $p_2 = (2, 5, 1, 6, 7, 8, 3, 4)$. In step 1, PMX will first randomly select a segment for these two solutions, e.g., loci 4 to 6 in this case. They are 4, 5, and 6 of p_1 and 6, 7, and 8 of p_2. In step 2, PMX will create three mappings for the selected subsolutions. Each mapping is associated with two subsolutions from the two current solutions at the same locus. This means that 4 of p_1 will be mapped to 6 of p_2, 5 of p_1 will be mapped to 7 of p_2, and 6 of p_1 will be mapped to 8 of p_2. The mappings now are $4 \leftrightarrow 6$, $5 \leftrightarrow 7$, and $6 \leftrightarrow 8$. Also, because the mappings $4 \leftrightarrow 6$ and $6 \leftrightarrow 8$ have the subsolution 6 in common, these two mappings can be concatenated to $4 \leftrightarrow 6 \leftrightarrow 8$. In step 3, PMX will exchange the selected segments of p_1' and p_2'. For p_1', 1, 2, 3, 7, and 8 are the other subsolutions at loci 1, 2, 3, 7, and 8 that must be checked to determine if they need to be changed. Because 7 and 8 match the mappings, the values that will be used to replace them are marked as "?." Similarly, for p_2', the two values that will be used to replace 5 at locus 2 and 4 at locus 8 are also marked as "?." In step 4, PMX will look for the mappings for the marked subsolutions to change p_1' and p_2'; thus, loci 7 and 8 in p_1' will be replaced by 5 and 4 because of the mappings $5 \rightarrow 7$ and $4 \rightarrow 6 \rightarrow 8$. Similarly, for p_2', loci 2 and 8 are replaced by 7 and 8. Finally, the new candidate solutions are $o_1 = (1, 2, 3, 6, 7, 8, 5, 4)$ and $o_2 = (2, 7, 1, 4, 5, 6, 3, 8)$. By using PMX, most structures of the searched solutions can be kept while making a few adjustments to them for later iterations.

As shown in Fig. 12.7, CX provides another way to transit the searched solutions of GA. The basic idea of CX is to find some useful relationships between the two searched solutions p_1 and p_2. In the first step, CX will start with the first subsolution of p_1 (i.e., $\langle 1 \rangle$) and then find the subsolution of p_2 (i.e., $\langle 2 \rangle$) at the same locus. This operator will then seek the subsolution $\langle 2 \rangle$ in p_1; the second subsolution of p_1 is a match in this case. At the same time, the subsolutions $\langle 1 \rangle$ and $\langle 2 \rangle$ will be added to the cycle set. The seeking process will be repeated until the subsolution sought in p_2 is the same as the

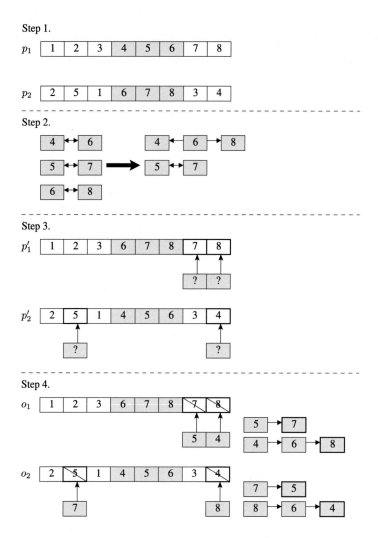

Figure 12.6 The basic idea of partially mapped crossover (PMX).

first subsolution in the cycle set. This example shows that when the third subsolution of p_1 (i.e., $\langle 3 \rangle$) is found, the seeking process will continue to seek the third subsolution of p_2 (i.e., $\langle 1 \rangle$), and it will end because the third subsolution of p_2 is in the cycle set and thus forms a cycle. Given the cycle set (1, 2, 5, 7, 3), in step 2, CX will mark the subsolutions of p_1 and p_2 in the cycle set by "X"; therefore, p_1' and p_2' will become (X, X, X, 4, X, 6, X, 8) and (X, X, X, 6, X, 8, X, 4), respectively, where "X" represents a subsolution that will not be exchanged with the other solution, while the unmarked subsolutions in solution p_1' will be exchanged with the other unmarked subsolutions in solution p_2'. This means that the fourth, sixth, and eighth subsolutions of these two solutions will be exchanged; that is, $\langle 4 \rangle$, $\langle 6 \rangle$, and $\langle 8 \rangle$ at these three subsolutions will be exchanged, and the new candidate solutions thus obtained in step 3 are $o_1 = (1, 2, 3, 6, 5, 8, 7, 4)$ and $o_2 = (2, 5, 1, 4, 7, 6, 3, 8)$.

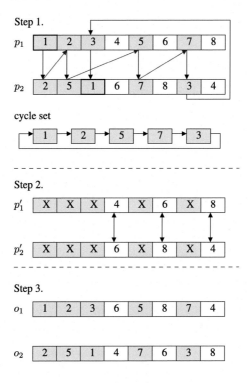

Figure 12.7 The basic idea of cycle crossover (CX).

As shown in Fig. 12.8, OX provides another way to keep the structure of the current solution for later iteration. Bear in mind that it is also referred to as OX1 in this book. In the first step, OX1 will randomly select a segment for these two solutions, just like the two–point crossover operator. In this example, the subsolutions 4–6 of p_1 and p_2 will be reserved for the next iteration; thus, (4, 5, 6) and (6, 7, 8) are marked accordingly in this step. In step 2, OX1 will find the visiting sequences of cities of these two searched solutions. These visiting sequences need to be readjusted starting from the first city. As a result, (1, 2, 3, 4, 5, 6, 7, 8) and (1, 6, 7, 8, 3, 4, 2, 5) are the visiting sequences of p_1 and p_2. In step 3, OX1 will mark all the

Step 1.

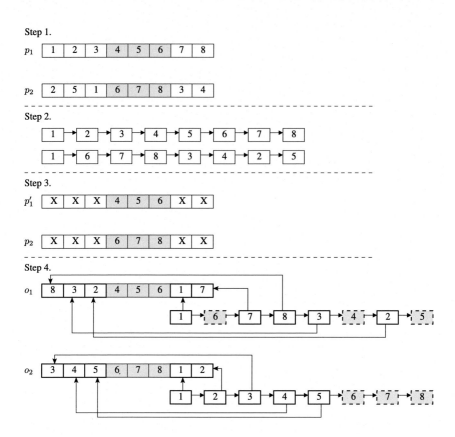

Figure 12.8 The basic idea of order crossover 1 (OX1).

subsolutions of p_1 and p_2 as "X" except (4, 5, 6) and (6, 7, 8) because these subsolutions have been marked in the first step. In step 4, OX1 will then use the visiting sequences to fill up the subsolutions marked "X" in step 3. This means that the visiting sequences of p_2 will be used to fill up the subsolutions marked "X" in p_1'. In this example, the fill up process will start at the seventh subsolution of p_1' and continue until the end of the solution is reached. The fill up process will then move back to the first subsolution and continue to fill all the subsolutions marked "X" up. Hence, (1, 7) are used to fill up the seventh and eighth subsolutions of p_1' while (8, 3, 2) are used to fill up the first, second, and third subsolutions. This is how the new candidate solution o_1 is created by OX1. Similarly, the same fill up process will be used to create the new candidate solution o_2.

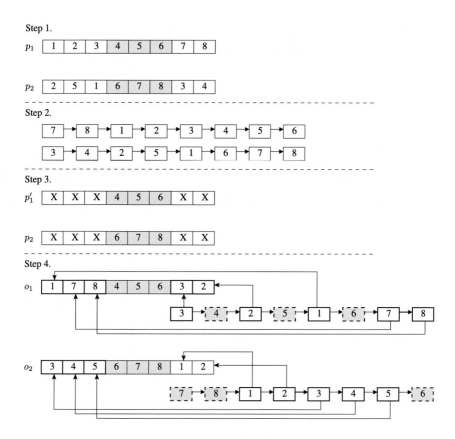

Figure 12.9 The basic idea of order crossover 2 (OX2).

As shown in Fig. 12.9, the first step of OX2 is the same as that of OX1. But in step 2, the procedure for finding the visiting sequences of cities of these two searched solutions is different from that of OX1. These visiting sequences will begin with locus 7, meaning that the visiting sequence of p_1 will be (7, 8), (1, 2, 3), and (4, 5, 6) while the visiting sequence of p_2 will be (3, 4), (2, 5, 1), and (6, 7, 8). Step 3 of OX2 is also the same as that of OX1 in the sense that it will mark all the subsolutions of p_1 and p_2 as "X" except (4, 5, 6) and (6, 7, 8). Step 4 of OX2 is very similar to that of OX1. Because the visiting sequences of p_1 and p_2 are different, the new solutions o_1 and o_2 will be different from the new solutions of OX1.

12.3. Implementation of GA for the TSP with different crossover operators

Above we discussed the basic ideas of different crossover operators and how different transition operators are designed. To make the implementation more flexible and to make it easier to add more crossover operators, we will first implement a simple version of GA for solving the TSP, as shown in Listing 12.1, in which the crossover operator has been declared as a `virtual` function. By implementing GA for solving the TSP in this way, the entire program, as shown in Listing 12.1, can be reused when adding a new crossover operator. This implies that all we have to do is to simply implement just this new crossover operator.

Listing 12.1: src/c++/3-tsp/main/ga.h: The implementation of GA for the TSP.

```
 1 #ifndef __GA_H_INCLUDED__
 2 #define __GA_H_INCLUDED__
 3
 4 #include <string>
 5 #include <fstream>
 6 #include <limits>
 7 #include "lib.h"
 8
 9 using namespace std;
10
11 class ga
12 {
13 public:
14
15     typedef vector<int> solution;
16     typedef vector<solution> population;
17     typedef vector<double> pattern;
18     typedef vector<pattern> instance;
19     typedef vector<double> dist_1d;
20     typedef vector<dist_1d> dist_2d;
21
22     ga(int num_runs,
23        int num_evals,
24        int num_patterns_sol,
25        string filename_ini,
26        string filename_ins,
27        int pop_size,
28        double crossover_rate,
29        double mutation_rate,
30        int num_players);
31
32     population run();
33
34 private:
35     population init();
36     vector<double> evaluate(const population &curr_pop, const dist_2d &dist);
37     void update_best_sol(const population &curr_pop, const vector<double> &curr_obj_vals,
        solution &best_sol, double &best_obj_val);
38     population tournament_select(const population &curr_pop, vector<double> &curr_obj_vals, int
        num_players);
```

```
39     virtual population crossover(const population& curr_pop, double cr);
40     population mutate(const population& curr_pop, double mr);
41
42 private:
43     const int num_runs;
44     const int num_evals;
45     const int num_patterns_sol;
46     const string filename_ini;
47     const string filename_ins;
48     const int pop_size;
49     const double crossover_rate;
50     const double mutation_rate;
51     int num_players;
52
53     population curr_pop;
54     vector<double> curr_obj_vals;
55     solution best_sol;
56     double best_obj_val;
57
58     instance ins_tsp;
59     solution opt_sol;
60     dist_2d dist;
61 };
62
63 inline ga::ga(int num_runs,
64               int num_evals,
65               int num_patterns_sol,
66               string filename_ini,
67               string filename_ins,
68               int pop_size,
69               double crossover_rate,
70               double mutation_rate,
71               int num_players
72               )
73     : num_runs(num_runs),
74     num_evals(num_evals),
75     num_patterns_sol(num_patterns_sol),
76     filename_ini(filename_ini),
77     filename_ins(filename_ins),
78     pop_size(pop_size),
79     crossover_rate(crossover_rate),
80     mutation_rate(mutation_rate),
81     num_players(num_players)
82 {
83     srand();
84 }
85
86 inline ga::population ga::run()
87 {
88     // double avg_obj_val = 0;
89     vector <double> avg_obj_val_eval(num_evals, 0.0);
90
91     for (int r = 0; r < num_runs; r++) {
92         int eval_count = 0;
93         double best_so_far = numeric_limits<double>::max();
94
95         // 0. initialization
96         curr_pop = init();
97
98         while (eval_count < num_evals) {
99             // 1. evaluation
100            curr_obj_vals = evaluate(curr_pop, dist);
101
102            update_best_sol(curr_pop, curr_obj_vals, best_sol, best_obj_val);  // please check
       to see if it is redundant
```

```
103
104            for (int i = 0; i < pop_size; i++) {
105                if (best_so_far > curr_obj_vals[i])
106                    best_so_far = curr_obj_vals[i];
107                if (eval_count < num_evals)
108                    avg_obj_val_eval[eval_count++] += best_so_far;
109            }
110
111            // 2. determination
112            curr_pop = tournament_select(curr_pop, curr_obj_vals, num_players);
113
114            // 3. transition
115            curr_pop = crossover(curr_pop, crossover_rate);
116            curr_pop = mutate(curr_pop, mutation_rate);
117        }
118
119        // avg_obj_val += best_obj_val;
120    }
121
122    // avg_obj_val /= num_runs;
123
124    for (int i = 0; i < num_evals; i++) {
125        avg_obj_val_eval[i] /= num_runs;
126        cout << fixed << setprecision(3) << avg_obj_val_eval[i] << endl;
127    }
128
129    cout << best_sol << endl;
130
131    return curr_pop;
132 }
133
134 inline ga::population ga::init()
135 {
136    // 1. parameters initialization
137    population curr_pop(pop_size, solution(num_patterns_sol));
138    curr_obj_vals.assign(pop_size, 0.0);
139    best_obj_val = numeric_limits<double>::max();
140    ins_tsp = instance(num_patterns_sol, pattern(2));
141    opt_sol = solution(num_patterns_sol);
142    dist = dist_2d(num_patterns_sol, dist_1d(num_patterns_sol));
143
144    // 2. input the TSP benchmark
145    if (!filename_ins.empty()) {
146        ifstream ifs(filename_ins.c_str());
147        for (int i = 0; i < num_patterns_sol ; i++)
148            for (int j = 0; j < 2; j++)
149                ifs >> ins_tsp[i][j];
150    }
151
152    // 3. input the optimum solution
153    string filename_ins_opt(filename_ins + ".opt");
154    if (!filename_ins_opt.empty()) {
155        ifstream ifs(filename_ins_opt.c_str());
156        for (int i = 0; i < num_patterns_sol ; i++) {
157            ifs >> opt_sol[i];
158            opt_sol[i]--;
159        }
160    }
161
162
163    // 4. initial solutions
164    if (!filename_ini.empty()) {
165        ifstream ifs(filename_ini.c_str());
166        for (int p = 0; p < pop_size; p++)
167            for (int i = 0; i < num_patterns_sol; i++)
```

```
168                    ifs >> curr_pop[p][i];
169        }
170    else {
171        for (int p = 0; p < pop_size; p++) {
172            for (int i = 0; i < num_patterns_sol; i++)
173                curr_pop[p][i] = i;
174            for (int i = num_patterns_sol-1; i > 0; --i) // shuffle the solutions
175                swap(curr_pop[p][i], curr_pop[p][rand() % (i+1)]);
176        }
177    }
178
179    // 5. construct the distance matrix
180    for (int i = 0; i < num_patterns_sol; i++)
181        for (int j = 0; j < num_patterns_sol; j++)
182            dist[i][j] = i == j ? 0.0 : distance(ins_tsp[i], ins_tsp[j]);
183
184    return curr_pop;
185 }
186
187 inline vector<double> ga::evaluate(const population &curr_pop, const dist_2d &dist)
188 {
189    vector<double> tour_dist(curr_pop.size(), 0.0);
190    for (size_t p = 0; p < curr_pop.size(); p++) {
191        for (size_t i = 0; i < curr_pop[p].size(); i++) {
192            const int r = curr_pop[p][i];
193            const int s = curr_pop[p][(i+1) % curr_pop[p].size()];
194            tour_dist[p] += dist[r][s];
195        }
196    }
197    return tour_dist;
198 }
199
200 inline void ga::update_best_sol(const population &curr_pop, const vector<double> &curr_obj_vals,
         solution &best_sol, double &best_obj_val)
201 {
202    for (size_t i = 0; i < curr_pop.size(); i++) {
203        if (curr_obj_vals[i] < best_obj_val) {
204            best_sol = curr_pop[i];
205            best_obj_val = curr_obj_vals[i];
206        }
207    }
208 }
209
210 inline ga::population ga::tournament_select(const population &curr_pop, vector<double> &
         curr_obj_vals, int num_players)
211 {
212    population tmp_pop(curr_pop.size());
213    for (size_t i = 0; i < curr_pop.size(); i++) {
214        int k = rand() % curr_pop.size();
215        double f = curr_obj_vals[k];
216        for (int j = 1; j < num_players; j++) {
217            int n = rand() % curr_pop.size();
218            if (curr_obj_vals[n] < f) {
219                k = n;
220                f = curr_obj_vals[k];
221            }
222        }
223        tmp_pop[i] = curr_pop[k];
224    }
225    return tmp_pop;
226 }
227
228 inline ga::population ga::crossover(const population& curr_pop, double cr)
229 {
230    population tmp_pop = curr_pop;
```

```
231
232    const size_t mid = tmp_pop.size()/2;
233    const size_t ssz = tmp_pop[0].size();
234    for (size_t i = 0; i < mid; i++) {
235        const double f = static_cast<double>(rand()) / RAND_MAX;
236        if (f <= cr) {
237            // 1. one-point crossover
238            const int xp = rand() % ssz;
239            solution& s1 = tmp_pop[i];
240            solution& s2 = tmp_pop[mid+i];
241            swap_ranges(s1.begin()+xp, s1.end(), s2.begin()+xp);
242
243            // 2. correct the solutions
244            for (size_t j = i; j < tmp_pop.size(); j += mid) {
245                solution& s = tmp_pop[j];
246
247                // find the number of times each city was visited,
248                // which can be either 0, 1, or 2; i.e., not visited,
249                // visited once, and visited twice
250                vector<int> visit(ssz, 0);
251                for (size_t k = 0; k < ssz; k++)
252                    ++visit[s[k]];
253
254                // put cities not visited in bag
255                vector<int> bag;
256                for (size_t k = 0; k < ssz; k++)
257                    if (visit[k] == 0)
258                        bag.push_back(k);
259
260                // correct cities visited twice, by replacing one of
261                // which by one randomly chosen from the bag, if
262                // necessary
263                if (bag.size() > 0) {
264                    random_shuffle(bag.begin(), bag.end());
265                    for (size_t n = 0, k = xp; k < ssz; k++) {
266                        if (visit[s[k]] == 2)
267                            s[k] = bag[n++];
268                    }
269                }
270            }
271        }
272    }
273
274    return tmp_pop;
275 }
276
277 inline ga::population ga::mutate(const population& curr_pop, double mr)
278 {
279    population tmp_pop = curr_pop;
280
281    for (size_t i = 0; i < tmp_pop.size(); i++){
282        const double f = static_cast<double>(rand()) / RAND_MAX;
283        if (f <= mr) {
284            int m1 = rand() % tmp_pop[0].size();        // mutation point
285            int m2 = rand() % tmp_pop[0].size();        // mutation point
286            swap(tmp_pop[i][m1], tmp_pop[i][m2]);
287        }
288    }
289
290    return tmp_pop;
291 }
292
293 #endif
```

12.3.1 Declaration of functions and parameters

Since the implementation of GA for solving the one-max problem $\mathbb{P}_1$ and the deceptive problem $\mathbb{P}_4$ in this book uses an integer vector to encode a solution and a solution vector to encode a population, lines 15–16 show that they can be reused by GA for solving the TSP. Lines 17–18 declare types to represent an instance of the TSP, while lines 19–20, just like the implementation of ACO shown in Listing 8.1, declare types to keep distances between all the cities as a lookup table. Because the implementation described herein is GA for the TSP, which is different from GA for the one-max and deceptive problems, the initialization (line 35) and evaluation (line 36) operators need to be modified. As mentioned before, the transition operator also needs to be modified to make it possible for GA to solve the TSP; therefore, the implementation described herein uses a simple crossover operator, named modified one-point crossover (MOPC), as shown in line 39, the basic idea of which is to use one-point crossover with a recovery mechanism to prevent it from creating infeasible solutions. As mentioned before, it is declared as a `virtual` function so that it can be easily customized, that is, replaced by a new crossover operator. Some parts of GA for the one-max and deceptive problems that can be adopted for the TSP are run(), as shown in line 32, and tournament_select(), as shown line 38. In this implementation, most of the declarations of variables are the same as for the one-max problem, as shown in Listing 7.1, except `ins_tsp`, `opt_sol`, and `dist`, which are variables for keeping the locations of cities, the optimal solution, and the distances between cities, as shown in lines 58–60.

12.3.2 The main loop

The main loop of GA for the TSP is shown in lines 86–132. To solve the TSP, for the operators update_best_sol() in line 102 and tournament_select() in line 112, we can almost apply the same operators as in the implementation of GA for solving the one-max problem, as described in Listing 7.1, except a few minor modifications. Because the structure of a solution to the TSP is different from that of a solution to the one-max problem, the initialization operator has to use the locations of cities (i.e., `ins_tsp`) to create the initial chromosomes and to calculate and save the distances between cities to the variable `dist`, as shown in line 96. The evaluate() function in line 100 and the crossover() function in line 115 also need to be modified to make GA capable of solving the TSP. In brief, the evaluate(), tournament_select(), update_best_sol(), crossover(), and mutation()

operators of this implementation will also be performed repeatedly for each generation until the termination criterion is met, as in the implementation of Listing 7.1.

12.3.3 Additional functions

As shown in lines 134–185, the initialization operator init() can be divided into five parts: initialize the parameters, input the TSP benchmark, input the optimal solution, create the initial solutions, and construct a distance matrix. First, the vector curr_pop is created to keep the population, as shown in line 137. Next, a vector curr_obj_vals and a scalar best_obj_val are created to keep track of the fitness value of each chromosome and the best fitness value so far, as shown in lines 138–139. Then, another three vectors ins_tsp, opt_sol, and dist are created to store the locations of cities, the optimal solution, and the distances between cities, as shown in lines 140–142. The second part is shown in lines 145–150. It reads the two-dimensional location information of each city from the input dataset and keeps them in the vector ins_tsp. The third part is shown in lines 153–160. It reads the optimal solution, that is, the tour routing order for cities, and keeps it in the vector opt_sol. The fourth part reads the initial solutions from the input file (as shown in lines 164–169) or creates the initial solution using a random process (as shown in lines 171–176). To randomly create the initial solution, this operator first assigns to each chromosome in the vector curr_pop the solution $\langle 0, 1, \ldots, n - 1 \rangle$, where $n = $ num_patterns_sol, and then shuffles it to disturb the subsolutions, as shown in lines 174–175. The fifth part, shown in lines 180–182, then calculates the distances between cities and keeps this information in the matrix dist.

As shown in lines 187–198, the evaluation operator needs to calculate the tour distance for all the solutions in the population. For each chromosome, it will calculate the distance from the starting city all the way to the last city and back to the starting city, as shown in lines 191–195. Compared to the goal of the one-max problem, which is to maximize the total value, the objective of the TSP is to minimize the total value. For the operator tournament_select(), only a minor change in line 218 is required, i.e., use "<" instead of ">." In other words, a new solution will be considered as a better solution if the fitness value of the new solution is less than the best fitness value of the selected chromosome for this tournament. As shown in lines 228–275, the MOPC can be divided into two major parts: a one-point crossover followed by a solution corrector, as depicted in Fig. 12.10. The first part consists of steps 1 and 2. In this part, the one-point crossover operator will be used to exchange parts of genes between pairs of chromosomes, e.g., p_1 and p_2. This process will, however, also create genes that are redundant in the child chromosomes, such as "3" and "4" for child o_1 and "5" and "6" for child o_2. The second part is composed of steps 3 and 4. In step 3, the crossover operator finds cities that are not in the child solutions o_1 and o_2. In step 4, these cities are used to randomly replace genes that are redundant in the child solutions. For example, in step 4, the seventh and eighth genes of o_1 are randomly replaced by "6" and "5" that are found in step 3. These steps show exactly how this crossover operator avoids creating an infeasible solution during the convergence process.

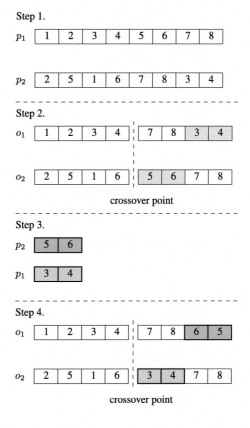

Figure 12.10 The basic idea of the modified one-point crossover (MOPC).

It is time now for the implementation of MOPC. The first part, as shown in lines 238–241, is the one-point crossover, which is responsible for exchanging genes—in the range of the crossover point to the end of the chromosome—in each pair of chromosomes if the random value f is less than or equal to the crossover rate cr. The second part, as shown in lines 244–270, first finds the number of times each city was visited, which can be 0, 1, or 2, i.e., not visited, visited once, and visited twice, and put cities not visited in bag after the one-point crossover is performed, as shown in lines 245–258. This is step 3 of Fig. 12.10. The crossover operator will then correct cities visited twice by replacing one of them (the one in the range swapped [line 241], to be more specific) by one randomly chosen from bag, if necessary, as shown in lines 263–269. This is step 4 of Fig. 12.10. As shown in lines 277–291, the mutation operator mutate() is designed specifically for the TSP. The basic idea of this operator is to randomly select two subsolutions (i.e., cities) from the solution (i.e., visiting or routing sequence) and then exchange both of them with a predefined probability. This implies that each solution has a certain chance to randomly swap two subsolutions to disrupt its visiting order to escape from a local optimum.

12.3.4 Adding new crossover operators

The other versions of crossover—PMX, CX, and OX2—are implemented in Listing 12.2, Listing 12.3, and Listing 12.4, respectively. As shown in these implementations, all we have to do is to realize these operators and reuse the other parts of Listing 12.1.

Listing 12.2: src/c++/3-tsp/main/gapmx.h: The implementation of PMX of GA for TSP.

```
 1 #ifndef __GAPMX_H_INCLUDED__
 2 #define __GAPMX_H_INCLUDED__
 3
 4 #include "ga.h"
 5
 6 class gapmx: public ga
 7 {
 8 public:
 9     using ga::ga;
10
11 private:
12     population crossover(const population& curr_pop, double cr) override;
13 };
14
15 inline population gapmx::crossover(const population& curr_pop, double cr)
16 {
17     population tmp_pop = curr_pop;
18
19     const int mid = curr_pop.size()/2;
20     for (int i = 0; i < mid; i++) {
21         const double f = static_cast<double>(rand()) / RAND_MAX;
22         if (f <= cr) {
```

```
23          // 1. select the two parents and offspring
24          const int p[2] = { i, i+mid };
25
26          // 2. select the mapping sections
27          size_t xp1 = rand() % (curr_pop[0].size() + 1);
28          size_t xp2 = rand() % (curr_pop[0].size() + 1);
29          if (xp1 > xp2)
30              swap(xp1, xp2);
31
32          // 3. swap the mapping sections
33          for (size_t j = xp1; j < xp2; j++)
34              swap(tmp_pop[p[0]][j], tmp_pop[p[1]][j]);
35
36          // 4. fix the duplicates
37          for (int k = 0; k < 2; k++) {
38              const int z = p[k];
39              for (size_t j = 0; j < curr_pop[0].size(); j++) {
40                  if (j < xp1 || j >= xp2) {
41                      int c = curr_pop[z][j];
42                      size_t m = xp1;
43                      while (m < xp2) {
44                          if (c == tmp_pop[z][m]) {
45                              c = curr_pop[z][m];
46                              m = xp1; // restart the while loop
47                          }
48                          else
49                              m++; // move on to the next
50                      }
51                      if (c != curr_pop[z][j])
52                          tmp_pop[z][j] = c;
53                  }
54              }
55          }
56      }
57  }
58
59    return tmp_pop;
60  }
61
62 #endif
```

The implementation of PMX is shown in lines 15–60 of Listing 12.2, which can be divided into four parts: select the parents, select the mapping sections, swap the mapping sections, and fix the duplicates. To be more specific, this operator first selects a pair of parents by saving the indices to them in the vector p, as shown in line 24. Next, it will select the mapping sections by randomly generating two numbers to indicate the beginning and the end of them, as shown in lines 27–30. Then, it will swap the selected genes (i.e., mapping sections) in parents p_1 and p_2, as shown in lines 33–34. Finally, it will recover genes that are outside the mapping sections but are a duplication of those inside the mapping sections, if necessary, as shown in lines 37–55. The check in line 40 excludes genes that are inside the mapping section. After that, it will find genes duplicated in the chromosome. As shown in lines 42–50, this operator will then try to find the mapping gene to replace the gene that is duplicated in the chromosome, such as

$5{\to}7$ for o_1 in step 4 of Fig. 12.6. The while loop is used to find mapping genes that may form a chain, such as the sequence $4{\to}6{\to}8$ for o_1 in step 4 of Fig. 12.6.

Listing 12.3: src/c++/3-tsp/main/gacx.h: The implementation of CX of GA for the TSP.

```
1 #ifndef __GACX_H_INCLUDED__
2 #define __GACX_H_INCLUDED__
3
4 #include "ga.h"
5
6 class gacx: public ga
7 {
8 public:
9     using ga::ga;
10
11 private:
12     population crossover(const population& curr_pop, double cr) override;
13 };
14
15 inline population gacx::crossover(const population& curr_pop, double cr)
16 {
17     population tmp_pop = curr_pop;
18
19     const int mid = curr_pop.size()/2;
20     const size_t ssz = curr_pop[0].size();
21     for (int i = 0; i < mid; i++) {
22         const double f = static_cast<double>(rand()) / RAND_MAX;
23         if (f <= cr) {
24             vector<bool> mask(ssz, false);
25             const int c1 = i;
26             const int c2 = i+mid;
27             int c = tmp_pop[c1][0];
28             int n = tmp_pop[c2][0];
29             mask[0] = true;
30             while (c != n) {
31                 for (size_t j = 0; j < ssz; j++) {
32                     if (n == tmp_pop[c1][j]) {
33                         if (mask[j])
34                             goto swap;
35                         c = n;
36                         n = tmp_pop[c2][j];
37                         mask[j] = true;
38                         break;
39                     }
40                 }
41             }
42     swap:
43             for (size_t j = 0; j < ssz; j++) {
44                 if (!mask[j])
45                     swap(tmp_pop[c1][j], tmp_pop[c2][j]);
46             }
47         }
48     }
49
50     return tmp_pop;
51 }
52
53 #endif
```

The implementation of CX is shown in lines 15–51 of Listing 12.3. In this operator, the variables c1 and c2 are used as the indices of the two parents (p_1 and p_2), as shown in lines 25–26. The variables c and n are used to hold the values of the genes of p_1 and p_2 at the same locus. As shown in lines 27–28, they initially hold the values of the first genes of p_1 and p_2, i.e., the values in tmp_pop[c1][0] and tmp_pop[c2][0]. As shown in lines 30–41, this operator will then seek both parents p_1 and p_2 together to find genes that form a cycle, like the cycle set described in Fig. 12.7. Moreover, we use the variable mask to keep track of genes that are in the cycle set. Finally, this operator will exchange genes at the same locus of the two parents but not in the cycle set, as shown in lines 43–46.

Listing 12.4: src/c++/3-tsp/main/gaox.h: The implementation of OX2 of GA for the TSP.

```
 1 #ifndef __GAOX_H_INCLUDED__
 2 #define __GAOX_H_INCLUDED__
 3
 4 #include "ga.h"
 5
 6 class gaox: public ga
 7 {
 8 public:
 9     using ga::ga;
10
11 private:
12     population crossover(const population& curr_pop, double cr) override;
13 };
14
15 inline population gaox::crossover(const population& curr_pop, double cr)
16 {
17     population tmp_pop = curr_pop;
18
19     const int mid = curr_pop.size()/2;
20     const size_t ssz = curr_pop[0].size();
21     for (int i = 0; i < mid; i++) {
22         const double f = static_cast<double>(rand()) / RAND_MAX;
23         if (f <= cr) {
24             // 1. create the mapping sections
25             size_t xp1 = rand() % (ssz + 1);
26             size_t xp2 = rand() % (ssz + 1);
27             if (xp1 > xp2)
28                 swap(xp1, xp2);
29
30             // 2. indices to the two parents and offspring
31             const int p[2] = { i, i+mid };
32
33             // 3. the main process of ox
34             for (int k = 0; k < 2; k++) {
35                 const int c1 = p[k];
36                 const int c2 = p[1-k];
37
38                 // 4. mask the genes between xp1 and xp2
39                 const auto& s1 = curr_pop[c1];
40                 const auto& s2 = curr_pop[c2];
41                 vector<bool> msk1(ssz, false);
```

```
42          for (size_t j = xp1; j < xp2; j++)
43              msk1[s1[j]] = true;
44          vector<bool> msk2(ssz, false);
45          for (size_t j = 0; j < ssz; j++)
46              msk2[j] = msk1[s2[j]];
47
48          // 5. replace the genes that are not masked
49          for (size_t j = xp2 % ssz, z = 0; z < ssz; z++) {
50              if (!msk2[z]) {
51                  tmp_pop[c1][j] = s2[z];
52                  j = (j+1) % ssz;
53              }
54          }
55      }
56  }
57 }
58
59    return tmp_pop;
60 }
61
62 #endif
```

The implementation of OX2 is shown in lines 15–60 of Listing 12.4. As far as the very first part is concerned, the implementation of OX2 is similar to all the other crossover operators, as shown in lines 25–31. It will first select the mapping sections for the parents p_1 and p_2. The variables c1 and c2 are used to represent the indices of p_1 and p_2 in the population. Similar to step 3 of Fig. 12.9, this operator has to check to see which genes in p_1' and p_2' cannot be changed again, as shown in lines 39–46. This means that the values of genes of offspring o_1 and o_2 that cannot be changed will be cloned from the values of genes in the same loci of their parents p_1' and p_2'. Finally, as shown in lines 49–54, this operator inserts genes from the other parent starting from the next locus of the mapping section, i.e., from xp2, to the end of the chromosome; it will then wrap around and insert genes again starting from the first gene of the chromosome until the start locus of the mapping section is reached.

12.4. Simulation results of GA for the TSP with different crossover operators

In this section, the TSP benchmark eil51 from TSPLIB (Reinelt, 1991) is used to evaluate the performance of different crossover operators of GA for the TSP. To better understand the performance of each crossover operator, it will be evaluated from different perspectives, which include the quality of end results, the impact of the crossover rate, and the computation time. As all simulations described in the previous chapters, all results are average values of 100 runs. Moreover, in this chapter, each run is carried out for 120,000 evaluations, while the number of chromosomes is set equal to 120, the crossover rate (c_r) is set to 20% (0.2), 40% (0.4), 60% (0.6), and 80% (0.8), and the mutation rate is set to 10% (0.1). Note that the mutation operator described in this chapter is different from that described in Section 7.2, in the sense that the mutation rate here is the probability of exchanging two subsolutions in the same chromosome (i.e., to be done once and only once) while the mutation rate in Section 7.2 is the probability of mutating each subsolution in a chromosome (i.e., to be done for each subsolution in the chromosome). Note that OX here represents the crossover operator OX2.

Table 12.1 compares the performance of these different crossover operators in terms of the total distance, where c_r denotes the crossover rate and m_r denotes the mutation rate. The results show that the crossover operator has a strong impact on the search performance of a metaheuristic algorithm even though the same parameter settings are used. Of course, the results further show that the crossover rate for the crossover operators will also affect the final results of GA for the TSP. For example, for OX, the results worsen when the crossover rate is increased from 0.2 to 0.8.

Table 12.1 Comparison between MOPC, PMX, CX, and OX with different crossover rates in terms of the total distance.

Crossover	MOPC	PMX	CX	OX
$c_r = 0.2$, $m_r = 0.1$	581.889	585.814	575.192	485.860
$c_r = 0.4$, $m_r = 0.1$	583.889	568.008	580.724	476.585
$c_r = 0.6$, $m_r = 0.1$	586.969	574.673	586.615	479.649
$c_r = 0.8$, $m_r = 0.1$	568.678	570.225	579.546	544.232

Figs. 12.11, 12.12, 12.13, and 12.14 further show the convergence status of these crossover operators. It can be seen that the convergence process of OX is faster than those of other crossovers from the very beginning. Fig. 12.11 shows that GA with OX yields better results than the other crossover operators when the crossover rate is set equal to 20% and that the final result of GA with OX is less than 500 in terms of the objective value, i.e., the total distance of a tour (solution). The results further show that it can find a better result by increasing the number of evaluations. Other test results can be found in Fig. 12.12, which shows that GA with OX also yields better results than the other crossover operators when the crossover rate is set equal to 40%. The final result of GA with OX is less than 500, and it can find better results by increasing the number of evaluations. Different from OX, the convergence curves of MOPC, PMX, and CX

are quite close to each other, and the results of CX are worse than those of MOPC and PMX, as shown in Fig. 12.12. If we change the crossover rate of these crossover operators to 60%, the results of GA with different crossover operators are very similar to those of GA with crossover rates of 20% and 40%, as shown in Fig. 12.13.

Fig. 12.14 compares the performance of MOPC, PMX, CX, and OX with the crossover rate set equal to 80%. MOPC, PMX, and CX achieve results that are similar to those obtained using other crossover rates. Table 12.1 also shows that the convergence speed of GA with OX at a crossover rate of 80% is worse than that of OX with the crossover rate set equal to 20%, 40%, or 60%. From these results, especially from the results of OX, we conclude that the crossover rate of 40% gives much better results than the other crossover rates. This illustrates that a suitable setting of the parameters is very important for metaheuristic algorithms.

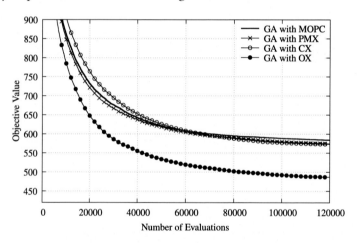

Figure 12.11 Simulation results of MOPC, PMX, CX, and OX with a crossover rate of 20% and a mutation rate of 10%.

Table 12.2 Computation time of MOPC, PMX, CX, and OX in seconds.

Crossover	MOPC	PMX	CX	OX
$c_r = 0.2$, $m_r = 0.1$	3.037	2.146	2.030	2.270
$c_r = 0.4$, $m_r = 0.1$	4.618	3.007	2.347	2.917
$c_r = 0.6$, $m_r = 0.1$	5.540	2.856	2.441	3.390
$c_r = 0.8$, $m_r = 0.1$	7.662	3.806	2.917	3.905

In addition to the end results of GA with different crossover operators, the computation time is also a critical issue. Table 12.2 compares

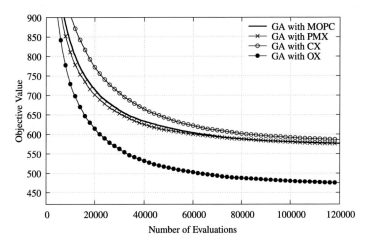

Figure 12.12 Simulation results of MOPC, PMX, CX, and OX with a crossover rate of 40% and a mutation rate of 10%.

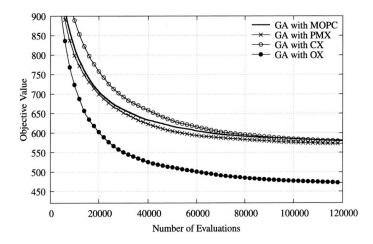

Figure 12.13 Simulation results of MOPC, PMX, CX, and OX with a crossover rate of 60% and a mutation rate of 10%.

these four crossover operators in terms of the computation time for the TSP benchmark eil51 with the crossover rate set equal to 20% to 80% and the mutation rate set equal to 10%. The results depicted in this table show that PMX, CX, and OX are faster than MOPC, especially when the crossover rate is enlarged to, say, $c_r = 0.8$. According to our observations, this is because MOPC requires a lot of computation time to recover infeasible solutions. The computation time of MOPC increases when increasing

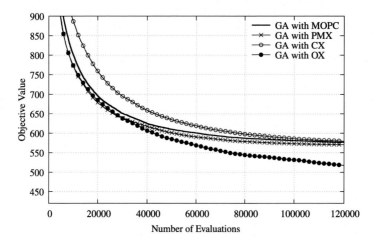

Figure 12.14 Simulation results of MOPC, PMX, CX, and OX with a crossover rate of 80% and a mutation rate of 10%.

the crossover rate because much more infeasible solutions will be created, and hence MOPC requires much more computation time to recover. The results show that a good transition operator, in the sense that it can avoid creating infeasible solutions during the convergence process, is very useful in reducing the computation time of metaheuristic algorithms for solving complex optimization problems.

12.5. Discussion

This chapter first presents three major reasons why we need to re-design a transition operator for metaheuristics that may be required by a different encoding method to avoid creating infeasible solutions or to enhance the search performance. Three well-known crossover operators and an MOPC operator are used as examples to show that the search strategy of GA in solving the TSP can be changed. The simulation results show that a good transition operator (e.g., OX) with suitable parameter settings may be employed to improve the quality of the end results. The results further show that a transition operator that creates infeasible solutions during the convergence process may reduce the search performance of a metaheuristic algorithm. According to our observations, for some optimization problems (especially scheduling problems), the original transition operator of metaheuristic algorithms often creates infeasible solutions during the convergence process; therefore, most studies using metaheuristic algorithms for scheduling problems either design a transition operator that will not create infeasible solutions or design a good recovery mechanism to solve this problem. In summary, the transition operator may have a strong impact on the search performance of a metaheuristic algorithm in terms of the end results and the computation time. This problem exists not only in GA, but in most metaheuristic algorithms. How to use domain knowledge of the optimization problem in question to enhance the performance of a transition operator is an interesting topic that is worth investigating to enhance the performance of metaheuristic algorithms. This chapter aims to explain how the transition operator may affect the performance and results of a metaheuristic algorithm. Of course, there are many ways to design and develop a better and more suitable transition operator for metaheuristic algorithms, as can be found in the discussions of SA in Section 5.4, TS in Section 6.4, GA in Section 7.4, ACO in Section 8.4, PSO in Section 9.4, and DE in Section 10.4. The strategies—using a historical memory to accumulate the searched information, using an adaptive or self-adaptive mechanism to adjust the transition operator, and using a predefined schedule to adjust the transition operator at different periods during the convergence process—provide some effective ways to generate new candidate solutions that are worthy of investigation.

Supplementary source code

Here are the hyperlinks to the directories where the source code described in this chapter resides.

https://github.com/cwtsaiai/metaheuristics_2023/src/c++/3-tsp/

CHAPTER THIRTEEN

Evaluation and determination operators

Although the evaluation and determination operators are two different operators of a metaheuristic algorithm, their design is typically inseparable, because the determination operator typically refers to information provided by the evaluation operator. Therefore, these two operators are discussed together in this chapter. In addition, the schema theorem and fitness landscape analysis are discussed to show two advanced analysis technologies for understanding the performance of a metaheuristic algorithm in solving an optimization problem.

13.1. Evaluation operator

Using an objective function $f()$ to get the objective value f_s of a solution s for an optimization problem $\mathbb{P}$ provides a *direct* way to measure the quality of a solution; thus, some studies used objective functions as evaluation operators of metaheuristic algorithms, such as simulated annealing (SA) and tabu search (TS), to evaluate the searched solutions during the convergence process. Based on the return values of the objective function for the searched solutions, a metaheuristic algorithm is then able to compare the qualities of searched solutions to determine the directions or regions that will be searched at later iterations. For example, when using SA to solve an optimization problem, the objective values f_s and f_v of the current solution s and the new candidate solution v will be compared in each iteration. The comparison result can then be used by SA to determine the better solution and the difference between s and v to decide to search around s or v in the next iteration.

Unlike most single-solution-based metaheuristic algorithms that only need to compare the current solution and the new candidate solution in terms of their objective values, population-based metaheuristic algorithms typically need to evaluate the qualities of multiple searched solutions at each iteration; thus, it might be difficult to discriminate "good solutions" from these solutions by their objective values in some situations. Note that if the objective values of searched solutions have to be transformed to another set of values for a metaheuristic algorithm, these indirect objective values are also called fitness values, and the function used to transform the objective values to fitness values is referred to as the fitness function in some metaheuristic algorithms. If a metaheuristic algorithm considers and compares only the objective values of searched solutions, it may always select the best one of them to the next iteration and the searches of this metaheuristic algorithm will converge very quickly. If a metaheuristic algorithm uses the ratio of the objective value of a searched solution to the objective values of all searched solutions, such as the roulette wheel selection operator of the genetic algorithm (GA), the probabilities of good and worse searched solutions selected for next iteration might be very close to each other. For these reasons, we need to find a better way to "evaluate the searched solutions" and to "determine the good search directions." A critical reason is that if the objective values from an evaluation operator for the searched solutions are too close to each other, then the determination operator may have trouble judging whether a solution is good or not compared to the other searched solutions. We may need to find a way to transform the objective values of the searched solutions from one space to another to make the measurement in question easier. For instance, suppose the objective values of the five chromosomes of GA at a generation are 1.1, 1.2, 2.4, 2.6, and 2.7, and the survival probabilities of each chromosome computed using roulette wheel selection (as shown in Eq. (7.1), $p_i = f_i / \sum_{i=1}^{m} f_i$, where f_i represents the objective value of a chromosome and m represents the number of chromosomes) are 11%, 12%, 24%, 26%, and 27%, respectively, for the maximization optimization problem. The survival probabilities of the fourth and fifth chromosomes are $p_4 = 2.6/(1.1 + 1.2 + 2.4 + 2.6 + 2.7) = 2.6/10.0 = 26\%$ and $p_5 = 2.7/10.0 = 27\%$, respectively, which mean that the survival probabilities of the last three chromosomes are eventually very close to each other. This example illustrates that this kind of survival probability for the determination mechanism is not useful in discriminating the searched solutions whose objective or fitness values are close to each other when a difference as small as 0.1 in the objective values eventually implies a very significant difference in the searched solutions. An intuitive way to solve this problem can be found in the design of tournament selection of GA, in which the objective value of each chromosome is transformed to rank. For the same example, the objective values of the five chromosomes 1.1, 1.2, 2.4, 2.6, and 2.7 can be transformed to the space called ranks 5, 4, 3, 2, and 1 as the fitness values for the maximization optimization problem. This transformation makes it easy for the selection (i.e., determination) operator to discriminate the searched solutions, and the last chromosome will have a high probability to be selected for the next generation.

Since the objective value f_s of a searched solution s does not contain information to depict the relationship and the difference between solution s and the other searched solutions, the searches of a metaheuristic algorithm might trend to similar directions or regions during the convergence process, thus reducing the search diversity. This means that the use of only the objective values of searched solutions in the determination operator of a metaheuristic algorithm may not be able to avoid searching similar directions or regions. This issue is apparent in some population-based metaheuristic algorithms because preventing multiple search directions from trending to particular directions is critical to keep a metaheuristic algorithm from falling into local optima; therefore, for some metaheuristic algorithms, the calculation method of the determination operator was even changed dynamically during the convergence process (Jin, 2005; Ratnaweera et al., 2004). These two reasons explain why for some metaheuristic algorithms, an *indirect* way for the determination operator to evaluate solutions was developed.

To enhance the performance of the fitness function, Jin (Jin, 2005) gave a brief review of approximate models of fitness functions for two major reasons: (1) the explicit fitness function does not exist and (2) the computation time of an explicit fitness function is very long. How to develop an approximate evaluation method for the determination operator is thus still a critical research topic for metaheuristic algorithms. A good lightweight fitness function may be useful to reduce the computation time of an evaluation operator while an approximate fitness function will be useful to reduce the risk of misleading the search direction for the determination operator. As shown in Fig. 13.1, a possible way to reduce the computation time of an evaluation operator is not to always employ the original fitness function $f()$ to evaluate a solution but also using the approximate fitness function $f_a()$ (Jin, 2005). Note that by the approximate fitness function, we mean a lightweight method to calculate the fitness function. For example, for the one-max problem, assuming that the solution is 1,000,000 bits long, it is obviously expensive to evaluate each solution. An approximate method using a sampling method can be used to enhance the performance of a fitness function. This approximate fitness function $f_{samples}(s)$ will randomly select only 0.1% of the subsolutions to decide the number of subsolutions that assumes the value "1." Then $f_{samples}(s) \times 1000$ will be the approximate fitness function value of solution s. Since $f_{samples}()$ can be used as the approximate fitness function $f_a() = f_{\text{samples}}()$, it will be used for all the solutions, but the best one based on the evaluation result of the approximate fitness function $f_a()$ will be reevaluated by $f()$. In this way, the computation time of the fitness function for all solutions can be significantly reduced if the approximate fitness function takes less time to evaluate a solution than the original fitness function.

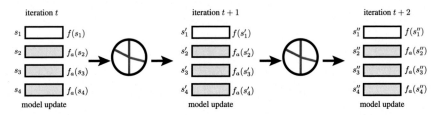

Figure 13.1 How to use the original and approximate fitness functions at the same generation during the convergence process.

Another possible way can also be found in (Jin, 2005). As shown in Fig. 13.2, it employs both the approximate and original fitness functions to measure the fitness values of all the solutions by repeatedly invoking these two functions until the termination condition is met, as follows: first the approximate fitness function for a certain number of iterations and then the original fitness function once. The example given in Fig. 13.2 shows that all the solutions will be evaluated by $f_a()$ at iterations t and $t+1$, and then the original fitness function will be employed to evaluate all the solutions once at iteration $t+2$. A different approach is to alternate between the original and approximate fitness functions to measure all the solutions with a fixed percentage or period. Some other studies (Jin et al., 2001; Nain and Deb, 2003) even attempted to dynamically adjust the frequency of the original and approximate fitness functions during the convergence process.

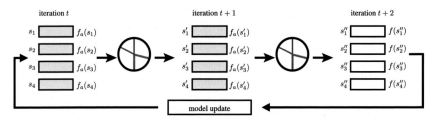

Figure 13.2 How to use the original and approximate fitness functions at different generations during the convergence process.

Different from changing the way to use the transition, evaluation, and determination operators for a metaheuristic algorithm to adjust the convergence process, another promising way is to modify the evaluation operator to make it possible to dynamically adjust the fitness function or calculate the fitness value during the convergence process. Dahal et al. (Dahal et al., 2008) discussed some variable fitness functions. A simple way can be found in (Remde et al., 2008). The fitness values of solutions of a metaheuristic algorithm are changed at different iterations by

$$f_v(s, t) = \sum_{i=1}^{n} s_i w_i(t),$$ (13.1)

where s is a solution consisting of n subsolutions, each of which, denoted s_i, is associated with a weight $w_i(t)$, as shown in Fig. 13.3. All the weights will be changed iteration by iteration with a predefined schedule for the weights.

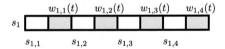

Figure 13.3 How the solution of a variable fitness function is encoded.

Sallam et al. (Sallam et al., 2017) further explained that the landscape information of a problem and the history of the fitness values of operators can also provide some insight for selecting a different transition operator to enhance the search performance of differential evolution (DE). In this study, to know which operator can improve the searched solutions, the search ability for each particular operator is calculated as

$$s_{op(i)} = \frac{op(i)^s}{op(i)_n},$$ (13.2)

where $op(i)^s$ represents the number of searched solutions whose fitness value can be improved by using the i-th operator $op(i)$, while $op(i)_n$ represents the total number of solutions by using this operator. With this information, DE can then pick the approximate search operator to enhance the search performance during the convergence process. Some similar ways to dynamically change the search strategies of metaheuristic algorithms have been discussed in several studies (Ratnaweera et al., 2004; Taherkhani and Safabakhsh, 2016) because they can provide better results than a fixed or single search strategy for solving complex optimization problems. A simple example can be found in (Ratnaweera et al., 2004), in which Ratnaweera et al. presented a dynamic adjustment method to adjust the acceleration coefficients of particles for particle swarm optimization (PSO) during the convergence process. This provides a better result than traditional PSO, which uses fixed acceleration coefficients. Since some of the designs can also be found in the determination operator of a metaheuristic algorithm, we will turn our discussion to the improved determination operator in the next section.

13.2. Determination operator

How to develop an effective or efficient way to improve the search ability of a metaheuristic algorithm for a particular optimization problem has become a critical research topic. The focus of this section is thus on the determination operator of a metaheuristic algorithm and ways to adjust the search strategy for SA, TS, GA, ant colony optimization (ACO), PSO, and DE.

13.2.1 Determination operator for single-solution-based metaheuristic algorithms

As mentioned in Chapter 5, the convergence process of SA is based on the probabilistic acceptance criterion P_a and the annealing schedule. Fig. 13.4 shows that the basic idea of the cooling schedule for the probabilistic acceptance criterion P_a of SA is to take into account the objective values of current and new solutions (f_1 and f_2) and the current temperature to compute the probability P_a of accepting the improved solution and whether to accept non-improved solutions at the next iteration. Since the annealing (cooling) schedule is an essential component of SA that is involved in the determination of the search strategy, several annealing schedules have been presented in previous studies (Luke, 2018; Schneider and Puchta, 2010; Suman and Kumar, 2006). Today, we can use some online tools to compare and display the difference between two different annealing schedules, such as Wolfram Alpha (Wolfram Alpha, 2009).

Fig. 13.5 shows three different annealing schedules (Luke, 2018; Schneider and Puchta, 2010) for adjusting the search strategy of SA for 1000 iterations. Annealing schedule 1 is based on Eq. (5.5) given in Section 5.1

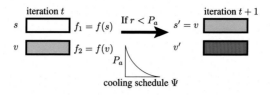

Figure 13.4 How the next solution to be searched is determined by simulated annealing.

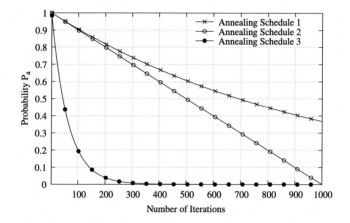

Figure 13.5 Comparison of three cooling schedules for simulated annealing.

and is defined as

$$\Psi_t = \Psi_{t-1} \times \alpha = \Psi_{\text{init}} \times \alpha^t, \tag{13.3}$$

where Ψ_{init} denotes the initial temperature and Ψ_t denotes the temperature at iteration t. Here, Ψ_{init} is set equal to 1.0 and α is set equal to 0.999. Annealing schedules 2 and 3 are defined, respectively, as

$$\Psi_t = \Psi_{\text{init}} - t \times \left(\frac{\Psi_{\text{init}} - \Psi_{\text{max}}}{n} \right), \tag{13.4}$$

$$\Psi_t = \Psi_{\text{init}} \times \left(\frac{\Psi_{\text{max}}}{\Psi_{\text{init}}} \right)^{\frac{t}{\Psi_{\text{max}}}}, \tag{13.5}$$

where Ψ_{max} denotes the temperature at the maximum iteration. In addition to the simple approach of annealing schedule 1, which updates the temperature by using a fixed proportional rate, annealing schedules 2 and 3 provide two more annealing schedules, linear and proportional, that make it possible for SA to construct the probability P_a in different ways. The

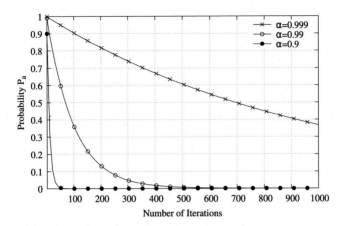

Figure 13.6 Comparison of three different values of α for cooling schedule 1 for simulated annealing.

linear annealing schedule will linearly decrease the probability of accepting the non-improved solution as the number of iterations increases; the proportional annealing schedule will, however, decrease P_a very quickly in the early period but very slowly in the later period.

Fig. 13.6 further shows that we can also change the search strategy of SA by adjusting the value of α for annealing schedule 1. In this example, the values of α are set equal to 0.999, 0.99, and 0.9. From these comparisons, it can be easily seen that P_a will decrease very quickly when the value of α is shrinked. Of course, these are not the only annealing schedules for SA; in fact, a lot of other annealing schedules have been presented (Luke, 2018; Schneider and Puchta, 2010; Suman and Kumar, 2006) (Section 5.4), which provide us different options to adjust the search strategy of SA.

TS is another kind of single-solution-based metaheuristic algorithm. Its determination operator also compares the objective values of current and new solutions (f_1 and f_2), just like SA. If the new solution v is better than the current solution s, it will also take into account the solutions in the tabu list to avoid searching the same solution or a particular region too frequently, as shown in Fig. 13.7. In the studies of Glover (Glover and Laguna, 1997) and others (Gendreau and Potvin, 2005; Glover and Kochenberger, 2003; Taillard, 2016), the short-term memory, long-term memory, diversification, and intensification have been discussed several times; they provide a systematic perspective for the design of a metaheuristic algorithm. The short-term memory can typically be regarded as the tabu list, the purpose of which is to prevent a metaheuristic algorithm from searching the solutions that have been recently searched again and again. The long-term memory can be regarded as a knowledge base of diversification rules to guide the search into previously unexplored regions in the solution space.

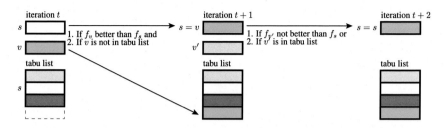

Figure 13.7 How the next solution to be searched is determined by tabu search.

The strategies of diversification and intensification, as the names suggest, are aimed to diversify and intensify the search; that is, to search more regions and to search a potential region more intensively. These two strategies are usually the "two ends of a scale," the balance point of which has always been an open issue for all the metaheuristics. The tabu list is an essential component for determining the search strategy of TS. In addition to the fixed length tabu list, varying the length of a tabu list during the convergence process and using multiple tabu lists are two other useful strategies that make it possible to keep more information to enhance the search performance of TS (Gendreau and Potvin, 2005).

13.2.2 Determination operator for population-based metaheuristic algorithms

As shown in Figs. 13.8 and 13.9, the roulette wheel selection and tournament selection operators are two representative selection operators that can be regarded as the determination operators of GA. The underlying ideas in the design of these selection operators have been discussed in Chapter 7.

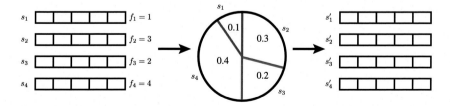

Figure 13.8 How roulette wheel selection works.

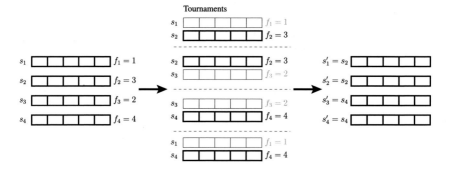

Figure 13.9 How tournament selection works.

Assuming that there are four solutions $s = \{s_1, s_2, s_3, s_4\}$, Fig. 13.8 shows that the roulette wheel selection operator will use the fitness values of these solutions to construct a probability model and then use this model to pick four solutions from s to get $s' = \{s'_1, s'_2, s'_3, s'_4\}$. As shown in Fig. 13.9, the tournament selection operator will randomly select t_o solutions (called players) and find the best of them as s'_i. This procedure is typically referred to as a tournament. In this example, four tournaments, each of which contains two solutions, will be performed by the tournament selection operator to get s'. Goldberg and Deb (Goldberg and Deb, 1991) gave a systematic comparison of the selection operators, including the time complexities. The results show that the time complexities of the roulette wheel and tournament selection operators are $O(n^2)$ and $O(n)$, respectively. This explains why roulette wheel selection is usually slower than tournament selection in practice. There are certainly other ways to enhance the performance of roulette wheel selection. For each new solution s'_i, the roulette wheel selection operator needs to compare a random number with the probability model to locate the correct slot from n slots, and this procedure will be performed n times; thus, it takes time $O(n^2)$. However, if a binary search is used to locate the correct slot, then its time complexity can be reduced to $O(n \log n)$. Even after this reduction, the time complexity of roulette wheel selection is still higher than that of tournament selection.

Although the tournament selection operator seems to perform better than the roulette wheel operator in terms of the time complexity and convergence speed, there still are some dilemmas that we have to take into consideration when using it for GA or other metaheuristic algorithms. A simple example can be found in Fig. 13.10, where it is assumed that GA uses only the tournament selection operator but not any other transition operators, such as crossover and mutation, to change the chromosomes. This example shows that a chromosome with the worst fitness value will be omitted by the tournament selection operator at every generation. This characteristic of the tournament selection operator will make all the chromosomes become the same after a certain number of generations. Although the chromosomes may be changed by the transition operators, a similar situation will still occur for some of the chromosomes in the population. From these observations, it can be easily seen that the tournament selection operator has the risk of leading GA into local optima, especially at later stages of the convergence process, because it will keep removing the chromosome with the worst fitness value during the convergence process.

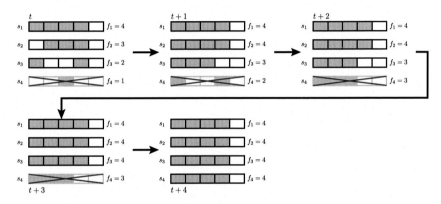

Figure 13.10 An example of tournament selection without any transition operators.

Another well-known selection operator is the ranking selection operator (Hancock, 1994). The linear and exponential ranking selection operators are two well-known ranking mechanisms. For linear ranking selection, the fitness values of all the chromosomes can be computed as follows:

$$f(s_i) = \begin{cases} r_s, & \text{if } s_i \text{ is the best solution of } s, \\ 2 - r_s, & \text{if } s_i \text{ is the worst solution } s, \\ r_s - \frac{2(\omega_i - 1)(r_s - 1)}{m - 1}, & \text{otherwise,} \end{cases} \qquad (13.6)$$

where r_s is a predefined value, ω_i is the rank of the i-th chromosome s_i, and m is the number of chromosomes in the population. If s_i is the best one in the population s in terms of the objective value, its fitness value will be r_s, which assumes a value in the range $(1, 2]$. Here, its value is set equal to 2. If s_i is the worst one, its fitness value will be $2 - r_s$. If r_s is set equal to 2, its fitness value will be $2 - 2 = 0$, which means that it has no chance to survive to the next generation. For all the other chromosomes, the ones ranked from 2 to $m - 1$, the fitness value will be calculated as defined above and will fall in the range $(0, 2)$. Suppose $m = 4$ and $r_s = 2$. Now, if the rank of s_i is 2, then its fitness value will be $2 - 2(2 - 1)(2 - 1)/(4 - 1) = 2 - 2/3 = 4/3$; if the rank of s_i is 3, then its fitness value will be $2 - 2(3 - 1)(2 - 1)/(4 - 1) = 2 - 4/3 = 2/3$. The fitness values of the best and worst chromosomes will be 2 and 0, respectively, again as defined above.

As for exponential ranking selection, it will adjust the fitness value of each solution s_i based on its rank ω_i defined by

$$f(s_i) = r_s^{(\omega_i - 1)}, \tag{13.7}$$

where r_s is a predefined value. Again, ω_i will be assigned 1 when s_i is the best chromosome, and it will be assigned m when s_i is the worst chromosome. For this example, provided that $r_s = 0.99$, the fitness value of the best solution will be $0.99^{(1-1)} = 0.99^0 = 1$. The fitness values of the second, third, and fourth best solutions will be $0.99^1 = 0.99$, $0.99^2 = 0.9801$, and $0.99^3 = 0.970299$, respectively. Based on these adjusted fitness values, the probability model of the ranking selection can be constructed, just like in roulette wheel selection. The probability p_i to select s_i is computed by

$$p_i = \frac{r_s^{(\omega_i - 1)}}{\sum_{i=1}^{m} r_s^{(\omega_i - 1)}}, \tag{13.8}$$

where m is the number of chromosomes in the population and the denominator $\sum_{i=1}^{m} r_s^{(\omega_i - 1)}$ is used to normalize the probabilities p_i so that $\sum_{i=1}^{m} p_i = 1$. A simple example is given in Fig. 13.11 to illustrate the exponential ranking selection. To increase the differences of the fitness values of

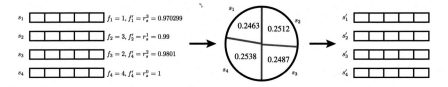

Figure 13.11 How ranking selection works.

the solutions, all we have to do is to decrease the value of r_s. For example, provided that $r_s = 0.5$, the fitness values of these four solutions will be 1 for s_4, 0.5 for s_2, 0.25 for s_3, and 0.125 for s_1.

Fig. 13.12 shows another representative selection operator for the so-called $(\mu + \lambda)$ and (μ, λ) evolution (overlapping generation) strategies (Al-rashdi and Sayyafzadeh, 2019; Hancock, 1994), where μ is the number of parents selected, $s_a = \{s_1, s_2, \ldots, s_\mu\}$, and λ is the number of offspring, $s_b = \{s'_1, s'_2, \ldots, s'_\lambda\}$. For $(\mu + \lambda)$, μ usually refers to the top μ current solutions and λ refers to the offspring generated by these parents via the transition operator. In this example, μ is set equal to 2 and λ is set equal to 4. After combining the parents and offspring, this selection operator (also called the truncation selection operator) can sort their fitness values to select the top m solutions for the next iteration.

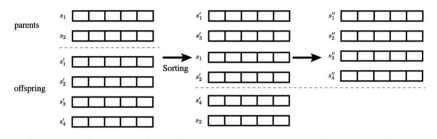

Figure 13.12 How evolution strategy selection works.

As a member of the family of evolutionary algorithms, DE (Das and Suganthan, 2011; Storn and Price, 1997) uses a similar tournament selection operator to select better searched solutions for the next generation. For the maximization problem, the determination operator of DE can be defined as follows:

$$s_i^{t+1} = \begin{cases} u_i, & \text{if } f(u_i) \geq f(s_i^t), \\ s_i^t, & \text{otherwise,} \end{cases} \tag{13.9}$$

where s_i^t denotes the i-th solution of generation t, s_i^{t+1} denotes the i-th solution of generation $t + 1$, u_i denotes the i-th offspring created by the mutation and crossover operators of DE, $f(s_i^t)$ denotes the fitness value of s_i^t, $f_i^u = f(u_i)$ denotes the fitness value of u_i, and $f()$ denotes the objective function to be maximized. In brief, the above equation states that u_i will replace s_i^t at generation $t + 1$ only if it is equally good or better, i.e., $f(u_i) \geq f(s_i^t)$. As shown in Fig. 13.13, this kind of selection operator is quite similar to the tournament selection operator of GA (Fig. 13.9). The main difference is that the players involved in the DE tournament selection mechanism are the m parents and the m offspring and each tournament involves only two players, whereas the players involved in the GA tournament selection mechanism are the $k \times m$ parents, where k is the number of players in each tournament. It can be easily seen from this example that this kind of selection operator can also be regarded as a special case of the $(\mu + \lambda)$ evolution strategy.

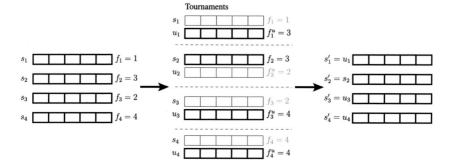

Figure 13.13 How the selection operator of DE works.

Fig. 13.14 gives an example to illustrate the basic idea of the solution construction operator of ant colony optimization. In this example, each solution s_i contains a set of subsolutions $s_{i,1}$, $s_{i,2}$, $s_{i,3}$, $s_{i,4}$, and $s_{i,5}$, each of which is constructed using the pheromone and distance information. Eq. (8.1) in Section 8.1 has been widely used to construct the probability model p to generate (pick) the candidate subsolutions one by one, as follows:

$$P^i_{\text{ACO},j,k} = \frac{[\tau_{j,k}]^\alpha [\eta_{j,k}]^\beta}{\sum_{g \in \mathbb{N}^i_j} [\tau_{j,g}]^\alpha [\eta_{j,g}]^\beta}, \tag{13.10}$$

where $P^i_{\text{ACO},j,k}$ denotes the transition probability of the i-th ant (i.e., the i-th solution) at node c_j (i.e., the j-th subsolution) choosing node c_k as the next node (i.e., the $(j + 1)$-th subsolution). Although the indices i, j, and k used in Eq. (13.10) are inconsistent with those used in Eq. (8.1), these two equations are eventually the same. For the definitions of $\tau_{j,k}$, $\tau_{j,g}$, $\eta_{j,k}$, $\eta_{j,g}$, α, β, and $\mathbb{N}^i_j$, the reader is referred to Section 8.1. This probability model is similar to that of the roulette wheel selection defined in Eq. (7.1) and denoted P_{GA} here for computing the probability of each solution. One of the most important differences between Eq. (13.10) (P_{ACO}) and Eq. (7.1) (P_{GA}) is that P_{ACO} is used to select the candidate "subsolution" of ACO but P_{GA} is used to select the candidate "solution" in most studies. P_{ACO} also takes into account the pheromone (i.e., search experience) and distance (i.e., domain knowledge of the problem) information, but P_{GA} only takes into account the fitness values of the solutions. As mentioned before, the time complexity of this kind of selection mechanism is $O(n^2)$ or $O(n \log n)$; thus, it is expected that it is computationally very expensive. That may be the reason why Dorigo and his colleagues (Dorigo and Gambardella, 1997) used the probability model of the ant colony system (ACS) to replace the probability model of the ant system (AS) (Dorigo et al., 1996) to reduce the computation time of this operator. The probability model of ACS can be easily stated as follows: Given that q_0 is a random number uniformly distributed in the range $[0, 1]$, each subsolution has a $100(1 - q_0)\%$ chance of being selected from the possible candidate subsolutions using Eq. (13.10) and a $100q_0\%$ chance of being the "best" possible candidate subsolution.

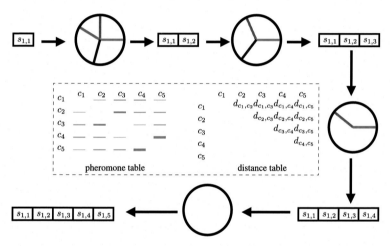

Figure 13.14 How ant colony optimization constructs its solution.

In addition to AS and ACS, there in fact exists a variety of other ant-based algorithms, such as elitist AS, Ant-Q, beam-ACO, rank-based ant system (AS_{rank}), and the max-min ant system (MMAS) (Blum, 2005a; Dorigo and Stützle, 2010). In most cases, each of them has its own determination operator. For example, AS_{rank} will first sort ants (i.e., searched solutions) based on the total lengths of tours, and then only the best $(w - 1)$ ants and the global best ant are allowed to deposit pheromone to the pheromone table. Another example is MMAS, in which the pheromone trail is limited to an interval $[\tau_{max}, \tau_{min}]$, and only the iteration-best ant and the best-so-far ant are allowed to add pheromone to its trail in the pheromone table.

The last metaheuristic algorithm we will discuss in this section is the so-called particle swarm optimization. Even though the functionality of the determination operator can be regarded as a local, global, and velocity update procedure, its convergence speed is typically very quick compared to other population-based metaheuristic algorithms. Since each particle will

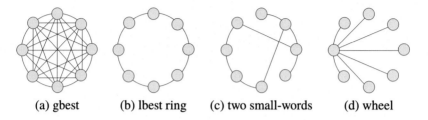

    (a) gbest   (b) lbest ring   (c) two small-words   (d) wheel

Figure 13.15 Four different topologies of particle swarm optimization.

be moved by using the information of its search experience (i.e., local best $p_{b,i}$), the best-so-far solution (i.e., global best g_b), and its original trajectory v_i^t, how to exchange the information has become a promising research topic. The topology of particles, therefore, is one of the hotspots in PSO research (Kennedy, 1999). A simple topology called gbest (Fig. 13.15(a)), in which each particle exchanges information with others, was widely used in many PSO studies. Kennedy (Kennedy, 1999) provided a brief discussion about different topologies, such as lbest ring, two small-words, and wheel topologies, as shown in Figs. 13.15(b–d), which represent different ways to exchange information between particles during the convergence process. In the gbest topology, each particle is attracted to the best solution found by any particle in the population. In the lbest topology, however, each particle is attracted to the best solution of its neighbors in the topology, where each particle has only two neighbors. From the examples of Figs. 13.15(c and d), it is easy to understand the two small-words and wheel topologies. These topologies are aimed to slow down the convergence speed of PSO because not all particles are directly attracted to the best-so-far solution at the same iteration. This means that each particle of the last three topologies will be attracted to the best neighbor solution instead of the best one of all particles so that not all particles will move toward the best particle too quickly.

    Another research trend is to modify the functionality of the determination operator of PSO by developing a better way for velocity update. The original velocity update can be found in Eq. (9.1) (Chapter 9). Two representative ways to adjust the search strategy of PSO are dynamically adjusting the inertial weight ω (Shi and Eberhart, 1999) and dynamically adjusting the acceleration coefficients c_1 and c_2 (Ratnaweera et al., 2004).

In (Taherkhani and Safabakhsh, 2016), some different ways to adjust the inertia weight were discussed, which can be divided into three categories: (1) constant and random inertial weights, (2) time-varying inertial weights, and (3) adaptive inertial weights. A representative example can be found in (Ratnaweera et al., 2004; Shi and Eberhart, 1999) in which the time-varying inertia weight (TVIW) ω will be adjusted linearly as the number of iterations increases, as follows:

$$\omega^t = \frac{t_{max} - t}{t_{max}}(\omega^{max} - \omega^{min}) + \omega^{max}, \qquad (13.11)$$

where t is the current iteration number, t_{max} is the maximum iteration number, and ω^{max} and ω^{min} are the initial and final values of ω, respectively. Now, the velocity update equation (Eq. (9.1)) can then be changed to Eq. (13.12) for adjusting the inertial weight linearly:

$$v_i^{t+1} = \omega^t v_i^t + c_1\varphi_1(p_{b,i}^t - s_i^t) + c_2\varphi_2(g_b^t - s_i^t). \qquad (13.12)$$

Ratnaweera and his colleagues (Ratnaweera et al., 2004) presented a way to dynamically adjust the acceleration coefficients, called the time-varying acceleration coefficient (TVAC), where the acceleration coefficients c_1 and c_2 are modified as follows:

$$c_1^t = (c_{1f} - a_{1i})\frac{t}{t_{max}} + c_{1i}, \qquad (13.13)$$

$$c_2^t = (c_{2f} - c_{2i})\frac{t}{t_{max}} + c_{2i}, \qquad (13.14)$$

where c_{1f}, c_{1i}, c_{1f}, and c_{1i} are four constants. These two acceleration coefficients can now be used to update the velocity as follows:

$$v_i^{t+1} = \omega v_i^t + c_1^t\varphi_1(p_{b,i}^t - s_i^t) + c_2^t\varphi_2(g_b^t - s_i^t). \qquad (13.15)$$

The simulation results described in (Ratnaweera et al., 2004) show that the strategies of dynamically adjusting the inertial weight and dynamically adjusting the acceleration coefficients are able to find a better result than the simple PSO, which uses fixed inertial weight and acceleration coefficients, for solving the function optimization problem $\mathbb{P}_6$.

13.3. Schema theorem

To better understand why the searches of most metaheuristic algorithms will trend to "good" solutions, in this section, we will use the schema theorem of Holland (Holland, 1975) to explain how a metaheuristic algorithm improves the search results gradually during the convergence process. In (Michalewicz, 1996), Michalewicz explained and emphasized that the schema theorem illustrates that some particular subsolutions $\mathbb{S}$ of all the chromosomes will be exponentially increasing during the convergence process if the fitness values of some chromosomes that contain these subsolutions $\mathbb{S}$ are above average.[1] From this perspective, the schema theorem (Holland, 1975; Michalewicz, 1996) can be regarded as follows:

> short, low-order, above-average schemata receive exponentially increasing trials in subsequent generations of a genetic algorithm.

Goldberg (Goldberg, 1989) further explained that these short, low-order, above-average schemata can be regarded as the building blocks of the search process, and the searches of GA can be regarded as a process of seeking near-optimal solutions by identifying and recombining good building blocks. The following discussions are heavily based on the descriptions of (Michalewicz, 1996) in which an extension is given to further explain the schema theorem. The one-max problem is used to explain how GA can find better results generation by generation. The example given in Fig. 13.16 shows that there are four solutions, s_1, s_2, s_3, and s_4, each of which has 11 genes. Schema $\mathbb{S}_1$ ($\star \star \star \star$ 1 1 0 $\star \star \star \star$) can be regarded as a template solution, where "$\star$" is the "don't care" symbol. In this case, the don't care symbols are located in loci $g_1, g_2, g_3, g_4, g_8, g_9, g_{10}$, and g_{11} and the fixed bits "110" are located in loci g_5, g_6, and g_7. Together, they represent the unique pattern of this schema $\mathbb{S}_1$, and apparently, two of the chromosomes (i.e., chromosomes s_1 and s_2) of the current population match the unique pattern of this schema $\mathbb{S}_1$ at all positions.

	g_1	g_2	g_3	g_4	g_5	g_6	g_7	g_8	g_9	g_{10}	g_{11}	
s_1	1	0	0	1	1	1	0	0	0	1	0	$f(s_1) = 8.4$
s_2	0	0	1	0	1	1	0	1	1	1	0	$f(s_2) = 8.3$
s_3	1	1	1	0	0	1	0	1	0	1	1	$f(s_3) = 2.1$
s_4	0	1	0	1	1	0	1	0	0	0	1	$f(s_4) = 1.1$
$\mathbb{S}_1$	*	*	*	*	1	1	0	*	*	*	*	

Figure 13.16 An example of schema $\mathbb{S}_1$.

Fig. 13.17 shows another example in which schema $\mathbb{S}_2$ (1 $\star \star \star \star$ 1 0 $\star$ 0 1 $\star$) matches chromosomes s_1 and s_3 of the current population. It is easily

---

[1] These kinds of subsolutions are also referred to as the schema.

seen that there are 2^r different chromosomes that will match a schema in solving the one-max problem, where r is the number of don't care bits.

	g_1	g_2	g_3	g_4	g_5	g_6	g_7	g_8	g_9	g_{10}	g_{11}	
s_1	1	0	0	1	1	1	0	0	0	1	0	$f(s_1) = 8.4$
s_2	0	0	1	0	1	1	0	1	1	1	0	$f(s_2) = 8.3$
s_3	1	1	1	0	0	1	0	1	0	1	1	$f(s_3) = 2.1$
s_4	0	1	0	1	1	0	1	0	0	0	1	$f(s_4) = 1.1$
$\mathbb{S}_2$	1	*	*	*	*	1	0	*	0	1	*	

Figure 13.17 An example of schema $\mathbb{S}_2$.

For example, the 2^6 different chromosomes that will match the schema $\mathbb{S}_2$ are listed as follows:

```
(1 0 0 0 0 1 0 0 0 1 0)
(1 0 0 0 0 1 0 0 0 1 1)
(1 0 0 0 0 1 0 1 0 1 0)
              ⋮
(1 1 1 1 1 1 0 1 0 1 0)
(1 1 1 1 1 1 0 1 0 1 1)
```

Since the selection, crossover, and mutation operators of GA will affect the contents of chromosomes during the convergence process, the schema theory was presented to explain the impact of these operators during the convergence process. The following notation is used to simplify the discussion of the schema theory that follows.

$o(\mathbb{S}_i)$ The *order* of schema $\mathbb{S}_i$. It represents the number of non-don't care bits in the schema. For example, $o(\mathbb{S}_1)$ is 3, because $\mathbb{S}_1 = (\star \star \star \star 1\ 1\ 0 \star \star \star \star)$ has three non-don't care bits, while $o(\mathbb{S}_2)$ is 5, because $\mathbb{S}_2 = (1 \star \star \star \star 1\ 0 \star 0\ 1 \star)$ has five non-don't care bits.

$\delta(\mathbb{S}_i)$ The defining *length* of the schema $\mathbb{S}_i$. It represents the distance between the first and last fixed bit positions. For example, $\delta(\mathbb{S}_1) = 7 - 5 = 2$, because the first and the last fixed bit positions of $\mathbb{S}_1$ are g_5 and g_7, i.e., $\mathbb{S}_1 = (\star \star \star \star 1\ 1\ 0 \star \star \star \star)$, while $\delta(\mathbb{S}_2) = 10 - 1 = 9$, because the first and last fixed bit positions of $\mathbb{S}_2$ are g_1 and g_{10}, i.e., $\mathbb{S}_2 = (\mathbf{1} \star \star \star \star 1\ 0 \star 0\ \mathbf{1} \star)$.

$\xi(\mathbb{S}_i, t)$ The number of chromosomes that matches schema $\mathbb{S}_i$ at generation t. For example, there are two chromosomes (s_1 and s_2) that match $\mathbb{S}_1$, and there are two (s_1 and s_3) that match $\mathbb{S}_2$; thus, both $\xi(\mathbb{S}_1, t)$ and $\xi(\mathbb{S}_2, t)$ at generation t are equal to 2.

$\varepsilon(\mathbb{S}_i, t)$ The total fitness value of the chromosomes that match schema $\mathbb{S}_i$ at generation t. For example, s_1 and s_2 match $\mathbb{S}_1$, thus, $\varepsilon(\mathbb{S}_1, t)$ is $8.4 + 8.3 = 16.7$, i.e., the sum of $f(s_1)$ and $f(s_2)$.

$\bar{\varepsilon}(\mathbb{S}_i, t)$ The average fitness value of the chromosomes that match schema $\mathbb{S}_i$ at generation t. For example, there are two chromosomes that match $\mathbb{S}_1$; thus, $\varepsilon(\mathbb{S}_1, t) = 16.7$. This implies that $\bar{\varepsilon}(\mathbb{S}_i, t) = \varepsilon(\mathbb{S}_1, t)/\xi(\mathbb{S}_i, t) = 16.7/2 = 8.35$.

$f(s, t)$ The total fitness value of the population at generation t, i.e., $\sum_{i=1}^{m} f(s_i, t)$, where m is the number of chromosomes in the population and $f(s_i, t)$ is the fitness value of the i-th chromosome s_i at generation t. For the example given in Fig. 13.16, $m = 4$; therefore, the total fitness value of the population is $\sum_{i=1}^{m} f(s_i) = 8.4 + 8.3 + 2.1 + 1.1 = 20.0$.

$\bar{f}(s, t)$ The average fitness value of the population at generation t, i.e., $\bar{f}(s, t) = f(s, t)/m$. Once again, for the example described in Fig. 13.16, $\bar{f}(s, t) = 20/4 = 5$.

p_c, p_m The crossover and mutation rates of GA.

Now, it is time to turn our attention to the impact of the search operators on schema $\mathbb{S}_i$ during the convergence process. In this discussion, we will first assume that some patterns (called schemata) are the "good subsolutions," which might appear in different chromosomes because of the search operators. Estimating the expected number of these schemata in the next generation is a critical step to understand the performance of a metaheuristic algorithm. To make the discussion as consistent as possible, we will use schema $\mathbb{S}_1$ given in Fig. 13.16 as an example to provide a brief introduction to the schema theory. In this example, the population size m is set equal to 4. Roulette wheel selection is used as the selection operator, while one-point crossover is used as the crossover operator. Since the selection, crossover, and mutation operators have a strong impact on the searches of GA during the convergence process, the following discussions will focus on how they affect schema $\mathbb{S}_i$.

13.3.1 Selection and fitness function for the schema

Since the selection operator will select suitable chromosomes for the next generation based on the fitness values of all the chromosomes, we need to take into account this information to estimate the impact of the selection operator on schema $\mathbb{S}_i$. To understand the impact of the selection operator on schema $\mathbb{S}_i$, we need the fitness values of the chromosomes that match schema $\mathbb{S}_i$ and the fitness values of all the chromosomes to estimate the survival probability of chromosomes that match schema $\mathbb{S}_i$ at the next generation $t + 1$. The number of chromosomes that match schema $\mathbb{S}_i$ at generation $t + 1$ can be estimated from the number of chromosomes that match schema $\mathbb{S}_i$ at generation t, as follows:

$$\xi(\mathbb{S}_i, t+1) = \xi(\mathbb{S}_i, t) \times \frac{\bar{\varepsilon}(\mathbb{S}_i, t)}{\bar{f}(s, t)}, \tag{13.16}$$

where $\bar{\varepsilon}(\mathbb{S}_i, t)$ and $\bar{f}(s, t)$ are as defined above. If the fitness values of chromosomes $s(\mathbb{S}_i, t)$ are "above average" compared to the fitness values of all the chromosomes, then the number of chromosomes that match $\mathbb{S}_i$ will be increased because they have a high probability to be selected by the selection operator for the next generation. By contrast, if the fitness values of chromosomes $s(\mathbb{S}_i, t)$ are "below average," the number of such chromosomes will be decreased because they are not as competitive as the other chromosomes. Let us use the case described in Fig. 13.16 as an example to predict the number of chromosomes that will match schema $\mathbb{S}_1$ at generation $t + 1$. It can be easily calculated for this example that $\xi(\mathbb{S}_1, t) = 2$, $\bar{\varepsilon}(\mathbb{S}_1, t) = \varepsilon(\mathbb{S}_1, t)/\xi(\mathbb{S}_1, t) = (8.4 + 8.3)/2 = 8.35$, $\bar{f}(s, t) = f(s, t)/m = (8.4 + 8.3 + 2.1 + 1.1)/4 = 5.0$, and the number of chromosomes that match schema $\mathbb{S}_1$ at generation $t + 1$ is

$$\xi(\mathbb{S}_1, t+1) = 2 \times \frac{8.35}{5.0} = 3.34. \tag{13.17}$$

This result shows that the number of chromosomes that match schema $\mathbb{S}_1$ will be increased from two to three or four at generation $t + 1$. This also implies that the ratio of $\bar{\varepsilon}(\mathbb{S}_1, t)$ (i.e., the average fitness value of the chromosomes that match schema $\mathbb{S}_1$) to $\bar{f}(s, t)$ (i.e., the average fitness value of the population) will have a strong impact on $\xi(\mathbb{S}_1, t+1)$. Let $\Delta = \bar{\varepsilon}(\mathbb{S}_1, t)/\bar{f}(s, t)$ be the ratio. If $\Delta > 1$, this means that schema $\mathbb{S}_1$ is "above average" and $\xi(\mathbb{S}_1, t + 1)$ will be increased at the next generation. On the other hand, if $\Delta < 1$, this means that schema $\mathbb{S}_1$ is "below average" and $\xi(\mathbb{S}_1, t+1)$ will be decreased at the next generation. If $\Delta = 1$, then $\xi(\mathbb{S}_1, t + 1)$ will be about the same as $\xi(\mathbb{S}_1, t)$.

13.3.2 Crossover for the schema

Now that we have discussed the selection and fitness functions for a schema, we can pay close attention to the impact of the crossover operator on schema $\mathbb{S}_i$. In this example, the one-point crossover operator is used as one of the transition operators of GA. As shown in Fig. 13.18, if the crossover point is at g_8, schema $\mathbb{S}_1$ of chromosome s_1 will be retained in chromosome s_1^c at the next generation. Fig. 13.19 shows another example for schema $\mathbb{S}_2$ of chromosome s_1. If the crossover point is again at g_8, then the structure of schema $\mathbb{S}_2$ will be broken, in the sense that some of the subsolutions of schema $\mathbb{S}_2$ will appear in s_1^c while some of them will appear in s_2^c, instead of in a single chromosome.

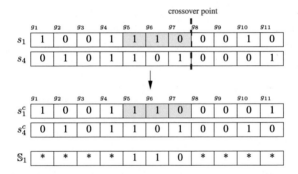

Figure 13.18 The impact of a crossover operator on schema $\mathbb{S}_1$.

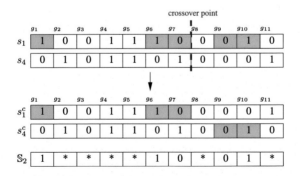

Figure 13.19 The impact of a crossover operator on schema $\mathbb{S}_2$.

These two examples illustrate that short and small schemata have a higher chance to be retained in the next and later generations, but it is just a surmise. To better understand the impact of the crossover operator on schema $\mathbb{S}_i$, the defining length of the schema $\delta(\mathbb{S}_i)$ plays a critical role in determining the probabilities of "destruction" and "survival." A large value of $\delta(\mathbb{S}_i)$ implies that the range of a schema is dispersed in the chromosome, while a small value of $\delta(\mathbb{S}_i)$ means that the range of a schema is concentrated in the chromosome. The schema theory defines the destruction probability of a schema $\delta(\mathbb{S}_i)$ as

$$s_c^d(\mathbb{S}_i) = \frac{\delta(\mathbb{S}_i)}{n-1} \tag{13.18}$$

and the survival probability of a schema $\delta(\mathbb{S}_i)$ as

$$s_c(\mathbb{S}_i) = 1 - \frac{\delta(\mathbb{S}_i)}{n-1}, \tag{13.19}$$

where n is the number of genes of a chromosome. In GA, typically, only part of the chromosomes will undergo crossover. This means that only $p_c \times m$ chromosomes have a chance to be changed by the crossover operator. For this reason, the survival probability of schema $\mathbb{S}_i$ can be defined as

$$s_c(\mathbb{S}_i) = 1 - p_c \times \frac{\delta(\mathbb{S}_i)}{n-1}. \tag{13.20}$$

For example, suppose the crossover rate p_c is set equal to 0.6. The survival probability of schema $\mathbb{S}_1$ will be

$$s_c(\mathbb{S}_1) = 1 - 0.6 \times \frac{2}{11-1} = 0.88 \tag{13.21}$$

and the survival probability of schema $\mathbb{S}_2$ will be

$$s_c(\mathbb{S}_2) = 1 - 0.6 \times \frac{9}{11-1} = 0.46. \tag{13.22}$$

It can be easily seen from the results that the survival probability of schema $\mathbb{S}_1$ is higher than that of schema $\mathbb{S}_2$.

13.3.3 Mutation for the schema

The mutation operator is another transition operator of GA that has a chance to change the value of a chromosome. As shown in Fig. 13.20, suppose the mutation rate $p_m = 0.01$; this implies that the value of each

gene has a 1% chance of being mutated, e.g., by inverting its value from
"1" to "0" or from "0" to "1." For this example, the mutation operator will
mutate the value of g_9 by inverting its value from "0" to "1." Besides, the
structure of $\mathbb{S}_1$ for chromosome s_1 will be retained for the next generation.
If the gene in g_5, g_6, or g_7 is mutated, the structure of $\mathbb{S}_1$ for chromosome
s_1 will be destructed (i.e., not retained for the next generation).

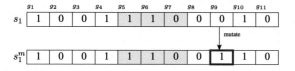

Figure 13.20 The impact of a mutation operator on schema $\mathbb{S}_1$.

Since mutations are considered to be independent from each other, the
survival probability of schema $\mathbb{S}_i$ for a mutation can be defined as

$$s_m(\mathbb{S}_i) = (1 - p_m)^{o(\mathbb{S}_i)}, \qquad (13.23)$$

where $o(\mathbb{S}_i)$ is the number of non-don't care bits of schema $\mathbb{S}_i$. The
value $s_m(\mathbb{S}_i)$ can be regarded as the probability of the mutation operator
not changing the value of genes for schema $\mathbb{S}_i$. Also, because $p_m \ll 1$, the
approximate survival probability of schema $\mathbb{S}_i$ can be defined as

$$s_m(\mathbb{S}_i) \approx 1 - o(\mathbb{S}_i) \times p_m. \qquad (13.24)$$

We can use schema $\mathbb{S}_1$ as an example to continue the discussion. Suppose the mutation rate p_m is set equal to 0.01 and $o(\mathbb{S}_1)$ is 3. Then the
approximate survival probability of schema $\mathbb{S}_1$ will be

$$s_m(\mathbb{S}_1) \approx 1 - 3 \times 0.01 = 0.97. \qquad (13.25)$$

13.3.4 A simple example of schema theory

Given Eqs. (13.16), (13.20) and (13.24), we can now have an integrated
view on the impact of selection, crossover, and mutation on the schema.
These equations can be combined to take into account the impact of selection, crossover, and mutation as follows:

$$\xi(\mathbb{S}_i, t+1) = \xi(\mathbb{S}_i, t) \times \frac{\bar{\varepsilon}(\mathbb{S}_i, t)}{\bar{f}(s, t)} \times \left[1 - p_c \times \frac{\delta(\mathbb{S}_i)}{n-1} - o(\mathbb{S}_i) \times p_m\right]. \qquad (13.26)$$

We can use this equation to estimate how many chromosomes will match schema $\mathbb{S}_i$ at the next generation $t + 1$. For schema $\mathbb{S}_1$, its survival probability will be

$$\xi(\mathbb{S}_1, t + 1) = 2 \times \frac{8.35}{5.0} \times \left[1 - 0.6 \times \frac{2}{10} - 3 \times 0.01 \right] = 2 \times 1.4195 = 2.839.$$

(13.27)

This means that the number of chromosomes that match schema $\mathbb{S}_1$ at generation $t + 1$ will be two or three, because $\xi(\mathbb{S}_1, t + 1) = \xi(\mathbb{S}_i, t) \times 1.4195 = 2 \times 1.4195 = 2.839$. However, at generation $t + 2$, $\xi(\mathbb{S}_1, t + 2) = \xi(\mathbb{S}_i, t) \times 1.4195^2 = 2 \times 2.01498025 = 4.0299605$, implying that the expected number of chromosomes that match schema $\mathbb{S}_1$ will increase from two at generation t to four at generation $t + 2$. The results state that "good schema" will be exponentially increased to different chromosomes during the convergence process, an echo of the assumption of the schema theory stating that the short, low-order, above-average schemata will be increasing exponentially in the population of GA.

13.4. Fitness landscape analysis

To better understand the search performance of a metaheuristic algorithm, several studies (Grefenstette, 1999; Malan and Engelbrecht, 2013) attempted to depict the relationships between candidate solutions, called fitness landscape analysis. The boundary between objective and fitness functions is unclear. Malan and Engelbrecht (Malan and Engelbrecht, 2013) mentioned that a broader notion of fitness and the fitness function can also involve the objective and the objective function for metaheuristic algorithms. In an early study (Wright, 1932), Wright presented a two-dimensional figure in which the contour lines are used to indicate the fitness values of chromosomes to show the fitness landscape for genetic evolution in 1932. As shown in Fig. 13.21, we can mimic Wright, that is, using valleys, peaks, ridges, and plateaus to display the kinds of fitness landscapes we have. In this example, the symbols "+" and "-" represent peaks and valleys, respectively. Colors with higher values represent higher fitness values, and vice versa.

An intuitive way to define the fitness landscape for a search algorithm is to define it as a directed graph in which two nodes (i.e., two candidate solutions) in the fitness landscape are neighbors if one node can be reached from the other node by a single step of the transition operator (Malan

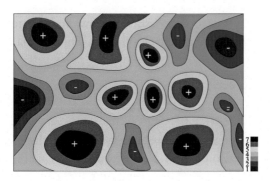

Figure 13.21 Example for illustrating the fitness landscape (Wright, 1932).

and Engelbrecht, 2013). A much more general and formal definition of landscape can be found in (Stadler, 2002) based on three assumptions:

- a solution s in the solution space,
- a candidate solution v that is the neighbor of s in the solution space, and
- a fitness function $f(s) : \mathbf{R}^n \to \mathbf{R}$.

Under these assumptions, v is a neighbor solution of s. For the discrete landscape, v may be created by using a transition operator for s once, such as the Hamming distance of these two solutions being one. For the continuous landscape, v can also be regarded as a neighbor solution of s, such as the Euclidean distance being less than a predefined threshold (Malan and Engelbrecht, 2013; Stadler, 2002).

The fitness landscape analysis itself is often very complex; however, it is very useful for a better understanding of the optimization problem in question before we eventually decide on a metaheuristic algorithm for solving it. Although this is not a perfect method for the fitness landscape analysis, many methods (e.g., the NK model (Kauffman and Weinberger, 1989) or local optima networks (Ochoa et al., 2014)) can be used to help us better understand the relationship between search algorithm and optimization problem and even find ways to develop better metaheuristic algorithms. Jones (Jones, 1995) pointed out that each operator in a metaheuristic algorithm has its own fitness landscape for the search, while Ochoa and Malan (Ochoa and Malan, 2019) pointed out that a fitness function can have many fitness landscapes. Malan and Engelbrecht (Malan and Engelbrecht, 2013) further pointed out that the information of fitness landscapes can be used to guide the searches of a metaheuristic algorithm to the right direction to move more easily toward the global optimum during the convergence process. Richter and Engelbrecht enlisted several experts to write a book (Richter and Engelbrecht, 2014) to provide a blueprint of the fitness landscape research domain that focused on principles of fitness landscape, measurement of search algorithm performance, problem hardness, visualization methods of fitness landscapes, and other relevant topics in fitness landscape analysis. In summary, from the above discussions, it can be easily seen that the information of fitness landscapes can be used to understand the "characteristics of an optimization problem," to know the "search behavior of a metaheuristic algorithm" in solving such an optimization problem, and to develop "a better metaheuristic algorithm" for such an optimization problem.

13.5. Discussion

This chapter provides a brief review of fitness (objective) functions in Section 13.1, determination operators in Section 13.2, the schema theorem in Section 13.3, and fitness landscape analysis in Section 13.4. The studies mentioned in this chapter show several possible ways to enhance the search performance of metaheuristic algorithms. From these studies, it can be easily seen that the fitness function and the determination operator can be modified in different ways, but we need to know the kind of landscape we confront to select an appropriate strategy for a metaheuristic algorithm. As shown in Table 13.1, if most of the fitness values of searched solutions are very close to each other, roulette wheel selection may not be useful to determine their quality because the probabilities for the next iteration will also be very close to each other. If the goal is to maximize the objective value, then the best chromosome is s_4 while the worst chromosome is s_1 in this example. The probabilities of s_4 and s_1 being selected for the next generation are $p_4 = 0.28 = 28\%$ and $p_1 = 0.22 = 22\%$, respectively, so the difference is only 6%. Because the survival probabilities of all the chromosomes are too close to each other, the selection operator may not always select the best or second best solution for the next generation.

Table 13.1 Number of possible routes for an n-city traveling salesman problem.

Fitness $f(s_i)$	$f(s_1)$	$f(s_2)$	$f(s_3)$	$f(s_4)$
	1.1	1.2	1.3	1.4
Probability p_i	p_1	p_2	p_3	p_4
	0.22	0.24	0.26	0.28

Tournament selection provides a solution for this situation because this kind of selection operator will simply compare the selected solutions. This means that better solutions will have a higher chance to be passed on to the next iteration in this situation. This solution, of course, is not a panacea. Every design of a metaheuristic algorithm has its pros and cons, and tournament selection is no exception. This kind of selection strategy may fall into local optima very quickly during the convergence process, especially when the population size is too small (i.e., most players in a tournament will be the same) or the number of players for each tournament is too high (i.e., the winner of different tournaments will be the same). As a result, the whole population will be composed of one and only one chromosome, that is, all chromosomes in the population are eventually the same.

The studies we discussed in this chapter also highlight several dynamic adjustment methods for the fitness function and the determination operator. A simple example can be found in (Grefenstette, 1999), in which the mutation rate is regarded as part of the subsolution to dynamically adjust the fitness function for dynamic fitness landscapes. Some representative examples for adjusting the convergence strategies by modifying the determination operator are dynamically adjusting the length of a tabu list during the convergence process (Gendreau and Potvin, 2005) and dynamically adjusting the inertial weight or the acceleration coefficients a_1 and a_2 (Bratton and Kennedy, 2007; Ratnaweera et al., 2004; Taherkhani and Safabakhsh, 2016).

This chapter shows several ways to modify the objective/fitness function and the determination operator in an attempt to provide an integrated viewpoint of them from the perspectives of different metaheuristic algorithms. The discussions also provide some niches to help us improve the search performance of a metaheuristic algorithm A from the concept of another metaheuristic algorithm B. For example, for ACO, we know that the solution construction operator is computationally very expensive. This implies that adopting the selection or determination methods from other metaheuristic algorithms will probably provide a useful way to enhance the search performance of ACO-based algorithms. Of course, we can even use the annealing schedule from SA to improve the determination operator of other metaheuristic algorithms, by giving them a better chance to escape from local optima. To further understand the impact of the evaluation and determination operators on a metaheuristic algorithm, a brief discussion on Holland's schema theorem was given in Section 13.3, which gives us not only some ideas on measuring the performance of metaheuristic algorithms in solving an optimization problem but also a simple way to understand why better searched solutions are selected for later iterations (or generations) during the convergence process. Unlike the schema theorem, which aims to understand the performance of a metaheuristic algorithm for solving an optimization problem from the perspective of search operators, it can be easily seen from the discussions given in Section 13.4 that some fitness landscape analysis tools are based on the observations from the perspective of relationships between solutions or between search algorithm, solution space landscape, and local optima to find a better metaheuristic algorithm for solving optimization problems. In summary, to design a new search strategy by modifying the fitness function and determination operators for solving optimization problems, the landscape information, domain knowledge, and characteristics of the fitness function and determination operators need to be considered together so that an appropriate search strategy can be selected or developed.

CHAPTER FOURTEEN

Parallel metaheuristic algorithm

14.1. The basic idea of the parallel metaheuristic algorithm

Although we already know that the basic idea of metaheuristic algorithms is to provide a faster way to find an approximate solution compared with traditional exclusive search, there is plenty of room to improve them. How to make a metaheuristic algorithm find the end result more quickly is a critical research direction because reducing the response time[1] is important in practice, especially for real-time systems. Since the rise of parallel and distributed computing technologies in the 1980s or even earlier, some studies (Bethke, 1976; Casotto et al., 1987; Crainic et al., 1995; Tanese, 1987) have attempted to apply metaheuristic algorithms to or even redesign them for this environment to further reduce the response time. Crainic et al. (Crainic et al., 1995) compare several parallel strategies for tabu search (TS) to reduce the response time. One is the master–slave architecture. Suppose the number of candidate solutions at each iteration is m and the number of slaves is $\ddot{p}$. Each slave will be used to perform the transition and evaluation operators for the $m/\ddot{p}$ of candidate solutions. Each slave will then use the surrogate function and the tabu list dispatched by the master to check the searched solutions and return the best solution. Several other studies (Bethke, 1976; Casotto et al., 1987; Tanese, 1987) show that simulated annealing (SA) and the genetic algorithm (GA) can also be applied to parallel and distributed computing environments to reduce the response time, for example on a multi-processor system with a shared-memory architecture. Such kinds of methods are collectively known as parallel metaheuristic algorithms. Cung et al. (Cung et al., 2002) show that the *speedup* of a parallel algorithm $\ddot{a}$ versus its sequential counterpart a for a problem $\mathbb{P}$ can be defined as

$$\mathbf{s}_{\ddot{a}}^{\mathbb{P}}(\ddot{p}) = \frac{\mathbf{T}_{a}^{\mathbb{P}}(p)}{\mathbf{T}_{\ddot{a}}^{\mathbb{P}}(\ddot{p})}, \tag{14.1}$$

---

[1] By most parallel computing technologies are able to reduce the "response time," we mean that they can reduce the time a system takes to react to the request of a user, but not the total amount of time.

where p denotes a single-processor machine, $\ddot{p}$ denotes the number of processors on a multi-processor machine, $T_a^{\mathbb{P}}(p)$ denotes the response time of the sequential algorithm a running on a single-processor machine for solving the problem $\mathbb{P}$, and $T_{\ddot{a}}^{\mathbb{P}}(\ddot{p})$ denotes the response time of the parallel algorithm $\ddot{a}$ running on a multi-processor machine using $\ddot{p}$ processors for solving the problem $\mathbb{P}$. To make it easier to understand the performance of a parallel algorithm $\ddot{a}$ using $\ddot{p}$ processors, the speedup can be normalized as

$$\eta_{\ddot{a}}^{\mathbb{P}}(\ddot{p}) = \frac{s_{\ddot{a}}^{\mathbb{P}}(\ddot{p})}{\ddot{p}}. \tag{14.2}$$

As some early studies (Crainic and Toulouse, 1998; Nowostawski and Poli, 1999) observed, parallelizing the computation of a metaheuristic algorithm may not only "reduce the response time," but also "find a better result" compared with the sequential version of a metaheuristic algorithm. In brief, because a metaheuristic algorithm typically takes a large amount of computation time to evaluate the objective or fitness values of searched solutions and other search operators, the response time of a metaheuristic algorithm can be significantly reduced if the computing power of a system can be increased by increasing the number of CPUs. Compared with the sequential version of a metaheuristic algorithm that may converge to particular regions or directions, thus falling into local optima easily, the parallel version of a metaheuristic algorithm provides a way that allows each processor (or computer node) to search a different subspace at the same time, thus reducing the chance to fall into the same regions or directions. Because of this, parallel metaheuristic algorithms are able to find better results than sequential metaheuristic algorithms.

Cantú-Paz (Cantú-Paz, 1998) provided an influential survey of parallel GA (PGA), which is divided into three categories: (1) global single-population master–slave, (2) single-population fine-grained, and (3) multiple-population coarse-grained. As the name suggests, global single-population master–slave has a global population that is maintained on the master, but the selection operation is distributed to the slaves. The single-population fine-grained structure is useful for massively parallelized computing environments, such as in grid computing. Although it also has only one population, the chromosomes are dispatched to different nodes. The selection and crossover of the chromosomes will be restricted to the neighbor nodes, i.e., a small neighborhood. The multiple-population coarse-grained structure divides the population into several subpopulations which will exchange the information they have after a certain number of generations. Since this is just like the situation in which there are many different populations, each of which lives on a different island and may migrate from one island to another, it is also called the multiple-deme or island model. To further classify PGA, another study (Nowostawski and Poli, 1999) took into account the implementation of the following four elements: (1) how the evaluation and transition operators are applied, (2) how many subpopulations GA contains, (3) how they exchange information when the number of subpopulations is more than one, and (4) how the determination operator is applied, e.g., globally or locally. Based on these concerns, PGA can be classified into eight classes: (1) a master–slave architecture to distribute the computation of the fitness function, (2) static subpopulations with migration, (3) static overlapping subpopulations without migration, (4) massive PGA, (5) dynamic subpopulations, (6) parallel steady-state GA, (7) parallel messy GA, and (8) hybrid methods.

In fact, the study of the parallelism of search operators is not just for GA; it is also found in other kinds of metaheuristic algorithms, since about the same time. The period around 2000 may be regarded as a turning point for parallel metaheuristic algorithms because the number of studies that attempted to use parallel computing technologies to enhance the performance of metaheuristic algorithms (Alba, 2005; Alba et al., 2013; Crainic and Toulouse, 1998; Cung et al., 2002; Nowostawski and Poli, 1999) grew significantly. In (Crainic and Toulouse, 1998), Crainic and Toulouse classified the parallel methods for metaheuristic algorithms into three different categories to differentiate their strategies.

Type 1: Metaheuristic algorithms of this type *parallelize the operations* within an iteration, which is also called low-level parallelism, the sole purpose of which is to reduce the response time of sequential metaheuristic algorithms, without improving the quality of the end results. The master–slave architecture for population-based metaheuristic algorithms is one of the methods in this category. For single-solution-based metaheuristic algorithms, one possible way to parallelize the computation is to decompose the computation of the evaluation operator, which is referred to as single-trial parallelism. Another possible way is called multiple parallelism, in which all computing nodes start with the same solution but perform the single-solution-based metaheuristic algorithm independently during the convergence process. The best of all the computing nodes is then selected as the final solution.

Type 2: Metaheuristic algorithms of this type decompose the optimization problem or solution space into a set of subproblems, which is also called parallelization by domain decomposition. This kind of parallel strategy is different in the sense that it will divide the solution space into a set of subspaces, which may change the search results compared with a sequential metaheuristic algorithm. This means that each slave will search the solutions in the subspace assigned to it independently and then it is the master that is responsible for combining the partial solutions from all the slaves to obtain the solution. For instance, we can randomly create an initial solution s for the traveling salesman problem (TSP). The n cities in the solution s can be divided into $\hat{n}$ subsets, each of which will be assigned to a computing node where the transition operator will be performed on the subset for a certain number of iterations. A synchronization method will then be employed on the master to integrate subsolutions from all the computing nodes to create a complete solution.

Type 3: Metaheuristic algorithms based on this parallel strategy use multiple threads to search for possible solutions simultaneously. This strategy is also called multiple search. It typically does not have a master computing node, meaning that all the computing nodes will perform the metaheuristic algorithm independently. The independent and cooperative strategies are two representative strategies. The basic idea of the independent strategy is that all the computing nodes will perform the metaheuristic algorithm to find a complete solution, without any exchange of information with other computing nodes during the convergence process. The best solution in these computing nodes will be selected at the end of the convergence process. The cooperative strategy can be categorized into single-population fine-grained or multiple-population coarse-grained. Computing nodes will exchange the searched information by using a predefined topology and communication method, such as broadcast and propagation.

Cung et al. in a later study (Cung et al., 2002) provided another classification of the following parallel metaheuristic algorithms: greedy randomized adaptive search procedure (GRASP), TS, SA, GA, and ant colony optimization (ACO). The parallel metaheuristic algorithms in this classification can be divided into single walk (fine-grained to medium-grained tasks) and multiple walks (medium-grained to coarse-grained tasks). The multiple walks also contain the independent and cooperative search threads. Because the development of parallel metaheuristic algorithms was very fast in the 1990s, Alba and another 40 researchers in the book (Alba, 2005) discussed many different kinds of metaheuristic algorithms and their possible applications. The focus is not only on how to parallelize the computation of TS, SA, GA, and ACO, but also on estimation of distribution algorithms (EDAs), scatter search (SS), hybrid metaheuristic algorithms, and so forth. This book provides a systematic overview and complete discussion of parallel metaheuristic algorithms and is a very useful roadmap for researchers who want to enter this research field. Alba et al. in a later study (Alba et al., 2013) provided a simple way to classify the parallel metaheuristic algorithms, by dividing these algorithms into trajectory-based (single-solution-based) parallel metaheuristic algorithms and population-based parallel metaheuristic algorithms. Each category contains three to four subcategories, and the differences between parallel metaheuristic algorithms are clearly explained. To discuss parallel metaheuristic algorithms, we also divide them into single-solution-based and population-based parallel metaheuristic algorithms, consistent with (Alba et al., 2013). The classification of single-solution-based parallel metaheuristic algorithms is mainly based on (Alba et al., 2013), while the classification of population-based parallel metaheuristic algorithms is based on (Cantú-Paz, 1998).

14.1.1 Single-solution-based parallel metaheuristic algorithms

The category of single-solution-based metaheuristic algorithms typically includes SA, TS, iterated local search (ILS), variable neighborhood search (VNS), greedy randomized adaptive search procedures (GRASP), and so forth. The parallel moves, parallel evaluation, and parallel multi-start models are three major parallel models of single-solution-based metaheuristic algorithms (Alba et al., 2013).

The parallel moves model: As shown in Fig. 14.1(a), the parallel moves model employs the master–slave architecture to parallelize the computations of a single-solution-based metaheuristic algorithm. The master ($\ddot{p}_0$) sends a copy of the current solution s to its slaves ($\ddot{p}_1$ and $\ddot{p}_2$) in the very beginning of each iteration, i.e., data 1 in Fig. 14.1(a). Each slave performs the single-solution-based metaheuristic algorithm independently and then returns the best solution to the master, i.e., info 1 and info 2 in Fig. 14.1(a). Fig. 14.1(b) provides a simple example to illustrate how the parallel moves model works. This example shows that the master will send a copy of its current solution s to its slaves, each of which will search for some possible candidate solutions independently, i.e., search for a certain number of iterations. These slaves will then return the best solutions they found, and

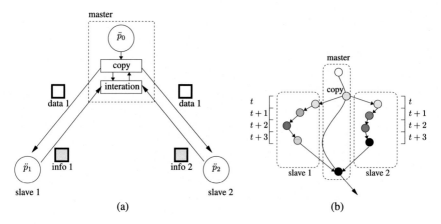

Figure 14.1 The parallel moves model. (a) The basic idea. (b) A simple example (Alba et al., 2013).

the master will pick the best of them as the current solution for the next iteration. It can be imagined and expected that this model makes it possible for a single-solution-based parallel metaheuristic algorithm to increase the number of solutions searched compared with a sequential single-solution-based metaheuristic algorithm with the same number of iterations. Since the number of searched solutions will be increased in the same number of iterations, the search diversity will also be increased; as a result, the quality of the end result of a single-solution-based parallel metaheuristic algorithm may also be improved with the same number of iterations (or the same response time) as a sequential single-solution-based metaheuristic algorithm.

The parallel evaluation model: Fig. 14.2(a) shows that this model is similar to the parallel moves model in the sense that it also employs the master–slave architecture to dispatch the computations from the master to its slaves. Unlike the parallel moves model, this model will divide the computations of a single-solution-based metaheuristic algorithm and assign them to the slaves, such as data 1 and data 2 in Fig. 14.2(a). After all the slaves return their results to the master, the results will be combined. Comparing parallel moves to parallel evaluation, it can be easily seen that each slave of the parallel moves model is responsible for carrying out a single-solution-based metaheuristic algorithm for a certain number of iterations, while each slave of the parallel evaluation model is responsible for carrying out only *a portion of the search tasks* of a single-solution-based metaheuristic algorithm each iteration, as shown in Fig. 14.2(b). The computations of

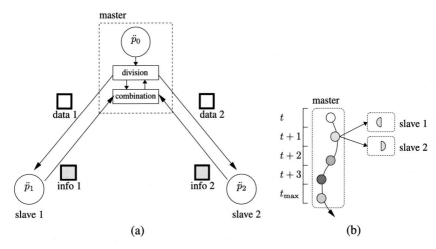

Figure 14.2 The parallel evaluation model. (a) The basic idea. (b) A simple example (Alba et al., 2013).

a single-solution-based metaheuristic algorithm on the master are divided into a set of subcomputations at each iteration, such as the computations of the evaluation operator for the solution $f(s)$. These subcomputations will be dispatched to slaves. The slaves will return the computation results to the master, and then the master will combine the results to get the actual result of $f(s)$. This approach can be used when the computations of the evaluation operator for the solution can be performed independently. For example, the objective value of a complete solution for the one-max problem can be divided into a number of computation tasks (i.e., accumulations of the "1's" for all the subsections) that can be computed independently. As shown in Fig. 14.3, in this example, a solution s has 12 subsolutions

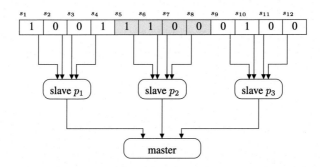

Figure 14.3 How the computation tasks are divided and distributed to different slaves.

and there are three slaves, each of which is responsible for accumulating the number of "1's" for a certain number of subsolutions. This means that each slave will accumulate the number of 1's for a particular segment of the solution, e.g., $\sum_{i=j}^{k} f(s_i)$, where s_i represents a subsolution in the segment containing the j-th to k-th subsolutions. After receiving the results from all the slaves, the single-solution-based parallel metaheuristic algorithm on the master can then get the actual result of $f(s)$ by combining the number of 1's of these subsolutions.

The parallel multi-start model: Different from the parallel moves and parallel evaluation models, in which all slave nodes cooperate with each other, the multi-start model launches a single-solution-based metaheuristic algorithm at each slave for a complete convergence process. As mentioned in (Alba et al., 2013), these single-solution-based metaheuristic algorithms are performed on different processors, and eventually different search strategies can be used:

- Heterogeneous or homogeneous search strategies. A heterogeneous search strategy means that each processor will use a different parameter setting or even run a different metaheuristic algorithm. A homogeneous search strategy means that each processor will run the same metaheuristic algorithm and use the same parameter settings.
- Independent or cooperative search strategies. An independent search strategy means that no search information will be exchanged by the processors during the convergence process. A cooperative search strategy means that search information will be exchanged by the processors during the convergence process.
- Each processor will start with the same or a different initial solution.

As shown in Fig. 14.4(a), all the processors (i.e., $\ddot{p}_1$ and $\ddot{p}_2$) of the multi-start model will independently and simultaneously perform their single-solution-based metaheuristic algorithms. Fig. 14.4(b) gives a simple example to show the case where each processor starts with a different initial solution and performs the single-solution-based metaheuristic algorithm to find the results independently.

14.1.2 Population-based parallel metaheuristic algorithms

The category of population-based metaheuristic algorithms typically includes GA, ACO, particle swarm optimization (PSO), differential evolution (DE), and so on. As far as this category is concerned, we can

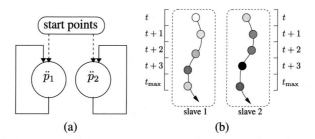

Figure 14.4 The parallel multi-start model. (a) The basic idea. (b) A simple example (Alba et al., 2013).

essentially use the classification method presented by Cantú-Paz (Cantú-Paz, 1998) for GA to further understand how the other metaheuristic algorithms parallelize their computation tasks. This classification method includes the following models for the classification of population-based parallel metaheuristic algorithms: master–slave, coarse-grained (i.e., island), fine-grained, and hierarchical.

The master–slave model: The master–slave architecture is typically not only used in single-solution-based parallel metaheuristic algorithms; it is also widely used in population-based parallel metaheuristic algorithms. As shown in Fig. 14.5(a), similar to other master–slave architectures, this architecture will dispatch all the computations (i.e., data 1 and data 2) from the master to its slaves. The slaves will then return the computed results

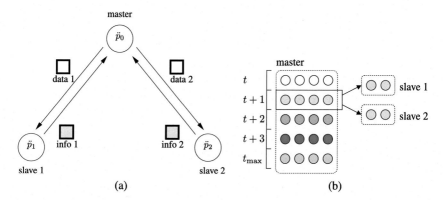

Figure 14.5 How the master–slave parallel model works. (a) The basic idea. (b) A simple example.

(i.e., info 1 and info 2) to the master. Fig. 14.5(b) further shows that the master–slave model of a population-based parallel metaheuristic algorithm is similar to the parallel moves and parallel evaluation models of a single-solution-based parallel metaheuristic algorithm. The main difference is that each slave of a population-based parallel metaheuristic algorithm is responsible for the computation tasks of one or more solutions, while each slave of a single-solution-based parallel metaheuristic algorithm is responsible for the computation tasks of one solution or a certain number of subsolutions. The computations dispatched to the slaves can be the evaluation or transition operators, which depend on the design of the master–slave architecture. Cantú-Paz (Cantú-Paz, 1998) also pointed out that the design of such an architecture has to take into account the following factors: synchronicity vs. asynchronicity and shared memory vs. distributed memory. The factor synchronicity vs. asynchronicity is important because slaves may run at different speed or have different computing power. For this reason, population-based parallel metaheuristic algorithms may not stop and wait for any slow slaves. Because of the parallel computing architecture and the communication costs between different processors, shared memory and distributed memory are two well-known ways to keep the searched solutions. If the overhead of communication is low, distributed memory will be a good choice in the sense that the population-based parallel metaheuristic algorithm will have more memory to use in this case. If the overhead of communication is high, shared memory is a possible way to reduce this kind of overhead of a parallel metaheuristic algorithm.

The coarse-grained model: As shown in Fig. 14.6(a), for the coarse-grained model, all the processors (also called islands) play the same role, and each of them has its own population and convergence process. As shown in Fig. 14.6(b), assuming that there are eight chromosomes in the population, compared with the master–slave model, which may have only a single population, the island model will usually divide the population into several subpopulations, each of which will contain two chromosomes and will be assigned to a different processor to perform the parallel metaheuristic algorithm independently for a certain number of iterations. The searched information will be exchanged between different islands. This process is referred to as migration. The information exchanged may be the best known searched solution of each island or other useful information. The communication topology of islands and the migration rate are two important factors of island models. As for the migration rate, the search will fall back to a sequential metaheuristic algorithm if the migration procedure is performed too often. It will also incur higher communication costs between

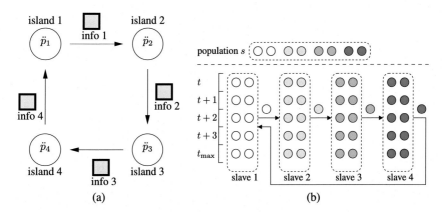

Figure 14.6 How the coarse-grained (island) model works. (a) The basic idea. (b) A simple example.

processors because a lot of chromosome(s) will be moved from one island to replace chromosome(s) on another. On the other hand, the search on each island will converge independently from each other if the interval between two migrations is too large. A similar concern exists for the communication topology. It is expected that the convergence will be accelerated if an island is able to connect to a large number of islands to exchange information because the "good" solution of an island will be spread to most of the islands, but the entire search of all the islands will trend to particular regions or directions in the solution space. For these two reasons, the communication topology and the migration rate are obviously two important research issues of this model because they affect the search performance. In general, the communication topology of the island model is like a ring. As shown in Fig. 14.6(a), processor $\ddot{p}_1$ will receive the result from one neighbor island ($\ddot{p}_4$ in this case) and send its searched result to the other neighbor island ($\ddot{p}_2$ in this case).

The fine-grained model: As shown in Fig. 14.7, the fine-grained model will typically restrict the communication of each processor to only a small number of processors. This model is suitable for some particular architectures, such as massively parallel computers. This kind of model is also often used in hierarchical models of parallel metaheuristic algorithms (Alba et al., 2013).

The hierarchical parallel model: Fig. 14.8 shows two different hierarchical GA models (also called hybrid models), which are discussed in

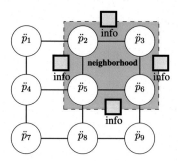

Figure 14.7 How the fine-grained model works.

(Cantú-Paz, 1998) and may also be used in other parallel metaheuristic algorithms (Alba et al., 2013). Hierarchical models can typically be di-

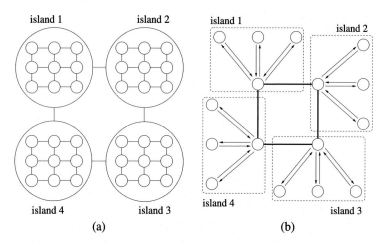

Figure 14.8 Two hierarchical parallel models. (Cantú-Paz, 1998). (a) The combination of coarse-grained and fine-grained models. (b) The combination of coarse-grained and master–slave models.

vided into two levels: the high level and the low level. Fig. 14.8(a) shows a hierarchical model that employs the coarse-grained (island) model as the high-level search architecture and the fine-grained model as the low-level search architecture. Fig. 14.8(b) shows another possible hierarchical model that also employs the coarse-grained (island) model as the high-level search architecture, but it uses the master–slave model as the low-level search architecture.

14.2. Implementation of parallel GA for the TSP

This section will use master–slave PGA (MSP-GA) as an example to show how a parallel metaheuristic algorithm parallelizes its computations during the convergence process. Listing 14.1 is an implementation of MSP-GA for solving the TSP, based on the sequential implementation shown in Listing 12.1. Moreover, the implementation of MSP-GA for solving the TSP described here is built on the multi-threading technologies of C++. Since there are many subtle differences between the sequential implementation and the parallel implementation, we decide to start anew for the parallel implementation, that is, we start with a copy of the code described in Listing 12.1, so that we are free to do all kinds of tests and optimizations without the burden of dealing with the differences in between. One thing to note is that instead of the modified one-point crossover (MOPC) of Listing 12.1, the parallel implementation uses the order crossover (OX) described in Listing 12.4.

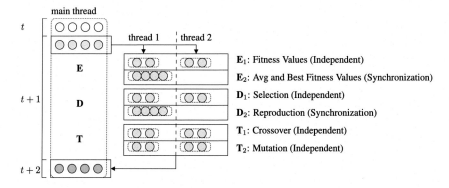

Figure 14.9 How MSP-GA is implemented as a multi-threading program.

Before we proceed to describe MSP-GA in detail, we will use Fig. 14.9 as an example to show how the multi-threading technologies are applied to GA for solving the TSP. In this figure, the main thread can be regarded as the master node, while threads 1 and 2 can be regarded as the slaves. GA includes the tasks of computing the fitness values, computing the average fitness values, finding the best-so-far solution and getting its fitness value, selection, reproduction, crossover, and mutation, which are performed by the evaluation, determination, and transition operators. This example also shows that although the computation tasks of the fitness function for calculating the fitness values (E_1) of all the chromosomes are "independent" of

each other, the computation tasks for computing the average fitness values
of all the chromosomes ($\mathbf{E}_2$), finding the best-so-far solution ($\mathbf{E}_2$), and getting its fitness value ($\mathbf{E}_2$) are "dependent" on each other, even though all of
them are computation tasks of the evaluation operator. The same situation
can be found for the determination operator, in which the tournament
selection ($\mathbf{D}_1$) operator is "independent" while the reproduction operator ($\mathbf{D}_2$) is "dependent." Since the computations of the crossover ($\mathbf{T}_1$) and
mutation ($\mathbf{T}_2$) operators involve either a pair of chromosomes or a single
chromosome, the computation tasks of these two operators are "independent."

The main reason why the computation tasks for computing average fitness values of all the chromosomes are dependent on each other is because
they need the fitness values of all the chromosomes. This is one of the
most important reasons why this procedure cannot be carried out before
all the slaves return their part of the fitness values. The computation tasks
to find the best-so-far solution and its fitness value are also dependent on
each other because there is no way for a slave to find the best-so-far solution and its fitness value without looking at those of the other slaves and
writing the best-so-far solution and its fitness value back to something vis-
ible to all the slaves. This will eventually lead to the *race condition* if two
or more slaves want to update the best-so-far solution and its fitness value
simultaneously. The same problem is also found for the reproduction operator. To solve this problem, MSP-GA has been designed from scratch to let
all the slaves perform the independent tasks simultaneously and let the slave
that finished the independent tasks *last* take the responsibility of completing
the dependent tasks, such as computing the average fitness values of all the
chromosomes, finding the best-so-far solution, getting its fitness value, and
performing the reproduction operator. This facilitates synchronization. After completing all the computation tasks, be it independent or dependent,
of these operators (evaluation, determination, and transition) of generation
$t + 1$, this search process of MSP-GA will be repeated for generation $t + 2$,
i.e., the next generation, until the predefined stop condition is met.

Listing 14.1: src/c++/3-tsp/main/pga.h: The implementation of island
model parallel GA for the TSP.

```
1 #ifndef __PGA_H_INCLUDED__
2 #define __PGA_H_INCLUDED__
3
4 #include <condition_variable>
5 #include <thread>
6 #include <chrono>
```

```
 7 #include <atomic>
 8
 9 #include <string>
10 #include <fstream>
11 #include <limits>
12 #include "lib.h"
13
14 using namespace std;
15
16
17
18 class pga
19 {
20 public:
21
22     typedef vector<int> solution;
23     typedef vector<solution> population;
24     typedef vector<double> pattern;
25     typedef vector<pattern> instance;
26     typedef vector<double> dist_1d;
27     typedef vector<dist_1d> dist_2d;
28
29     pga(int num_runs,
30         int num_evals,
31         int num_patterns_sol,
32         string filename_ini,
33         string filename_ins,
34         int pop_size,
35         double crossover_rate,
36         double mutation_rate,
37         int num_players,
38         int num_threads);
39
40     population run();
41
42     static void thread_run(population& curr_pop,
43                            population& tmp_curr_pop,
44                            vector<double>& curr_obj_vals,
45                            const dist_2d& dist,
46                            int thread_idx,
47                            int num_threads,
48                            int nevals,
49                            double& best_obj_val,
50                            solution& best_sol,
51                            pattern& avg_obj_val_eval,
52                            int thread_pop_size,
53                            double crossover_rate,
54                            double mutation_rate,
55                            int num_players,
56                            std::mutex& cv_m,
57                            std::condition_variable& cv,
58                            std::atomic<int>& icnt
59                            );
60
61 private:
62     population init();
63     static double evaluate(const solution& curr_sol, const dist_2d& dist);
64     static void tournament_select(const population& curr_pop, const vector<double>&
       curr_obj_vals, population& tmp_curr_pop, int num_players, int thread_idx, int
       thread_pop_size);
65     static void crossover_ox(population& curr_pop, population& tmp_curr_pop, double cr, int
       thread_idx, int thread_pop_size);
66     static void mutate(population& curr_pop, double mr, int thread_idx, int thread_pop_size);
67
68 private:
```

```
69     const int num_runs;
70     const int num_evals;
71     const int num_patterns_sol;
72     const string filename_ini;
73     const string filename_ins;
74     const int pop_size;
75     const double crossover_rate;
76     const double mutation_rate;
77     const int num_players;
78
79     population curr_pop;
80     vector<double> curr_obj_vals;
81     solution best_sol;
82     double best_obj_val;
83
84     instance ins_tsp;
85     solution opt_sol;
86     dist_2d dist;
87
88     const int num_threads;
89 };
90
91 inline pga::pga(int num_runs,
92                 int num_evals,
93                 int num_patterns_sol,
94                 string filename_ini,
95                 string filename_ins,
96                 int pop_size,
97                 double crossover_rate,
98                 double mutation_rate,
99                 int num_players,
100                int num_threads
101                )
102     : num_runs(num_runs),
103       num_evals(num_evals),
104       num_patterns_sol(num_patterns_sol),
105       filename_ini(filename_ini),
106       filename_ins(filename_ins),
107       pop_size(pop_size),
108       crossover_rate(crossover_rate),
109       mutation_rate(mutation_rate),
110       num_players(num_players),
111       num_threads(num_threads)
112 {
113     srand();
114 }
115
116 inline pga::population pga::run()
117 {
118     vector<double> avg_obj_val_eval(num_evals, 0.0);
119     std::mutex cv_m;
120     std::condition_variable cv;
121     std::atomic<int> icnt(num_threads);
122
123     for (int r = 0; r < num_runs; r++) {
124         // 0. initialization
125         population tmp_curr_pop = curr_pop = init();
126
127         vector<thread> t;
128         for (int i = 0; i < num_threads; i++) {
129             t.push_back(thread(thread_run,
130                                ref(curr_pop),
131                                ref(tmp_curr_pop),
132                                ref(curr_obj_vals),
133                                ref(dist),
```

```
134                               i, // thread_idx
135                               num_threads,
136                               num_evals,
137                               ref(best_obj_val),
138                               ref(best_sol),
139                               ref(avg_obj_val_eval),
140                               curr_pop.size()/num_threads,
141                               crossover_rate,
142                               mutation_rate,
143                               num_players,
144                               ref(cv_m),
145                               ref(cv),
146                               ref(icnt)
147                               ));
148
149           cpu_set_t cpuset;
150           CPU_ZERO(&cpuset);
151           CPU_SET(i, &cpuset);
152           int rc = pthread_setaffinity_np(t[i].native_handle(), sizeof(cpu_set_t), &cpuset);
153           if (rc != 0) {
154               std::cerr << "Error calling pthread_setaffinity_np: " << rc << "\n";
155           }
156       }
157
158       for (int i = 0; i < num_threads; i++)
159           t[i].join();
160   }
161
162   for (int i = 0; i < num_evals; i++) {
163       avg_obj_val_eval[i] /= num_runs;
164       cout << fixed << setprecision(3) << avg_obj_val_eval[i] << endl;
165   }
166
167   cout << best_sol << endl;
168
169   return curr_pop;
170 }
171
172 inline pga::population pga::init()
173 {
174     // 1. parameters initialization
175     population curr_pop(pop_size, solution(num_patterns_sol));
176     curr_obj_vals.assign(pop_size, 0.0);
177     best_obj_val = numeric_limits<double>::max();
178     ins_tsp = instance(num_patterns_sol, pattern(2));
179     opt_sol = solution(num_patterns_sol);
180     dist = dist_2d(num_patterns_sol, dist_1d(num_patterns_sol));
181
182     // 2. input the TSP benchmark
183     if (!filename_ins.empty()) {
184         ifstream ifs(filename_ins.c_str());
185         for (int i = 0; i < num_patterns_sol ; i++)
186             for (int j = 0; j < 2; j++)
187                 ifs >> ins_tsp[i][j];
188     }
189
190     // 3. input the optimum solution
191     string filename_ins_opt(filename_ins + ".opt");
192     if (!filename_ins_opt.empty()) {
193         ifstream ifs(filename_ins_opt.c_str());
194         for (int i = 0; i < num_patterns_sol ; i++) {
195             ifs >> opt_sol[i];
196             opt_sol[i]--;
197         }
198     }
```

```
199
200
201     // 4. initial solutions
202     if (!filename_ini.empty()) {
203         ifstream ifs(filename_ini.c_str());
204         for (int p = 0; p < pop_size; p++)
205             for (int i = 0; i < num_patterns_sol; i++)
206                 ifs >> curr_pop[p][i];
207     }
208     else {
209         for (int p = 0; p < pop_size; p++) {
210             for (int i = 0; i < num_patterns_sol; i++)
211                 curr_pop[p][i] = i;
212             for (int i = num_patterns_sol-1; i > 0; --i) // shuffle the solutions
213                 swap(curr_pop[p][i], curr_pop[p][rand() % (i+1)]);
214         }
215     }
216
217     // 5. construct the distance matrix
218     for (int i = 0; i < num_patterns_sol; i++)
219         for (int j = 0; j < num_patterns_sol; j++)
220             dist[i][j] = i == j ? 0.0 : distance(ins_tsp[i], ins_tsp[j]);
221
222     return curr_pop;
223 }
224
225 inline void pga::thread_run(population& curr_pop,
226                             population& tmp_curr_pop,
227                             vector<double>& curr_obj_vals,
228                             const dist_2d& dist,
229                             int thread_idx,
230                             int num_threads,
231                             int num_evals,
232                             double& best_obj_val,
233                             solution& best_sol,
234                             pattern& avg_obj_val_eval,
235                             int thread_pop_size,
236                             double crossover_rate,
237                             double mutation_rate,
238                             int num_players,
239                             std::mutex& cv_m,
240                             std::condition_variable& cv,
241                             std::atomic<int>& icnt
242                             )
243 {
244     // 0. declaration
245     const int num_gens = num_evals / curr_pop.size();
246     const int s_idx = thread_idx * thread_pop_size;
247     const int e_idx = s_idx + thread_pop_size;
248
249     int gen_count = 0;
250
251     while (gen_count < num_gens) {
252         // 1. evaluation
253         for (int i = s_idx; i < e_idx; i++)
254             curr_obj_vals[i] = evaluate(curr_pop[i], dist);
255         {
256             std::unique_lock<std::mutex> lk(cv_m);
257             if (--icnt)
258                 cv.wait(lk);
259             else {
260                 // 1.1. sync
261                 size_t z = 0;
262                 auto avg = &avg_obj_val_eval[gen_count * curr_pop.size()];
263                 for (size_t i = 0; i < curr_pop.size(); i++) {
```

```
264                    if (curr_obj_vals[i] < best_obj_val)
265                        best_obj_val = curr_obj_vals[z = i];
266                    avg[i] += best_obj_val;
267                }
268                best_sol = curr_pop[z];
269                // 1.2. continue
270                icnt = num_threads;
271                cv.notify_all();
272            }
273        }
274
275        // 2. determination
276        tournament_select(curr_pop, curr_obj_vals, tmp_curr_pop, num_players, thread_idx,
       thread_pop_size);
277        {
278            std::unique_lock<std::mutex> lk(cv_m);
279            if (--icnt)
280                cv.wait(lk);
281            else {
282                // 2.1. sync
283                curr_pop = tmp_curr_pop;
284                // 2.2. continue
285                icnt = num_threads;
286                cv.notify_all();
287            }
288        }
289
290        // 3. transition
291        crossover_ox(curr_pop, tmp_curr_pop, crossover_rate, thread_idx, thread_pop_size);
292        mutate(curr_pop, mutation_rate, thread_idx, thread_pop_size);
293
294        ++gen_count;
295    }
296
297    // avg_obj_val += best_obj_val;
298 }
299
300 inline double pga::evaluate(const solution& curr_sol, const dist_2d& dist)
301 {
302    double tour_dist = 0.0;
303    for (size_t i = 0; i < curr_sol.size(); i++) {
304        const int r = curr_sol[i];
305        const int s = curr_sol[(i+1) % curr_sol.size()];
306        tour_dist += dist[r][s];
307    }
308
309    return tour_dist;
310 }
311
312 inline void pga::tournament_select(const population& curr_pop, const vector<double>&
       curr_obj_vals, population& tmp_curr_pop, int num_players, int thread_idx, int
       thread_pop_size)
313 {
314    int s_idx = thread_idx * thread_pop_size;
315    int e_idx = s_idx + thread_pop_size;
316
317    for (int i = s_idx; i < e_idx; i++) {
318        int k = rand() % curr_pop.size();
319        double f = curr_obj_vals[k];
320        for (int j = 1; j < num_players; j++) {
321            int n = rand() % curr_pop.size();
322            if (curr_obj_vals[n] < f) {
323                k = n;
324                f = curr_obj_vals[k];
325            }
```

```
326
327              }
328              tmp_curr_pop[i] = curr_pop[k];
329         }
330 }
331
332 inline void pga::crossover_ox(population& curr_pop, population& tmp_curr_pop, double cr, int
        thread_idx, int thread_pop_size)
333 {
334     const int mid = thread_pop_size / 2;
335     const int s_idx = thread_idx * thread_pop_size;
336     const int m_idx = thread_idx * thread_pop_size + mid;
337     const int e_idx = thread_idx * thread_pop_size + thread_pop_size;
338     const size_t ssz = curr_pop[0].size();
339     for (int i = s_idx; i < m_idx; i++) {
340         const double f = static_cast<double>(rand()) / RAND_MAX;
341         if (f <= cr) {
342             // 1. create the mapping sections
343             size_t xp1 = rand() % (ssz + 1);
344             size_t xp2 = rand() % (ssz + 1);
345             if (xp1 > xp2)
346                 swap(xp1, xp2);
347
348             // 2. indices to the two parents and offspring
349             const int p[2] = { i, i+mid };
350
351             // 3. the main process of ox
352             for (int k = 0; k < 2; k++) {
353                 const int c1 = p[k];
354                 const int c2 = p[1-k];
355
356                 // 4. mask the genes between xp1 and xp2
357                 const auto& s1 = curr_pop[c1];
358                 const auto& s2 = curr_pop[c2];
359                 vector<bool> msk1(ssz, false);
360                 for (size_t j = xp1; j < xp2; j++)
361                     msk1[s1[j]] = true;
362                 vector<bool> msk2(ssz, false);
363                 for (size_t j = 0; j < ssz; j++)
364                     msk2[j] = msk1[s2[j]];
365
366                 // 5. replace the genes that are not masked
367                 for (size_t j = xp2 % ssz, z = 0; z < ssz; z++) {
368                     if (!msk2[z]) {
369                         tmp_curr_pop[c1][j] = s2[z];
370                         j = (j+1) % ssz;
371                     }
372                 }
373             }
374         }
375     }
376     for (int i = s_idx; i < e_idx; i++)
377         curr_pop[i] = tmp_curr_pop[i];
378 }
379
380 inline void pga::mutate(population& curr_pop, double mr, int thread_idx, int thread_pop_size)
381 {
382     const int s_idx = thread_idx * thread_pop_size;
383     const int e_idx = thread_idx * thread_pop_size + thread_pop_size;
384     for (int i = s_idx; i < e_idx; i++){
385         const double f = static_cast<double>(rand()) / RAND_MAX;
386         if (f <= mr) {
387             int m1 = rand() % curr_pop[0].size();        // mutation point
388             int m2 = rand() % curr_pop[0].size();        // mutation point
389             swap(curr_pop[i][m1], curr_pop[i][m2]);
```

```
390        }
391    }
392 }
393
394 #endif
```

14.2.1 Declaration of functions and parameters

As shown in lines 22–27 and lines 69–86 of Listing 14.1, most of the declarations are the same as those for simple GA for solving the TSP in Listing 12.1. The data member num_threads defined in line 88 is for the number of threads to be used by MSP-GA. The declaration of the thread_run() operator (lines 42–59) of MSP-GA for the TSP is the entry point of the thread (slave). It is declared as a static member function as far as this implementation is concerned, and it requires a lot of additional information from the run() operator, namely, thread_idx, num_threads, nevals, thread_pop_size, cv_m, cv, and icnt. The parameters thread_idx and num_threads are for the thread index (a number between 0 and num_threads − 1) and the number of threads to be created by MSP-GA during the convergence process. The parameters nevals and thread_pop_size are used to represent the number of evaluations to be carried out and the number of chromosomes each thread will work on. The parameters cv_m, cv, and icnt are declared to control the threads. More precisely, cv_m is the mutex for the condition variable cv, which is eventually a queue for the slaves that have finished their tasks to wait for the last slave to finish its task and synchronize the data, while icnt is the number of slaves that can be considered as a reference count so that when the count goes down to 0, this means that this particular slave is the last one finishing its tasks and thus is responsible for the task of synchronization, that is, for moving the data around all the slaves. As mentioned before, because thread_run() is implemented as a static member function, the operators evaluate(), tournament_select(), crossover_ox(), and mutate() are also implemented as a static member function. Note that this is not necessarily the only way to implement a thread using the multi-threading technologies of C++. Rather, it can be implemented as a non-static member function. We will leave this to the reader.

14.2.2 The main loop

The main loop is composed of run() and thread_run(), which are shown in lines 116–170 and lines 225–298, respectively. The functionality of these two functions is basically the same as that of simple GA for solving the TSP except that it is the member function thread_run() that calls the functions tournament_select(), crossover_ox(), and mutate() instead of the member function run() because these operators will be carried out by each thread instead of by run(). The discussion that follows will focus on the main differences between simple GA and MSP-GA to show how MSP-GA works. In run(), the parameters cv_m, cv, and icnt, shown in lines 119–121, will be first declared and initialized if possible. The init() operator will then be called to create the population and initialize the other parameters of MSP-GA, as shown in line 125. In line 127, a vector of threads named t is defined and initialized as an empty vector. In lines 128–147, the push_back method is used to add num_threads threads to the vector t. Lines 149–155 assign each newly created thread to a particular CPU so that it will be run on that CPU throughout the lifetime of the thread. The main reason is that the scheduler of an operating system may move a thread from one CPU to another, and this may slow down the performance of a thread. Hence, in this implementation, each newly created thread will be assigned to a particular CPU and be run on that CPU throughout the lifetime of the thread. Lines 158–159 will call join() for each thread to wait for it to terminate.

Although thread_run() is not the main function of MSP-GA, it plays an essential role. The member function run() is responsible for the creation of threads that will be run simultaneously for the search of solutions via the call to thread().[2] Therefore, we will discuss it in this subsection. The main structure of this function can be divided into four parts: declaration, evaluation, determination, and transition. The variables num_gens, s_idx, e_idx, and gen_count are declared and initialized in lines 245–249. Among them, num_gens represents the number of generations that will be carried out by MSP-GA and its threads. Since only the number of evaluations to be performed by MSP-GA is given, dividing it by the number of chromosomes (i.e., curr_pop.size()) gives the number of generations to be performed. The variables s_idx and e_idx are used to specify the range of chromosomes[3] that the thread indexed by thread_idx is to work on, where thread_idx represents the index of a thread (i.e., this thread is the thread_idx-th thread) and thread_pop_size represents the number of chromosomes to be worked on by a thread. As shown in Fig. 14.10, suppose the size of the population (curr_pop.size()) is set equal to four and the number

---

[2] Eventually, it is also responsible for waiting for each thread to terminate.
[3] That is, chromosomes in the range of s_idx (inclusive) to e_idx (exclusive), or symbolically, [s_idx, e_idx).

of threads is set equal to two. Then the size of the population of each thread (i.e., `thread_pop_size`) will be two. In other words, this means that the population will be divided into subpopulation 1 (SP1) and SP2 for thread 1 and thread 2, respectively. For the first thread (i.e., `thread_idx = 0`), its `s_idx` and `e_idx` will assume the values 0 and 2, which bound the chromosomes controlled by the first thread. For the second thread (i.e., `thread_idx = 1`), `s_idx` and `e_idx` will assume the values 2 and 4, which bound the chromosomes controlled by the second thread. This indexing mechanism is developed to isolate the chromosomes for each thread as much as possible so that no copy of the chromosomes for each thread is required while at the same time minimizing the need of dealing with the race condition. Unfortunately, even with this indexing mechanism, it is still not possible to fully avoid the race condition problem. For instance, one thread is updating a chromosome in the population `curr_pop`, while another thread is reading the chromosome from the population `curr_pop` at the same time. Hence, some mechanisms are needed to make sure that all the threads have finished their tasks before they fall asleep to wait for the data to be synchronized. As far as the implementation of MSP-GA described here is concerned, this can be done by using a condition variable that acts as a ready queue so that each thread that has finished its task will simply fall asleep and wait on the ready queue, except the last one, which is responsible for synchronizing the data for all the threads. Once the last one completes its task of synchronizing the data, all the threads will simply move on to the next stage simultaneously.

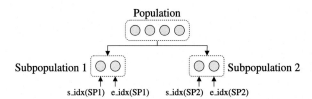

Figure 14.10 How to divide the population into a set of subpopulations for each thread.

The main loop of the thread thread_run() is shown in lines 251–295, which will be repeatedly performed for a certain number of generations. The very first operator in the main loop is the evaluation operator, as shown in lines 253–273, the tasks of which can be divided into two stages: (1) compute the fitness values of all the chromosomes it gets assigned and (2) update the average fitness values, the best-so-far solution, and its fitness value. As mentioned in Fig. 14.9, the fitness values of all the chromosomes can be computed independently and thus simultaneously. As shown in lines 253–254, each thread will compute the fitness values of all the chromosomes it gets assigned. After that, as shown in lines 256–258, each thread will proceed to decrease the island count (i.e., the value of icnt) by one on line 257 to see if it should fall asleep to wait for all the other threads to complete their tasks (line 258) or if it should take care of synchronizing the data for all the threads (lines 261–268). This makes it easy to ensure that all the threads have completed their computation of the fitness values and thus all the data are essentially ready for the second stage (i.e., the stage responsible for synchronizing the data for all the threads) to perform its task. More precisely, as shown in lines 259–272, MSP-GA will update the average fitness values, the best-so-far solution, and its fitness value. As shown in Fig. 14.9, these update procedures are eventually dependent on each other because two or more threads may want to update the best-so-far solution and its fitness value simultaneously, thus causing the race condition. A simple solution to this problem, as the implementation of MSP-GA shows, is to let the thread ($\mathcal{T}_L$) that finishes the computation of the fitness values last do the update, that is, update (1) the average fitness values (line 266) and (2) the best-so-far solution (line 268) and its fitness value (line 265). After the update, the thread $\mathcal{T}_L$ will restore the value of icnt back to the number of threads (line 270) before it resumes the execution of all the threads that have been waiting on the ready queue for the synchronization to be done so that they can now move on to the next stage (line 271). However, before that, it is worth mentioning that the braces surrounding the block starting with the declaration lk are to limit the scope of the lock.

The determination operator of MSP-GA contains two parts, namely, tournament selection and chromosomes reproduction, as shown in lines 276–288. As shown in Fig. 14.9 and discussed before, the tournament selection operator can be regarded as composed of a set of independent computation tasks; thus, all the threads are able to select their offspring via a call to the tournament selection operator simultaneously, as shown in line 276. What each thread has to do after the call to the tournament selection operator returns is exactly the same as with the evaluation operator. That is, each thread will proceed to decrease the island count (i.e., the value of icnt) by one on line 279 to see if it should fall asleep to wait for all the other threads to complete their tasks (line 280) or it should take care of synchronizing the data for all the threads (line 283). Once again, the thread ($\mathcal{T}_L$) that finishes the tournament selection task last is responsible for synchronizing the data for all the threads, which in this case is simply copying the offspring (i.e., tmp_curr_pop) created by all the threads to the parents (i.e., curr_pop) so that it is ready for the next generation, as shown in line 283. After copying, the thread T_L will as for the evaluation operator restore the value of icnt back to the number of threads (line 285) before it resumes the execution of all the threads that have been waiting on the ready queue for the synchronization to be done (line 286) so that they can now move on to the next stage (i.e., crossover and mutation, shown in lines 291–292). More precisely, the crossover and mutation operators on lines 291–292 can be regarded as that each is composed of a set of independent computation tasks; as such, they can be carried out by all the threads simultaneously. Finally, in the end of this loop, gen_count will be increased by one to let the thread of MSP-GA move on to the next generation, as shown in line 294.

14.2.3 Additional functions

As shown in lines 172–223, since the initialization operator of MSP-GA "init()" is performed in the main thread, its implementation is similar to the implementation of init() for simple GA, as shown in Listing 12.1. The implementation of the evaluation operator "evaluate()" on lines 300–310 is also very similar to the implementation of evaluate() for the simple GA shown in Listing 12.1. The main difference, however, is that the implementation of evaluate() for the simple GA uses a two-level nested loop to compute the fitness values of all the chromosomes of the population (the outer loop is for each chromosome and the inner [nested] loop is for each gene of a chromosome), while the implementation of evaluate() for MSP-GA uses a loop to compute the fitness value of one chromosome. The main reason is that each thread of MSP-GA will evaluate a subset of the chromosomes; therefore, the evaluate() operator needs to be modified so that it computes the fitness value of a chromosome instead of the fitness values of all the chromosomes in a single call. The tournament_select() operator of MSP-GA also needs to be modified a bit for the same reason so that all the threads can be run simultaneously without causing a race condition. In other words, each thread will be performed on the part of the chromosomes for which it is responsible, as shown in lines 312–330. The very first change is the addition of two new parameters thread_idx and thread_pop_size. The former is the thread index, while the latter is the number of chromosomes assigned to this thread. On lines 314–315, two variables s_idx and e_idx are then defined to control the range of chromosomes the tournament selection operator of this thread gets accessed, that is, the half-open range [s_idx, e_idx) specifies the range of chromosomes in `curr_pop` and corresponding objective values in `curr_obj_vals`. Another modification is found in line 317, where the lower and upper bounds of the outer loop have been changed accordingly from [0, `curr_pop.size()`) to [`s_idx, e_idx`).

Similar modifications can also be found in the crossover and mutation operators for the same reason. The lower and upper bounds of the loop have to be changed accordingly from [0, `curr_pop.size()`) to [`s_idx, e_idx`), to ensure that they access only the portion of chromosomes they get assigned. As shown in lines 332–378, OX crossover is used in this implementation. Still, a few modifications are needed to make it work for MSP-GA. First, the additional parameters `thread_idx` and `thread_pop_size` are added to specify the thread index and the number of chromosomes each thread is allowed to get accessed. The parameter `thread_pop_size` instead of `curr_pop.size()` of simple GA is used to calculate `mid` (line 334). The variables `s_idx`, `m_idx`, and `e_idx` in lines 335–337 are then defined and initialized to specify the range of chromosomes that gets assigned to the thread indexed by `thread_idx` and the index to the middle one based on the values of `thread_idx`, `thread_pop_size`, and `mid`. For the mutation operator, as shown in lines 380–392, the additional parameters `thread_idx` and `thread_pop_size` are again used to specify the thread index and the number of chromosomes each thread is allowed to get accessed. The variables `s_idx` and `m_idx` in lines 382–383 are then defined and initialized to specify the range of chromosomes that gets assigned to the thread indexed by `thread_idx` based on the values of `thread_idx` and `thread_pop_size`.

14.3. Simulation results of parallel GA for the TSP

To better understand the performance of MSP-GA, we compare it with simple GA (as described in Chapter 12) for solving the benchmarks eil51 and pr1002 of the TSP. Also, the population size is set equal to 120, 240, 480, and 960 for both simple GA and MSP-GA. The number of evaluations is set equal to 120,000. The crossover and mutation rates are set equal to 0.4 and 0.1, respectively. As shown in Table 14.1, we first compare the performance of simple GA (GA-OX) and MSP-GA with four threads ($T_h = 4$) for the benchmark eil51. The parameters $\mathcal{A}$, m, and t in this table represent the algorithm, the number of chromosomes, and computation time in seconds. The terms "Real," "User," and "System" stand for the response time, the user CPU time, and the system CPU time. Note that these three times come from the Linux command "time." This table shows that all the response times of MSP-GA with different numbers of chromosomes are longer than those of simple GA. According to our observations, this may be because the problem size is too small. This means that multi-threading technologies are not particularly useful in this case.

Table 14.1 Comparison between GA-OX and MSP-GA with different numbers of chromosomes for eil51 in terms of computation time.

$\mathcal{A}$	GA-OX			MSP-GA ($T_h = 4$)		
m/t	Real	User	System	Real	User	System
120	12.005	11.778	0.197	20.993	40.100	25.211
240	12.287	12.058	0.198	19.509	37.750	24.203
480	12.511	12.268	0.207	19.603	38.738	24.213
960	12.688	12.452	0.199	20.393	41.232	24.917

We then compared the performance of simple GA (GA-OX) and MSP-GA with four threads ($T_h = 4$) for the benchmark pr1002 (Table 14.2). All the "Real" times indicate that MSP-GA is much faster than simple GA. When the population size is set equal to 120, MSP-GA can reduce the response time[4] by only about 50.96%; however, when the population size is set equal to 960, MSP-GA can reduce the response time by about 54.35%.[5] Interestingly, MSP-GA takes a longer "User" time than GA-OX. This implies that increasing the number of threads also incurs some overhead. Multi-threading and the master–slave parallel model provide ways to accelerate the response time of GA, but not the computation time.

Table 14.2 Comparison between GA-OX and MSP-GA with a different number of chromosomes for pr1002 in terms of computation time.

$\mathcal{A}$	GA-OX			MSP-GA ($T_h = 4$)		
m/t	Real	User	System	Real	User	System
120	172.530	171.508	0.498	84.614	251.297	22.631
240	178.507	177.478	0.442	82.626	253.037	19.963
480	185.202	184.053	0.579	82.782	262.370	16.221
960	195.929	194.952	0.303	89.440	278.752	25.729

Fig. 14.11 compares simple GA and MSP-GA for solving the TSP benchmarks eil51 and pr1002 in an alternative way. It can be easily seen that

[4] $(84.614 - 172.530)/172.530 \times 100\% = 50.96\%.$
[5] $(89.440 - 195.929)/195.929 \times 100\% = 54.35\%.$

for eil51, both the response time (Fig. 14.11(a)) and the computation time (Fig. 14.11(b)) MSP-GA takes are longer than for simple GA. As expected, for pr1002, the response time (Fig. 14.11(c)) MSP-GA takes is shorter than for simple GA, even though the computation time (Fig. 14.11(d)) it takes is longer than for simple GA. Note that in these comparisons, the number of evaluations is the same for all simulations, and the time simple GA with different numbers of chromosomes takes is almost the same for both benchmarks. Similar results were found for MSP-GA for both benchmarks.

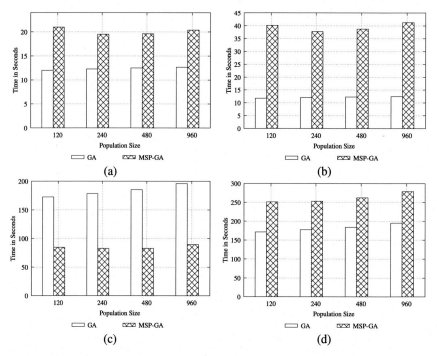

Figure 14.11 Simulation results of simple GA and MSP-GA for solving the TSP benchmarks eil51 and pr1002. (a) The real response time simple GA and MSP-GA take for eil51. (b) The total computation time simple GA and MSP-GA take for eil51. (c) The real response time simple GA and MSP-GA take for pr1002. (d) The total computation time simple GA and MSP-GA take for pr1002.

Table 14.3 compares simple GA and MSP-GA with four threads ($T_h = 4$) in terms of the end results. Since the search strategies of simple GA and MSP-GA for solving the TSP are the same and the only difference is that simple GA uses one thread while MSP-GA uses four threads during the convergence process, the end results (i.e., total distance of tours) of these two GAs are, as expected, almost the same. These results match the assumptions and goals of the master–slave parallel model of GA.

Table 14.3 Comparison between GA-OX and MSP-GA with different numbers of chromosomes for eil51 and pr1002 in terms of the quality.

Benchmark	eil51		pr1002	
$m/\mathcal{A}$	GA-OX	MSP-GA ($T_h = 4$)	GA-OX	MSP-GA ($T_h = 4$)
120	474.879	473.798	2,864,484.970	2,857,568.046
240	475.807	476.169	3,341,313.360	3,337,738.613
480	482.986	484.919	3,928,694.245	3,932,310.887
960	531.092	532.126	4,560,967.142	4,557,417.999

To evaluate the performance in terms of the number of threads used, Table 14.4 and Fig. 14.12 show the results of MSP-GA with different numbers of threads ($T_h = 2$, 4, 6, and 8) and different numbers of chromosomes ($m = 120$, 240, 480, and 960). The results show that the "User" time (i.e., the computation time) for all simulations increases as the number of threads used by MSP-GA increases. The "Real" time (i.e., the response time) decreases as the number of threads used by MSP-GA increases. This means that using a larger number of threads is a possible way to speed up MSP-GA; however, there is no guarantee that the speedup will be linearly proportional to the number of threads for MSP-GA.

Table 14.4 Comparison of MSP-GA with different numbers of threads for pr1002 in terms of response time.

$\mathcal{A}$	MSP-GA ($m = 120$)			MSP-GA ($m = 240$)		
T_h/t	Real	User	System	Real	User	System
2	118.267	208.218	10.252	118.585	210.665	10.329
4	84.614	251.297	22.631	82.626	253.037	19.963
6	71.963	268.179	49.812	69.019	267.232	46.388
8	63.840	274.813	74.857	61.152	275.227	73.726

$\mathcal{A}$	MSP-GA ($m = 480$)			MSP-GA ($m = 960$)		
T_h/t	Real	User	System	Real	User	System
2	121.724	217.393	10.973	130.210	235.012	10.627
4	82.782	262.370	16.221	89.440	278.752	25.729
6	68.222	270.414	47.138	70.428	286.385	47.142
8	59.852	277.917	72.680	61.014	291.679	75.089

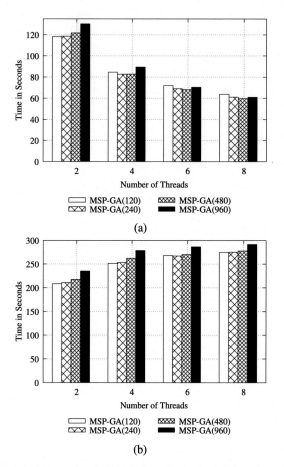

Figure 14.12 Simulation results of MSP-GA for solving the TSP benchmark pr1002 with different numbers of threads. (a) The real response time of MSP-GA with different numbers of threads. (b) The total computation time of MSP-GA with different numbers of threads.

14.4. Discussion

This chapter provides a brief review of parallel metaheuristic algorithms from different perspectives. They provide alternative ways to enhance the performance of metaheuristic algorithms in the sense that not only the response time is reduced, but also the end results are improved. Parallel metaheuristic algorithms have been classified in different ways. Alba et al. (Alba et al., 2013) divide them into single-solution-based and population-based parallel metaheuristic algorithms. To evaluate the performance of parallel metaheuristic algorithms, in this chapter we used master–slave PGA for solving the TSP as an example to show how the performance of parallel metaheuristic algorithms is enhanced. The simulation results show that when designing parallel metaheuristic algorithms, several factors must be taken into account, concerning not only the computer system but also the parallel metaheuristic algorithm itself. Regarding the use of multi-threading technologies for the implementation of a metaheuristic algorithm, the very first thing we have to do is to figure out whether or not the computation tasks of each operator of the metaheuristic algorithm to be performed by each thread are "independent" of each other. Computation tasks that are independent of each other can be performed in parallel by all the threads, while for the computation tasks that are dependent on each other, we do not have the luxury to do the same. Rather, what we have to do in this case is to avoid the so-called race condition. This is more easily said than done. Luckily, for the implementation of GA for the TSP described in this chapter, it can be easily achieved by having one of the threads—the one that finished the computation tasks last—move the data around (from one thread to another). After finishing the movement of data, all this thread has to do is to resume the execution of all the other threads that have been waiting for this thread to finish its tasks. This process is repeated until all computation tasks are finished. Hence, to speed up parallel algorithms, the independent tasks must be maximized while minimizing the dependent ones.

One of the most important research trends in parallel metaheuristic algorithms is to apply them to developed platforms, such as cloud computing platforms and GPUs, because some cloud computing platforms (e.g., Hadoop and Spark) and Nvidia CUDA provide an API that makes it easy for researchers to realize parallel metaheuristic algorithms on multiple processors or computers. Some modifications to parallel metaheuristic algorithms are typically needed to make it possible to apply them in these promising environments. For cloud computing platforms, parallel metaheuristic algorithms need to be parallelized based on the MapReduce model. An intuitive way to apply parallel metaheuristic algorithms to cloud computing platforms is to use the master–slave architecture, because the MapReduce model can be regarded as a master–slave architecture in one way or another. This, however, does not mean that other kinds of parallel metaheuristic algorithms cannot be realized in such cloud computing environments. We actually believe that most of them can be applied to cloud computing platforms or GPU systems. One of the essential issues for most parallel metaheuristic algorithms to apply them to real-time systems is to reduce the response time during the convergence process. However, this incurs overhead. In summary, although new and promising environments are continuously presented, some open issues still exist in parallel environments. The communication time, the topology of different processors or computers, memory usage, load balance, and management are the major open issues in these new environments; therefore, we need to take these issues into account when designing parallel metaheuristic algorithms for new environments.

Supplementary source code

Here are the hyperlinks to the directories where the source code described in this chapter resides.

https://github.com/cwtsaiai/metaheuristics_2023/src/c++/3-tsp/

CHAPTER FIFTEEN

Hybrid metaheuristic and hyperheuristic algorithms

15.1. The basic idea of the hybrid metaheuristic algorithm

After using metaheuristic algorithms to solve an optimization problem, most researchers ask themselves which algorithm is best for other optimization problems. The answer to this question can be found in an early study (Wolpert and Macready, 1997), in which Wolpert and Macready pointed out that for a search algorithm $\mathcal{A}$ that outperforms another search algorithm $\mathcal{B}$ in solving an optimization problem $\mathbb{P}_a$, there always exists another optimization problem $\mathbb{P}_b$ for which $\mathcal{B}$ will beat $\mathcal{A}$. This means that every search algorithm has its pros and cons and consequently a search algorithm may perform well only for a specific kind of optimization problem. A possible way to enhance the search performance of metaheuristics by integrating two or more (meta)heuristic algorithms, called hybrid metaheuristics, has been discussed for years. The basic idea is to leverage the strong points of different search algorithms to provide a comprehensive search strategy in solving optimization problems. An intuitive way to integrate two search algorithms is to use a population-based metaheuristic algorithm to identify promising areas in the solution space for the so-called "global search" while using a single-solution-based metaheuristic algorithm or local search algorithm to fine-tune the solutions searched by the population-based metaheuristic algorithm for the so-called "local search." Because such a combination takes into account the diversification (also called global search and exploration) and intensification (also called local search and exploitation), such a hybrid metaheuristic algorithm can typically give better results than single metaheuristic algorithms in solving complex optimization problems.

Because a combination of multiple (meta)heuristic algorithms as a hybrid metaheuristic algorithm may provide better results than single metaheuristic algorithms, most researchers in this field are interested in developing high-performance hybrid metaheuristic algorithms. In (Blum et al., 2011), Blum and his colleagues raised three important questions that a researcher needs to take into account before developing a new hybrid meta-heuristic algorithm: (1) What is the goal of optimization? (2) Is there still room to improve? (3) And which type of hybrid metaheuristic algorithm can find better results for the optimization problem in question? The first question is asked to clarify whether the goal in the development of a hybrid metaheuristic algorithm is to accelerate the "response time" or to improve the "quality of the end result" when solving a particular optimization problem before developing a new hybrid metaheuristic algorithm. The second question is to make sure that there exists a good metaheuristic algorithm or a heuristic algorithm that works well for the optimization problem in question within a reasonable time. We do not need a new hybrid metaheuristic algorithm when there is a good existing metaheuristic algorithm that can be used for solving the optimization problem in question. It is, however, necessary to develop a new hybrid metaheuristic algorithm if there is still plenty of room to improve, especially with respect to the quality of the end result. The third question is to make sure which combination of meta-heuristic algorithms or metaheuristic algorithm and heuristic algorithms can provide a better result than a metaheuristic algorithm by itself can pro-vide. Before we combine two or more (meta)heuristic algorithms into one metaheuristic algorithm, we must know that there is no guarantee that the performance of a metaheuristic algorithm or a hybrid algorithm under de-velopment will be improved by adding more search algorithms. This means that we need to take into account the side effects of adding more search algorithms, because the impact may not always be positive. This explains why we need to make sure that the side effect is acceptable before we use this new hybrid heuristic algorithm to replace the original search algorithm we have.

Since there are different ways to integrate different metaheuristic and heuristic algorithms, these hybrid metaheuristic algorithms were classified in some studies. In an early study (Sinha and Goldberg, 2003), Sinha and Goldberg presented a classification method to classify hybrid architectures of evolutionary algorithms into the following categories: pipeline hybrids, asynchronous hybrids, hierarchical hybrids, and embedded hybrids. Al-though this taxonomy is originally for genetic and evolutionary algorithms, it provides a roadmap for understanding most possible ways to integrate the genetic algorithm (GA) and other search algorithms. In the same study, Sinha and Goldberg also showed that preprocessor, postprocessor, and staged hybrids are three possible ways to combine evolutionary and other search algorithms in the category of pipeline hybrids, as shown in

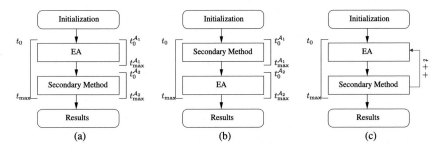

Figure 15.1 Different ways of hybrids (Sinha and Goldberg, 2003). (a) Preprocessor hybrids. (b) Postprocessor hybrids. (c) Staged hybrids.

Fig. 15.1. In these examples, as shown in Fig. 15.1(a and b), the first two classes take a different order to perform the evolutionary and other search algorithms during the convergence process, where $\mathcal{A}_1$ and $\mathcal{A}_2$ represent the first and second search algorithms, t_0 and t_{max} represent the first and last generations (or iterations), $t_0^{\mathcal{A}_i}$ and $t_{max}^{\mathcal{A}_i}$ represent the first and last generations (or iterations) of the i-th search algorithm, and t represents the current generation. This implies that the results of the first search algorithm $\mathcal{A}_1$ will be the initial seeds of the second search algorithm $\mathcal{A}_2$. It is just like performing a search algorithm $\mathcal{A}_1$ and then passing its results to another search algorithm $\mathcal{A}_2$ as soon as the first search algorithm $\mathcal{A}_1$ converges. In the first hybridization, $\mathcal{A}_1$ is the evolutionary algorithm that is used to find promising regions in the solution space before the search continues with another search algorithm $\mathcal{A}_2$. In the second hybridization, $\mathcal{A}_1$ will be a secondary method that is used to find a good initial population for the evolutionary algorithm $\mathcal{A}_2$. As shown in Fig. 15.1(c), the third class shows another way to integrate different search algorithms. It will, however, perform these two search algorithms $\mathcal{A}_1$ and $\mathcal{A}_2$ repeatedly during the convergence process. Sinha and Goldberg pointed out some possible secondary methods that can be integrated into an evolutionary algorithm, e.g., local search methods, simulated annealing (SA), artificial neural networks, fuzzy logic, metaheuristic algorithms, and intelligent search algorithms.

In a later study (Lozano and García-Martínez, 2010), Lozano and García-Martínez provided a survey on the hybridizations of metaheuristic algorithm with evolutionary algorithms and classified them into three categories: (1) collaborative hybrid metaheuristic algorithms, (2) integrative hybrid metaheuristic algorithms, and (3) metaheuristic algorithms with evolutionary intensification and diversification components.

In collaborative hybrid metaheuristic algorithms, several metaheuristic algorithms are performed in parallel and information is exchanged during the convergence process. In this class, one possibility is to divide the population into a set of subpopulations in such a way that each gets assigned to a metaheuristic algorithm to search for a good solution simultaneously. After all the collaborative metaheuristic algorithms complete their searches, the subpopulations can be merged back together as a population. Using the migration mechanism to move a chromosome from one subpopulation to another is also a possible way to exchange information between subpopulations. Another possibility is that the hybrid metaheuristic algorithm first performs the evolutionary algorithm and then uses the local search algorithm to fine-tune the results of the evolutionary algorithm. In this way, the output of a search algorithm is used as the input of another search algorithm.

The second class consists of the integrative hybrid metaheuristic algorithms. In this class, one possibility is to let one of the metaheuristic algorithms, say A_1, play the role of master while letting the other metaheuristic algorithm, say A_2, play the role of subordinate of A_1. The memetic algorithm can be regarded as an example of this kind of hybridization in which the evolutionary algorithm acts as the search algorithm A_1 to search for good solutions in the whole solution space first and then the local search algorithm A_2 takes care of fine-tuning the solutions obtained by A_1.

The third class is composed of the metaheuristic algorithms with evolutionary intensification and diversification components. These algorithms are very similar to integrative hybrid metaheuristic algorithms because both integrate two different search algorithms into a single search algorithm. In this way, $A_1.O_1$ and $A_1.O_2$, which are essentially two different metaheuristic algorithms by themselves, represent two different operators of a metaheuristic algorithm A_1. The main difference between integrative hybrid metaheuristic algorithms and metaheuristic algorithms with evolutionary intensification and diversification components is that in the former the search algorithm A_2 is added to the main search algorithm A_1, while in the latter the search algorithm A_2 is used to replace a particular operator of another search algorithm A_1.

Ting et al. (Ting et al., 2015) divided the hybrid metaheuristics into two categories: collaborate and integrative hybridizations. Collaborate hybridization consists of three different ways to integrate the metaheuristic algorithms: multi-stage, sequential, and parallel. The multi-stage and sequential hybridizations are similar to the preprocessor and staged hybridizations of (Sinha and Goldberg, 2003), respectively. Parallel hybridization is very interesting in the sense that two or more search algorithms are performed simultaneously on the same population. Integrative hybridization includes two possible approaches: full manipulation and partial manipulation. The main difference between these two approaches is that the hybrid metaheuristic algorithm will change the entire population or only a portion of the entire population.

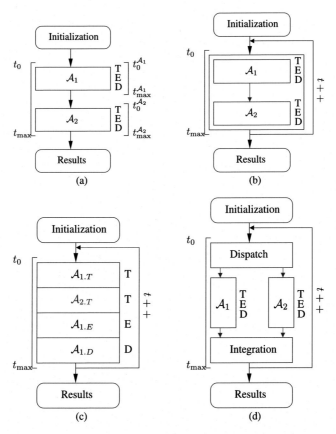

Figure 15.2 The frameworks of hybrid metaheuristics. (a) Multi-stage. (b) Sequential. (c) Embedded. (d) Parallel.

The classifications given by the aforementioned studies (Blum et al., 2010, 2011; Lozano and García-Martínez, 2010; Sinha and Goldberg, 2003; Ting et al., 2015) can be summarized as four different frameworks to integrate different (meta)heuristics into a hybrid metaheuristic algorithm (Fig. 15.2): (1) multi-stage, (2) sequential, (3) embedded, and (4) parallel frameworks. In the multi-stage framework, the hybrid metaheuristic algorithm performs two or more (meta)heuristics in multiple stages. In the sequential framework, some hybrid metaheuristic algorithms perform two or more (meta)heuristics repeatedly at each iteration. In the embedded framework, we use a search algorithm A_1 as the main search algorithm and the other search algorithm A_2 as the auxiliary search algorithm. They can be integrated by either adding A_2 to A_1 or using A_2 to replace an operator of A_1 for searching good solutions at each iteration. In the parallel framework, the (meta)heuristics are performed simultaneously for the same solution space or for different solution spaces.

In summary, since several successful cases have shown that hybrid metaheuristics provide a better way to solve optimization problems than metaheuristics, they can be found in several studies (Blum et al., 2008, 2011; Talbi, 2014). Based on studies given in (Blum et al., 2008, 2010, 2011; Lozano and García-Martínez, 2010; Sinha and Goldberg, 2003; Talbi, 2014; Ting et al., 2015), the major advantage of hybrid metaheuristic algorithms is that they are able to improve the quality of the end result provided that the integration is suitable. The most critical factor for success is that hybrid metaheuristic algorithms can leverage the strength of different (meta)heuristics. However, most hybrid metaheuristics are computationally more expensive than metaheuristics, because most hybrid metaheuristics are more complicated than metaheuristics. For this reason, Blum and his colleagues (Blum et al., 2011) recommended that researchers in this field make sure that the goal for solving an optimization problem is either reducing the response time or improving the quality of the end result before a new hybrid metaheuristic algorithm is developed.

15.2. The basic idea of the hyperheuristic algorithm

Among the hybrid metaheuristics, there is a special case that integrates two or more (meta)heuristic algorithms as a search algorithm, called hyperheuristic algorithm. Since the way the underlying algorithms are integrated and the way the framework works are different from most of the hybrid metaheuristics, we will turn our attention to hyperheuristic algorithms in this section. In (Burke et al., 2003), Burke et al. defined hyperheuristic algorithms as follows:

"A hyper-heuristic can be (often is) a (meta-)heuristic and it can operate on (meta-)heuristics."

This quote implies that a hyperheuristic algorithm will use a (meta)heuristic algorithm as the high-level heuristic (HLH) algorithm to drive a set of low-level (meta)heuristics in solving an optimization problem. The three simple examples given in Fig. 15.3 show the differences between metaheuristic algorithms, hybrid metaheuristic algorithms, and hyperheuristic algorithms. As shown in Fig. 15.3(a), the general framework of a metaheuristic algorithm is to repeatedly perform the transition (T), evaluation (E), and determination (D) operators of a search algorithm during the convergence process. As shown in Fig. 15.3(b), the so-called hybrid metaheuristic algorithms, which integrate multiple (meta)heuristic algorithms as a search algorithm, repeatedly perform the transition, evaluation, and determination operators of the underlying metaheuristic algorithms to find the result during the convergence process. By leveraging the strength of different (meta)heuristic algorithms, they typically find better results than metaheuristic algorithms alone. There exist different variants of the hybrid metaheuristics, which can be found in Section 15.1; the example given here, however, is a general framework of the hybrid metaheuristic algorithm. As shown in Fig. 15.3(c), the so-called hyperheuristic algorithms also integrate multiple (meta)heuristic algorithms into a search algorithm. The main characteristic of hyperheuristic algorithms, however, is that it is not required to perform all the (meta)heuristic algorithms they have at each iteration during the convergence process. Instead, a hyperheuristic algorithm first puts all the (meta)heuristic algorithms in a candidate pool as the low-level search algorithms (i.e., the top-right side of Fig. 15.3(c)). It will then use an HLH as the low-level heuristic (LLH) selection mechanism to pick a suitable LLH algorithm to find the solution. At the end of each iteration, it will use a predefined acceptance criterion to determine whether the search results of the current LLH algorithm will be accepted or not.

Fig. 15.4 gives three additional examples to further explain how metaheuristic algorithms, hybrid metaheuristic algorithms, and hyperheuristic algorithms work during the convergence process. As shown in Fig. 15.4(a), metaheuristic algorithms use one and only one metaheuristic algorithm to find better results in the same iteration or generation. As shown in Fig. 15.4(b), different from the search strategy of metaheuristic algorithms, hybrid metaheuristic algorithms adopt two or more (meta)heuristics to search in the same iteration. As shown in Fig. 15.4(c), the search strategy of hyperheuristic algorithms is similar to that of metaheuristic algorithms; only a metaheuristic algorithm is used to search in the same iteration.

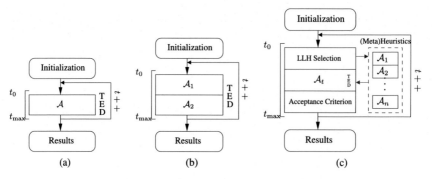

Figure 15.3 Differences between metaheuristic algorithms, hybrid metaheuristic algorithms, and hyperheuristic algorithms. (a) Metaheuristic algorithm. (b) Hybrid metaheuristic algorithm. (c) Hyperheuristic algorithm.

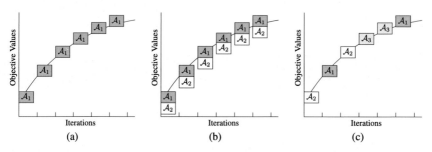

Figure 15.4 The search strategies of metaheuristic algorithms, hybrid metaheuristic algorithms, and hyperheuristic algorithms during the convergence process. (a) Metaheuristic algorithm. (b) Hybrid metaheuristic algorithm. (c) Hyperheuristic algorithm.

The main difference between metaheuristic algorithms and hyperheuristic algorithms is that the searches of metaheuristic algorithms in different iterations are guided by the same metaheuristic algorithm, while the searches of hyperheuristic algorithms in different iterations may be guided by different (meta)heuristic algorithms. This example also shows that the search strategy of hyperheuristics is to use only one (meta)heuristic algorithm in the same iteration while hybrid metaheuristic algorithms use two or more (meta)heuristic algorithms in the same iteration.

A critical reason for integrating a set of (meta)heuristic algorithms in such a way into a search algorithm during the convergence process can be found in (Cowling et al., 2001). Cowling et al. pointed out that a hyperheuristic can use only the performance indicators "without domain knowledge" to determine and choose appropriate low-level search algorithms to solve an optimization problem. Bai and Kendall (Bai and Kendall,

2005) stated that once an HLH algorithm is developed, a different optimization problem can also be solved by replacing the set of LLH algorithms and the objective function. This means that a hyperheuristic algorithm requires neither domain knowledge nor changes to the HLH algorithm for solving an unknown optimization problem. As such, the hyperheuristic algorithm can be regarded as a non-problem-specific search algorithm. Burke et al. (Burke et al., 2013) further stated that the definition of hyperheuristic algorithm can be extended as follows:

"A search method or learning mechanism for selecting or generating heuristics to solve computational search problems."

To realize the hyperheuristic algorithm, Bai and Kendall (Bai and Kendall, 2005) presented a simple and intuitive framework, called SA hyperheuristic (SAHH), that uses SA as the HLH to determine whether or not to accept the solution found by the LLH and pass it on to the next iteration. As shown in line 5 of Algorithm 11, SAHH will first randomly

---

Algorithm 11 Simulated annealing hyperheuristic.

1 Set the initial temperature Ψ	I
2 Randomly create the initial solution s	I
3 $f_s =$ Evaluate(s)	E
4 **While** the termination criterion is not met	
5 Randomly select a heuristic $\mathcal{A}_t \in \mathcal{A}$	D
6 $L_h = 0$	
7 **While** $L_h < L_{\max}$	
8 $L_h{+}{+}$	
9 $v = \mathcal{A}_t(s)$	T $\mapsto$ E
10 $f_v =$ Evaluate(v)	E
11 If $r < P_a$	D
12 $s = v$	
13 $f_s = f_v$	
14 Endif	
15 **End**	
16 Update Ψ according to the annealing schedule	D
17 **End**	
18 **Output** s	O

---

select a heuristic algorithm $\mathcal{A}_t$ from the candidate set $\mathcal{A}$, which contains 12 LLHs for solving the planogram optimization problem. Then L_h is reset to 0 before the selected LLH is performed. The inner loop, as shown in lines 7–15 of Algorithm 11, will perform the selected LLH, as shown in line 9, to transit the solution s to a new solution v and then use the mechanism of SA to determine whether or not to accept the new solution v. After performing the selected LLH for $L_{\max}$ iterations, this hyperheuristic algorithm will update the temperature Ψ and then randomly select another LLH.

Ouelhadj and Petrovic (Ouelhadj and Petrovic, 2010) presented an agent-based cooperative hyperheuristic framework to obtain better results than with traditional sequential hyperheuristic algorithms. If the solution s_b found by the LLH $\mathcal{A}_1$ is better than the current global best solution, then this framework will accept the solution s_b. Otherwise, if the solution s_w found by the LLH $\mathcal{A}_1$ is worse than the current global best solution, the cooperative agent of this framework will decide whether or not to accept this solution s_w by using the acceptance criteria of improving only, all moves, tabu search, SA, and great deluge. In (Han and Kendall, 2003), Han and Kendall attempted to integrate GA and tabu search into a hyperheuristic algorithm. This GA-based hyperheuristic algorithm has two distinguishing features. The first is that it is an indirect search algorithm, which implies that the searched result of a chromosome is not a solution for the problem in question; rather, each gene of the chromosome eventually encodes an index to the LLH algorithms. A solution found by this GA-based hyperheuristic algorithm is the sequence to perform the LLH algorithms. The second is that each subsolution is associated with a counter t_i to avoid the present LLH from being performed again within a predefined number of iterations t_{max}. For example, if an LLH $\mathcal{A}_i$ has been performed at iteration t, this algorithm will first set the counter t_i for $\mathcal{A}_i$ to t_{max} ($t_i = t_{max}$) and then t_i will be decreased by 1 ($t_i = t_i - 1$) after each iteration. A similar concept can also be found in (Burke et al., 2007) in which Burke et al. used the tabu list instead to record the LLHs that have been performed to avoid performing them again within a short time. In (Tsai et al., 2014), Tsai et al. presented a simple hyperheuristic framework for solving the workflow scheduling and map-task scheduling optimization problems for cloud computing platforms. The main characteristic of this hyperheuristic algorithm is that the acceptance operator uses a diversity detection operator and an improvement detection operator to avoid the hyperheuristic algorithm from using the same LLH for a long time during the convergence process. The simulation results show that the hyperheuristic algorithm is able to find better results than traditional deterministic algorithms, such as max-min and min-min, and other metaheuristics alone, such as SA, GA, particle swarm optimization, and ant colony optimization.

Since many hyperheuristic algorithms have been presented, each with its own way to perform the LLH algorithms, Ouelhadj et al. (Özcan et al., 2008) divided them into two categories—deterministic and non-

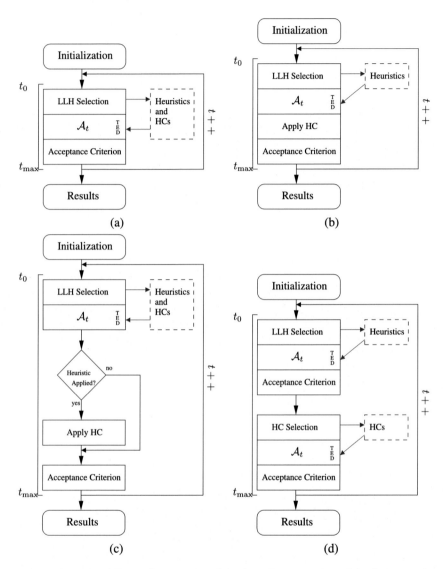

Figure 15.5 Four different frameworks of the hyperheuristic algorithm (Özcan et al., 2008). (a) F_A. (b) F_B. (c) F_C. (d) F_D.

deterministic—from the perspective of the heuristic scheduler over a set of heuristics. Deterministic means that the LLH algorithms are performed in terms of a predefined schedule or deterministic rule (e.g., round–robin strategy), while non–deterministic means that the hyperheuristic algorithm chooses the next LLH algorithm based on a probability distribution or selection mechanism during the convergence process. In the same study, Ouelhadj et al. further defined four different frameworks of hyperheuristic algorithms from the perspective of the low-level search algorithm selection mechanism, as shown in Fig. 15.5. The hyperheuristic algorithm framework F_A described in Fig. 15.5(a) will simply perform an LLH algorithm selected from the candidate set, which contains heuristic and hill climbing (HC) algorithms in this case, during the convergence process. The framework F_B described in Fig. 15.5(b) will perform an LLH algorithm selected from the candidate set, which contains only heuristic algorithms now, before performing a predefined HC algorithm to fine-tune the solution found by the selected heuristic algorithm. This guarantees that the HC algorithm will be performed at each step of the hyperheuristic algorithm. The framework F_C described in Fig. 15.5(c) gives another way to perform the HC algorithm after the heuristic algorithm, which is similar to F_B described in Fig. 15.5(b). The difference is that F_B guarantees that HC will be performed by the hyperheuristic algorithm at each iteration whereas F_C provides no such guarantee, meaning that F_C may skip HC at some iterations. Fig. 15.5(d) shows a two-stage framework, which will first perform the heuristic algorithm and then the HC algorithm, by selecting each of them out of a different pool. Ouelhadj et al. also noted that the LLH selection may adjust its selection strategy based on the feedback from the HC selection at the previous step.

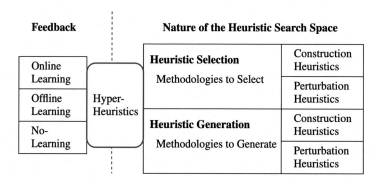

Figure 15.6 A classification of hyperheuristic approaches (Burke et al., 2010).

Another critical classification of hyperheuristics can also be found in (Burke et al., 2010), which provides a roadmap to researchers who want to enter this research field. As shown in Fig. 15.6, Burke et al. (Burke et al., 2010) classified hyperheuristic algorithms into two major categories:

(1) feedback and (2) nature of the heuristic search space. Hyperheuristic algorithms in the "feedback" category can be regarded as learning algorithms based on some sort of feedback from the convergence process. Hence, this category contains three subcategories: online learning, offline learning, and no-learning. Online learning means that the hyperheuristic algorithm will adjust its search strategy based on the searched results, while offline learning means that the hyperheuristic algorithm will make a search plan based on the training instances, prior knowledge, or domain knowledge. We can use a training method to make a good schedule for the low-level search algorithms of the hyperheuristic algorithm. No-learning means that no feedback will be used during the convergence process. As for the nature of the heuristic search space, two major categories exist: heuristic selection and heuristic generation. Heuristic selection means that the hyperheuristic algorithm will choose (or select) (meta)heuristics from the candidate set. Heuristic generation means that the hyperheuristic algorithm will generate some (meta)heuristics from the components of (meta)heuristics in the candidate set. In this figure, construction heuristics means that the transition operator will construct a solution from partial solutions step by step, while perturbation heuristics means that the transition operator will construct a solution by refining a complete solution.

Compared to metaheuristics and hybrid metaheuristics, the major advantage of a hyperheuristic algorithm is that it can find good results without domain knowledge at all or with just a little. The strong point is that it can be regarded as a non-problem-specific algorithm when a researcher wants to use it to solve different optimization problems. Some hyperheuristic algorithms are computationally cheaper than hybrid heuristic algorithms because they perform one and only one LLH algorithm at each iteration, such as the SA hyperheuristic of (Bai and Kendall, 2005) and the hyperheuristic algorithm of (Tsai et al., 2014), whereas hybrid heuristic algorithms perform two or more heuristic algorithms at each iteration. Of course, there is no free lunch. For the hyperheuristic algorithms, there are some open issues that we need to take into account. The first issue is that they are much more difficult to realize than metaheuristics and hybrid metaheuristics, because all the heuristic algorithms in the candidate set need to be realized. Another issue is that being a non-problem-specific algorithm means that some of the computations during the convergence process are in fact redundant because not all the selected LLHs are suitable for the optimization problem in question. In summary, the hyperheuristic algorithm is a feasible solution to improve the end results of metaheuristics without the need of domain knowledge at all or with just a little. It is like a cocktail, which is a mixture of different drinks; as such, its unique characteristic is that it is not unique. Fortunately, such a mixture provides us with advanced solutions to many optimization problems.

15.3. Implementation of the hybrid heuristic algorithm for the TSP

In this section, a hybridization of GA and SA called the hybrid GA (HGA) is used as an example. As shown in Fig. 15.7, the basic idea of HGA is to use GA as the major search algorithm while using SA as a search operator. This means that GA in HGA will play the role of global search while SA in HGA will play the role of local search, to fine-tune the search results of GA. The implementation of this algorithm can be regarded as the embedded framework of hybrid metaheuristic algorithms, as shown in Fig. 15.2(c).

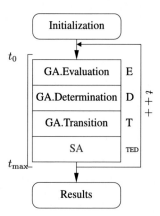

Figure 15.7 The embedded hybrid metaheuristic algorithm used in the implementation of HGA.

Both GA and SA are performed in each iteration. Of course, an iteration of this hybrid metaheuristic algorithm can also be regarded as a generation of GA if SA is seen as an operator of GA. The main feature of this hybrid metaheuristic algorithm is that GA is used for the search of all the chromosomes while SA is used for the search of a chromosome produced by GA. This design is aimed to avoid using SA for too many chromosomes in the same iteration, which will decrease the number of searches of GA during the convergence process. Two parameters, α and β, are used to determine when to start SA and the number of evaluations to be performed by SA in each iteration.

The implementation of HGA is composed of GA and SA, as shown in Listings 15.1 and 15.2, respectively. Since the implementation of HGA described here (Listing 15.1) is for solving the traveling salesman problem (TSP), it is based on the implementation of GA given in Listing 12.1, the implementation of OX crossover given in Listing 12.4, and the implementation of SA given in Listing 5.1. Note that we have made a few modifications to both GA and SA so that HGA (Listing 15.1) will call SA at the right time for a predefined number of iterations and so that SA (Listing 15.2) can be used not only as a standalone program but also as an operator of HGA for solving the TSP.

Listing 15.1: src/c++/3-tsp/main/hga.h: A hybrid GA for the TSP.

```
 1 #ifndef __HGA_H_INCLUDED__
 2 #define __HGA_H_INCLUDED__
 3
 4 #include <string>
 5 #include <fstream>
 6 #include <limits>
 7 #include "lib.h"
 8 #include "sa.h"
 9
10 using namespace std;
11
12 class hga
13 {
14 public:
15     typedef vector<int> solution;
16     typedef vector<solution> population;
17     typedef vector<double> pattern;
18     typedef vector<pattern> instance;
19     typedef vector<double> dist_1d;
20     typedef vector<dist_1d> dist_2d;
21
22     hga(int num_runs,
23         int num_evals,
24         int num_patterns_sol,
25         string filename_ini,
26         string filename_ins,
27         int pop_size,
28         double crossover_rate,
29         double mutation_rate,
30         int num_players,
31         int sa_num_evals,      // parameters of SA
32         double sa_time2run,    // parameters of SA
33         double max_temp,       // parameters of SA
34         double min_temp        // parameters of SA
35         );
36
37     population run();
38
39 private:
40     population init();
41     vector<double> evaluate(const population &curr_pop, const dist_2d &dist);
```

```
42    double evaluate(const solution &curr_sol, const dist_2d &dist);
43    void update_best_sol();
44    population tournament_select(const population &curr_pop, vector<double> &curr_obj_vals, int
      num_players);
45    population crossover_ox(const population& curr_pop, double cr);
46    population mutate(const population& curr_pop, double mr);
47    population saf();
48
49 private:
50    int num_runs;
51    int num_evals;
52    int num_patterns_sol;
53    string filename_ini;
54    string filename_ins;
55
56    population curr_pop;
57    vector<double> curr_obj_vals;
58
59    solution best_sol;
60    double best_obj_val;
61
62    dist_2d dist;
63    instance ins_tsp;
64    solution opt_sol;
65
66    int pop_size;
67    double crossover_rate;
68    double mutation_rate;
69    int num_players;
70    int sa_num_evals;    // parameters of SA
71    double sa_time2run; // parameters of SA
72    double max_temp;     // parameters of SA
73    double min_temp;     // parameters of SA
74
75    int eval_count = 0;
76    double best_so_far;
77    vector<double> avg_obj_val_eval;
78 };
79
80 inline hga::hga(int num_runs,
81                 int num_evals,
82                 int num_patterns_sol,
83                 string filename_ini,
84                 string filename_ins,
85                 int pop_size,
86                 double crossover_rate,
87                 double mutation_rate,
88                 int num_players,
89                 int sa_num_evals,
90                 double sa_time2run,
91                 double max_temp,
92                 double min_temp
93                 )
94    : num_runs(num_runs),
95      num_evals(num_evals),
96      num_patterns_sol(num_patterns_sol),
97      filename_ini(filename_ini),
98      filename_ins(filename_ins),
99      pop_size(pop_size),
100     crossover_rate(crossover_rate),
101     mutation_rate(mutation_rate),
102     num_players(num_players),
103     sa_num_evals(sa_num_evals),
104     sa_time2run(sa_time2run),
105     max_temp(max_temp),
```

```
106        min_temp(min_temp)
107    {
108        srand();
109    }
110
111    inline hga::population hga::run()
112    {
113        double avg_obj_val = 0;
114        avg_obj_val_eval = vector<double>(num_evals, 0.0);
115
116        for (int r = 0; r < num_runs; r++) {
117            eval_count = 0;
118            best_so_far = numeric_limits<double>::max();
119
120            // 0. initialization
121            curr_pop = init();
122
123            while (eval_count < num_evals) {
124                // 1. evaluation
125                curr_obj_vals = evaluate(curr_pop, dist);
126                update_best_sol();
127
128                // 2. determination
129                curr_pop = tournament_select(curr_pop, curr_obj_vals, num_players);
130
131                // 3. transition
132                curr_pop = crossover_ox(curr_pop, crossover_rate);
133                curr_pop = mutate(curr_pop, mutation_rate);
134
135                if (eval_count > num_evals * sa_time2run)
136                    curr_pop = saf();
137            }
138            avg_obj_val += best_obj_val;
139        }
140        avg_obj_val /= num_runs;
141
142        for (int i = 0; i < num_evals; i++) {
143            avg_obj_val_eval[i] /= num_runs;
144            cout << fixed << setprecision(3) << avg_obj_val_eval[i] << endl;
145        }
146        cout << best_sol << endl;
147
148        return curr_pop;
149    }
150
151    inline hga::population hga::saf()
152    {
153        population tmp_pop = curr_pop;
154
155        sa sa_search(1,
156                     sa_num_evals,
157                     curr_pop[0].size(),
158                     "",
159                     max_temp,
160                     min_temp,
161                     dist,
162                     true);
163
164        int n = 0;
165        double v = curr_obj_vals[0];
166        for (size_t i = 1; i < curr_pop.size(); i++) {
167            if (curr_obj_vals[i] < v) {
168                n = i;
169                v = curr_obj_vals[i];
170            }
```

```
171    }
172
173    tmp_pop[n] = sa_search.run(curr_pop[n], eval_count, avg_obj_val_eval, best_obj_val, best_sol
       );
174
175    return tmp_pop;
176 }
177
178 inline hga::population hga::init()
179 {
180    // 1. parameters initialization
181    population curr_pop(pop_size, solution(num_patterns_sol));
182    ins_tsp = instance(num_patterns_sol, pattern(2, 0.0));
183    opt_sol = solution(num_patterns_sol, 0);
184    dist = dist_2d(num_patterns_sol, dist_1d(num_patterns_sol, 0.0));
185    curr_obj_vals = vector<double>(pop_size, 0.0);
186    best_obj_val = numeric_limits<double>::max();
187
188    // 2. input the TSP benchmark
189    if (!filename_ins.empty()) {
190        ifstream ifs(filename_ins.c_str());
191        for (int i = 0; i < num_patterns_sol ; i++)
192            for (int j = 0; j < 2; j++)
193                ifs >> ins_tsp[i][j];
194    }
195
196    // 3. input the optimum solution
197    string filename_ins_opt(filename_ins + ".opt");
198    if (!filename_ins_opt.empty()) {
199        ifstream ifs(filename_ins_opt.c_str());
200        for (int i = 0; i < num_patterns_sol ; i++) {
201            ifs >> opt_sol[i];
202            --opt_sol[i];
203        }
204    }
205
206    // 4. initial solutions
207    if (!filename_ini.empty()) {
208        ifstream ifs(filename_ini.c_str());
209        for (int p = 0; p < pop_size; p++)
210            for (int i = 0; i < num_patterns_sol; i++)
211                ifs >> curr_pop[p][i];
212    }
213    else {
214        for (int p = 0; p < pop_size; p++) {
215            for (int i = 0; i < num_patterns_sol; i++)
216                curr_pop[p][i] = i;
217            for (int i = num_patterns_sol-1; i > 0; --i)    // shuffle the solutions
218                swap (curr_pop[p][i], curr_pop[p][rand() % (i+1)]);
219        }
220    }
221
222    // 5. construct the distance matrix
223    for (int i = 0; i < num_patterns_sol; i++)
224        for (int j = 0; j < num_patterns_sol; j++)
225            dist[i][j] = i == j ? 0.0 : distance(ins_tsp[i], ins_tsp[j]);
226
227    return curr_pop;
228 }
229
230 inline vector<double> hga::evaluate(const population &curr_pop, const dist_2d &dist)
231 {
232    vector<double> tour_dist(curr_pop.size(), 0.0);
233    for (size_t p = 0; p < curr_pop.size(); p++)
234        tour_dist[p] = evaluate(curr_pop[p], dist);
```

```
235     return tour_dist;
236 }
237
238 inline double hga::evaluate(const solution& curr_sol, const dist_2d &dist)
239 {
240     double tour_dist = 0.0;
241     for (size_t i = 0; i < curr_sol.size(); i++) {
242         const int r = curr_sol[i];
243         const int s = curr_sol[(i+1) % curr_pop[0].size()];
244         tour_dist += dist[r][s];
245     }
246
247     if (best_so_far > tour_dist)
248         best_so_far = tour_dist;
249     if (eval_count < num_evals)
250         avg_obj_val_eval[eval_count++] += best_so_far;
251
252     return tour_dist;
253 }
254
255 inline void hga::update_best_sol()
256 {
257     for (size_t i = 0; i < curr_pop.size(); i++) {
258         if (curr_obj_vals[i] < best_obj_val) {
259             best_sol = curr_pop[i];
260             best_obj_val = curr_obj_vals[i];
261         }
262     }
263 }
264
265 inline hga::population hga::tournament_select(const population &curr_pop, vector<double> &
        curr_obj_vals, int num_players)
266 {
267     population tmp_pop(curr_pop.size());
268     for (size_t i = 0; i < curr_pop.size(); i++) {
269         int k = rand() % curr_pop.size();
270         double f = curr_obj_vals[k];
271         for (int j = 1; j < num_players; j++) {
272             int n = rand() % curr_pop.size();
273             if (curr_obj_vals[n] < f) {
274                 k = n;
275                 f = curr_obj_vals[k];
276             }
277         }
278         tmp_pop[i] = curr_pop[k];
279     }
280     return tmp_pop;
281 }
282
283 inline hga::population hga::crossover_ox(const population& curr_pop, double cr)
284 {
285     population tmp_pop = curr_pop;
286
287     const int mid = curr_pop.size()/2;
288     const size_t ssz = curr_pop[0].size();
289     for (int i = 0; i < mid; i++) {
290         const double f = static_cast<double>(rand()) / RAND_MAX;
291         if (f <= cr) {
292             // 1. create the mapping sections
293             size_t xp1 = rand() % (ssz + 1);
294             size_t xp2 = rand() % (ssz + 1);
295             if (xp1 > xp2)
296                 swap(xp1, xp2);
297
298             // 2. indices to the two parents and offspring
```

```
299              const int p[2] = { i, i+mid };
300
301              // 3. the main process of ox
302              for (int k = 0; k < 2; k++) {
303                  const int c1 = p[k];
304                  const int c2 = p[1-k];
305
306                  // 4. mask the genes between xp1 and xp2
307                  const auto& s1 = curr_pop[c1];
308                  const auto& s2 = curr_pop[c2];
309                  vector<bool> msk1(ssz, false);
310                  for (size_t j = xp1; j < xp2; j++)
311                      msk1[s1[j]] = true;
312                  vector<bool> msk2(ssz, false);
313                  for (size_t j = 0; j < ssz; j++)
314                      msk2[j] = msk1[s2[j]];
315
316                  // 5. replace the genes that are not masked
317                  for (size_t j = xp2 % ssz, z = 0; z < ssz; z++) {
318                      if (!msk2[z]) {
319                          tmp_pop[c1][j] = s2[z];
320                          j = (j+1) % ssz;
321                      }
322                  }
323              }
324          }
325      }
326
327      return tmp_pop;
328  }
329
330  inline hga::population hga::mutate(const population& curr_pop, double mr)
331  {
332      population tmp_pop = curr_pop;
333      for (size_t i = 0; i < tmp_pop.size(); i++){
334          const double f = static_cast<double>(rand()) / RAND_MAX;
335          if (f <= mr) {
336              const int m1 = rand() % tmp_pop[0].size();  // mutation point
337              const int m2 = rand() % tmp_pop[0].size();  // mutation point
338              swap(tmp_pop[i][m1], tmp_pop[i][m2]);
339          }
340      }
341      return tmp_pop;
342  }
343
344  #endif
```

Listing 15.2: src/c++/3-tsp/main/sa.h: A modified SA used in hybrid GA for the TSP.

```
1  #ifndef __SA_H_INCLUDED__
2  #define __SA_H_INCLUDED__
3
4  #include <string>
5  #include <limits>
6  #include "lib.h"
7
8  using namespace std;
9
10 class sa
11 {
12 public:
```

```
13    typedef vector<int> solution;
14    typedef vector<double> dist_1d;
15    typedef vector<dist_1d> dist_2d;
16
17    sa(int num_runs,
18       int num_evals,
19       int num_patterns_sol,
20       string filename_ini,
21       double max_temp, // sa-specific parameter
22       double min_temp, // sa-specific parameter
23       dist_2d dist,
24       bool continue_flag
25       );
26
27    solution run(solution& sol_orig, int& eval_count_orig, vector<double>& avg_obj_val_eval_orig
      , double& best_obj_val_orig, solution& best_sol_orig);
28
29 private:
30    solution init();
31    double evaluate(const solution &s, const dist_2d &dist);
32    solution transit(const solution& s);
33    // begin the sa functions
34    bool determine(double tmp_obj_val, double obj_val, double temperature);
35    double annealing(double temperature);
36    // end the sa functions
37
38 private:
39    int num_runs;
40    int num_evals;
41    int num_patterns_sol;
42    string filename_ini;
43
44    // begin the sa parameters
45    double max_temp;
46    double min_temp;
47    // end the sa parameters
48
49    dist_2d dist;
50    bool continue_flag;
51
52    solution best_sol;
53    double best_obj_val;
54    solution curr_sol;
55    double curr_temp;
56 };
57
58 inline sa::sa(int num_runs,
59               int num_evals,
60               int num_patterns_sol,
61               string filename_ini,
62               double max_temp, // sa-specific parameter
63               double min_temp, // sa-specific parameter
64               dist_2d dist,
65               bool continue_flag
66               )
67    : num_runs(num_runs),
68      num_evals(num_evals),
69      num_patterns_sol(num_patterns_sol),
70      filename_ini(filename_ini),
71      max_temp(max_temp),
72      min_temp(min_temp),
73      dist(dist),
74      continue_flag(continue_flag)
75 {
76    srand();
```

```
77  }
78
79  inline sa::solution sa::run(solution& sol_orig, int& eval_count_orig, vector<double>&
        avg_obj_val_eval_orig, double& best_obj_val_orig, solution& best_sol_orig)
80  {
81      const int eval_count_saved = eval_count_orig;
82
83      for (int r = 0; r < num_runs; r++) {
84          int eval_count = eval_count_saved;
85
86          // 0. Initialization
87          curr_sol = init();
88
89          best_obj_val = best_obj_val_orig;
90          best_sol = best_sol_orig;
91          curr_sol = sol_orig;
92
93          double obj_val = evaluate(curr_sol, dist);
94
95          while (eval_count < num_evals + eval_count_saved) {
96              // 1. Transition
97              solution tmp_sol = transit(curr_sol);
98
99              // 2. Evaluation
100             double tmp_obj_val = evaluate(tmp_sol, dist);
101
102             // 3. Determination
103             if (determine(tmp_obj_val, obj_val, curr_temp)) {
104                 curr_sol = tmp_sol;
105                 obj_val = tmp_obj_val;
106             }
107
108             if (obj_val < best_obj_val) {
109                 best_sol = curr_sol;
110                 best_obj_val = obj_val;
111             }
112
113             curr_temp = annealing(curr_temp);
114             avg_obj_val_eval_orig[eval_count++] += best_obj_val;
115         }
116
117         eval_count_orig = eval_count;
118
119         if (best_obj_val < best_obj_val_orig) {
120             best_obj_val_orig = best_obj_val;
121             best_sol_orig = best_sol;
122             sol_orig = best_sol;
123         }
124         else
125             sol_orig = curr_sol;
126     }
127
128     return sol_orig;
129  }
130
131
132 inline sa::solution sa::init()
133 {
134     curr_temp = max_temp;
135
136     if (!continue_flag) {
137         curr_sol = solution(num_patterns_sol);
138         for (int i = 0; i < num_patterns_sol; i++)
139             curr_sol[i] = i;
140         for (int i = num_patterns_sol-1; i > 0; --i)     // shuffle the solutions
```

```
141            swap(curr_sol[i], curr_sol[rand() % (i+1)]);
142      }
143
144      return curr_sol;
145  }
146
147  inline sa::solution sa::transit(const solution& s)
148  {
149      solution t(s);
150      const int i = rand() % s.size();
151      const int j = rand() % s.size();
152      swap(t[i], t[j]);
153      return t;
154  }
155
156  inline double sa::evaluate(const solution &s, const dist_2d &dist)
157  {
158      double tour_dist = 0.0;
159      for (size_t i = 0; i < s.size(); i++){
160          const int c = s[i];
161          const int n = s[(i+1) % s.size()];
162          tour_dist += dist[c][n];
163      }
164      return tour_dist;
165  }
166
167  inline bool sa::determine(double tmp_obj_val, double obj_val, double temperature)
168  {
169      const double r = static_cast<double>(rand()) / RAND_MAX;
170      const double p = exp((tmp_obj_val - obj_val) / temperature);
171      return r > p;
172  }
173
174  inline double sa::annealing(double temperature)
175  {
176      return 0.9 * temperature;
177  }
178  #endif
```

15.3.1 Declaration of functions and parameters

For HGA (Listing 15.1), most parameter declarations are the same as those for GA, as described in Listings 12.1 and 12.4. The member function saf() is declared in line 47 for HGA to call SA. Four additional parameters specific to SA—the number of iterations for SA, time to launch SA during the convergence process, maximum temperature, and minimum temperature—namely, `sa_num_evals`, `sa_time2run`, `max_temp`, and `min_temp`, are defined in lines 70–73.

15.3.2 The main loop of HGA

As shown in lines 111–149, the implementation of the main function of HGA is similar to that described in Listing 12.1, except for lines 135–136, where a call to saf() is made when it is time to do so. This implies that HGA will still perform the evaluation, determination, and transition operators of GA, but it will use saf() as a transition operator in the end of each iteration to fine-tune a chromosome if needed. Line 135 is used to determine when to launch SA. This means that HGA will launch SA after `sa_time2run` percent of evaluations have been performed during the convergence process, which is also referred to as α in this chapter.

15.3.3 Additional functions of HGA

As shown in Listing 15.1, most functions of HGA are the same as those described in Listings 12.1 and 12.4 except saf() shown in lines 151–176. The purpose of saf() is to improve the best solution found by the GA in this iteration, if possible. This means that HGA will find the best chromosome in the current population, as shown in lines 164–171, and use SA to improve it, as shown in line 173. The variables (1) `curr_pop[n]`, (2) `eval_count`, (3) `avg_obj_val_eval`, (4) `best_obj_val`, and (5) `best_sol` of HGA will also be passed to SA for continuing the search of HGA. These variables represent (1) the best solution in the current population in terms of the objective value, (2) the number of evaluations to be performed by SA (also called β in this chapter), (3) the array of average objective values for each evaluation, (4) the objective value of the best-so-far solution, and (5) the best-so-far solution, respectively.

15.3.4 The declaration of functions and parameters of SA

Two additional vector types, `dist_1d` and `dist_2d` (lines 14–15), and two additional parameters, `dist` and `continue_flag` (lines 49–50), of Listing 15.2 are added to SA (Listing 5.1). The former is a two-dimensional array (or a matrix to be more specific) that is used to pass the distances between all the cities to SA to speed up its computation of the distance of a solution. The latter is a flag used to determine how the initial solution is created for SA. If its value is false, the initial solution will be randomly created as a brand-new solution so that SA can be run as a standalone program to find a TSP tour; otherwise, the initial solution will be taken from the program that calls SA, that is, HGA in this case. As shown in line 31, the declaration of function evaluate() in Listing 15.2 is a little bit different from that in Listing 5.1 because of the addition of a second parameter named `dist` that is used to pass the distances between all the cities from HGA to SA.

15.3.5 The main loop of SA

The main loop of SA is shown in lines 79–129 of Listing 15.2, which takes five parameters from HGA, as mentioned in Section 15.3.3: `sol_orig`, `eval_count_orig`, `avg_obj_val_eval_orig`, `best_obj_val_orig`, and `best_sol_orig`. To simplify the discussion of this function, we will discuss only the differences between Listing 15.2 and Listing 5.1. To keep track of the evaluation count, two variables, `eval_count_saved` (line 81) and `eval_count` (line 84), are used. The former is used to save the evaluation count so far while the latter is used to keep track of the evaluation performed by SA and as an index into a one-dimensional array for keeping track of the average objective values (line 114). In lines 89–91, the variables `best_obj_val`, `best_sol`, and `curr_sol` are initialized to the objective value of the best-so-far solution, the best-so-far solution, and the best solution found by GA so far. Another difference is found in line 95, where `num_evals` in Listing 5.1 is replaced by `num_evals + eval_count_saved` because the evaluation count for SA is started from `eval_count_saved` and the number of evaluations to be performed by SA is `num_evals`. A similar modification can also be found in line 114, where `avg_obj_val_eval[eval_count++]` in Listing 5.1 is replaced by `avg_obj_val_eval_orig[eval_count++]` because this is how the average objective values are passed back to `avg_obj_val_eval` of HGA. Finally, in lines 114–128, the evaluation count, objective value of the best-so-far solution, the best-so-far solution, and the current solution are passed back to HGA before the searches are switched back to GA.

15.3.6 Additional functions of SA

As shown in lines 132–145, in init(), in case SA is asked to create the initial seed by itself, it will put the ID of all the cites into a solution first and then do a random shuffle on the solution to randomize the order in which the cities are visited in a tour. Since this implementation is for solving the TSP, the transit() operator that was originally designed for the one-max problem has to be modified so that it works for the solution of the TSP instead of just flipping a subsolution of the solution of a one-max problem. As shown in lines 147–154, it will first randomly select two subsolutions (i.e., cities) from a solution and then swap their loci to create a new routing path for the TSP. Another modification is given in lines 156–165, where the evaluate() function is modified to speed up its calculation of the distance of a solution for the TSP.

15.4. Simulation results of the hybrid heuristic algorithm for the TSP

To evaluate the performance of HGA, we compare it with GA for solving the TSP benchmark eil51 with different parameter settings. All the algorithms employ OX as the crossover operator. Here, GA represents simple GA and HGA represents hybrid GA that calls saf(). As the simulations described in the other chapters of this book, all results are the average values of 100 runs, each of which is carried out for 120,000 evaluations, while the number of chromosomes is set equal to 120, the crossover rate (c_r) is set equal to 40% (0.4), the mutation rate is set equal to 10% (0.1), and the number of players for the tournament selection operator is set equal to three.

The parameters α and β represent the time to start SA and the number of evaluations SA will perform when it is called by HGA. Table 15.1 shows the impact the values of β may have on the performance of GA and HGA. The results show that GA and HGA take almost the same computation time. This is very likely because all of them use the same number of evaluations during the convergence process. The results also show that the parameter setting $\alpha = 0.1$ and $\beta = 10$ for HGA gives better results than GA and HGA with other values of β. The results show that as far as this version of HGA is concerned, a small number of evaluations that SA performs for each call of HGA give a better result than a large number of evaluations that SA performs for each call of HGA.

Table 15.1 Comparison of GA and HGA with $\alpha = 0.1$ and different values of β.

	Total distance	Running time (in seconds)
GA	474.879	12.005
HGA ($\beta = 10$)	472.062	13.907
HGA ($\beta = 20$)	474.764	13.654
HGA ($\beta = 40$)	475.417	12.920
HGA ($\beta = 60$)	475.585	12.893
HGA ($\beta = 80$)	476.344	11.490
HGA ($\beta = 100$)	481.647	11.042

Tables 15.2 and 15.3 further show the impact the values of α and β may have on the performance of GA and HGA. In this case, β is set equal to 10 in Table 15.2 and to 100 in Table 15.3, while α is set equal to 0.1, 0.2, 0.4, 0.6, and 0.8. It is *not* easily seen from Table 15.2 that the results of

HGA get worse as the value of α increases for $\beta = 10$; however, it can be easily seen from Table 15.3 that the results of HGA will get better as the value of α increases for $\beta = 100$ in most cases. The results also show that the end results will be influenced by both α and β, not just one. How to find a set of good parameters is thus a critical research issue for such hybrid metaheuristic algorithm.

Table 15.2 Comparison of GA and HGA with $\beta = 10$ and different values of α.

	Total distance	Running time (in seconds)
GA	474.879	12.005
HGA ($\alpha = 0.1$)	472.062	13.907
HGA ($\alpha = 0.2$)	474.622	13.596
HGA ($\alpha = 0.4$)	475.633	13.231
HGA ($\alpha = 0.6$)	472.153	12.879
HGA ($\alpha = 0.8$)	475.619	12.495

Table 15.3 Comparison of GA and HGA with $\beta = 100$ and different values of α.

	Total distance	Running time (in seconds)
GA	474.879	12.005
HGA ($\alpha = 0.1$)	481.647	11.042
HGA ($\alpha = 0.2$)	479.173	11.194
HGA ($\alpha = 0.4$)	475.622	11.486
HGA ($\alpha = 0.6$)	471.546	11.687
HGA ($\alpha = 0.8$)	474.997	11.970

Figs. 15.8 and 15.9 show the convergence results of HGA with different values of α and β in terms of the quality of the end results. Fig. 15.8 shows that HGA performs SA for only "10" evaluations in the end of each iteration for $\beta = 10$ with different timing to launch SA. It can be easily seen that the convergence speeds of GA and HGA are quite similar from evaluation 1 to 20,000, as shown in Fig. 15.8(a). The results of Fig. 15.8(b) show the detailed convergence curves of GA and HGA from evaluation 60,000 to 120,000. It can be easily seen that the convergence of HGA with $\alpha = 0.1$ is faster than those of GA and HGA with other values of α. Different from the results of Fig. 15.8, in which SA is performed for only "10" evaluations in the end of each iteration, Fig. 15.9 shows the results of performing SA

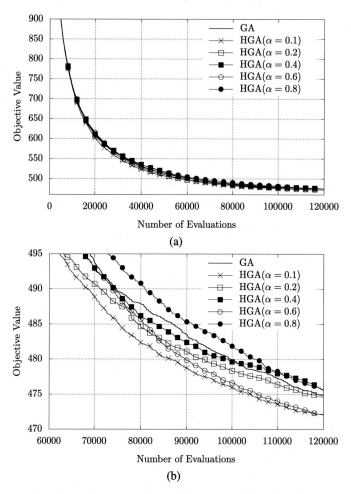

Figure 15.8 Convergence results of GA and HGA with $\beta = 10$ and different values of α. (a) The convergence curves of GA and HGA. (b) The details of Fig. 15.8(a).

for "100" evaluations with $\beta = 100$ with different timing to launch SA to see the impact of α. Fig. 15.9(a) shows that the convergence speeds of GA and HGA are quite similar from evaluation 1 to 10,000. Fig. 15.9(b) further shows that the convergence of HGA with $\alpha = 0.4$ is much faster than those of HGA with different values of α before evaluation 80,000. After that, the results of HGA with $\alpha = 0.6$ are better than those of GA and HGA with other values of α. This means that we still need to take into account the impact of other parameters of HGA, e.g., β.

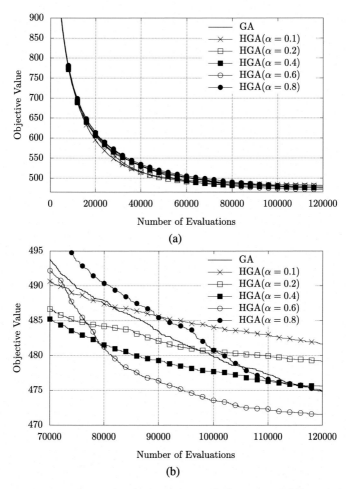

Figure 15.9 Convergence results of GA and HGA with $\beta = 100$ and different values of α. (a) The convergence curves of GA and HGA. (b) The details of Fig. 15.9(a).

15.5. Discussion

This chapter briefly reviews hybrid metaheuristics and hyperheuristics. A simple hybrid metaheuristic algorithm is also given to show that hybridization of metaheuristic algorithms is an alternative way to improve the quality of the end results of metaheuristics. However, both hybrid metaheuristics and hyperheuristics may incur extra computation time, which is obviously the price that you have to pay for the quality. Hybrid metaheuristics and hyperheuristics are, of course, more difficult to implement than other simple metaheuristics because they integrate two or more meta-heuristic algorithms, which are independent of the framework of hybrid metaheuristics and hyperheuristics. The good news is that the implementation for most metaheuristics can be found on the Internet and in many research societies. In fact, you may find implementations in different programming languages, such as C, C++, Java, Python, and MATLAB®. This makes it much easier to realize hybrid metaheuristics or hyperheuristics today. As for the computation time issue, cloud computing and GPU are two promising technologies that may be able to speed up the response time for hybrid metaheuristics and hyperheuristics. The simulation results also show that the improvement HGA makes for GA is not that significant in solving the TSP benchmark eil51. We also applied GA and HGA to the TSP benchmark pr1002 for 1,200,000 evaluations, and the improvement was about 2.323% $((1,136,091.594 - 1,163,109.149)/1,163,109.149 \times 100\% = 2.323\%)$. Although the results of GA and HGA are far away from the best known solution (259,045), we can still see that HGA essentially improves the quality of the end result. According to our observation, this is due to the fact that SA, which plays the role of a local search operator of HGA, is not very suitable for the TSP because no domain knowledge is used to fine-tune the solution obtained by GA. An intuitive way is to use a (meta)heuristic algorithm (e.g., 2-opt) which takes into account the domain knowledge for solving the TSP to replace SA so that it may increase the improvement of HGA, which will be discussed in the next chapter. From the discussions of this chapter, it is possible to imagine that if many metaheuristic algorithms can be put in the candidate pool and used as the low-level search algorithms for hyperheuristic algorithm, the improvement for solving the TSP may be significant compared with the combination of GA and SA because some of these low-level search algorithms may be particularly suitable for solving the TSP.

Supplementary source code

Here are the hyperlinks to the directories where the source code described in this chapter resides.

https://github.com/cwtsaiai/metaheuristics_2023/src/c++/3-tsp/

CHAPTER SIXTEEN

Local search algorithm

16.1. The basic idea of local search

Several local search (LS) algorithms (Feo and Resende, 1995; Lourenço et al., 2003; Mladenović and Hansen, 1997; Talbi, 2009; Voudouris and Tsang, 1999, 1996) have been presented and are widely used. Most of them are very simple and easy to implement and can typically find a "good solution" for the problem in question. Suppose we have an optimization problem where we have to minimize the objective value of the solution. Its fitness landscape is shown in Fig. 16.1. Further suppose that the starting point is s^1. LS will find the local optimum s^2 at the first valley in the solution space with a sufficient number of iterations because LS will search the neighbors of s^1 and move toward a region where it can find a better solution. This characteristic makes it possible for LS to find a solution that is better than s^1, and this process will be repeated until it cannot find a better solution from the neighbors, thus getting stuck at a local optimum, i.e., s^2 in this case.

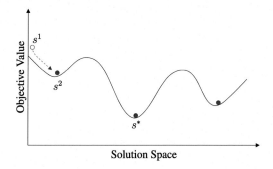

Figure 16.1 A local search algorithm.

Although LS may fall into a local optimum quickly, it can still be used to find an approximate solution to the optimization problem in question if such a search algorithm can be integrated with other global search algorithms. Algorithm 12 gives a general framework of LS algorithms. It is similar to hill climbing (HC) described in Algorithm 2 of Section 3.2.1 because HC can be seen as a special case of an LS algorithm.

The main difference between LS and HC is shown in lines 4 and 6 of Algorithm 12, where the DeterministicNeighborSelection() operator replaces the NeighborSelection() operator of Algorithm 2 in line 4 and the Improve() operator replaces the determination operator of Algorithm 2 in lines 6–9. The reason to use the DeterministicNeighborSelection() operator for LS is that the new candidate solutions can be created by either a random process or a deterministic process and here it is a deterministic process that is used for LS; thus, the name DeterministicNeighborSelection() is used to differentiate this operator for LS, HC, and other single-solution-based metaheuristic algorithms. The reason to use the Improve() operator for LS is that there are several determination mechanisms that can be used in the Improved() operator for the LS algorithm, such as first improvement, best improvement, random selection, and even possibilities to accept a non-improving neighbor.

Algorithm 12 Local search.

1 Randomly create the initial solution s	[I]
2 $f_s = $ Evaluate(s)	[E]
3 **While** the termination criterion is not met	
4 $v = $ DeterministicNeighborSelection(s)	[T]
5 $f_v = $ Evaluate(v)	[E]
6 $s, f_s = $ Improve(s, v, f_s, f_v)	[D]
7 **End**	
8 **Output** s	[O]

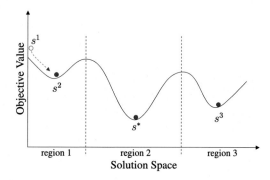

Figure 16.2 A local search algorithm with the perspective of solution space.

As shown in Fig. 16.2, the end result of an LS algorithm is typically very sensitive to the starting point in the solution space. This example shows that unless the starting point is in region 2, a general LS algorithm will not be able to find the global optimum, i.e., s^*. This does not mean that there is no way to improve the end result of LS. In addition to tabu search (TS) and simulated annealing (SA), several LS algorithms—greedy randomized adaptive search procedure (GRASP) (Feo and Resende, 1995), variable neighborhood search (VNS) (Mladenović and Hansen, 1997), guided LS (GLS) (Voudouris and Tsang, 1999, 1996), and iterated LS (ILS) (Lourenço et al., 2003)—have become popular and are widely used in many studies on solving complex optimization problems. Talbi pointed out in his book (Talbi, 2009) that since the 1980s, many other LS algorithms have been presented, the strategies of which can be divided into three categories: (1) iterating from different solutions, (2) changing the landscape of the problem, and (3) accepting non-improving neighbors. As shown in Fig. 16.3, the basic idea of these three kinds of advanced LS algorithms is typically to prevent the search from getting stuck at a local optimum.

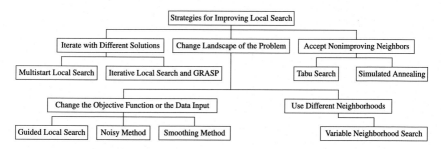

Figure 16.3 The classification of improved local search algorithms (Talbi, 2009).

16.1.1 Iterating with different solutions

For the first category, one possible way for iterating with different solutions is obviously to use different initial solutions for the LS algorithm so that it may jump from one region to another in the solution space. ILS (Lourenço et al., 2003) can be regarded as one of the LS algorithms in the category of iterating from different solutions. As shown in Fig. 16.4, the basic idea of ILS is to allow LS to search neighbors in different regions, called perturbation. For example, the searches will be moved from region 1 to region 2 instead of limiting to only the nearest neighbors of region 1 in which the current solution resides. The searches near solution s_1 of LS can then be shifted to another region, say solution s_2, although s_1 is at a local minimum of region 1. In addition to the concept of perturbation, the main search strategy is still to search solutions nearby the current solution iteratively for a certain number of iterations until a region that has a better solution is found, such as from s^2 to s^*.

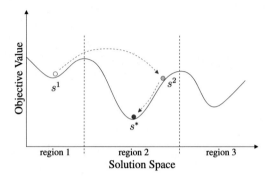

Figure 16.4 An example of iterated local search.

Algorithm 13 Iterated local search.

1 Randomly create the initial solution s	[I]
2 $s = \text{LocalSearch}(s)$	[T]
3 $f_s = \text{Evaluate}(s)$	[E]
4 **While** the termination criterion is not met	
5 $\quad v = \text{Perturbation}(s, H)$	[T]
6 $\quad v' = \text{LocalSearch}(v)$	[T]
7 $\quad f_{v'} = \text{Evaluate}(v')$	[E]
8 $\quad s, f_s = \text{Acceptance}(s, v', f_s, f_{v'}, H)$	[D]
9 **End**	
10 **Output** s	[O]

Algorithm 13 shows that ILS will first call the simple local search algorithm (e.g., Algorithm 12) to improve the solution, as shown in line 2. It will then repeatedly perform the Perturbation(), LocalSearch(), Evaluate(), and Acceptance() operators during the convergence process, as shown in lines 5–8. The Perturbation() operator allows ILS to search not only neighbors in the same region of the current solution s; therefore, it can be regarded as a transition operator that makes it possible for ILS to search for candidate solutions in different regions. If each new search is considered as a modification of the current solution s to generate a new candidate solution v, the modification done by Perturbation() will be larger than the modification done by LocalSearch(). This means that the difference or distance between s and v (solution generated by Perturbation()), $d(s, v)$, is larger than that between v and v' (solution generated by LocalSearch()), $d(v, v')$, thus making it possible for ILS to escape local optima or particular regions of the search space (Stützle and Ruiz, 2018). Lourenço et al. (Lourenço et al., 2003) pointed out that if the modification by Perturbation() is too small, it will quickly fall back to the local minimum of the region in which the solution s resides, whereas if the modification is too large, the search will become a random search. The search history H can be used to keep track of solutions searched in the previous iterations to prevent ILS from searching the same region again shortly. This means that the Perturbation() operator will create a new solution v based on both the solution s and the history of search H. In this way, ILS will search not only the region in which a particular solution resides but also other regions. As such, it is able to find a better solution than a simple LS does. The Acceptance() operator can be used as in SA so that ILS will consider the search history H and accept only a better solution or a worse solution from time to time. The search history H can be considered as temperature Ψ and the probability for acceptance can be defined as $\exp(f_s - f_{v'})/\Psi$ when the new candidate solution v' is worse than the current solution s for solving a minimization optimization problem. Note that f_s and $f_{v'}$ denote, respectively, the objective or fitness values of the solutions s and v'. Although ILS is simple and easy to implement compared with most metaheuristic algorithms, it makes LS capable of escaping from local optima.

16.1.2 Changing the landscape of the problem

The second category—changing the landscape of the problem—includes changing the objective function and using different neighborhoods. Changing the objective function, of course, is a possible way to prevent LS from searching the same position for many iterations. An intuitive method is to reduce the objective and/or fitness value for a particular solution s_i if LS cannot find a better solution from the neighbors of s_i for many iterations.

In (Voudouris and Tsang, 1996), Voudouris and Tsang presented another advanced LS algorithm, called GLS, which can be regarded as in the category of changing the objective function or the input data of the problem. As shown in Fig. 16.5, GLS will try to change the objective function to guide the search of GLS to another region in the solution space. This example shows that the basic idea of GLS is to add penalties to the searched solution s^1 if it has been searched by GLS for a certain number of times. Fig. 16.5(a) shows that the objective value of solution s^1 of GLS at iteration t will be calculated using the original objective function. GLS will then add penalties to the searched solution s^1 to lower its superiority if GLS

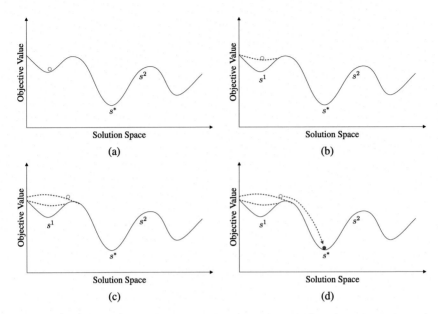

Figure 16.5 An example of guided local search. (a) Solution s^1 at $t = 1$. (b) Solution s^1 at $t = 2$. (c) Solution s^1 at $t = 3$. (d) Solution s^1 at $t = 4$, which may move to s^*.

searches it again in a short period of time, as shown in Fig. 16.5(b). After adding penalties to the solution s^1 for a while, the superiority of solution s^2 is no longer significant, as shown in Fig. 16.5(c). GLS may now be able to escape the local optimum and move to s^*, as shown in Fig. 16.5(d). To realize this concept, Voudouris and Tsang proposed three major equations for adding penalties to each feature of the searched solution. For a minimum optimization problem, the objective function of GLS is defined as

$$h(s) = f(s) + \lambda \sum_{i=1}^{n} p_i I_i(s). \tag{16.1}$$

That is, besides the original objective function for the optimization problem $f(s)$, it adds a second term $\lambda \sum_{i=1}^{n} p_i I_i(s)$ to the original objective function, where λ is a parameter for controlling the impact of penalty, p_i is the penalty associated with the i-th feature $\mathbf{f}_i$, and $I_i(s)$ is an indicator function to determine whether the feature $\mathbf{f}_i$ is present in the solution s, which is defined as

$$I_i(s) = \begin{cases} 1, & \text{solution } s \text{ has feature } \mathbf{f}_i, \\ 0, & \text{otherwise.} \end{cases} \tag{16.2}$$

The utility u_i is defined as

$$u_i(s, \mathbf{f}_i) = I_i(s)\frac{c_i}{1 + p_i}, \tag{16.3}$$

where c_i is the cost of feature $\mathbf{f}_i$.

Algorithm 14 Guided local search.

1 Randomly create the initial solution s	$\boxed{\text{I}}$
2 Associate each feature $\mathbf{f}_i$ with a penalty $p_i = 0$ and a utility u_i	
3 **While** the termination criterion is not met	
4 $\quad h = $ EvaluateWithPenalties(s) using Eqs. (16.1) and (16.2)	$\boxed{\text{E}}$
5 $\quad v = $ LocalSearch(s, h)	$\boxed{\text{T}}$
6 $\quad s = v$	
7 $\quad$ Update all utilities u using Eq. (16.3)	$\boxed{\text{D}}$
8 $\quad$ Update the penalties if u_i is maximum, $p_i = p_i + 1$	$\boxed{\text{D}}$
9 **End**	
10 **Output** s	$\boxed{\text{O}}$

Algorithm 14 shows the details of GLS. In brief, it will first associate each feature $\mathbf{f}_i$ with a penalty p_i and a utility u_i, as shown in line 2. The main convergence process is shown in lines 3–9. GLS will first evaluate the solution s using Eqs. (16.1) and (16.2), as shown in line 4. It will then employ the LS operator to find a better solution based on the modified objective function value h and the solution s, as shown in line 5. The new candidate solution v will then be used to replace the current solution s in line 6 before GLS moves on to updating all the utilities using Eq. (16.3) in line 7 and the penalty for utility u_i with features that have the maximum utility value by $p_i = p_i + 1$ in line 8. The penalty parameter p_i of GLS plays the role of counting how many times the feature $\mathbf{f}_i$ is penalized.

Using different neighbors is another well-known way to change the landscape of a solution space to enhance the search performance of LS. The landscape of the solution space of an optimization problem can be regarded as the neighbor structure of feasible solutions that a search algorithm sees. As such, a certain number of landscapes, $\mathcal{L} = \{\mathcal{L}_1, \mathcal{L}_2, \ldots, \mathcal{L}_m\}$, of a solution space can be generated to represent the same optimization problem. For instance, by saying that solutions s_1 and s_2 are neighbors of a search algorithm $\mathcal{A}$, it means that the search for solution s_1 may jump to solution s_2 when applying the search operator $\oplus$ of search algorithm $\mathcal{A}$ to solution s_1 once in $\mathcal{L}_1$. Solutions s_1 and s_2 are thus said to be neighbors in $\mathcal{L}_1$. In case landscape $\mathcal{L}_2$ is used, applying the search operator $\oplus$ to solution s_1 once will essentially generate s_3 instead of s_2, which means that s_1 and

s_3 are neighbors in $\mathcal{L}_2$. Hence, this kind of search strategy will prevent LS from searching only particular neighbors of the current solution so that it will not get stuck at a particular region because different neighbors may guide the search to different regions or even improve the search result. As shown in Fig. 16.6, VNS (Mladenović and Hansen, 1997) can be regarded as a representative search algorithm in the category of "changing the neighborhood." The basic idea is to change the landscape of the solution space of the search algorithm to prevent the search algorithm from easily falling into local optima. In this example, two neighborhood structures (i.e., neighborhood 1 and neighborhood 2) will be created before the convergence process of VNS. If VNS first uses neighborhood structure 1, it may get stuck at solution s^1, but the search procedure of VNS may change to neighborhood structure 2. By using this new neighborhood structure, VNS may escape from the local optimum of s^1 to the local optimum of s^2, and it may then be able to find s^3 when VNS changes back to neighborhood structure 1 again.

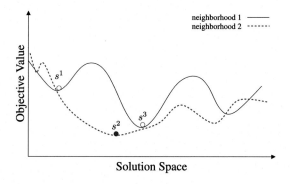

Figure 16.6 An example of variable neighborhood search.

Algorithm 15 shows that VNS is quite similar to the simple LS (i.e., Algorithm 12). That is, after the initial solution is randomly created and evaluated, VNS first creates a set of neighborhood structures, as shown in line 3. VNS will then randomly select a neighborhood structure $\mathcal{N}_k$ for the NeighborSelection() operator, as shown in lines 5–7. In this way, VNS will not always search the same direction. For instance, as the example described in Fig. 16.6 shows, after s^1, VNS will not move to the right-hand side if the same neighborhood structure 1 is used, but VNS may move to the right-hand side if neighborhood structure 2 is used; thus, VNS is able to enhance the performance of a simple LS algorithm.

Algorithm 15 Variable neighbor search.

1 Randomly create the initial solution s	$\boxed{\text{I}}$
2 $f_s =$ Evaluate(s)	$\boxed{\text{E}}$
3 Create a set of neighborhood structures $\mathcal{N}$	$\boxed{\text{D}}$
4 **While** the termination criterion is not met	
5 $\quad \mathcal{N}_k = NeighborhoodSelection(s, \mathcal{N}_k)$	$\boxed{\text{T}}$
6 $\quad v =$ NeighborSelection($s, \mathcal{N}_k$)	$\boxed{\text{T}}$
7 $\quad f_v =$ Evaluate(v)	$\boxed{\text{E}}$
8 $\quad s, f_s =$ Improve(s, v, f_s, f_v)	$\boxed{\text{D}}$
9 **End**	
10 **Output** s	$\boxed{\text{O}}$

16.1.3 Accepting non-improving neighbors

The last category is accepting non-improving neighbors. Two representative search algorithms in this category are SA and TS, which have been discussed in Chapters 5 and 6. Either accepting solutions worse than the current solution (or best-so-far solution) from time to time or searching particular solutions makes it possible for LS to escape from local optima.

16.1.4 k-opt

In addition to the aforementioned LS algorithms, k-opt (Croes, 1958), Lin–Kernighan (LK) (Lin and Kernighan, 1973), and Lin–Kernighan heuristic (LKH) (Helsgaun, 2000, 2009) are other kinds of LS algorithms that are widely used for many combinatorial optimization problems. The basic idea of k-opt can be summarized as follows:

1. Apply some transformations to the partial subsolutions of s_s to make them the new subsolutions of s'_s.
2. Compute the objective values of s_s and s'_s. The search algorithm will accept these new subsolutions if the objective value of the new solution s'_s is better than the objective value of the current solution.

Fig. 16.7 shows a simple example of 2-opt. Suppose that the 2-opt algorithm will select four cities, c_1, c_2, c_3, and c_4, and take into account two edges, $e_{1,3}$ (the edge from city c_1 to city c_3) and $e_{2,4}$ (the edge from city c_2 to city c_4), as shown in Fig. 16.7(a). As shown in Fig. 16.7(b), edges $e_{1,3}$ and $e_{2,4}$ will be replaced by $e_{1,2}$ (the edge from city c_1 to city c_2) and $e_{3,4}$ (the edge from city c_3 to city c_4). This implies that the 2-opt algorithm will accept these two new edges if $d_{1,2} + d_{3,4} < d_{1,3} + d_{2,4}$, where $d_{i,j}$ denotes the distance between cities c_i and c_j. As shown in Fig. 16.7, besides replacing edges $e_{1,3}$ and $e_{2,4}$ by edges $e_{1,2}$ and $e_{3,4}$, the direction of edge $e_{3,2}$ (before 2-opt) will be reversed from $c_3 \rightarrow c_2$ to $c_2 \rightarrow c_3$ (i.e., after 2-opt, edge $e_{3,2}$

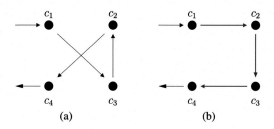

Figure 16.7 An example of 2-opt. (a) Before 2-opt. (b) After 2-opt.

will become edge $e_{2,3}$) for solving the traveling salesman problem (TSP). As shown in Fig. 16.8, in case there are cities between c_2 and c_3, the routing direction of these cities will be changed from $c_z \rightarrow c_y \rightarrow c_x$ to $c_x \rightarrow c_y \rightarrow c_z$ if edges $e_{1,3}$ and $e_{2,4}$ are replaced by edges $e_{1,2}$ and $e_{3,4}$.

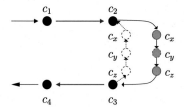

Figure 16.8 The example of 2-opt given in Fig. 16.7 with cities between c_2 and c_3.

Unlike the 2-opt algorithm, LK (Lin and Kernighan, 1973) is built on the concept of "variable depth" so that a different number of edge changes can be made to reduce the computation time. An improved version of LK, called LKH, can be found in (Helsgaun, 2000, 2009), in which the problem in question is partitioned into a certain number of smaller problems to reduce the complexity of the problem. In addition to 2-opt, 3-opt, and 4-opt, LKH also uses sequential 5-opt and non-sequential 4-opt to further improve the quality of the end result.

16.2. Metaheuristic algorithm with local search

Using LS to improve solution(s) found by a metaheuristic algorithm is an alternative way to improve the end results, which can also be regarded as a hybrid metaheuristic algorithm. Different from other hybridizations, this kind of hybrid metaheuristic algorithm will search the region in which the current solution resides via the deterministic rule of an LS algorithm to find a better solution iteratively until the best solution of this region, i.e., local optimum, is found. Using LS as another search algorithm after a metaheuristic algorithm converges and embedding LS as a search operator of a metaheuristic algorithm are two representative ways to integrate LS and a metaheuristic algorithm. Because LS makes it possible for a metaheuristic algorithm to improve the quality of the end result, some metaheuristic algorithms have been designed in such a way that LS acts as one of its search operators, such as an ant colony system with 3-opt (Dorigo and Gambardella, 1997).

Another way to combine a metaheuristic algorithm and an LS algorithm is to integrate two different (meta)heuristic algorithms. In such a combination, one of the search algorithms is a population-based metaheuristic algorithm while the other is a single-solution-based metaheuristic algorithm or a heuristic algorithm. This kind of combination can be regarded as a hybrid metaheuristic algorithm. The population-based algorithm in this hybrid metaheuristic algorithm typically plays the role of global search, while the single-solution-based metaheuristic algorithm or the heuristic algorithm typically plays the role of LS. Because this combination makes it possible for a metaheuristic algorithm to find better results than a metaheuristic algorithm by itself, it has been a research hotspot since the 1990s. A well-known example that was mentioned in Section 12.1 is the genetic k-means algorithm (GKA) (Krishna and Narasimha Murty, 1999). This hybrid metaheuristic algorithm combines a genetic algorithm (GA) and k-means, in which GA plays the role of global search while k-means plays the role of LS. Another example that can also be found in (Mantawy et al., 1999) combines GA, TS, and SA as a hybrid metaheuristic algorithm for solving the unit commitment problem.[1] As far as this hybrid algorithm is concerned, GA plays the role of major search algorithm for global search. First, the initial population is randomly generated. At each iteration of this hybrid metaheuristic algorithm, it will first compute the fitness values of all the chromosomes, and then it will use TS to create a number of new candidate solutions v that are the neighbors of the current population s. The crossover and mutation operators of GA are then used to transit the solutions v. The last operator of this hybrid metaheuristic algorithm is SA, which is used to determine whether or not to accept the solutions v for the next generation. Applying an LS algorithm to a metaheuristic algorithm,

[1] The unit commitment problem is an optimization problem aimed at finding a solution that minimizes the total operating costs of the power generating units during the scheduling time horizon.

of course, is found not only in the studies of GA but also in the studies of other metaheuristics. In the studies of (Higashi and Iba, 2003; Mahi et al., 2015), particle swarm optimization (PSO) was used as the major search algorithm while LS (i.e., a mutation operator or 3-opt) was used to fine-tune the results found by PSO. Combining PSO and LS is a way to improve the end results of PSO, which also reveals that LS is able to enhance the performance of metaheuristics in terms of quality of the end results.

In addition to the aforementioned combinations of metaheuristic algorithms with LS, a promising hybrid metaheuristic algorithm, called the memetic algorithm (MA) (Moscato, 1989; Neri and Cotta, 2012), was presented recently. The basic idea is to use a population-based metaheuristic algorithm as the major search algorithm and a set of LS algorithms as the search operator to find a better end result. Algorithm 16 gives a general framework of MA, which is similar to GA, i.e., Algorithm 7 (Section 7.1). This means that lines 1–7 of MA are similar to GA. However, after performing the evaluation, selection, and transition operators, MA uses the LocalSearch() operator to fine-tune the searched solutions v_m to the new candidate solutions s_m and then evaluates the fitness values of these new solutions, as shown in lines 8–9. An important difference between MA and other hybrid metaheuristic algorithms is that it will select the new solutions for the next iteration from the current solutions s and new candidate solutions s_m. From the perspective of GA, the solutions for the next generation of MA are selected from parents s and its offspring s_m. Another difference is that MA will check to see if it converges to a local optimum. If it converges to a local optimum (e.g., the average of Hamming distances between searched solutions is less than a predefined threshold), it will perform a Restart() operator to prevent MA from searching solutions that have been searched for many times during the convergence process. An intuitive way to implement the Restart() operator is to keep a fraction of the current solutions and replace some current solutions by new random solutions (Neri and Cotta, 2012).

16.3. Implementation of GA with 2-opt for the TSP

A simple method that combines GA with 2-opt, called GA+2-opt, for solving the TSP is used here to illustrate the performance of LS for metaheuristic algorithms. The basic idea of this hybrid metaheuristic algorithm is to use GA as the major search algorithm while using 2-opt as a search operator to improve the results at the end of each generation, that is,

Algorithm 16 Memetic algorithm.

1 Set the crossover and mutation rate	I
2 Randomly create the initial population s	I
3 **While** the termination criterion is not met	
4 $\quad f = \text{Fitness}(s)$	E
5 $\quad v = \text{Selection}(s, f)$	D
6 $\quad v_c = \text{Crossover}(v)$	T
7 $\quad v_m = \text{Mutation}(v_c)$	T
8 $\quad s_m = \text{LocalSearch}(v_m)$	T
9 $\quad f_m = \text{Fitness}(s_m)$	E
10 $\quad s = \text{Selection}(s, s_m, f, f_m)$	D
11 $\quad$ **If** $\textit{Converged}(s)$	D
12 $\qquad s = \text{Restart}(s)$	T
13 **End**	
14 **Output** s	O

after all the other search operators of GA, e.g., the crossover and mutation operators, have been performed. The implementation details of GA+2-opt are shown in Listing 16.1.

Listing 16.1: src/c++/3-tsp/main/gals.h: GA+2-opt for the TSP.

```
1 #ifndef __GALS_H_INCLUDED__
2 #define __GALS_H_INCLUDED__
3
4 #include <string>
5 #include <fstream>
6 #include <limits>
7 #include "lib.h"
8
9 using namespace std;
10
11 class gals
12 {
13 public:
14     typedef vector<int> solution;
15     typedef vector<solution> population;
16     typedef vector<double> pattern;
17     typedef vector<pattern> instance;
18     typedef vector<double> dist_1d;
19     typedef vector<dist_1d> dist_2d;
20
21     gals(int num_runs,
22          int num_evals,
23          int num_patterns_sol,
24          string filename_ini,
25          string filename_ins,
26          int pop_size,
27          double crossover_rate,
28          double mutation_rate,
29          int num_players,
30          int num_ls,
31          int ls_flag);
32
33     population run();
```

```
34
35 private:
36     population init();
37     vector<double> evaluate(const population &curr_pop, const dist_2d &dist);
38     double evaluate(const solution& curr_sol, const dist_2d &dist);
39     void update_best_sol(const population &curr_pop, const vector<double> &curr_obj_vals,
       solution &best_sol, double &best_obj_val);
40     population tournament_select(const population &curr_pop, vector<double> &curr_obj_vals, int
       num_players);
41     population crossover_ox(const population& curr_pop, double cr);
42     population mutate(const population& curr_pop, double mr);
43     solution   two_opt(const solution& curr_sol, const dist_2d &dist, int num_ls);
44     population two_opt(const population& curr_pop, const dist_2d &dist, int num_ls);
45
46 private:
47     int num_runs;
48     int num_evals;
49     int num_patterns_sol;
50     string filename_ini;
51     string filename_ins;
52
53     double best_obj_val;
54     solution best_sol;
55
56     double avg_obj_val;
57     vector<double> avg_obj_val_eval;
58
59     population curr_pop;
60     vector<double> curr_obj_vals;
61
62     dist_2d dist;
63     instance ins_tsp;
64     solution opt_sol;
65
66     int pop_size;
67     double crossover_rate;
68     double mutation_rate;
69     int num_players;
70
71     int num_ls;
72     int ls_flag;
73
74     int eval_count;
75     double best_so_far;
76 };
77
78 inline gals::gals(int num_runs,
79                   int num_evals,
80                   int num_patterns_sol,
81                   string filename_ini,
82                   string filename_ins,
83                   int pop_size,
84                   double crossover_rate,
85                   double mutation_rate,
86                   int num_players,
87                   int num_ls,
88                   int ls_flag
89                   )
90     : num_runs(num_runs),
91       num_evals(num_evals),
92       num_patterns_sol(num_patterns_sol),
93       filename_ini(filename_ini),
94       filename_ins(filename_ins),
95       pop_size(pop_size),
96       crossover_rate(crossover_rate),
```

```
 97        mutation_rate(mutation_rate),
 98        num_players(num_players),
 99        num_ls(num_ls),
100        ls_flag(ls_flag)
101 {
102     srand();
103 }
104
105 inline gals::population gals::run()
106 {
107     avg_obj_val = 0;
108     avg_obj_val_eval = vector<double>(num_evals, 0.0);
109
110     for (int r = 0; r < num_runs; r++) {
111         eval_count = 0;
112         best_so_far = numeric_limits<double>::max();
113
114         // 0. initialization
115         curr_pop = init();
116
117         while (eval_count < num_evals) {
118             // 1. evaluation
119             curr_obj_vals = evaluate(curr_pop, dist);
120             update_best_sol(curr_pop, curr_obj_vals, best_sol, best_obj_val);
121
122             // 2. determination
123             curr_pop = tournament_select(curr_pop, curr_obj_vals, num_players);
124
125             // 3. transition
126             curr_pop = crossover_ox(curr_pop, crossover_rate);
127             curr_pop = mutate(curr_pop, mutation_rate);
128
129             if (ls_flag == 0)
130                 curr_pop = two_opt(curr_pop, dist, num_ls);
131             else {
132                 curr_obj_vals = evaluate(curr_pop, dist);
133                 int n = 0;
134                 double v = curr_obj_vals[0];
135                 for (size_t i = 1; i < curr_pop.size(); i++) {
136                     if (curr_obj_vals[i] < v) {
137                         n = i;
138                         v = curr_obj_vals[i];
139                     }
140                 }
141                 curr_pop[n] = two_opt(curr_pop[n], dist, num_ls);
142             }
143             update_best_sol(curr_pop, curr_obj_vals, best_sol, best_obj_val);
144         }
145         avg_obj_val += best_obj_val;
146     }
147
148     avg_obj_val /= num_runs;
149
150     for (int i = 0; i < num_evals; i++) {
151         avg_obj_val_eval[i] /= num_runs;
152         cout << fixed << setprecision(3) << avg_obj_val_eval[i] << endl;
153     }
154     cout << best_sol << endl;
155
156     return curr_pop;
157 }
158
159 inline gals::population gals::init()
160 {
161     // 1. parameters initialization
```

```
162    population curr_pop(pop_size, solution(num_patterns_sol));
163    curr_obj_vals.assign(pop_size, 0.0);
164    best_obj_val = numeric_limits<double>::max();
165    ins_tsp = instance(num_patterns_sol, pattern(2));
166    opt_sol = solution(num_patterns_sol);
167    dist = dist_2d(num_patterns_sol, dist_1d(num_patterns_sol));
168
169    // 2. input the TSP benchmark
170    if (!filename_ins.empty()) {
171        ifstream ifs(filename_ins.c_str());
172        for (int i = 0; i < num_patterns_sol ; i++)
173            for (int j = 0; j < 2; j++)
174                ifs >> ins_tsp[i][j];
175    }
176
177    // 3. input the optimum solution
178    string filename_ins_opt(filename_ins + ".opt");
179    if (!filename_ins_opt.empty()) {
180        ifstream ifs(filename_ins_opt.c_str());
181        for (int i = 0; i < num_patterns_sol ; i++) {
182            ifs >> opt_sol[i];
183            opt_sol[i]--;
184        }
185    }
186
187    // 4. initial solutions
188    if (!filename_ini.empty()) {
189        ifstream ifs(filename_ini.c_str());
190        for (int p = 0; p < pop_size; p++)
191            for (int i = 0; i < num_patterns_sol; i++)
192                ifs >> curr_pop[p][i];
193    }
194    else {
195        for (int p = 0; p < pop_size; p++) {
196            for (int i = 0; i < num_patterns_sol; i++)
197                curr_pop[p][i] = i;
198            for (int i = num_patterns_sol-1; i > 0; --i)     // shuffle the solutions
199                swap(curr_pop[p][i], curr_pop[p][rand() % (i+1)]);
200        }
201    }
202
203    // 5. construct the distance matrix
204    for (int i = 0; i < num_patterns_sol; i++)
205        for (int j = 0; j < num_patterns_sol; j++)
206            dist[i][j] = i == j ? 0.0 : distance(ins_tsp[i], ins_tsp[j]);
207
208    return curr_pop;
209 }
210
211 inline vector<double> gals::evaluate(const population &curr_pop, const dist_2d &dist)
212 {
213    vector<double> tour_dist(curr_pop.size(), 0.0);
214    for (size_t p = 0; p < curr_pop.size(); p++)
215        tour_dist[p] = evaluate(curr_pop[p], dist);
216    return tour_dist;
217 }
218
219 inline double gals::evaluate(const solution& curr_sol, const dist_2d &dist)
220 {
221    double tour_dist = 0.0;
222    for (size_t i = 0; i < curr_sol.size(); i++) {
223        const int r = curr_sol[i];
224        const int s = curr_sol[(i+1) % curr_pop[0].size()];
225        tour_dist += dist[r][s];
226    }
```

```
227
228    if (best_so_far > tour_dist)
229        best_so_far = tour_dist;
230    if (eval_count < num_evals)
231        avg_obj_val_eval[eval_count++] += best_so_far;
232
233    return tour_dist;
234 }
235
236 inline void gals::update_best_sol(const population &curr_pop, const vector<double> &
        curr_obj_vals, solution &best_sol, double &best_obj_val)
237 {
238    for (size_t i = 0; i < curr_pop.size(); i++) {
239        if (curr_obj_vals[i] < best_obj_val) {
240            best_sol = curr_pop[i];
241            best_obj_val = curr_obj_vals[i];
242        }
243    }
244 }
245
246 inline gals::population gals::tournament_select(const population &curr_pop, vector<double> &
        curr_obj_vals, int num_players)
247 {
248    population tmp_pop(curr_pop.size());
249    for (size_t i = 0; i < curr_pop.size(); i++) {
250        int k = rand() % curr_pop.size();
251        double f = curr_obj_vals[k];
252        for (int j = 1; j < num_players; j++) {
253            int n = rand() % curr_pop.size();
254            if (curr_obj_vals[n] < f) {
255                k = n;
256                f = curr_obj_vals[k];
257            }
258        }
259        tmp_pop[i] = curr_pop[k];
260    }
261    return tmp_pop;
262 }
263
264 inline gals::population gals::crossover_ox(const population& curr_pop, double cr)
265 {
266    population tmp_pop = curr_pop;
267
268    const int mid = curr_pop.size()/2;
269    const size_t ssz = curr_pop[0].size();
270    for (int i = 0; i < mid; i++) {
271        const double f = static_cast<double>(rand()) / RAND_MAX;
272        if (f <= cr) {
273            // 1. create the mapping sections
274            size_t xp1 = rand() % (ssz + 1);
275            size_t xp2 = rand() % (ssz + 1);
276            if (xp1 > xp2)
277                swap(xp1, xp2);
278
279            // 2. indices to the two parents and offspring
280            const int p[2] = { i, i+mid };
281
282            // 3. the main process of ox
283            for (int k = 0; k < 2; k++) {
284                const int c1 = p[k];
285                const int c2 = p[1-k];
286
287                // 4 mask the genes between xp1 and xp2
288                const auto& s1 = curr_pop[c1];
289                const auto& s2 = curr_pop[c2];
```

```
290                 vector<bool> msk1(ssz, false);
291                 for (size_t j = xp1; j < xp2; j++)
292                     msk1[s1[j]] = true;
293                 vector<bool> msk2(ssz, false);
294                 for (size_t j = 0; j < ssz; j++)
295                     msk2[j] = msk1[s2[j]];
296
297                 // 5. replace the genes that are not masked
298                 for (size_t j = xp2 % ssz, z = 0; z < ssz; z++) {
299                     if (!msk2[z]) {
300                         tmp_pop[c1][j] = s2[z];
301                         j = (j+1) % ssz;
302                     }
303                 }
304             }
305         }
306     }
307
308     return tmp_pop;
309 }
310
311 inline gals::population gals::mutate(const population& curr_pop, double mr)
312 {
313     population tmp_pop = curr_pop;
314
315     for (size_t i = 0; i < tmp_pop.size(); i++){
316         const double f = static_cast<double>(rand()) / RAND_MAX;
317         if (f <= mr) {
318             int m1 = rand() % tmp_pop[0].size();        // mutation point
319             int m2 = rand() % tmp_pop[0].size();        // mutation point
320             swap(tmp_pop[i][m1], tmp_pop[i][m2]);
321         }
322     }
323
324     return tmp_pop;
325 }
326
327 inline gals::population gals::two_opt(const population& curr_pop, const dist_2d &dist, int
        num_ls)
328 {
329     population tmp_pop(curr_pop.size());
330     for (size_t i = 0; i < curr_pop.size(); i++)
331         tmp_pop[i] = two_opt(curr_pop[i], dist, num_ls);
332     return tmp_pop;
333 }
334
335 inline gals::solution gals::two_opt(const solution& curr_sol, const dist_2d &dist, int num_ls)
336 {
337     int r_start = rand() % curr_pop.size();
338     int ls_count = 0;
339
340     solution tmp_sol = curr_sol;
341     double tmp_sol_dist = evaluate(tmp_sol, dist);
342     for (size_t i = r_start; i < tmp_sol.size()-1; i++) {
343         for (size_t k = i+1; k < tmp_sol.size(); k++) {
344             solution tmp_sol_2opt = tmp_sol;
345             for (size_t b = i, e = k; b <= k; ++b, --e)
346                 tmp_sol_2opt[b] = tmp_sol[e];
347
348             double tmp_sol_2opt_dist = evaluate(tmp_sol_2opt, dist);
349             if (tmp_sol_dist > tmp_sol_2opt_dist) {
350                 tmp_sol = tmp_sol_2opt;
351                 tmp_sol_dist = tmp_sol_2opt_dist;
352             }
353
```

```
354          if (++ls_count == num_ls) goto done;
355       }
356    }
357 done:
358    return tmp_sol;
359 }
360
361 #endif
```

16.3.1 Declaration of functions and parameters

As shown in lines 30–31, in addition to the parameters for GA, GA+2-opt requires two additional parameters, namely, num_ls and ls_flag; the former specifies the number of LSs to be performed, while the latter specifies whether to fine-tune the whole population (when ls_flag is set equal to 0) or just the current best solution (when ls_flag is set equal to 1). The data members num_ls and ls_flag are declared in lines 71–72 to hold the values of these additional parameters. To evaluate the performance of 2-opt with a limited number of exchanges, the function two_opt() is declared in lines 43–44. The function is declared twice because it is overloaded so that it can take as input either a solution or a population, as the first parameter of two_opt() shows. In this implementation, the function two_opt() declared in line 44 that takes as input a population will eventually call the two_opt() function declared in line 43 that takes as input a solution for each solution in the population to fine-tune the entire population.

16.3.2 The main loop

Since the implementation is also based on that given in Listing 12.1, the main difference is that 2-opt has been added, as shown in lines 129–142. In this implementation, there are two possibilities to use LS for the meta-heuristic algorithm. One is to use 2-opt to fine-tune all of the current solutions (when ls_flag is equal to 0), as shown in lines 129–130, while the other is to use 2-opt to fine-tune only the current best solution (when ls_flag is set equal to 1), as shown in lines 131–142.

16.3.3 Additional functions

As shown in lines 327–333, this function will repeatedly call the two_opt() function in lines 335–359 to improve all the solutions. As shown in lines 335–359, the purpose of this two_opt() function is to reduce the total distance of a tour by adjusting the order of some edges of the solution. It will first randomly select a city, as shown in line 337, and then use this city as the starting point of the LS. Because 2-opt is an LS method, in this implementation, it will check a certain number of possible combinations, as shown in lines 342–343. The reason to check only a certain number of possible combinations instead of a percentage of the number of cities is to avoid increasing the computation time for a problem of large size. An alternative approach is obviously to limit the number of edges exchanged by the LS method. Fig. 16.9 gives a simple example to illustrate the basic idea of the implementation of this operator. This example is an extension of the example given in Fig. 16.7. Again, there are four cities (c_1, c_2, c_3, and c_4) and three edges ($e_{1,3}$, $e_{3,2}$, and $e_{2,4}$), the directions of which are shown in the part denoted "before 2-opt." The part denoted "after 2-opt" shows that the direction of the edge from c_3 to c_2 is reversed so that it is now from c_2 to c_3. Besides, the next city from c_1 will be c_2, and the next city from c_3 will be c_4. In fact, all we have to do is to reverse the cities residing between the first and last cities of the selected subsolution, e.g., c_3 and c_2 of subsolution $c_1 \rightarrow c_3 \rightarrow c_2 \rightarrow c_4$, as shown in lines 345–346. After that, this operator will compute the distance of the new subsolution (e.g., $c_1 \rightarrow c_2 \rightarrow c_3 \rightarrow c_4$) and use this to replace the current subsolution (e.g., $c_1 \rightarrow c_3 \rightarrow c_2 \rightarrow c_4$) if the new distance is shorter than that of the current subsolution, as shown in lines 348–352.

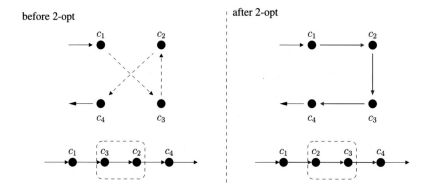

Figure 16.9 Example illustrating the implementation of 2-opt.

As shown in line 354, this operator will stop exchanging the edges, i.e., reversing the cities between the start and end cities of the selected subsolution, when the limit is reached.

16.4. Simulation results of GA with 2-opt for the TSP

Empirical analysis was conducted on a PC with Intel Xeon E5-2620v4 CPUs (2.10GHz, 20 MB cache, and eight cores) and 16 GB of memory running Fedora 26 with Linux 4.12.11-300.fc26.x86_64. The programs were written in C++ and compiled using g++. To evaluate the performance of LS for metaheuristic algorithms, GA, GA with 2-opt for all solutions (GA+2-opt-w), and GA with 2-opt for the current best solution (GA+2-opt-b) for solving the TSP benchmark eil51 are compared in this section. The number L_{max} in "GA+2-opt-w (L_{max})" and "GA+2-opt-b (L_{max})" denotes the number of LSs to be employed for each solution at each generation. For example, GA+2-opt-w (100) denotes that each chromosome will be fine-tuned by performing 100 LSs at each generation. Note that all GAs in this chapter use OX as the transition operator with the crossover and mutation rates set equal to 0.4 and 0.1, respectively.

Table 16.1 Comparison of GA and GA with 2-opt for the whole population or the current best chromosome with different parameter settings of L_{max}.

Algorithm	Total distance	Running time (in seconds)
GA	474.879	12.005
GA+2-opt-w (100)	461.853	6.791
GA+2-opt-w (200)	470.439	6.556
GA+2-opt-w (400)	465.170	6.556
GA+2-opt-w (600)	461.614	6.716
GA+2-opt-w (800)	457.180	6.917
GA+2-opt-w (1000)	454.176	6.746
GA+2-opt-b (100)	453.338	8.455
GA+2-opt-b (200)	454.007	8.157
GA+2-opt-b (400)	452.805	7.848
GA+2-opt-b (600)	451.800	7.632
GA+2-opt-b (800)	451.655	7.525
GA+2-opt-b (1000)	452.837	7.472

Table 16.1 provides a detailed comparison between GA, GA+2-opt-w, and GA+2-opt-b with different parameter settings. It can be easily seen that GA with 2-opt is faster than GA. According to our observation, because all GAs described here perform the same number of evaluations and 2-opt takes much less time in generating a candidate solution than GA does, the results of GA with 2-opt depicted in these comparisons show that they are

all faster than GA. These comparisons further show that GA with LS is able
to find a better result than GA alone. For example, the simulation results
of GA+2-opt-w with $L_{max} = 100, 200, 400, 600, 800$, and 1000 are better
than those of GA alone. Compared with GA+2-opt-w, GA+2-opt-b uses
a different fine-tune strategy that attempts to improve only the current best
chromosome, which makes it possible to find better results than GA alone
and GA+2-opt-w with the same number of evaluations. The results show
that GA+2-opt-b beats GA alone for $L_{max} = 100, 200, 400, 600, 800$, and
1000 in terms of the total distance.

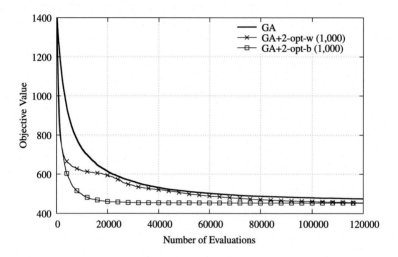

Figure 16.10 Convergence curves of GA, GA+2-opt-w, and GA+2-opt-b.

Fig. 16.10 shows the convergence curves of GA, GA+2-opt-w, and
GA+2-opt-b. The results show that the convergence of both versions of
GA with 2-opt is faster than that of GA from the early stage to the end
of the convergence process. This means that GA with 2-opt makes it possible for GA to find better results in the early stage of the convergence
process. Fig. 16.11 further shows that the end results of these GAs with
2-opt are very similar to each other, and the convergence of GA+2-opt-
b is faster than that of GA and GA+2-opt-w. The results show that the
frequency of using the LS method is a critical issue when combining
population-based metaheuristic algorithm with LS for solving an optimization problem because its setting may strongly impact the performance of
such a hybrid metaheuristic algorithm. The results show that such a hybrid
GA (GA with 2-opt) is able to find better results than the hybrid GA (GA

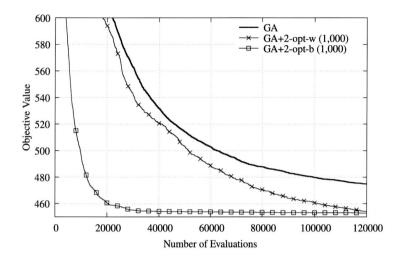

Figure 16.11 Partial enlargement of Fig. 16.10.

with SA) that we mentioned in Section 16.2. The reason is obvious. The design of 2-opt involves domain knowledge of the TSP; thus, GA with 2-opt can significantly improve the end result of GA alone for solving the TSP.

16.5. Discussion

This chapter briefly reviews LS algorithms and metaheuristic algorithms with LS algorithms. GA with 2-opt is used as a simple example to illustrate how to realize a metaheuristic algorithm with an LS algorithm to improve the end result. Even though an LS algorithm makes it possible for metaheuristic algorithms to improve the end results in most cases, an LS algorithm may not always be able to enhance the search performance of a metaheuristic algorithm. The result of GA with SA in Section 15.4 is about 471.546, whereas the result of GA with 2-opt is about 451.655. Hence, we have to take into account whether this is acceptable or not for solving a specific optimization problem before applying it to a metaheuristic algorithm during the convergence process. The good news is that several heuristic LS algorithms have been presented (e.g., LKH), which make it possible for a metaheuristic algorithm to avoid unnecessary LSs, thus reducing the overhead of LS of a metaheuristic algorithm.

Supplementary source code

Here are the hyperlinks to the directories where the source code described in this chapter resides.

https://github.com/cwtsaiai/metaheuristics_2023/src/c++/3-tsp/

CHAPTER SEVENTEEN

Pattern reduction

17.1. The basic idea of pattern reduction

How to reduce the computation time of metaheuristic algorithms is a critical research issue because the optimization problems we are facing have become much more complex. Chapter 14 provides some solutions to reduce the "response time," but not the "computation time," of a metaheuristic algorithm. Different from the parallelization methods that eventually require much more computing resources to speed up metaheuristic algorithms, several other acceleration methods (Elkan, 2003; Lai et al., 2008; Liu and Motoda, 1998; Wold et al., 1987; Xu and Wunsch-II, 2008) have been presented to remove "computations" during the convergence process. Dimensional reduction (Liu and Motoda, 1998; Wold et al., 1987) and removal of meaningless computations during searches (Elkan, 2003; Lai et al., 2008) are two ways to reduce the "computation time" of a metaheuristic algorithm.

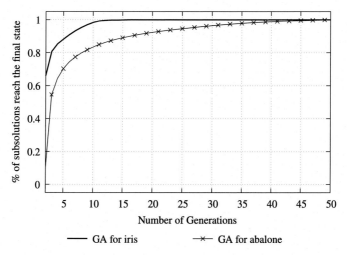

Figure 17.1 An example showing the subsolutions that reach the final state during the convergence process.

In this chapter, an efficient method for reducing the computation time of metaheuristic algorithms, called pattern reduction (PR) (Tsai, 2009; Tsai

et al., 2007), is introduced. The basic idea of PR is motivated by the observation that *some subsolutions of a solution* for most metaheuristic algorithms
will arrive at the final state more quickly than the others and will remain
intact until the end of the convergence process for solving optimization
problems. As shown in Fig. 17.1, we use the genetic algorithm (GA) with
one-iteration k-means for solving data clustering problems as an example
to demonstrate the percentage of subsolutions that reach the final state at
different times during the convergence process. It can be easily seen from
this figure that more than 95% of the subsolutions reach the final state after 10 generations during the convergence process of GA for dataset iris.[1]
A similar situation can be found in GA for dataset abalone,[2] in which at
least 80% of the subsolutions reach the final state after 10 generations. In
Fig. 17.1, by "a subsolution reaches the final state," we mean that the solution will no longer change its membership in later generations; thus, any
further computations done on the subsolution are eventually redundant.

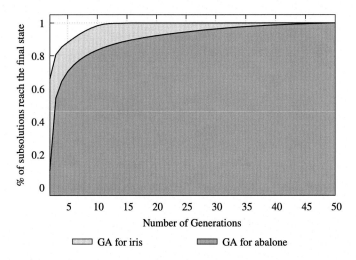

Figure 17.2 An alternative way to display the results of Fig. 17.1 to show that many of
the computations are eventually redundant during the convergence process.

Fig. 17.2 shows that the areas below the convergence curves in Fig. 17.1
indicate that a lot of computations are eventually redundant during the
convergence process of GA for solving clustering problems. This example
also shows that metaheuristic algorithms for solving different optimiza-

[1] https://archive.ics.uci.edu/ml/datasets/iris.
[2] https://archive.ics.uci.edu/ml/datasets/abalone.

tion problems may encounter similar situations. It can be easily seen from Figs. 17.1 and 17.2 that the computations repeatedly performed on the subsolutions of a solution are eventually redundant during the convergence process. This phenomenon can be easily found in metaheuristic algorithms for solving optimization problems, especially for combinatorial optimization problems. This implies that we may be able to reduce the computation time of a metaheuristic algorithm if an efficient mechanism can be found to detect and remove these redundant computations during the convergence process. This mechanism works under the following assumptions: (1) all subsolutions arrive at the final state at different times during the convergence process, (2) the mechanism (PR) is able to find computations that are eventually redundant during the convergence process, and (3) the mechanism (PR) is efficient enough so that it takes less time than the time it can save.

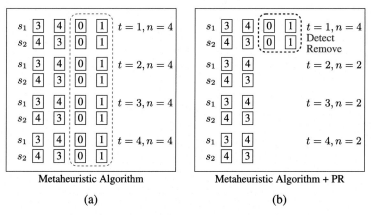

Figure 17.3 The main difference between metaheuristic algorithms and metaheuristic algorithms with PR. (a) The computation costs of a metaheuristic algorithm. (b) The computation costs of a metaheuristic algorithm with PR.

Fig. 17.3 gives an example to show the main difference between a metaheuristic algorithm and a metaheuristic algorithm with PR. In this example, the metaheuristic algorithm will search two solutions (s_1 and s_2), each of which has four subsolutions at a time during the convergence process, that is, the size of each solution will remain intact. Suppose the number of computations it takes is $2 \times 4 \times 4 = 32$, as shown in Fig. 17.3(a). Fig. 17.3(a) further shows that the last two subsolutions of these two solutions are the same at iteration $t = 1$, i.e., $s_{1,3} = s_{2,3} = 0$ and $s_{1,4} = s_{2,4} = 1$. If these two subsolutions of solutions s_1 and s_2 remain intact until the end of the convergence process, as in this case, the computations that the transition, evaluation, and determination operators performed on these two subsolutions of the two solutions will be redundant. How to detect and remove these redundant computations has thus been the main research issues of PR. Fig. 17.3(b) shows that if PR is able to detect and remove these redundant computations during the convergence process, the total computation time can be significantly reduced. The example shows that PR is able to detect and remove the redundant computations on the third and fourth subsolutions at iteration $t = 1$ so that no more computations are done on them at later iterations and consequently the number of computations that it can save is about $2 \times 2 \times 3 = 12$, that is, the total number of computations can be reduced from 32 to 20.

We will mention some of the issues that we have to pay special attention to before applying PR to metaheuristic algorithms to solve optimization problems. First, we need to take into account the risk of removing subsolutions during the convergence process. The answer to this issue can be very simple or very complex. The one-max problem provides us a simple answer. For a population-based metaheuristic algorithm, suppose that the population size is set equal to 10 (i.e., $m = 10$). The first subsolution of each of the 10 initial solutions has a 50/50 chance of being "0" or "1." If a random procedure is used to create these 10 initial seeds, it may end up creating five whose first subsolutions are "1" and five whose first subsolutions are "0." From a statistical perspective, the probability of all first subsolutions being the same is 2^{-m}, where m is the number of solutions. In this case, $m = 10$, so it is $2^{-10} = 1/1024$. It can be easily seen that for $m = 20$, the probability of all the first subsolutions being the same is $2^{-20} = 1/1{,}048{,}576$. The probability of occurrence of this event would certainly be very small if a random procedure is used by a metaheuristic algorithm to search and transit the search directions. Although a metaheuristic algorithm itself will control or adjust the search solutions during the convergence process, the probability of a subsolution at the same locus being the same will still be very small. If all the subsolutions at the same locus are the same, of course, it may be this kind of subsolution that would provide a positive impact on the metaheuristic algorithm, for example increasing the objective value of the search solution. This implies that if a subsolution s_x with value x can provide a better result than a subsolution s_y with value y, it will have a much higher probability to be selected and passed on to the next iteration. This assumption conforms to the schema theorem (Holland, 1975), which states that the short and low-order schemata of GA with above-average fitness will be increasing exponentially in later generations. For these reasons, removal of such a subsolution may be useful for reducing the redundant computations of a metaheuristic algorithm during the convergence process.

Since removing a portion of subsolutions during the convergence process may negatively affect the end results of a metaheuristic algorithm, another issue that we have to take into account is the kind of optimization problem that requires such an acceleration method. One obvious example is real-time application because this kind of application expects a good solution that can be found in a "reasonable time" instead of an exact solution that may take an enormous amount of time to obtain. How fast a metaheuristic algorithm can get such an approximate solution is the main focus of this kind of application. Although PR is able to detect and remove redundant computations of a metaheuristic algorithm for solving combinatorial optimization problems, another critical research issue is how to make it possible for PR to solve optimization problems in a continuous space. This research issue is very important because how to detect subsolutions that have the same value at the same locus of different solutions is a key issue in PR, but the values of subsolutions at the same locus may not be exactly the same. The subsolutions $s_{1,1} = 0.0001$ and $s_{2,1} = 0.00011$, which are the same or different for PR in solving an optimization problem, can be used as a simple example. A possible solution can be found in (Tsai et al., 2012), which states that if the difference between these two subsolutions $0.00001 = 0.00011 - 0.0001$ will not affect the end results of the solutions, then even though these two subsolutions are not exactly the same, they can eventually be considered the same. The basic idea of this detection method is to first check to see if the "differences between subsolutions in the same locus" are smaller than a predefined threshold and then check to see if these differences will have an impact on the quality of the solution. With this kind of detection mechanism, PR can then remove such subsolutions during the convergence process of a metaheuristic algorithm.

Algorithm 17 gives a framework for metaheuristics with PR that is similar to the framework for metaheuristics (Algorithm 3 in Section 4.1) except lines 3 and 8–9. The framework described herein shows that PR can be embedded in most metaheuristic algorithms. In this framework, s^r represents subsolutions that are detected by PR and thus will remain intact at later iterations during the convergence process, while f_r represents the fitness value of subsolutions s^r. Line 3 shows that s^r is set equal to an empty set while f_r is set equal to 0 before entering the convergence process. Line 6 shows that a solution is composed of two parts, uncompressed subsolutions

---

Algorithm 17 Framework for metaheuristics with pattern reduction.

---

1 s = Initialization(i) $\boxed{\text{I}}$
2 f_s = Evaluation(s) $\boxed{\text{E}}$
3 $s^r = \emptyset, f_r = 0$ $\boxed{\text{I}}$
4 **While** the termination criterion is not met
5   v = Transition(s, s^r) $\boxed{\text{T}}$
6   f_v = Evaluation(v, s^r, f_r) $\boxed{\text{E}}$
7   s, f_s = Determination(s, v, f_v, f_s) $\boxed{\text{D}}$
8   r, f_r = Detection(s, s^r, f_s, f_r) $\boxed{\text{D}}$
9   s, s^r, f_s = Remove(s, s^r, r, f_s, f_r) $\boxed{\text{T}}{\mapsto}\boxed{\text{D}}$
10 **End**
11 **Output** s $\boxed{\text{O}}$

---

v and compressed subsolutions s^r, and the fitness or objective value of s^r is represented by f_r to increase the speed of the evaluation operator. After the operators of a metaheuristic algorithm are finished, the Detection() operator of PR, as shown in line 8, will be performed to detect subsolutions that will no longer change; thus, any further computations performed on them are eventually redundant at later iterations. After that, the Remove() operator will remove them from the solution s, that is, $s^r = s^r \cup r$ and $s = s \setminus s^r$. The fitness or objective value f_s of each solution will then be modified based on f_r to reduce the number of computations performed by the evaluation operator. Algorithm 17 shows that PR is not only simple and easy to implement, but it can also significantly reduce the computation time of a metaheuristic algorithm during the convergence process, by removing most of the redundant computations, especially for complex or large-scale optimization problems, such as data clustering (Chiang et al., 2011; Tsai et al., 2015, 2012), image processing (Tsai et al., 2009, 2013), and the traveling salesman problem (Tsai et al., 2010). Even more importantly, it can be applied to different kinds of metaheuristic algorithms, such as simulated annealing, tabu search, GA (Tsai et al., 2010), ant colony optimization (Tsai et al., 2013), particle swarm optimization (Tsai et al., 2015), and even the iterative deterministic algorithm (Chiang et al., 2011; Tsai et al., 2007).

Since the underlying concept of PR is motivated by the observation that most metaheuristics contain subsolutions that will remain intact at later iterations, a critical question is, how are they detected by PR? In (Tsai, 2009), the detection strategies of PR have been summarized into the following three categories:

- **Time-oriented**: This strategy is usually suitable for single-solution-based metaheuristic algorithms because it aims to check each subsolu-

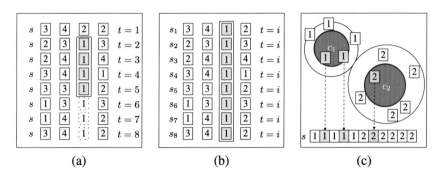

Figure 17.4 The basic idea of different detection operators. (a) Time-oriented detection. (b) Space-oriented detection. (c) Problem-specific detection.

tion at the same locus for a certain number of iterations to see if any further computations done on the subsolution are redundant. Here is how it works. As shown in Fig. 17.4(a), suppose $s = \{s_1, s_2, \ldots, s_n\}$ is a solution of a single-solution-based metaheuristic algorithm, where s_i is a subsolution of s. The basic idea of the detection operator of PR for a single-solution-based metaheuristic algorithm is to detect subsolutions that have not been changed for a predefined number of iterations. It has been an important research issue regarding *how long* is long enough to be certain that a subsolution will not be changed anymore because different metaheuristic algorithms for different optimization problems have different convergence characteristics; therefore, there exists no single good solution for determining this factor. A simple way to determine this predefined threshold for the detection of redundant subsolutions is to observe the characteristic of the convergence process of a metaheuristic algorithm in question. Another way is to use the concept of probe to remove some of the subsolutions if they will not affect the end results of a metaheuristic algorithm.

- **Space-oriented**: As for population-based metaheuristic algorithms, the so-called population s represents a set of solutions $s = \{s_1, s_2, \ldots, s_m\}$, each of which $(s_i = \{s_{i,1}, s_{i,2}, \ldots, s_{i,n}\})$ contains a set of subsolutions, where $s_{i,j}$ denotes the j-th subsolution of s_i. Fig. 17.4(b) gives a simple example to illustrate that when the values of the j-th subsolution of all the solutions are the same, it can be assumed that they have arrived at the final state. As mentioned before, the probability that subsolutions in the same locus of all the solutions are the same is very low. The probability of all the first subsolutions being the same is 2^{-m}, where

m is the number of solutions. The occurrence of such a situation to any subsolution during the convergence process is likely due to the determination operator that makes such a subsolution in all the solutions survive the selection and continue its life. This means that the chance is very small that it is caused by a random procedure. It is the reason that any further computations done on these subsolutions very likely will be redundant. Hence, they will be detected and removed by PR as the so-called redundant pattern s^r to reduce the computation time of a metaheuristic algorithm.

- **Problem-specific**: Different from time-oriented and space-oriented strategies, the design of a detection operator in this case can take into account the domain knowledge to find the subsolutions that have a small chance to change their values during the convergence process. As shown in Fig. 17.4(c), when applying a metaheuristic algorithm to a data clustering problem, the clusters to which all the data objects belong can be encoded as a solution $s = \{s_1, s_2, \ldots, s_n\}$, that is, each subsolution s_i encodes the cluster to which a data object is assigned. A data object near its centroid of a centroid-based clustering algorithm has a very small chance to move from one group to another. Based on this observation, we can then design a detection operator that uses this information to find data that may incur computations that are redundant at later iterations. This example shows that there are two clusters, denoted c_1 and c_2, that can be calculated by using the positions of all the data and how they are divided into these two clusters. As is well known, data near a centroid have a small chance to be moved to another group, especially in the later stage of the convergence process, because most clustering algorithms will converge and the moving distance of all centroids will be decreased gradually iteration by iteration; as such, we can assume that the two data objects near the centroid c_1 and the one data object near the centroid c_2 have reached the final state so that they will not be moved to another cluster during the convergence process. In this way, many computations done on these three data objects, say, determining to which cluster the data belong, can be skipped at later iterations. This means that a detection operator can be used to find these three data items during the convergence process so that a metaheuristic algorithm is able to avoid these redundant computations, thus reducing the computation time.

The Remove() operator can be designed in two different ways to cut down the redundant computations. The first one is to use a mask vector to inform the transition and evaluation operators that any further computations on the subsolutions in s^r will be redundant so that these operators can simply ignore the computations done on the subsolutions in s^r. The second one is to design the Remove() operator in such a way that it will compress the redundant subsolutions s^r into a single subsolution so that the number of computations can be reduced.

Suppose the time complexity of a metaheuristic algorithm is $O(mn\ell)$, where m denotes the number of solutions, n denotes the number of subsolutions, and ℓ denotes the number of iterations. In the ideal case, PR can reduce the time complexity of a metaheuristic algorithm from $O(mn\ell)$ to $O(nm)$ (Tsai, 2009). This can be easily proved by assuming that $\Delta \in [0, 1]$ is the percentage of subsolutions to be retained at each iteration, that is, $1 - \Delta$ is the percentage of subsolutions to be removed by PR at each iteration, as follows:

$$\sum_{i=0}^{\ell-1} \Delta^i mn = mn \sum_{i=0}^{\ell-1} \Delta^i \leq mn \sum_{i=0}^{\infty} \Delta^i = mn \frac{1}{1 - \Delta} = O(mn). \tag{17.1}$$

The time complexity of a metaheuristic algorithm with PR is now bound from above by $O(mn\ell)$ and from below by $O(mn)$. This explains why PR is able to save a large percentage of computation time of a metaheuristic algorithm.

17.2. Implementation of PREGA for clustering problems

Listing 17.1 gives a so-called pattern reduction enhanced genetic algorithm (PREGA) as a simple example to illustrate how PR works exactly. This implementation will also record the state of each subsolution iteration by iteration so that it can be easily checked to see when it reaches the final state during the convergence process. The basic idea of PREGA is to first find the so-called "redundant patterns" in the solutions and then keep other search operators informed so that any further computations on these common patterns will be avoided. As far as this implementation is concerned, the detection operator will first find the common patterns and then use a mask vector (i.e., pr_mask[m][n], where m is the population size and n is the number of subsolutions) to inform the search operators to avoid these redundant computations.

Since this implementation of PREGA is for the data clustering problem $\mathbb{P}_9$ (Section 12.1), the detection operator for this version of PR uses

the so-called problem-specific strategy discussed in Section 17.1 and illustrated in Fig. 17.4(c). This means that the detection() operator of PREGA
will be used to first detect and then remove patterns (i.e., subsolutions)
that are close to their centroids during the convergence process. Although
this implementation can be regarded as an extended version of Listing 7.1
in Section 7.2, it requires a few changes. The init() and evaluate() operators need to be modified for the data clustering problem. The crossover()
and mutate() operators also need to be modified to avoid changing the
redundant patterns that have already been compressed by PREGA. The
okm() operator is a one-iteration k-means, which is used as the local operator to fine-tune the solutions found by GA. The detection() operator
is used to detect the common subsolutions, and the detection strategy is
problem-specific as far as this implementation is concerned. Note that the
implementation described here is a simple example of PR. Many more
parts of PREGA need to be redesigned and optimized, e.g., the way the
means are calculated (Chiang et al., 2011), if we want to further reduce the
redundant computations of PREGA during the convergence process. Here
is an example command line for invoking PREGA for the data clustering
problem.

```
$ ./search prega 100 1000 150 4 3 "" "iris.data" 20 0.6 0.1 80
  0.1 0
```

The first term "prega" means that PREGA will be used by the `search`
command as the underlying search algorithm. The second term states that
the underlying search algorithm will be carried out for 100 runs, while the
third term says that 1000 evaluations (20 chromosomes for 50 generations)
will be performed per run because the number of chromosomes (i.e., the
population size) is set equal to 20. The fourth term in this case is the
number of data objects, that is, each solution contains 150 subsolutions.
The fifth and sixth terms are the number of dimensions of each object and
the number of clusters of this dataset, respectively. The seventh term (i.e.,
the first string) is the name of the file where the initial solutions reside.
A filename that is an empty string means that all the solutions will be
generated randomly. The eighth term is the name of the dataset to be
benchmarked. In this case, it is "iris.data." The three terms that follow,
20, 0.6, and 0.1, denote the number of chromosomes, the crossover rate,
and the mutation rate of PREGA, respectively. The last three terms, 80,

0.1, and 0, are the time to start the PR mechanism for GA (i.e., after 80 evaluations in this case), the reduction rate of PR, and a switch for turning the tracking mechanism for the state of all the subsolutions on or off, with "1" indicating that it is on and "0" indicating that it is off, respectively. The state of all the subsolutions during the convergence process will be recorded only when the tracking mechanism is on.

Listing 17.1: src/c++/6-clustering/main/prega.h: The implementation of PREGA for the clustering problem.

```
 1 #ifndef __PREGA_H_INCLUDED__
 2 #define __PREGA_H_INCLUDED__
 3
 4 #include <string>
 5 #include <fstream>
 6 #include <limits>
 7 #include "lib.h"
 8
 9 using namespace std;
10
11 class prega
12 {
13 public:
14     typedef vector<int> solution;
15     typedef vector<solution> population;
16     typedef vector<bool> sol_mask;
17     typedef vector<sol_mask> pop_mask;
18     typedef vector<double> pattern;
19     typedef vector<pattern> instance;
20     typedef vector<instance> instances;
21     typedef vector<double> obj_val_type;
22     typedef vector<population> population_conv; // pra
23
24     prega(int num_runs,
25           int num_evals,
26           int num_patterns_sol,
27           int num_dims,
28           int num_clusters,
29           string filename_ini,
30           string filename_ins,
31           int pop_size,
32           double crossover_rate,
33           double mutation_rate,
34           int num_players,
35           int num_detections,
36           double reduction_rate,
37           int pra);
38
39     population run();
40
41     // begin the ga functions
42     population crossover(population& curr_pop, double cr, pop_mask& pr_mask);
43     population mutate(population& curr_pop, double mr, pop_mask& pr_mask);
44     population okm(population& curr_pop, pop_mask& pr_mask, int eval);
45     pop_mask detection(population& curr_pop, pop_mask& pr_mask, int eval);
46     population select(population& curr_pop, obj_val_type& curr_obj_vals);
47     population tournament_select(population& curr_pop, obj_val_type& curr_obj_vals, int
        num_players);
48     // end the ga functions
```

```
49
50 private:
51     void init();
52     obj_val_type evaluate(const population& curr_pop);
53     void update_best_sol(const population& curr_pop, const obj_val_type& curr_obj_vals);
54
55     void redundant_analysis_1(const int curr_gen);
56     void redundant_analysis_2(const int curr_gen, const int curr_run);
57
58 private:
59     int num_runs;
60     int num_evals;
61     int num_patterns_sol;
62     int num_dims;
63     int num_clusters;
64     string filename_ini;
65     string filename_ins;
66
67     double avg_obj_val;
68     obj_val_type avg_obj_val_eval;
69     double best_obj_val;
70     solution best_sol;
71
72     instance ins_clustering;
73     instances centroids;
74
75     // for population-based algorithms-begin
76     population curr_pop;
77     obj_val_type curr_obj_vals;
78     // for population-based algorithms-end
79
80     // ga parameters-begin
81     int pop_size;
82     double crossover_rate;
83     double mutation_rate;
84     int num_players;
85     // ga parameters-end
86
87     // pr-begin
88     instance centroid_dist;
89     pop_mask pr_mask;
90     int num_detections;
91     double reduction_rate;
92     // pr-end
93
94     // pra: redundant analysis-begin
95     int pra;
96     int curr_gen;
97     population_conv pop_conv;
98     population_conv final_state;
99     instance  avg_final_state;
100    pattern  all_final_state;
101    // pra: redundant analysis-end
102 };
103
104 prega::prega(int num_runs,
105              int num_evals,
106              int num_patterns_sol,
107              int num_dims,
108              int num_clusters,
109              string filename_ini,
110              string filename_ins,
111              int pop_size,
112              double crossover_rate,
113              double mutation_rate,
```

```
114              int num_players,
115              int num_detections,
116              double reduction_rate,
117              int pra
118              )
119   : num_runs(num_runs),
120     num_evals(num_evals),
121     num_patterns_sol(num_patterns_sol),
122     num_dims(num_dims),
123     num_clusters(num_clusters),
124     filename_ini(filename_ini),
125     filename_ins(filename_ins),
126     pop_size(pop_size),
127     crossover_rate(crossover_rate),
128     mutation_rate(mutation_rate),
129     num_players(num_players),
130     num_detections(num_detections),
131     reduction_rate(reduction_rate),
132     pra(pra)
133 {
134     srand();
135 }
136
137 prega::population prega::run()
138 {
139     avg_obj_val = 0.0;
140     obj_val_type avg_obj_val_eval(num_evals, 0.0);
141     obj_val_type pr_kcounter(num_evals, 0.0);
142
143     if (pra) {
144         avg_final_state = instance(num_runs, pattern(num_evals/pop_size, 0.0));
145         all_final_state = pattern(num_evals/pop_size, 0.0);
146     }
147
148     for (int r = 0; r < num_runs; r++) {
149         int eval_count = 0;
150         double best_so_far = numeric_limits<double>::max();
151
152         // 0. initialization
153         init();
154
155         while (eval_count < num_evals) {
156             // 1. evaluation
157             curr_obj_vals = evaluate(curr_pop);
158             update_best_sol(curr_pop, curr_obj_vals);
159
160             if (pra) {
161                 redundant_analysis_1(curr_gen++);
162                 for (int p = 0; p < pop_size; p++) {
163                     for (int i = 0; i < num_patterns_sol; i++)
164                         if (pr_mask[p][i])
165                             pr_kcounter[p+eval_count]++;
166                 }
167             }
168
169             for (int i = 0; i < pop_size; i++) {
170                 if (best_so_far > curr_obj_vals[i])
171                     best_so_far = curr_obj_vals[i];
172                 if (eval_count < num_evals)
173                     avg_obj_val_eval[eval_count++] += best_so_far;
174             }
175
176             // 2. determination
177             curr_pop = tournament_select(curr_pop, curr_obj_vals, num_players);
178
```

```
179            // 3. transition
180            curr_pop = crossover(curr_pop, crossover_rate, pr_mask);
181            curr_pop = mutate(curr_pop, mutation_rate, pr_mask);
182            curr_pop = okm(curr_pop, pr_mask, eval_count);
183
184            if (eval_count > num_detections)
185                pr_mask = detection(curr_pop, pr_mask, (eval_count + pop_size) / pop_size);
186        }
187
188        avg_obj_val += best_obj_val;
189
190        if (pra)
191            redundant_analysis_2(curr_gen, r);
192    }
193
194    avg_obj_val /= num_runs;
195
196    cout << fixed << setprecision(3);
197    for (int i = 0; i < num_evals; i++) {
198        avg_obj_val_eval[i] /= num_runs;
199        cout << avg_obj_val_eval[i];
200        if (pra)
201            cout << ", " << pr_kcounter[i] / (num_runs * num_patterns_sol);
202        cout << endl;
203    }
204
205    return curr_pop;
206 }
207
208 void prega::init()
209 {
210    curr_pop = population(pop_size, solution(num_patterns_sol));
211    ins_clustering = instance(num_patterns_sol, pattern(num_dims, 0.0));
212    centroids = instances(pop_size, instance(num_clusters, pattern(num_dims, 0.0)));
213    obj_val_type curr_obj_vals(pop_size, 0.0);
214    best_obj_val = numeric_limits<double>::max();
215    centroid_dist = instance(pop_size, pattern(num_clusters, 0.0));
216    pr_mask = pop_mask(pop_size, sol_mask(num_patterns_sol, false));
217
218    // pra: redundant analysis-begin
219    if (pra) {
220        pop_conv = population_conv((num_evals/pop_size), population(pop_size, solution(
num_patterns_sol, 0)));
221        final_state = population_conv(num_runs, population(pop_size, solution(num_patterns_sol,
0)));
222        curr_gen = 0;
223    }
224    // pra: redundant analysis-end
225
226    // 1. input the dataset to be clustered
227    if (!filename_ins.empty()) {
228        ifstream ifs(filename_ins.c_str());
229        for (int i = 0; i < num_patterns_sol; i++)
230            for (int j = 0; j < num_dims; j++)
231                ifs >> ins_clustering[i][j];
232    }
233
234    // 2. create the initial solutions
235    if (!filename_ini.empty()) {
236        ifstream ifs(filename_ini.c_str());
237        for (int p = 0; p < pop_size; p++)
238            for (int i = 0; i < num_patterns_sol; i++)
239                ifs >> curr_pop[p][i];
240    }
241    else {
```

```
242         for (int p = 0; p < pop_size; p++)
243             for (int i = 0; i < num_patterns_sol; i++)
244                 curr_pop[p][i] = rand() % num_clusters;
245     }
246 }
247
248 prega::obj_val_type prega::evaluate(const population& curr_pop)
249 {
250     obj_val_type sse(pop_size, 0.0);
251
252     for (int p = 0; p < pop_size; p++) {
253         // 1. assign each pattern to its cluster
254         vector<int> count_patterns(num_clusters, 0);
255         for (int k = 0; k < num_clusters; k++)
256             for (int d = 0; d < num_dims; d++)
257                 centroids[p][k][d] = 0.0;
258         for (int k = 0; k < num_clusters; k++) {
259             for (int i = 0; i < num_patterns_sol; i++) {
260                 if (curr_pop[p][i] == k) {
261                     for (int d = 0; d < num_dims; d++)
262                         centroids[p][k][d] += ins_clustering[i][d];
263                     count_patterns[k]++;
264                 }
265             }
266         }
267
268         // 2. compute the centroids (means)
269         for (int k = 0; k < num_clusters; k++)
270             for (int d = 0; d < num_dims; d++)
271                 centroids[p][k][d] /= count_patterns[k];
272
273         // 3. compute sse
274         for (int i = 0; i < num_patterns_sol; i++) {
275             const int k = curr_pop[p][i];
276             for (int d = 0; d < num_dims; d++)
277                 sse[p] += pow(ins_clustering[i][d] - centroids[p][k][d], 2);
278         }
279     }
280
281     return sse;
282 }
283
284 prega::population prega::tournament_select(population& curr_pop, obj_val_type& curr_obj_vals,
        int num_players)
285 {
286     population tmp_pop(curr_pop.size());
287     pop_mask tmp_pr_mask(pop_size, sol_mask(num_patterns_sol, false));
288     for (size_t i = 0; i < curr_pop.size(); i++) {
289         int k = rand() % curr_pop.size();
290         double f = curr_obj_vals[k];
291         for (int j = 1; j < num_players; j++) {
292             int n = rand() % curr_pop.size();
293             if (curr_obj_vals[n] < f) {
294                 k = n;
295                 f = curr_obj_vals[k];
296             }
297         }
298         tmp_pop[i] = curr_pop[k];
299         tmp_pr_mask[i] = pr_mask[k];
300     }
301     pr_mask = tmp_pr_mask;
302     return tmp_pop;
303 }
304
305 void prega::update_best_sol(const population& curr_pop, const obj_val_type& curr_obj_vals)
```

```
306 {
307     for (size_t i = 0; i < curr_pop.size(); i++) {
308         if (curr_obj_vals[i] < best_obj_val) {
309             best_obj_val = curr_obj_vals[i];
310             best_sol = curr_pop[i];
311         }
312     }
313 }
314
315 prega::population prega::crossover(population& curr_pop, double cr, pop_mask& pr_mask)  // one-
        point crossover
316 {
317     const size_t mid = curr_pop.size()/2;
318     for (size_t i = 0; i < mid; i++) {
319         const double f = static_cast<double>(rand()) / RAND_MAX;
320         if (f <= cr) {
321             int xp = rand() % curr_pop[0].size();        // crossover point
322             for (size_t j = xp; j < curr_pop[0].size(); j++) {
323                 if (!pr_mask[i][j])
324                     swap(curr_pop[i][j], curr_pop[mid+i][j]);
325             }
326         }
327     }
328     return curr_pop;
329 }
330
331 prega::population prega::mutate(population& curr_pop, double mr, pop_mask& pr_mask)
332 {
333     for (size_t i = 0; i < curr_pop.size(); i++) {
334         const double f = static_cast<double>(rand()) / RAND_MAX;
335         if (f <= mr) {
336             const int m = rand() % curr_pop[0].size();   // mutation point
337             if (!pr_mask[i][m])
338                 curr_pop[i][m] = rand() % num_clusters;
339         }
340     }
341     return curr_pop;
342 }
343
344 prega::population prega::okm(population& curr_pop, pop_mask& pr_mask, int eval)
345 {
346     for (size_t p = 0; p < curr_pop.size(); p++) {
347         // 1. one iteration k-means
348         vector<int> count(num_clusters, 0);
349         for (int i = 0; i < num_patterns_sol; i++) {
350             if (!pr_mask[p][i] || eval > num_evals/2) {
351                 double dist = numeric_limits<double>::max();
352                 int c = 0;
353                 for (int k = 0; k < num_clusters; k++) {
354                     double dist_tmp = distance(ins_clustering[i], centroids[p][k]);
355                     if (dist_tmp < dist) {
356                         dist = dist_tmp;
357                         c = k;
358                     }
359                 }
360                 curr_pop[p][i] = c;
361                 ++count[c];
362                 if (centroid_dist[p][c] < dist)
363                     centroid_dist[p][c] = dist;
364             }
365         }
366
367         // 2. randomly assign a pattern to each empty cluster
368         for (int k = 0; k < num_clusters; k++) {
369             if (count[k] == 0) {
```

```
370                     const int i = rand() % num_patterns_sol;
371                     curr_pop[p][i] = k;
372                     if (pr_mask[p][i])
373                         pr_mask[p][i] = false;
374                 }
375         }
376     }
377     return curr_pop;
378 }
379
380 prega::pop_mask prega::detection(population& curr_pop, pop_mask& pr_mask, int eval)
381 {
382     double rr = eval * reduction_rate;
383     for (size_t p = 0; p < curr_pop.size(); p++) {
384         vector<double> dist(num_clusters);
385         for (int i = 0; i < num_clusters; i++)
386             dist[i] = centroid_dist[p][i] * rr;
387         for (int i = 0; i < num_patterns_sol; i++) {
388             if (!pr_mask[p][i]) {
389                 double dist_tmp = distance(ins_clustering[i], centroids[p][curr_pop[p][i]]);
390                 if (dist_tmp < dist[curr_pop[p][i]])
391                     pr_mask[p][i] = true;
392             }
393         }
394     }
395     return pr_mask;
396 }
397
398 void prega::redundant_analysis_1(const int curr_gen)
399 {
400     pop_conv[curr_gen] = curr_pop;
401 }
402
403 void prega::redundant_analysis_2(const int curr_gen, const int curr_run)
404 {
405     // 1. find the generation at which each subsolution reaches the final state for each run
406     for (int i = 0; i < pop_size; i++)
407         for (int j = 0; j < num_patterns_sol; j++) {
408             for (int k = curr_gen-1 ; k > 1 ; k--) {
409                 if (pop_conv[k][i][j] != pop_conv[k-1][i][j]) {
410                     final_state[curr_run][i][j] = k;
411                     break;
412                 }
413             }
414         }
415     }
416
417     // 2. accumulate the information obtained above for each run
418     for (int k = 0 ; k < curr_gen ; k++) {
419         int final_state_count = 0;
420         for (int i = 0; i < pop_size; i++) {
421             for (int j = 0; j < num_patterns_sol; j++) {
422                 if (final_state[curr_run][i][j] <= k)
423                     final_state_count++;
424             }
425         }
426         avg_final_state[curr_run][k] = final_state_count / pop_size;
427         all_final_state[k] += avg_final_state[curr_run][k];
428     }
429
430     // 3. compute the percentage of subsolutions that reach the final state for all runs
431     if (curr_run == num_runs-1) {
432         cout << "rd: ";
433         for (int k = 0 ; k < curr_gen ; k++)
```

```
434            cout << fixed << setprecision(3) << all_final_state[k] / (num_runs *
        num_patterns_sol) << ", ";
435         cout << endl;
436      }
437 }
438
439 #endif
```

17.2.1 Declaration of functions and parameters

Different from the implementation of GA for the one-max problem, the implementation described herein has to read the dataset from the file; thus, two new data types—instance and instances—are added for the input data objects and centroids, as shown in lines 18–20. Also, these two vectors are declared in lines 72 and 73. The input data objects in this chapter also represent the data nodes. The variable ins_clustering[n][d] is a two-dimensional vector (i.e., a vector of vectors), where n is the number of input data objects and d is the number of dimensions of each input data object. The variable centroids[m][k][d] is a three-dimensional vector, where m is the number of chromosomes, k is the number of clusters, and d is the number of dimensions of each centroid. As shown in line 30, the filename of the input dataset will be received as a parameter, while the variable for the filename is declared in line 65. The crossover() and mutate() operators will receive an additional piece of information (i.e., pr_mask[m][n]) from the main function, as shown in lines 42–43.

Compared to Listing 7.1, two additional operators, okm() and detection(), are added in PREGA, as shown in lines 44 and 45. As shown in lines 55–56, redundant_analysis_1() and redundant_analysis_2() record the state of all the subsolutions and check if a subsolution has reached the final state, respectively. As shown in lines 88–91, four additional variables, namely, centroid_dist, pr_mask, num_detections, and reduction_rate, are declared in PREGA for the detection() operator of PR. The variable centroid_dist is used to store the maximum distance between a data object and its centroid to make it easier to compute the radius of each cluster. This information will be used by PREGA to determine whether or not to remove a pattern during the convergence process. The values in the variables reduction_rate and num_detections are predefined for the determination of how many patterns will be removed and how fast they will be removed by PREGA during the convergence process. As shown in lines 95–100, six variables, namely, pra, curr_gen, pop_conv, final_state, avg_final_state, and all_final_state, are declared for redundant_analysis_1() and redundant_analysis_2(). The variable pra is used to enable or disable redundant_analysis_1() and redundant_analysis_2(). The variable curr_gen indicates the current generation, while the variable pop_conv[t][i][j] records the value of the j-th subsolution of the i-th solution at the t-th iteration. The variable final_state[t][i][j] is used to keep track of when the j-th subsolution of the i-th solution will reach the final state. The variables all_final_state[t] and avg_final_state[r][t] are used to keep track of the number of subsolutions that have reached the final state at the t-th iteration and the number of subsolutions that have reached the final state at the t-th iteration of the r-th run, respectively.

17.2.2 The main loop

As shown in lines 137–206, the main loop of PREGA is similar to Listing 7.1 except for a few modifications. First, as shown in lines 180–182, the main loop needs to pass pr_mask on to the crossover(), mutate(), and okm() operators. Second, the main loop also passes the evaluation number t to the okm() operator to reduce the number of redundant computations. Third, as shown in lines 184–185, the detection() operator will be called by PREGA for a certain number of times as far as this implementation is concerned, but it can be changed depending on the detection strategy and policy a PR-based algorithm uses. Fourth, in lines 143–146, avg_final_state and all_final_state are initialized. In lines 160–161, PREGA will call redundant_analysis_1() to keep track of the state of all the subsolutions at each iteration. Line 191 shows that PREGA will then call redundant_analysis_2() at the end of each run to check when each subsolution reaches its final state.

17.2.3 Additional functions

Compared to that in Listing 7.1, the init() operator here has to initialize four additional vectors, namely, ins_clustering, centroids, centroid_dist, and pr_mask, for PREGA, as shown in lines 211–216, and it also has to initialize three additional variables, namely, pop_conv, final_state, and curr_gen, for redundancy analysis, as shown in lines 220–222, first. It will then read the dataset from the file to ins_clustering, as shown in lines 227–232. Moreover, this operator will create the initial solutions, as shown in lines 235–245. Since each solution is encoded as $s_i = \{s_{i,1}, s_{i,2}, \ldots, s_{i,n}\}$, where n is the number of subsolutions, and each subsolution $s_{i,j}$ is randomly created from an integer value in the range $[0, k-1]$, where k is the number of clusters, this implies that each subsolution indicates to which cluster an input datum belongs. Because this implementation is for data clustering, the evaluate() operator of PREGA will use the sum of squared errors (SSE)[3] to measure the objective value of each solution, which can be divided into three steps: (1) assign each input datum to its cluster of each searched solution, (2) compute the centroids of each searched solution, and (3) compute the SSE of all the searched solutions. As shown in lines 254–266, for each solution, this operator will first initialize the variables count_patterns and centroids and then assign each input datum to the cluster to which it belongs. The variable count_patterns is used to keep track of the number of input data objects in each cluster, while the variable centroids is used to keep track of the value of the input data objects in a cluster. As shown in lines 269–271, this operator will then compute the centroids by dividing each centroid in centroids by the number of input data objects in that cluster. Finally, this operator will compute the SSE, as shown in lines 274–278. As shown in lines 323–324 and 337–338, the crossover() and mutate() operators will check pr_mask to see if the value associated with a subsolution is "0" or "1." If the value associated with a solution is "1," the crossover() and mutate() operators will bypass the transition for the subsolution.

As shown in lines 344–378, the okm() operator of PREGA will use a one-iteration k-means to fine-tune the solutions found by GA. According to our observation, this operator can be regarded as the domain knowledge. Without this operator, the convergence of simple GA will be very poor for the data clustering problem. This operator is composed of two steps: (1) assign the input data objects to the nearest cluster and (2) check to see if a cluster is empty. As shown in lines 349–365, this operator will first assign each input datum to the nearest cluster. As shown in line 350, this operator will check pr_mask to see if an input datum is marked as a redundant pattern. For an input datum that is not marked as a redundant pattern, it will be fine-tuned by this operator. Since the number of fine-tunes on all the input data objects performed by okm() will be reduced as the number of input data objects marked and frozen by PREGA increases, they will be unmarked after $t_{max}/2$ evaluations to unfreeze these frozen data objects again, as shown in line 350. This means that all the input data objects will be allowed to be fine-tuned by this operator. As shown in lines 362–363, in this step, this operator will also keep track of the distance between the furthest input datum and its cluster centroid, i.e., the so-called radius of each cluster. As shown in lines 368–375, the okm() operator will

[3] The definition of SSE can be found in the problem definition $\mathbb{P}_9$ in Section 12.1.

check to see if a cluster is empty. If a cluster is empty (i.e., has no input data object), this operator will then randomly select an input datum and assign it to this empty cluster.

As shown in lines 380–396, the detection() operator will first compute the reduction rate, which is defined as `eval × reduction_rate`, that is, it is proportional to the number of evaluations during the convergence process. The variable `rr` is used to determine if an input datum is redundant. This is based on the observation described in (Chiang et al., 2011), which states that an input datum that is close to its centroid will have a small chance to change its membership, that is, to be moved from one cluster to another. This means that if the distance between an input datum and its centroid is less than the threshold, it will be marked as a redundant pattern, and then any further computations on this subsolution will be avoided by the crossover(), mutate(), and okm() operators.

Fig. 17.5 shows a simple example to illustrate how redundant_analysis_1() (lines 398–401) and redundant_analysis_2() (lines 403–437) work. Suppose that s_1 and s_2 are two solutions that will be searched by a metaheuristic algorithm for eight iterations. Suppose further that each of them contains two subsolutions, such as $s_{1,1}$ and $s_{1,2}$ of s_1. Each subsolution can be either in the final state or not in the final state. The iteration at which a subsolution reaches its final state for the first time is defined as the time point at which the subsolution reaches its final state. To find the time points of all the subsolutions, we first assume that all the subsolutions reach their final states at the last iteration. We then record the states of all the subsolutions iteration by iteration until the final iteration is reached. The time point at which a subsolution reaches the final state can be easily found by looking at the value of the subsolution iteration by iteration from the last iteration back until the very first iteration bumped on the way back (say, iteration $n-1$), where the value of the subsolution is different from the value of the subsolution at the final state. Obviously, the iteration before (namely, iteration n) will be the time point at which the subsolution reaches its final state. In this way, the state of each subsolution can be recorded and checked so that we can know roughly when each subsolution reaches the final state during the convergence process and the percentage of subsolutions that reach the final state each iteration, just like the example shown in Fig. 17.1.

As shown in lines 398–401, the redundant_analysis_1() operator is used to save the values of all the subsolutions for each iteration to a pool. As shown in lines 403–437, the redundant_analysis_2() operator is used to an-

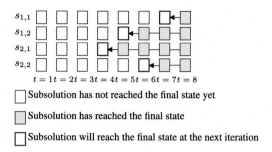

$s_{1,1}$

$s_{1,2}$

$s_{2,1}$

$s_{2,2}$

$t = 1 \quad t = 2 \quad t = 3 \quad t = 4 \quad t = 5 \quad t = 6 \quad t = 7 \quad t = 8$

☐ Subsolution has not reached the final state yet

☐ Subsolution has reached the final state

☐ Subsolution will reach the final state at the next iteration

Figure 17.5 The basic ideas of different detection operators.

alyze when a subsolution reaches the final state during the convergence process which can be divided into three steps: (1) finding the iteration at which a subsolution reaches the final state, as shown in lines 406–415, (2) counting the number of subsolutions that reach the final state at each iteration, as shown in lines 418–428, and (3) computing the average percentage of subsolutions that reach the final state each iteration of all runs, as shown in lines 431–436.

17.3. Simulation results of PREGA for clustering problems

Two datasets from the UCI Machine Learning Repository (Dua and Graff, 2017)—iris and abalone—were used for the simulations. The dataset iris has 150 input data objects, each of which has four attributes, and consists of three groups. The dataset abalone has 4177 input data objects, each of which has eight attributes, and consists of 29 groups.

To evaluate the performance of PREGA, let $\beta \in \{D, T\}$ denote either the SSE ($\beta = D$) or the computation time ($\beta = T$). Further, let Δ denote the enhancement of ϕ (PREGA) with respect to ψ (GA) as a percentage. The performance of PREGA, Δ_β, can now be defined as follows:

$$\Delta_\beta = \frac{\phi_\beta - \psi_\beta}{\psi_\beta} \times 100\%. \tag{17.2}$$

To compare GA and PREGA, we first set the launch time l_t and the reduction rate r_r of PREGA equal to 80 and 0.1, respectively. As shown in Table 17.1, PREGA is not faster than GA, but it gives a better result than GA for iris. A similar situation is observed in the comparison between

GA and PREGA in solving abalone. For abalone, PREGA can reduce the computation time by about 23.517%, which is more than for iris.

Table 17.1 Comparison of GA and PREGA for solving iris and abalone.

DS	Algorithm	n	t_{max}	m	SSE	Δ_D	T	Δ_T
iris	GA	150	20	20	78.942		0.216	
iris	PREGA	150	20	20	78.941	−0.001	0.244	12.963
abalone	GA	4177	50	20	642.245		89.320	
abalone	PREGA	4177	50	20	629.646	−1.962	68.315	−23.517

T: Running time (in seconds).

Table 17.2 shows the impact different launch times l_t have on clustering abalone. It can be easily seen that PREGA will take a longer running time when we postpone the time to launch the detection and remove operators of PREGA. The end results may also be worse when the launch time l_t is large compared with the results of GA, but not always. Note that this analysis result provides us with some information regarding the impact this parameter may have on the performance of PREGA; however, we are not arguing that a larger launch time l_t will always negatively affect the end results of all clustering algorithms.

Table 17.2 Comparison of PREGA with different launch schedules for solving abalone.

Launch time	SSE	Running time (in seconds)
40	644.944	65.536
60	656.275	66.831
80	629.646	68.315
100	647.112	70.237
120	633.348	71.768
160	637.401	72.870
180	637.870	74.472
200	653.046	75.947

Table 17.3 further shows the impact different reduction rates r_r may have on clustering abalone. It can be easily seen that the end results get worse when the reduction rate is increased, which shows that removing too many data objects belonging to a cluster may negatively affect the end results. On the other hand, a smaller reduction rate r_r means that the selection method for removing the data objects will become more conservative. The results

show that a smaller reduction rate r_r will eventually improve the end results of PREGA, making them even better than those of GA, while at the same time reducing its computation time.

Table 17.3 Comparison of PREGA with different reduction rates for solving abalone.

Reduction rate	SSE	Running time (in seconds)
0.01	576.500	74.407
0.02	577.019	71.338
0.06	614.011	68.484
0.08	637.519	66.982
0.1	629.646	68.315
0.2	630.631	66.704
0.4	639.884	66.683
0.6	635.602	67.170

Fig. 17.6 shows that the convergence processes of PREGA and GA for solving iris are very similar to each other. If we look closely at the convergence processes of PREGA and GA, as shown in Fig. 17.7, it can be easily seen that the convergence speed of PREGA is still very similar to that of GA.

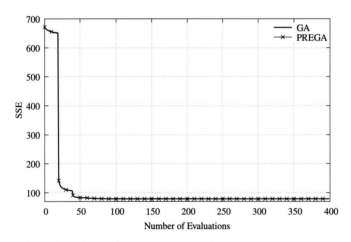

Figure 17.6 Comparison of GA and PREGA for iris.

To better understand the convergence of PREGA and GA, we also show the convergence of PREGA and GA for solving abalone. As shown

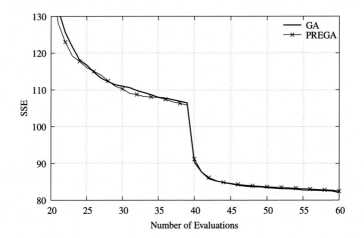

Figure 17.7 Details of Fig. 17.6.

in Fig. 17.8, PREGA with different settings and GA are very similar to each other because the interval of SSE is too large to see the details. In Fig. 17.9, the interval of SSE is made much smaller, i.e., in the range of 550 to 800. It can be easily seen from Figs. 17.8 and 17.9 that the SSEs of GA and PREGA-based algorithms strongly improve from 14,000 down to about 750 from evaluation 1 to evaluation 100. The main reason is that one-iteration k-means can provide a good fine-tuning for the searched solutions of GA and PREGA. During the period from evaluation 100 to evaluation 500, the results of PREGA with different parameter settings are worse than those of GA. According to our observation, this is due to the fact that most data objects are marked and fixed by PREGA, and they are not changed by any transition operator (e.g., crossover, mutation, and one-iteration k-means). To solve this issue, the data objects are unmarked after $t_{max}/2$ evaluations for the one-iteration k-means operator. As a consequence, PREGA can strongly improve after 500 evaluations so that it may be able to find a better result than GA alone.

In summary, these two simulations show that although PR is able to eliminate computations on the subsolutions that are essentially redundant, it may also negatively affect the end results in some cases. It is an alternative way to accelerate the computation time of metaheuristic algorithms, especially for large-scale or complex optimization problems.

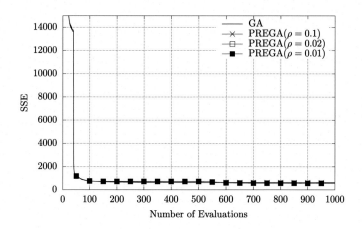

Figure 17.8 Comparison of GA and PREGA for abalone.

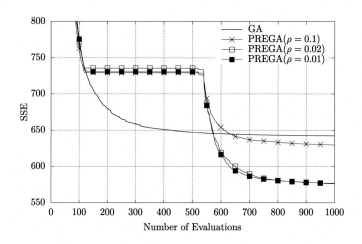

Figure 17.9 Details of Fig. 17.8.

17.4. Related work

In addition to PR, several studies (Baluja, 1994; Geem, 2009; Geem et al., 2001; Larraanaga and Lozano, 2001; Michalski, 2000; Mühlenbein and Paaß, 1996) attempted to use the search experience to construct probabilistic models to develop new metaheuristic algorithms to enhance the search performance. Most of them are designed to improve the end results, not to reduce the computation time of a metaheuristic algorithm.

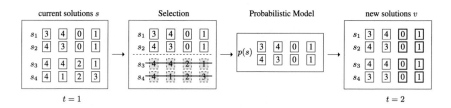

Figure 17.10 The basic idea of estimation of distribution algorithms.

The estimation of distribution algorithm (EDA) (Larraanaga and Lozano, 2001; Pelikan et al., 2000) is one of the most well-known alternatives. An interesting characteristic of EDAs is that they are neither crossover nor mutation operators; instead, the new solutions are sampled by the probabilistic model $p(x)$ that is constructed based on the current searched solutions. As shown in Fig. 17.10, EDA first sorts the current solutions by their fitness or objective values and then removes some of them. In this example, only solutions s_1 and s_2 are retained. EDA will then use these two selected solutions (i.e., s_1 and s_2) to construct its probabilistic model. An intuitive and simple way to construct this probabilistic model is called the univariate marginal distribution algorithm (UMDA) (Mühlenbein, 1997). UMDA estimates the joint probability distribution of the selected solutions that are used to construct the probabilistic model $p(x)$ to create the new solutions as follows:

$$p(x) = \prod_{i=1}^{n} p(x_i), \qquad (17.3)$$

where n is the number of subsolutions of EDA and $p(x_i)$ is estimated from the marginal frequencies as follows:

$$p(x_i) = \frac{\sum_{j=1}^{n_s} \delta_j(X_i = x_i|s')}{n_{s'}}, \qquad (17.4)$$

where $n_{s'}$ is the number of selected solutions s' and $\delta_j(X_i = x_i|s')$ can be computed as

$$\delta_j(X_i = x_i|s') = \begin{cases} 1, & \text{if } X_i = x_i \text{ in the } j\text{-th case of } s' \\ 0, & \text{otherwise.} \end{cases} \qquad (17.5)$$

By using these equations, the probabilistic model $p(x)$ of EDA can be easily constructed. For example, suppose the problem size of the one-max problem is four (i.e., the solution length is set equal to four). If there are two solutions $\langle 1, 1, 0, 0 \rangle$ and $\langle 0, 0, 1, 1 \rangle$, the probabilistic model $p(x) = \prod_{i=1}^{4} p(x_i)$, where $p(X_i = 1) = 0.5$, for $i = 1, \ldots, 4$. In addition to UMDA, a variety of EDAs have been presented, which include EDA without dependencies, with bivariate dependencies, with multi-variate dependencies, and with mixture models (Larraanaga and Lozano, 2001).

At the time that EDAs were presented, several studies (Geem, 2009; Geem et al., 2001; Michalski, 2000) were conducted to construct a probabilistic model or a historical (elitism) list based on "good searched solutions" to develop new metaheuristic algorithms. In (Michalski, 2000), Michalski presented a non-Darwinian-type evolution model called the learnable evolution model (LEM). The solutions of LEM can be divided into two groups: the high-performance group (H-group) and the low-performance group (L-group). The new solutions are created using the H-group descriptions while at the same time against the L-group descriptions. Another well-known metaheuristic algorithm called harmony search (HS), which also uses a probabilistic model to create the new solutions, can be found in (Geem, 2009; Geem et al., 2001). HS uses the searched solutions to construct the probabilistic model and then use it to generate the new solutions. In (Geem, 2009), Geem showed several successful cases of using HS to solve different optimization problems. In summary, PR, EDA, and HS can be regarded as metaheuristic algorithms which attempt to construct probabilistic models or historical (elitism) lists to enhance their search performance. The simulation results of these search algorithms show that such mechanisms may provide a much more efficient or effective way to develop metaheuristic algorithms. The main difference is that EDA and HS are designed to find better results than other evolutionary algorithms (EAs), while PR is designed to accelerate the computation time of EAs and other metaheuristic algorithms. According to our observation, most of them will hit the premature problem because the search diversity will degrade quicker than for other traditional metaheuristic algorithms. For this reason, how to maintain or increase the search diversity of this kind of metaheuristic algorithm has been a critical research issue.

17.5. Discussion

This chapter gives a brief review on PR, which provides an alternative way to accelerate metaheuristic algorithms by sacrificing a bit the quality of the end results. PR is very useful for solving large-scale and complex optimization problems. A brief review on the relevant work of using probabilistic models to enhance the performance of metaheuristic algorithms is given, which includes the estimation of distribution algorithm, the learnable evolution model, and HS. According to our observation, most of these search algorithms (including all the PR-based algorithms) converge faster than GA, which makes it possible for them to fall into local optima easily; therefore, some useful mechanisms to increase their search diversity are needed, such as the multi-start strategy. Even though these kinds of search algorithms may easily fall into local optima, they still provide us with alternative ways to enhance the search performance of metaheuristic algorithms, by reducing the computation time, improving the end result, or both. The good news is that by combining this kind of search algorithm and the multi-start strategy, we can not only reduce the computation time of metaheuristic algorithms, but also improve the end results; see, e.g., (Tsai et al., 2015, 2013).

Supplementary source code

Here are the hyperlinks to the directories where the source code described in this chapter resides.

https://github.com/cwtsaiai/metaheuristics_2023/src/c++/6-clustering/

CHAPTER EIGHTEEN

Search economics

18.1. The basic idea of search economics

The so-called metaheuristic algorithms are designed to find approximate solutions within a reasonable time. Most of them are intelligent enough to guess the right search directions for finding good solutions during the convergence process. Such search strategies can be used to replace exhaustive search (ES) to avoid systematically generating and checking every possible solution. This explains why most metaheuristic algorithms are able to find "good results" faster than ES.

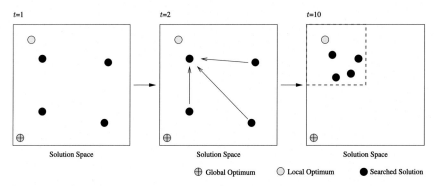

Figure 18.1 An illustration of the convergence of a metaheuristic algorithm in the solution space.

As shown in Fig. 18.1, in most cases, searches of a metaheuristic algorithm are doomed to converge to particular regions after a certain number of iterations. This figure also shows that a metaheuristic algorithm may be misled by a particular local optimum and most of the searches will converge to particular regions. In this situation, the search diversity of the metaheuristic algorithm will be reduced during the convergence process. This explains why retaining the search diversity is an important way to avoid searches converging to particular regions, i.e., falling into local optima. To deal with the problem that searches of a metaheuristic algorithm will converge to particular regions after a certain number of iterations, adding mechanisms to keep the diversity or balance the diversification and intensification of searches are two promising research directions. Another way to make the searches of a metaheuristic algorithm more precise during the convergence process is to use the so-called fitness landscape (Jin, 2005; Malan and Engelbrecht, 2013; Ochoa and Malan, 2019; Richter and Engelbrecht, 2014). One intuitive way is to keep a set of searched solutions, their objective values, and their positions in the solution space and then use this supplementary information to guide later searches during the convergence process. A major problem is that we cannot keep every searched solution; otherwise, it is going to be something just like adding all the searched solutions to an unlimited tabu list of a tabu search (TS), which will require a tremendous amount of memory space and an enormous number of checks to generate new candidate solutions. Another problem is that most metaheuristic algorithms do not have enough information about the fitness landscape to guide the searches during the convergence process.

Because of the aforementioned issues, a novel metaheuristic algorithm was presented in (Tsai, 2015). The unique characteristics of search economics (SE) can be summarized as follows. (1) It first divides the solution space into a set of subspaces (called regions) and then aims to *depict the landscape* by using some of the representative searched solutions. (2) Then it uses the *expected value of a region* to replace the objective and/or fitness value of a solution to determine later search directions. This means that SE searches the solution space based on the "potential" of each region; that is, to use the potential of each region to lead the search directions, just like investment selection in a market.

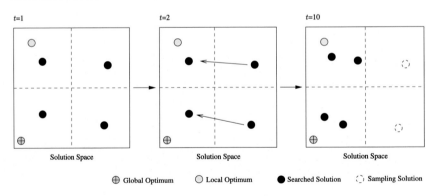

Figure 18.2 An illustration of the convergence of SE in the solution space.

As shown in Fig. 18.2, SE first divides the entire solution space into a set of regions at iteration $t = 1$. Each searcher s_i of SE will trend to a region that has high potential to get a solution that is better than its current solution.

Because SE considers additional information during the convergence process, searches will not always stay at the same region. Different from most metaheuristic algorithms that use only the objective value to evaluate the searched solution, SE uses additional information—to guide the searches, just like using the so-called return on investment (ROI) to measure every search during the convergence process—which includes: (1) investment status, (2) potential, and (3) solution space landscape of each region. These three pieces of information can also be regarded as the past, future, and present, respectively. The investment information contains the number of times a region has been searched and the number of times a region has not been searched by SE. SE avoids searching the same region for a long time while at the same time attempting to search the region that it has not searched for a long time to increase the search diversity. SE will also compute the potentials of solutions in different regions to provide another piece of information to determine whether or not to move the search from one region to another. SE accumulates the information of searched solutions from the very beginning, which are condensed into the so-called sampling solutions. The *expected value* of each region will then be computed, which is composed of the investment status and potential as well as the landscape information of each region that can be used by the evaluation and determination operators to make the search plan at later iterations. With these additional pieces of information and the objective value of searched solutions, each search of SE will be more meaningful than when using only the objective values. The ultimate goal of SE is to obtain better results by investing more computation resources[1] because it will attempt to take down some "representative searched solutions" to depict the landscape of the solution space to avoid meaningless searches and to use the expected value to make each search meaningful. In brief, the design concept of SE is that "you get whatever solution for which you actually pay."

As shown in Algorithm 18, SE is composed of the Initialization(), ResourceArrangement(), VisionSearch(), and MarketingResearch() operators. The Initialization() operator initializes the parameters of SE and creates the initial seeds, that is, it is similar to the Initialization() operator of most metaheuristic algorithms. The ResourceArrangement() operator divides the solution space into a set of regions. The VisionSearch() operator searches the approximate solution based on the expected value of each

---

[1] By "investing more computation resources" we mean that it will search more candidate solutions by having it run for more times or increasing the number of solutions searched at each iteration.

region. Different from `VisionSearch()`, which is aimed at searching the solutions, the `MarketingResearch()` operator is responsible for depicting the landscape of the entire solution space and recording the search circumstances of each region. These pieces of information will be passed to the `VisionSearch()` operator at the next iteration.

Algorithm 18 Search economics.

1 $s = $ Initialization()	$\boxed{\text{I}}$
2 $m = $ ResourceArrangement(s)	$\boxed{\text{I}}$
3 **While** the termination criterion is not met	
4 $s = $ VisionSearch(s, m)	$\boxed{\text{T}} \mapsto \boxed{\text{E}} \mapsto \boxed{\text{D}}$
5 $m = $ MarketingResearch(s, m)	$\boxed{\text{D}}$
6 **End**	
7 **Output** s	$\boxed{\text{O}}$

Compared to most metaheuristic algorithms, several additional variables are required by SE for keeping the information during the convergence process, which is summarized as follows:

s, s_i The set of solutions, i.e., $s = \{s_1, s_2, \dots, s_n\}$, where s_i is the i-th solution and n is the number of solutions. Note that s_i in SE is also called the searcher.

r, r_j The set of regions in the solution space, i.e., $r = \{r_1, r_2, \dots, r_h\}$, where r_j is the j-th region and h is the number of regions.

r_j^b The best-so-far solution of region r_j.

$m, m_{j,k}$ The set of sampling solutions for all the regions, i.e., $m = \{m_{1,1}, \dots, m_{1,w}, \dots, m_{h,1}, \dots, m_{h,w}\}$, where $m_{j,k}$ is the k-th sampling solution in the j-th region, h is the number of regions, and w is the number of sampling solutions in each region.

$v^i, v_{j,k}^i$ The set of candidate solutions of the i-th searcher, which is defined as the transition operator of s_i and all the sampling solutions in m, that is, $v_{j,k}^i = s_i \otimes m_{j,k}$, with $\otimes$ being the transition operator to exchange the information of them.

e, e_j^i The set of expected values, i.e., e_j^i is the expected value for the i-th searcher in the j-th region.

t_j^a The number of times the j-th region has been searched.

t_j^b The number of times the j-th region has not been searched.

18.1.1 The resource arrangement operator

As shown in Algorithm 19, the resource arrangement operator will first divide the solution space into h regions, i.e., subspaces. It will also randomly create w solutions for each region, called the sampling solutions m. This means that each sampling solution $m_{j,k}$ will be restricted to the j-th region. This operator will then randomly create n solutions s_i, where n is not necessarily equal to h because SE is designed in such a way that the computation resources can be dynamically added or reduced. The i-th searcher will be created and restricted to the j-th region, e.g., $j = (i \mod h)$. For example, if the solution space is divided into four regions and there are 10 searchers, then the first two regions will be assigned three searchers while the last two regions will be assigned two searchers.

Algorithm 19 ResourceArrangement(s).

```
 1  r = Divide the solution space into h regions
 2  For j = 1 to h
 3        For k = 1 to w
 4              m_{j,k} = Random(r_j)
 5        End
 6  End
 7  For i = 1 to n
 8        s_i = Random(r)
 9        Assign(s_i, m)
10  End
```

In (Tsai, 2015), we used the first $\log_2 h$ subsolutions to classify the solution space into h regions for solving the one-max problem. If h is set equal to 4, the solutions in the first region r_1 will be $\{00XX\}$, where X indicates that the value of the subsolution can be either 0 or 1. The solutions in regions r_2, r_3, and r_4 will be $\{01XX\}$, $\{10XX\}$, and $\{11XX\}$, respectively. In (Tsai and Liu, 2018), another division method was presented that uses the remainder of the sum of the values of all the subsolutions of a solution (i.e., the mod function) to divide the solution space and decide to which region a searcher is to be assigned. If h is set equal to 4, for the solutions $\{1357\}$ and $\{1247\}$, the remainders of the sums of the values of these two solutions are, respectively, $(1 + 3 + 5 + 7) \pmod 4 = 16 \pmod 4 = 0$ and $(1 + 2 + 4 + 7) \pmod 4 = 14 \pmod 4 = 2$. This means that they will be classified to regions r_1 and r_3, respectively. In (Tsai and Liu, 2019), we attempted to fix a portion of the subsolutions to represent a region of the solution space. If the solution space is to be divided into two regions, the first region can be represented by $\{\phi\phi XX\}$, while the second region can be represented by $\{XX\phi\phi\}$, where ϕ stands for a subsolution that cannot be changed by the transition operator and X stands for a subsolution that can be changed by the transition operator. In addition to these division methods (for dividing the solution space), the clustering algorithm (e.g., k-means), of course, can also be used to classify the searched solutions into a set of groups, which can then be used to divide the solution space.

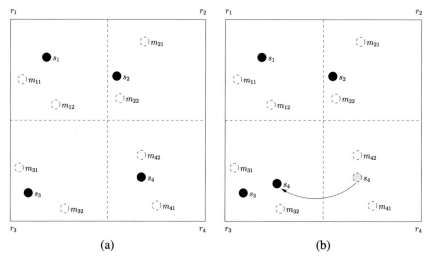

Figure 18.3 The relationships between solution s_i, sampling node $m_{j,k}$, and subregion r_j of SE. (a) At iteration t. (b) At iteration $t+1$.

How to deal with the issue of search perhaps being biased to particular regions has become an important research topic. As shown in Fig. 18.3, this issue typically will not occur because a searcher will not always stay in the same region. This means that SE will determine if a searcher should stay in the same region (i.e., intensification) or move to another region (i.e., diversification) based on the expected value of the searcher for different regions. As a result, the searcher s_4 in this figure is moved to region r_3. Because the expected value is composed of the objective values of searched solutions, the objective values of new probe solutions, and the number of times each region is searched, in (Tsai and Fang, 2021), Tsai and Fang pointed out that a searcher that moves to a region r^* can be due to the fact that the objective value of the best-so-far solution in the region r^* is better than the others, the average objective value of the probe solutions in the region r^* is better than those in the other regions, or the region r^* has not been searched for a long time.

18.1.2 The vision search operator

As shown in Algorithm 20, the vision search operator is composed of `Transition()`, `ExpectedValue()`, and `Determination()`, which play the roles of transition, evaluation, and determination, just like the counterparts of most metaheuristic algorithms. This means that if the `Transition()` or `Determination()` operator of other metaheuristic algorithms performs very well for an optimization problem $\mathbb{P}$ we are facing, then it can be adopted by SE for solving this problem, perhaps with a few modifications. For example, the one-point crossover operator of GA can be adopted as the `Transition()` operator of SE for solving the one-max problem (Tsai, 2015), while the mutation operator of DE can be used as the `Transition()` operator of SE for solving the single-objective real-parameter optimization problem (Tsai and Liu, 2020).

Algorithm 20 VisionSearch(s, m).

1 $v = \text{Transition}(s, m)$
2 $e = \text{ExpectedValue}(v, m, t^a_i, t^b)$
3 $s = \text{Determination}(v, e)$

As shown in Fig. 18.4, the main difference between most GA- and DE-based algorithms and SE is that the former will exchange information between two or more current searched solutions while the latter will exchange information between searchers s_i and the sampling solutions $m_{j,k}$ to generate probe solutions $v^i_{j,k}$ of different regions for the `ExpectedValue()` operator to determine whether a searcher s_i should stay in the current region or move to another region.

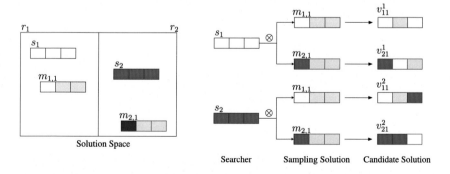

Figure 18.4 An illustration of the design of the transition operator of SE.

A representative characteristic of VisionSearch() is that it uses the expected value e_j^i to evaluate the possibility of region r_j to allow the Determination() operator to determine if the searcher s_i should stay in the current region or move to another region. The expected value for the ExpectedValue() operator is defined by

$$e_j^i = T_j V_j^i M_j, \tag{18.1}$$

where T_j represents the *investment status* of region r_j (information accumulated from the previous searches), V_j^i represents the *potential* of searcher s_i moving to region r_j (information from new candidate solutions), and M_j represents the *solution space landscape* and also how good the best-so-far solution at r_j is (information of the current best solution). T_j is defined as

$$T_j = \mathbb{N}\left(\frac{t_j^b}{t_j^a}\right), \tag{18.2}$$

where $\mathbb{N}(\cdot)$ indicates that the value of T_j is normalized by the normalization function $\mathbb{N}$ so that it is in the range $[0, 1]$. T_j provides two pieces of information: how many times the j-th region has been searched (i.e., t_j^a) and how long the j-th region has not been searched (i.e., t_j^b). The ratio of these two pieces of information allows SE to avoid getting stuck in a particular region for a long time. V_j^i is defined as

$$V_j^i = \mathbb{N}\left(\frac{\sum_{k=1}^{w} f(v_{j,k}^i)}{w}\right), \tag{18.3}$$

where $\mathbb{N}(\cdot)$ again indicates that the value of V_j^i is normalized by a normalization function. It is used to evaluate the temporary candidate solutions $v_{j,k}^i$ that can be regarded as the potential of the i-th searcher to be moved to the j-th region. M_j, which represents how good the best-so-far searched solution of the j-th region is, can be defined as

$$M_j = \mathbb{N}\left(\frac{f(r_j^b)}{\sum_{j=1}^{h} \sum_{k=1}^{w} f(m_{j,k})}\right). \tag{18.4}$$

The design of the Determination() operator is similar to that of the Transition() operator in the sense that it can be replaced by the Determination() operator of other metaheuristic algorithms, perhaps with a few changes. This means that the Determination() operator of SE for choosing a region r_j to which the searcher s_i will be moved in the next iteration based on the expected values of s_i can essentially be a determination operator of most metaheuristic algorithms. For example, the tournament selection operator of GA can be used with the expected values of a searcher s_i to different regions to decide to which region the searcher s_i will be moved at the next iteration.

18.1.3 The marketing research operator

As shown in Algorithm 21, the marketing research operator will use a certain number of sampling solutions to depict the landscape of each region, and it will also use t^a and t^b to keep track of the investment status of each region. It can be regarded as a simple realization method because it will keep only a few pieces of information of the searched solutions from the very beginning of the convergence process.

Algorithm 21 MarketingResearch(s, m).

---
1 $m = $ Update(s, m)
2 $t^a = $ Accumulation1(s, m)
3 $t^b = $ Accumulation2(s, m)

An alternative method to keep more information of the searched solutions in a region is to use the following three pieces of information: (1) the centroid of the searched solutions to keep track of the center of the information, (2) the radius to trace the scope of the searched solutions, and (3) the standard deviation to keep track of the degree of the scatter of the searched solutions. It is easily seen that increasing the number of centroids to represent the searched solutions in a region will certainly increase the accuracy of landscape depiction. Of course, there are other ways that can be used to depict the landscape. The computation costs will be a critical issue in the design of this operator.

18.2. Implementation of SE for the one-max problem

To explain the design of SE and how it works, SE for solving the one-max problem is used as an example. As shown in Fig. 18.5, to simplify SE and reduce the number of probe solutions $v^i_{j,k}$ generated at each iteration, the transition operator is modified. The main difference is that the searcher s_i will not exchange the information with the sampling solutions $m_{j,k}$ of all the regions. This means that the searcher s_i will randomly choose one, and

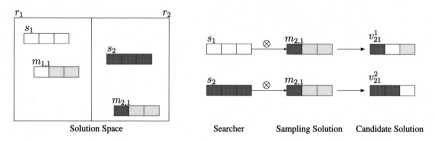

Figure 18.5 An illustration of the design of the transition operator of SE for solving the one-max problem.

only one, sample from each region to generate the probe solution $v^i_{j,k}$. The calculation of the expected value for the ExpectedValue() operator has also been simplified; it is defined as

$$e^i_j = \tilde{T}_j \tilde{M}_j, \tag{18.5}$$

where $\tilde{T}_j$ and $\tilde{M}_j$ represent, respectively, the *investment status* and *up-to-date sample solutions* of the j-th region and are defined as

$$\tilde{T}_j = \left(\frac{t^b_j}{t^a_j}\right), \tag{18.6}$$

$$\tilde{M}_j = \left(\frac{f(r^b_j)}{\sum_{j=1}^h \sum_{k=1}^w f(\tilde{m}_{j,k})}\right). \tag{18.7}$$

Different from M_j of Eq. (18.4), which is calculated based on the current sample solutions $m_{j,k}$, $\tilde{M}_j$ is calculated based on the up–to–date sample solutions $\tilde{m}_{j,k}$, which is the best one of the current sample solutions $m_{j,k}$ and probe solutions $v^i_{j,k}$. Note that the normalization functions for $\tilde{T}_j$ and $\tilde{M}_j$ were also removed to simplify the calculation of expected values. Listing 18.1 gives a simple example to illustrate how to apply SE to the one-max problem.

Listing 18.1: src/c++/1-onemax/main/se.h: The implementation of SE for the one-max problem.

```
1 #include <iostream>
2 #include <ctime>
3 #include <cmath>
4 #include <vector>
5
```

```
 6 using namespace std;
 7
 8 class se {
 9 public:
10     typedef vector<int> i1d;
11     typedef vector<i1d> i2d;
12     typedef vector<i2d> i3d;
13     typedef vector<i3d> i4d;
14     typedef vector<double> d1d;
15     typedef vector<d1d> d2d;
16     typedef vector<d2d> d3d;
17
18 public:
19     se(int num_runs,
20         int num_evals,
21         int num_bits_sol,
22         string filename_ini,
23         int num_searchers,
24         int num_regions,
25         int num_samples,
26         int num_players,
27         int scatter_plot
28         );
29
30     void run();
31
32 private:
33     // 1. initialization
34     void init();
35
36     // 2. resource arrangement
37     void resource_arrangement();
38
39     // 3. vision search
40     void vision_search(int eval);
41     void transit();
42     virtual void compute_expected_value(int eval);
43     int  evaluate_fitness(const i1d& sol);
44     void vision_selection(int player, int eval);
45
46     // 4. marketing survey
47     void marketing_survey(int& best_so_far);
48
49 private:
50     // 0.1 environment settings
51     int num_runs;
52     int num_evals;
53     int num_bits_sol;
54     string filename_ini;
55     int num_searchers;
56     int num_regions;
57     int num_samples;
58     int num_players;
59     int scatter_plot;
60
61     // 0.2 search results
62     d1d avg_obj_val_eval;
63     int best_so_far;
64     i1d best_sol;
65
66     // 0.3 search algorithm
67     i2d searcher_sol;       // [searcher, num_bits_sol]
68     i3d sample_sol;         // [region, sample, num_bits_sol]
69     i2d sample_sol_best;    // [region, num_bits_sol]
70     i4d sampleV_sol;        // [searcher, region, sample, num_bits_sol]
```

```
71
72      d1d searcher_sol_fitness;
73      d2d sample_sol_fitness;
74      d1d sample_sol_best_fitness;
75      d3d sampleV_sol_fitness;
76
77      d1d ta;
78      d1d tb;
79
80      d2d expected_value;
81      d1d T_j;
82      d1d M_j;
83
84      i1d searcher_region_id; // [searcher], region to which a searcher is assigned
85      i2d id_bits;            // region id: tabu bits
86      int num_id_bits;        // number of tabu bits
87
88      int eval_count;         // evaluation count
89  };
90
91  inline se::se(int num_runs,
92                int num_evals,
93                int num_bits_sol,
94                string filename_ini,
95                int num_searchers,
96                int num_regions,
97                int num_samples,
98                int num_players,
99                int scatter_plot
100               )
101     : num_runs(num_runs),
102       num_evals(num_evals),
103       num_bits_sol(num_bits_sol),
104       filename_ini(filename_ini),
105       num_searchers(num_searchers),
106       num_regions(num_regions),
107       num_samples(num_samples),
108       num_players(num_players),
109       scatter_plot(scatter_plot)
110 {
111     srand();
112 }
113
114 inline void se::run()
115 {
116     avg_obj_val_eval.assign(num_evals, 0.0);
117
118     for (int r = 0; r < num_runs; r++) {
119         init();                          // 1. initialization
120         resource_arrangement();          // 2. resource arrangement
121         while (eval_count < num_evals) {
122             int eval_cc = eval_count;
123             vision_search(eval_cc);         // 3. vision search
124             marketing_survey(best_so_far); // 4. marketing survey
125             for (int i = eval_cc; i < min(eval_count, num_evals); i++)
126                 avg_obj_val_eval[i] += best_so_far;
127         }
128     }
129
130     cout << fixed << setprecision(3);
131     for (int i = 0; i < num_evals; i++)
132         cout << avg_obj_val_eval[i] / num_runs << endl;
133 }
134
135 // 1. initialization
```

```
136  inline void se::init()
137  {
138      // set aside arrays for searchers, samples, and sampleV
139      searcher_sol.assign(num_searchers, i1d(num_bits_sol, 0));
140      sample_sol.assign(num_regions, i2d(num_samples, i1d(num_bits_sol, 0)));
141      sample_sol_best.assign(num_regions, i1d(num_bits_sol, 0));
142      sampleV_sol.assign(num_searchers, i3d(num_regions, i2d(num_samples*2, i1d(num_bits_sol, 0)))
         );
143
144      searcher_sol_fitness.assign(num_searchers, 0.0);
145      sample_sol_fitness.assign(num_regions, d1d(num_samples, 0.0));
146      sample_sol_best_fitness.assign(num_regions, 0.0);
147      sampleV_sol_fitness.assign(num_searchers, d2d(num_regions, d1d(num_samples*2, 0.0)));
148
149      best_sol.assign(num_bits_sol, 0);
150      best_so_far = 0;
151      eval_count = 0;
152
153      for (int i = 0; i < num_searchers; i++)
154          for (int j = 0; j < num_bits_sol; j++)
155              searcher_sol[i][j] = rand() % 2;
156  }
157
158  // 2. resource arrangement
159  inline void se::resource_arrangement()
160  {
161      // 2.1. initialize searchers and regions
162      num_id_bits = log2(num_regions);
163      searcher_region_id.assign(num_searchers, 0);
164      id_bits.assign(num_regions, i1d(num_id_bits, 0));
165      // region_id_bits();
166      for (int i = 0; i < num_regions; i++) {
167          int n = i;
168          int j = num_id_bits;
169          while (n > 0) {
170              id_bits[i][--j] = n % 2;
171              n /= 2;
172          }
173      }
174
175      // 2.1.1 assign searchers to regions
176      for (int i = 0; i < num_searchers; i++) {
177          // assign_search_region(i, i % num_regions);
178          const int r = i % num_regions;
179          searcher_region_id[i] = r;
180          for (int j = 0; j < num_id_bits; j++)
181              searcher_sol[i][j] = id_bits[r][j];
182      }
183
184      // 2.1.2 initialize sample solutions
185      for (int i = 0; i < num_regions; i++)
186          for (int j = 0; j < num_samples; j++) {
187              for (int k = 0; k < num_id_bits; k++)
188                  sample_sol[i][j][k] = id_bits[i][k];
189              for (int k = num_id_bits; k < num_bits_sol; k++)
190                  sample_sol[i][j][k] = rand() % 2;
191          }
192
193      // 2.2. initialize investment and how long regions have not been searched
194      ta.assign(num_regions, 0.0);
195      tb.assign(num_regions, 1.0);
196      for (int i = 0; i < num_searchers; i++) {
197          int r = searcher_region_id[i];
198          ta[r]++;
199          tb[r] = 1.0;
```

```
200      }
201
202      // 2.3. initialize expected values (ev)
203      expected_value.assign(num_searchers, d1d(num_regions, 0.0));
204      T_j.assign(num_regions, 0.0);
205      M_j.assign(num_regions, 0.0);
206  }
207
208  // 3. vision search
209  inline void se::vision_search(int eval)
210  {
211      // 3.1 construct V (searcher X sample)
212      if (eval > 0) transit();
213
214      // 3.2 compute the expected value of all regions of searchers
215      compute_expected_value(eval);
216
217      // 3.3 select region to which a searcher will be assigned
218      vision_selection(num_players, eval);
219  }
220
221  // 3.1 construct V (searcher X sample)
222  inline void se::transit()
223  {
224      // 3.1.1 exchange information between searchers and samples;
225      for (int i = 0; i < num_searchers; i++) {
226          for (int j = 0; j < num_regions; j++) {
227              for (int k = 0; k < num_samples; k++) {
228                  const int x = rand() % (num_bits_sol + 1);
229                  const int m = k << 1;
230
231                  for (int l = 0; l < num_id_bits; l++) {
232                      sampleV_sol[i][j][m][l] = id_bits[j][l];
233                      sampleV_sol[i][j][m+1][l] = id_bits[j][l];
234                  }
235
236                  for (int l = num_id_bits; l < num_bits_sol; l++) {
237                      if (l < x) {
238                          sampleV_sol[i][j][m][l] = searcher_sol[i][l];
239                          sampleV_sol[i][j][m+1][l] = sample_sol[j][k][l];
240                      }
241                      else {
242                          sampleV_sol[i][j][m][l] = sample_sol[j][k][l];
243                          sampleV_sol[i][j][m+1][l] = searcher_sol[i][l];
244                      }
245                  }
246              }
247          }
248      }
249
250      // 3.1.2 randomly change one bit of each solution in sampleV_sol
251      for (int i = 0; i < num_searchers; i++) {
252          for (int j = 0; j < num_regions; j++) {
253              for (int k = 0; k < 2*num_samples; k++) {
254                  int m = rand() % num_bits_sol; // bit to mutate
255                  if (m >= num_id_bits)
256                      sampleV_sol[i][j][k][m] = !sampleV_sol[i][j][k][m];
257              }
258          }
259      }
260  }
261
262  // 3.2 expected value for onemax problem
263  inline void se::compute_expected_value(int eval)
264  {
```

```
265     // 3.2.1 fitness value of searchers and sampleV_sol (new candidate solutions)
266     if (eval == 0) {
267         // 3.2.1a fitness value of searchers
268         for (int i = 0; i < num_searchers; i++)
269             searcher_sol_fitness[i] = evaluate_fitness(searcher_sol[i]);
270     }
271     else {
272         // 3.2.1b fitness value of sampleV_sol (new candidate solutions)
273         for (int i = 0; i < num_searchers; i++) {
274             int j = searcher_region_id[i];
275             if (scatter_plot == 1)
276                 cout << eval_count << " " << j-0.1*i+0.1*num_regions/2 << " " << i << " ";
277             for (int k = 0; k < num_samples; k++) {
278                 int n = rand() % 2*num_samples;
279                 int f = evaluate_fitness(sampleV_sol[i][j][n]);
280
281                 if (f > searcher_sol_fitness[i]) {
282                     searcher_sol[i] = sampleV_sol[i][j][n];
283                     searcher_sol_fitness[i] = f;
284                 }
285
286                 if (f > sample_sol_fitness[j][k]) {
287                     sample_sol[j][k] = sampleV_sol[i][j][n];
288                     sample_sol_fitness[j][k] = f;
289                 }
290             }
291         }
292         if (scatter_plot == 1)
293             cout << endl;
294     }
295
296     // 3.2.2 fitness value of samples
297     if (eval == 0) //
298         for (int j = 0; j < num_regions; j++)
299             for (int k = 0; k < num_samples; k++)
300                 sample_sol_fitness[j][k] = evaluate_fitness(sample_sol[j][k]);
301
302     double total_sample_fitness = 0.0; // f(m_j)
303     for (int j = 0; j < num_regions; j++) {
304         double rbj = 0.0;
305         int b = -1;
306
307         for (int k = 0; k < num_samples; k++) {
308             total_sample_fitness += sample_sol_fitness[j][k];
309             // update fbj
310             if (sample_sol_fitness[j][k] > rbj) {
311                 b = k;
312                 rbj = sample_sol_fitness[j][k];
313             }
314         }
315         if (b >= 0) {
316             sample_sol_best_fitness[j] = rbj;
317             sample_sol_best[j] = sample_sol[j][b];
318         }
319     }
320
321     // 3.2.3 M_j
322     for (int j = 0; j < num_regions; j++)
323         M_j[j] = sample_sol_best_fitness[j] / total_sample_fitness;
324
325     // 3.2.4 T_j
326     for (int j = 0; j < num_regions; j++)
327         T_j[j] = ta[j] / tb[j];
328
329     // 3.2.5 compute the expected_value
```

```
330     for (int i = 0; i < num_searchers; i++) {
331         for (int j = 0; j < num_regions; j++)
332             expected_value[i][j] = T_j[j] * M_j[j];
333     }
334 }
335
336 // subfunction: 3.2.1 fitness value
337 inline int se::evaluate_fitness(const i1d& sol)
338 {
339     eval_count++;
340     return accumulate(sol.begin(), sol.end(), 0);
341 }
342
343 // 3.3 select region to which a searcher will be assigned
344 inline void se::vision_selection(int player, int eval)
345 {
346     for (int j = 0; j < num_regions; j++)
347         tb[j]++;
348
349     // find index of the best vij
350     for (int i = 0; i < num_searchers; i++) {
351         int j = rand() % num_regions;
352         double ev = expected_value[i][j];
353         for (int p = 0; p < num_players-1; p++) {
354             int c = rand() % num_regions;
355             if (expected_value[i][c] > ev) {
356                 j = c;
357                 ev = expected_value[i][j];
358             }
359         }
360
361         // assign searcher i to region j
362         searcher_region_id[i] = j;
363
364         // update ta[j] and tb[j];
365         ta[j]++;
366         tb[j] = 1;
367     }
368 }
369
370 // 4. marketing survey
371 inline void se::marketing_survey(int& best_so_far)
372 {
373     for (int j = 0; j < num_regions; j++)
374         if (tb[j] > 1)
375             ta[j] = 1.0;
376
377     for (int i = 0; i < num_searchers; i++)
378         if (searcher_sol_fitness[i] > best_so_far) {
379             best_so_far = searcher_sol_fitness[i];
380             best_sol = searcher_sol[i];
381         }
382
383     for (int j = 0; j < num_regions; j++)
384         for (int k = 0; k < num_samples; k++)
385             if (sample_sol_fitness[j][k] > best_so_far) {
386                 best_so_far = sample_sol_fitness[j][k];
387                 best_sol = sample_sol[j][k];
388             }
389 }
```

18.2.1 Declaration of functions and parameters

In addition to the declaration of init() in line 34, the declarations of functions of SE can be found in lines 37–47. The declaration of the resource arrangement operator, resource_arrangement(), is given in line 37. As far as this implementation is concerned, the vision search operator, shown in lines 40–44, is composed of several functions, namely, vision_search(), transit(), compute_expected_value(), evaluate_fitness(), and vision_selection(). The declaration of the marketing survey operator is given in line 47. As shown in lines 51–88, the variable declarations can be divided into three parts: environment settings, search results, and search algorithm. To make it easier for the audience of the book to understand these declarations, the names used in the description of the algorithm and the names used in the implementation are summarized in Table 18.1. The three additional variables searcher_region_id, id_bits, and num_id_bits are declared to realize SE for solving the one-max problem. The variable searcher_region_id declared in line 84 is used to keep track of the region to which a searcher s_i is assigned. The variable id_bits declared in line 85 marks the region by a subset of solutions, while the variable num_id_bits declared in line 86 represents the number of identity bits. To keep track of the convergence status of each evaluation, the counter eval_count is declared in line 88.

18.2.2 The main loop

As shown in lines 241–260 of src/c++/1-onemax/main/main.cpp, the main() function of this implementation takes as input the parameter settings from the user and plays the role of driving the run() function defined in src/c++/1-onemax/main/se.h. As shown in lines 114–133, the implementation of run() is similar to the implementations of other metaheuristic algorithms described in this book that will carry out SE for a certain number of runs and evaluations and then show the average results.

18.2.3 Additional functions

As shown in lines 136–156, the variables for searchers, sampling solutions, and others are initialized in the init() function. The solution of each searcher will then be randomly created. As shown in lines 159–206, the ResourceArrangement() operator can be divided into three parts. First, all the relevant variables are initialized in lines 162–191, namely, num_id_bits, which is needed for dividing the solution space into a set of regions, in

Table 18.1 A cross-reference table of variables used in the description and implementation of SE.

Algorithm	Implementation	Line	Description
runs	num_runs	51	number of runs (rounds)
t_{max}	num_evals	52	number of evaluations per run
players	num_players	58	number of players for the tournament selection
$f(\cdot)$	avg_obj_val_eval	62	average best objective value of each evaluation
	best_so_far	63	best-so-far objective value
	best_sol	64	best-so-far searched solution
n	num_bits_sol	53	number of subsolutions of each solution
m	num_searchers	55	number of searchers (solutions)
h	num_regions	56	number of regions
w	num_samples	57	number of sampling solutions at each region
s	searcher_sol	67	set of searchers (solutions)
m	sample_sol	68	set of sampling solutions
r^b	sample_sol_best	69	best sampling solution of each region
v	sampleV_sol	70	set of candidate solutions
$f(s_i)$	searcher_sol_fitness	72	objective values of searchers
$f(m_{j,k})$	sample_sol_fitness	73	objective values of sampling solutions
$f(r_j^b)$	sample_sol_best_fitness	74	objective value of best sampling solutions
$f(v_{j,k}^i)$	sampleV_sol_fitness	75	objective values of candidate solutions
t^a	ta	77	investment status of each region
t^b	tb	78	investment status of each region
e	expected_val	80	expected value of SE
$\tilde{T}_j$	T_j	81	investment status of region r_j
$\tilde{M}_j$	M_j	82	quality of best-so-far solution at region r_j

line 162, id_bits in line 164, the assignment of searchers to regions and the replacement of the first $\log_2(h)$ subsolutions by id_bits in lines 176–182, and the creation and assignment of sampling solutions to regions based on the value of id_bits, as shown in lines 185–191. Second, t^a (ta) and t^b (tb) for each region are initialized, as shown in lines 194–200. Third, the variables e (expected_val), $\tilde{T}_j$ (T_j), and $\tilde{M}_j$ (M_j) are initialized, as shown in lines 203–205. Fig. 18.6 gives a simple example to illustrate how id_bits is used to assign each searcher s_i to a region. In this example, the number of id_bits is set equal to 2, meaning that the first two subsolutions of all solutions in each region are fixed. As far as this example is concerned, the template of s_2 is <01XX>, where X can assume the value of 0 or 1. In other words, this means that the solutions of s_2 can be <0100>, <0101>, <0110>, or <0111>.

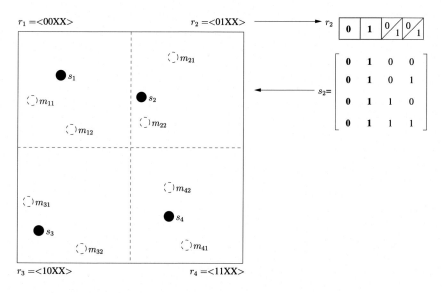

Figure 18.6 An illustration of the resource_arrangement() operator.

The vision_search() operator of SE is composed of transit(), compute_expected_value(), and vision_selection(), as shown in lines 209–219. As far as this implementation is concerned, the transit() operator can be divided into two parts: the modified crossover operator and the mutation operator, as shown in lines 222–260. Although the information exchange method of SE is very similar to the modified crossover operator of GA, the main difference is that the modified crossover operator of GA allows a searcher s_i to exchange its information with another searcher s'_i, whereas SE allows a searcher s_i to exchange its information with all the sampling solutions $m_{j,k}$ that may reside in the same or different regions, as shown in lines 225–248. Fig. 18.7 gives a simple example to illustrate how this implementation is realized. In this example, s_2 is <0111>. The transit() operator will exchange the searcher s_2 with a sampling solution $m_{j,k}$ to create temporary candidate solutions $v_{j,k}^2$. In this example, the sampling solution $m_{4,1}$ is randomly selected by the searcher s_2 to generate a new temporary candidate solution. The mutation operator will then be used to randomly change a subsolution of all the candidate solutions $v_{j,k}^i$, as shown in lines 251–259.

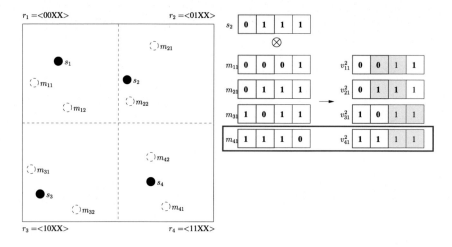

Figure 18.7 An illustration of the `transition()` operator.

As shown in lines 263–334, the `compute_expected_value` operator can be divided into five parts for computing $f(s_i)$, $v_{j,k}^i$, $f(v_{j,k}^i)$, $m_{j,k}$, $f(m_{j,k})$, $\tilde{T}_j$, $\tilde{M}_j$, and e. The first part is used to pick the candidate solution $v_{j,k}^i$ (`sampleV_sol`) and compute the objective values of each searcher s_i, $f(s_i)$ (`searcher_sol_fitness`), and candidate solution $v_{j,k}^i$, $f(v_{j,k}^i)$ (`sampleV_sol_fitness`), as shown in lines 266–294. At the first iteration, SE will only compute the objective values of all the searchers, as shown in lines 266–270. At the later iterations, SE will only compute the objective values of all the candidate solutions, as shown in lines 271–294. In lines 278–279, it will randomly select a possible solution generated by `transit()` as the candidate solution $v_{j,k}^i$ and then compute its objective value. It is designed like this because the objective values of all the searchers will be updated by the candidate solutions, as shown in lines 281–284; thus, the computations of the objective values of all the searchers are not really necessary after the first iteration. The second part is used to update the fitnesses of $m_{j,k}$ (`sample_sol_fitness`) and the best sampling solution r^b (`sample_sol_best`) in the j-th region, as shown in lines 297–319.

Lines 297–300 show that the objective values of the sampling solutions $m_{j,k}$ will be computed at the first iteration. The objective value of the best sampling solution r^b in the j-th region will then be updated by the comparison between the candidate solution r_j^b and the sampling solution $m_{j,k}$, as shown in lines 310–318. The third part will compute $\tilde{M}_j$ using Eq. (18.7), as shown in lines 322–323, while the fourth part will compute $\tilde{T}_j$ using Eq. (18.6), as shown in lines 326–327. The fifth part is used to compute the expected value e using Eq. (18.5), as shown in lines 330–333. As shown in lines 344–368, the tournament selection operator with the expected value is employed in this implementation to determine to which region each searcher will be moved. The last operator of this implementation is `marketing_survey()`, as shown in lines 371–389. As far as this implementation is concerned, this operator will update t^a and t^b for each region at the end of each iteration and the best-so-far solution. Bear in mind that the landscape depiction in this implementation is based on the sampling solutions m updated in `compute_expected_value()`, as shown in lines 286–289. However, when the landscape depiction becomes more complex than this implementation, it is more suitable to realize it in `marketing_survey()`.

18.3. Simulation results of SE for the one-max problem

To evaluate the performance of SE, in this section, we compare it with four other metaheuristic algorithms; namely, hill climbing (HC), simulated annealing (SA), TS, and the genetic algorithm (GA), for solving the one-max problem. All tests were carried out for 100 runs, and 1000 evaluations were performed in each run for the one-max problem.

Table 18.2 Number of evaluations needed by HC, SA, TS, GA, and SE to find the optimal solution with $t_{max} = 1000$.

Case	HC	SA	TS	GA-r	GA-t	SE-4r4s	SE-2r1s	SE-1r1s
$n = 10$	59	79	33	[9.77]	692	477	151	68
$n = 20$	145	148	96	[17.29]	873	[19.98]	227	153
$n = 40$	277	306	325	[30.20]	[39.95]	[39.73]	549	328
$n = 60$	486	460	374	[42.50]	[59.37]	[59.00]	845	539
$n = 80$	786	576	600	[53.94]	[77.41]	[77.10]	[79.98]	658
$n = 100$	965	826	901	[64.56]	[93.82]	[94.23]	[99.94]	741

Table 18.2 shows the results of SE for solving the one-max problem of sizes $n = 10$, $n = 20$, $n = 40$, $n = 60$, $n = 80$, and $n = 100$. In this table, a value in brackets (i.e., [*value*]) shows not only the end result of a metaheuristic algorithm but also the fact that the metaheuristic algorithm cannot reach the optimal solution during the convergence process, that is, within 1000 evaluations in this case. In this table, SE-*hrms* indicates an SE with h regions and m searchers. It can be easily seen that higher numbers of regions and searchers are associated with a lower probability that SE will be able to find the optimal solution using only 1000 evaluations. This situation can be mitigated by decreasing the number of regions and searchers, say, to one region and one searcher, during the convergence process. Although the results show that SE-1r1s is much faster than HC, SA, TS, GA-r, and GA-t in most cases, our observation shows that the results of SE-1r1s are not stable, which means that the results may be different from simulation to simulation. The convergence speed of SE-1r1s in solving the one-max problem is typically somewhere between those of SA and TS. The results for SE with different parameter settings do not mean that using multiple searchers and sampling solutions is not useful; on the contrary, it will eventually increase the diversity of the search directions of SE. This implies that the convergence of SE will be slower than that of HC, SA, and TS when SE divides the solution space into two or more regions or SE uses two or more searchers during the convergence process. This characteristic makes the searches of SE unlikely to fall into local optima at early iterations, especially when solving optimization problems. As shown in Table 18.3, in fact, SE with a larger number of searchers or regions can also find the optimal solution of the one-max optimization problem in most cases. The only issue is that SE requires a much larger number of evaluations to find the optimal solution. In summary, SE is suitable for solving large-scale or complex optimization problems.

Table 18.3 Number of evaluations needed by SE to find the optimal solution with $t_{max} = 5000$.

Case	SE-4r4s	SE-2r1s	SE-1r1s
$n = 10$	477	151	68
$n = 20$	1285	227	153
$n = 40$	2261	549	328
$n = 60$	2365	845	539
$n = 80$	4237	1055	658
$n = 100$	4805	1597	741

To make it easier to understand the performance of SE, Fig. 18.8 shows the convergence results of SE for solving the one-max problem of size $n =$ 100 with the following parameter settings: $h = 1$, $m = 1$ (1r1s); $h = 2$, $m = 1$ (2r1s); and $h = 4$, $m = 4$ (4r4s). The y-axis is used to represent the objective value. This figure shows that the convergence of SE with more searchers and regions (e.g., 4r4s) is much slower than that of SE with the other two parameter settings. An interesting thing to note, as this figure shows, is that the objective values of the initial solutions of SE-4r4s are better than those of the others, which may be due to the fact that it generates many more initial solutions than the other two.

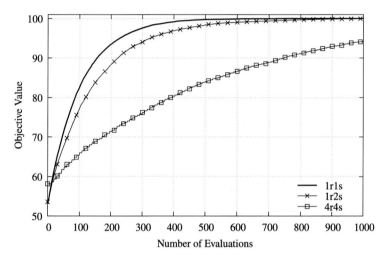

Figure 18.8 Convergence of SE with different numbers of searchers and regions for the one-max problem of size $n = 100$ and $t_{max} = 1000$.

In Fig. 18.9, the value of t_{max} was increased to 5000 to further see the convergence curves of SE. The results show that all three SEs are able to find the optimal solution for the one-max problem of size $n = 100$ in most runs. Figs. 18.10 and 18.11 show the convergence results of SE-4r4s and SE-1r1s for solving the one-max problem of sizes $n = 10$, $n = 20$, $n = 40$, $n = 60$, $n = 80$, and $n = 100$. It can be easily seen that the convergence of SE-4r4s for solving these six one-max problems is much slower than that of SE-1r1s, but SE-4r4s may still find the optimal solution after "a sufficient number of evaluations." Fig. 18.12 shows the convergence results of SE-1r1s for 1000 evaluations. It can be easily seen from this figure that the convergence curves are very similar to those of SE-4r4s for 5000 evaluations, as shown

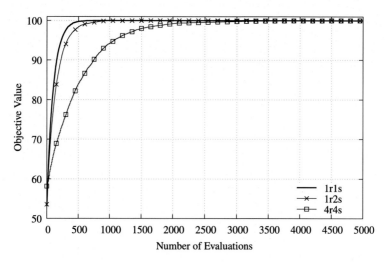

Figure 18.9 Convergence of SE with different numbers of searchers and regions for the one-max problem of size $n = 100$ and $t_{max} = 5000$.

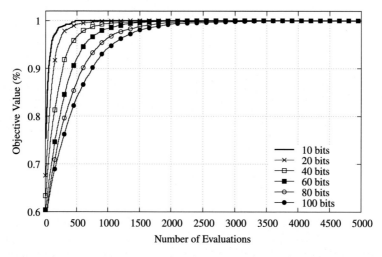

Figure 18.10 Convergence of SE with four regions and four searchers for the one-max problem of sizes $n = 10, n = 20, n = 40, n = 60, n = 80$, and $n = 100$.

in Fig. 18.10. This shows that the convergence speed of SE can be adjusted while retaining its convergence characteristic.

To better understand how the searchers move during the convergence process, we have to keep track of the movements of all searchers. This can be easily achieved by uncommenting line 276 and line 293 of Listing 18.1.

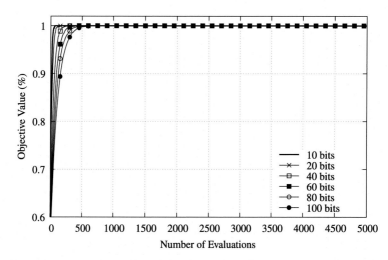

Figure 18.11 Convergence of SE with one region and one searcher for the one-max problem of sizes $n = 10$, $n = 20$, $n = 40$, $n = 60$, $n = 80$, and $n = 100$.

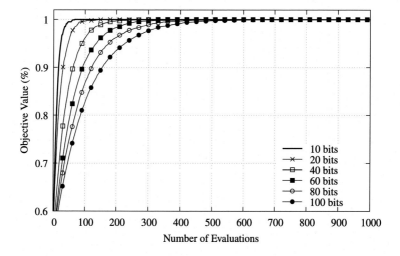

Figure 18.12 An enlargement of a portion of Fig. 18.11.

Fig. 18.13 shows that in case the solution space is divided into two regions (r_1 and r_2) and there is only one searcher, the search will spend about half of the time in each region. This implies that the searcher does not easily get stuck in a particular region. Fig. 18.14 further shows the case of four searchers and four regions r_1, r_2, r_3, and r_4, the first two subsolutions of which are $\{00\}$, $\{01\}$, $\{10\}$, and $\{11\}$. It can be easily seen that most of

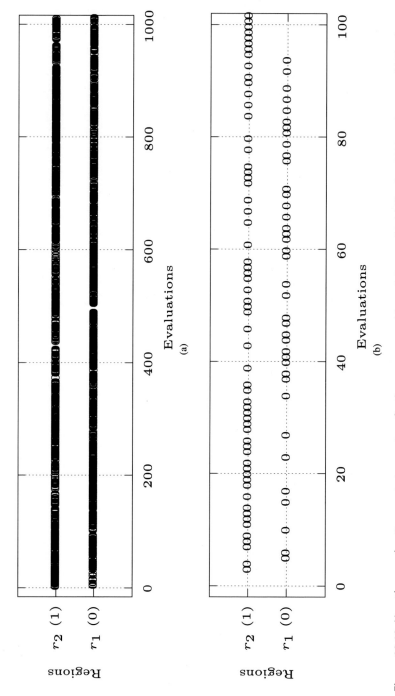

Figure 18.13 How the searcher 0 moves in two regions for the one-max problem of size $n = 10$. (a) The first 1000 evaluations. (b) The first 100 evaluations.

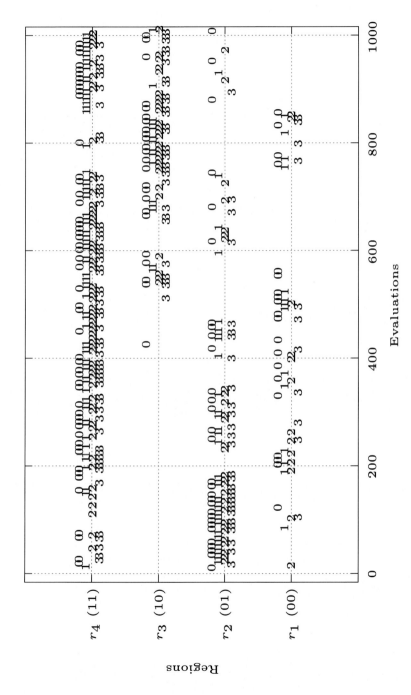

Figure 18.14 How the searchers 0, 1, 2, and 3 move in four regions for the one-max problem of size $n = 10$ in the first 1000 evaluations.

the searchers will move to region r_4 very frequently compared with the other regions. Most of the searchers prefer moving to regions r_2 and r_3 rather than region r_1. These phenomena are due to the fact that the first two subsolutions {11} contribute many more objective values than {01}, {10}, and {00}; and {01} and {10} contribute many more objective values than {00} to a complete solution. This means that the expected value of region r_4 is larger than those of regions r_3 and r_2, while the expected values of regions r_3 and r_2 are larger than that of region r_1. From this perspective, it can be easily understood why the searchers move to region r_4 very frequently. These phenomena reveal that the searches will tend to regions that have higher expected values while some searches will move to a region that has not been searched for a long time even though it has a worse expected value. With this kind of movement strategy, SE can keep the search diversity and thus does not easily fall into local optima during the convergence process.

18.4. Discussion

SE is a relatively young metaheuristic algorithm first presented in (Tsai, 2015) for solving the one-max problem. Later studies (Tsai, 2016; Tsai et al., 2017) showed that it can also be used to solve the wireless sensor deployment optimization problem and the wireless sensor clustering problem. In (Tsai and Liu, 2018), Tsai and Liu showed that SE is able to solve the service-to-interface assignment problem in an internet of things environment while in (Tsai and Fang, 2021), Tsai and Fang showed that it can also be used to find a better setting of the hyperparameters of a deep neural network in predicting the number of passengers at a metro station. In addition to the combinatorial optimization problem in a discrete solution space, a recent study (Tsai and Liu, 2020) showed that SE is able to solve the single-objective real-parameter optimization problem in a continuous solution space. The main differences between SE and metaheuristic algorithms can be summarized as follows:

- **The expected value**: The basic idea of SE is inspired by the concept of *return on investment* (ROI) from economics to prevent the searches from having a bias caused by the objective value. It uses the so-called expected value to evaluate the potential (i.e., the objective values of searched solutions, the objective values of new candidate solutions, and the number of times each subspace is searched) of different regions (i.e., subspaces of the solution space), that is, it uses the expected value to replace the objective value for determining later search directions.

- **Different way to exchange information**: The transition operator of vision search provides a different way to exchange the information between a solution and sampling solutions at different regions, which is useful in keeping the search diversity of SE.

- **The landscape description**: SE also attempts to keep the landscape information for the determination operator so that it has additional information to provide an accurate search to find better results.

In summary, SE still has some open issues that need to be solved, such as high computation costs and the way to depict the landscape of the solution space. The good news is that SE is able to find "better results" than many traditional metaheuristic algorithms for some optimization problems. Since more and more attention is paid to the depiction of the landscape of a solution space, it has become a critical research topic of metaheuristic algorithms in recent years. SE, with this concept in the core of its design, may become a research hotspot in the near future.

Supplementary source code

Here are the hyperlinks to the directories where the source code described in this chapter resides.

https://github.com/cwtsaiai/metaheuristics_2023/src/c++/1-onemax/

CHAPTER NINETEEN

Advanced applications

Since we know the basic ideas, theory, and implementation of metaheuristic algorithms discussed in this book, we are now well prepared to move on to the applications of these algorithms. In this chapter we will use five examples to illustrate how to apply metaheuristic algorithms to different kinds of advanced optimization problems. The examples, which will be discussed in Section 19.1 to Section 19.5, include (1) data clustering, (2) cluster-head selection (CHS) in a wireless sensor network (WSN), (3) traffic light control, (4) hyperparameter optimization (HPO) in a deep neural network (DNN), and (5) DNN pruning. This is just like looking for a solution to a new problem. To make it easier to understand these examples, the discussion of each example is ordered as follows: (1) problem description and definition, (2) solution encoding, and (3) how to apply metaheuristic algorithms to the optimization problem or application. Because each example can be regarded as a standalone system or a module of a system, in Section 19.6 we will provide a high-level view for a possible design of an intelligent system to explain the relationship between module and system by using metaheuristic algorithms as modules of such a system. We hope that these examples will provide the reader with some hints and inspiration to find ways to modify or develop appropriate versions of metaheuristic algorithms for various problems, applications, and systems.

19.1. Data clustering

19.1.1 Problem description and definition

To apply a metaheuristic algorithm to a new optimization problem or use it as a module of an application, the very first thing we have to do is to understand the problem, especially its definition. For a general definition of a data clustering problem, the reader is referred to $\mathbb{P}_9$ in Section 11.1. The goal of data clustering is typically to divide the input data objects into a number of groups based on some predefined similarity metrics. (See Fig. 19.1.) An ideal data clustering solution is found by minimizing the distances between objects in the same group while at the same time maximizing the distances between objects in different groups. From this perspective, a possible way

435

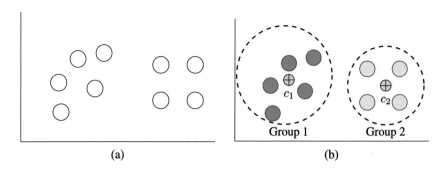

Figure 19.1 The data clustering problem. (a) Input data. (b) Clustering results.

to design an objective function for a data clustering problem is obviously to minimize the intra-cluster distance and maximize the inter-cluster distance simultaneously. One way to simplify such an objective function is to use the so-called sum of squared errors (SSE) (Definition 14) to minimize only the total distance between objects and their cluster centroids. In addition to choosing an objective function, how to calculate the similarity between a pair of objects and the similarity between an object and its centroid also needs to be taken into account when solving a data clustering problem. The Euclidean distance, Manhattan distance, Minkowski distance, and cosine similarity can be used to calculate the similarity between a pair of objects and the similarity between an object and its centroid, while the single-linkage, complete-linkage, average-linkage, centroid, and medoid distances can be used to compute the distance between clusters (Han et al., 2012). Because the data clustering problem (Jain et al., 1999; Xu and Wunsch-II, 2005) is a well-known optimization problem, a large number of data clustering algorithms from different perspectives have been presented since the last century.

Algorithm 22 k-means.

---

1 Randomly create an initial solution s $\boxed{\text{I}}$
2 **While** the termination criterion is not met
3      $c = \text{Update}(s)$ $\boxed{\text{T}}$
4      $s = \text{Assign}(s, c)$ $\boxed{\text{E}} \mapsto \boxed{\text{D}} \mapsto \boxed{\text{T}}$
5 **End**
6 **Output** s $\boxed{\text{O}}$

Algorithm 22 is the most well-known clustering algorithm, named k-means (MacQueen, 1967). Line 1 shows that it will first randomly divide all the objects into k non-empty groups s. Lines 3–4 show that it will repeatedly perform the "update" and "assign" operators to compute the centroids c (i.e., means) of all the clusters of the current partition and then assign each object to the group to which it belongs, that is, to the centroid to which it is closest. Because k-means requires the number of clusters to be given and is not able to handle non-linearly separable data objects (Chiang et al., 2011), several density-based clustering methods have been presented, such as density-based spatial clustering of applications with noise (DBSCAN) (Ester et al., 1996) and ordering points to identify the clustering structure (OPTICS) (Ankerst et al., 1999). The data clustering problem can be found in many different research topics and domains, and it has been investigated for years; thus, many data clustering algorithms have been presented using different approaches, such as partitioning, hierarchical, density-based, and grid-based clustering algorithms (Han et al., 2012). The data clustering problem can also be considered as an optimization problem. As a consequence, a metaheuristic algorithm with applicable modifications can be used in solving the data clustering problem. From some early studies (Bandyopadhyay and Maulik, 2002; Krishna and Narasimha Murty, 1999; Omran et al., 2002; van der Merwe and Engelbrecht, 2003), it is easy to find the footprints of metaheuristic algorithms for the data clustering problem, and most of them provide a better result than k-means.

19.1.2 Solution encoding

Once we are familiar with the problem in question and its definition, the next thing we have to deal with is how to encode a solution. An intuitive way is to encode a solution based on the structure of a feasible solution of the problem definition. The goal of the data clustering problem is to classify a set of n objects $x = \{x_1, x_2, \ldots, x_n\}$ into k groups. From this aspect, a simple way is to encode the solution as $s_i = \{s_{i,1}, s_{i,2}, \ldots, s_{i,n}\}$, where each subsolution $s_{i,j} \in \{1, 2, \ldots, k\}$ represents to which cluster the j-th object belongs. This encoding schema, also called the cluster ID, was used in a genetic algorithm (GA) for solving the data clustering problem (Krishna and Narasimha Murty, 1999). The goal of using this encoding schema is to search for the cluster to which each object belongs.

As mentioned in Section 11.1, the solution encoding of a metaheuristic algorithm for data clustering can also be by centroid or even as a hybrid. The reason to use a different solution encoding schema can be that we have some important concerns for solving the data clustering problem. In (Bandyopadhyay and Maulik, 2002), Bandyopadhyay and Maulik encoded a solution of GA for a data clustering problem as $s_i = \{s_{i,1}, s_{i,2}, \ldots, s_{i,k \times d}\}$, where k is the number of clusters and d is the number of dimensions. This can eventually reduce the size of a solution, especially for large-scale data clustering problems, because this kind of encoding schema typically assumes that $n \gg k$ and $n \gg d$, so $n \gg k \times d$. For example, to divide 100,000,000 four-dimensional input data objects into three groups, a feasible solution using cluster ID as the solution encoding schema will take 100,000,000 $\times$ 2 bits, while a feasible solution using centroid as the solution encoding schema will take only 3 $\times$ 4 real numbers and is essentially independent of the number of input data objects. It is not just the size of this solution encoding schema that will be much smaller than when using cluster ID as the number of input objects increases, but the solution space and goal will also be changed from using a metaheuristic algorithm to search for the group to which each object belongs to searching for the cluster center with which each object is associated. From these explanations, it is not hard to understand that when selecting a solution encoding schema, one has to take into account many things, such as size of the input data and characteristics of the optimization problem, application, system, and computer.

19.1.3 Metaheuristic algorithm for data clustering

Algorithm 23 gives a simple example to illustrate how to apply a metaheuristic algorithm to the data clustering problem. As far as this algorithm is concerned, a solution is encoded as a set of centroids $s_i = \{s_{i,1}, s_{i,2}, \ldots, s_{i,k \times d}\}$, where the first d subsolutions denote the first centroid, the second d subsolutions (i.e., the $(d+1)$-th to $2d$-th subsolutions) denote the second centroid, and so on. The metaheuristic algorithm we give here will first randomly create the initial population $s = \{s_1, s_2, \ldots, s_m\}$, where m is the number of solutions, as shown in line 1.

Algorithm 23 Metaheuristic algorithm for data clustering.

1 Randomly create the initial population s	I
2 $f_s = \text{Evaluation}(s)$	E
3 **While** the termination criterion is not met	
4 $v = \text{Transition}(s)$	T
5 $v' = k\text{-means}(v)$	T
6 $f_v = \text{Evaluation}(v')$	E
7 $s = \text{Determination}(v', s, f_v, f_s)$	D
8 **End**	
9 **Output** s	O

The solutions in the current population will then be evaluated, as shown in lines 2 and 6. The transition operator used here, as shown in line 5, depends on the kind of problem we want to solve and the kind of solution encoding schema we use. In case we want to use GA as the metaheuristic algorithm for solving the data clustering algorithm, we should consider the solution encoding schema we use now. Since the solution in this case is encoded as a set of centroids, each subsolution of which is a real number instead of a binary number or an integer, the bit-flip mutation operator might not be useful for such an encoding. To solve this problem, in (Bandyopadhyay and Maulik, 2002), the mutation operator for the j-th dimension of the i-th centroid with a fixed probability is defined as follows:

$$s_{i,j} = \begin{cases} s_{i,j} + \delta \times (\max(s_{i,j}) - s_{i,j}) & \text{if } \delta \geq 0, \\ s_{i,j} + \delta \times (s_{i,j} - \min(s_{i,j})) & \text{otherwise,} \end{cases} \quad (19.1)$$

where δ denotes a random number uniformly distributed in the range $[-R, +R]$, with R being calculated by the clustering metric (i.e., the sum of the Euclidean distances between the objects and their respective cluster centroids), $\min(s_{i,j})$ denotes the minimum value of the j-th dimension, and $\max(s_{i,j})$ denotes the maximum value of the j-th dimension. We can also use particle swarm optimization (PSO) as the metaheuristic algorithm for solving the data clustering algorithm. In this case, the position of the i-th particle of PSO can be regarded as a candidate solution of the data clustering problem that represents a set of centroids, and the velocity v_i and position s_i update (see also Eqs. (9.1) and (9.2) in Section 9.1) can be used as the transition operator to generate new candidate solutions. An interesting thing can also be found in some studies (Krishna and Narasimha Murty, 1999; Tsai et al., 2015) in which the k-means algorithm is used as a search operator (a local search operator) to fine-tune the new candidate solutions generated by GA or PSO, as shown in line 5. This shows that using domain-specific knowledge to develop the search operator is a useful way to improve the quality of the end result when using a metaheuristic algorithm as a clustering algorithm for solving the data clustering problem.

19.2. Cluster-head selection

19.2.1 Problem description and definition

Generally, wireless sensors are loaded with a non-rechargeable and small-capacity battery. As a consequence, the energy (lifetime) of such a network is limited. How to prolong the lifetime of a sensor has become a critical problem of WSN (Tsai et al., 2016; Yetgin et al., 2017). The CHS problem (Abbasi and Younis, 2007; Afsar and Tayarani-N, 2014) is one of the most well-known optimization problems in WSNs because a good plan to assign appropriate nodes as the cluster heads (CHs) will significantly prolong the lifetime of WSNs while at the same time making the packet transmissions between sensors, sink nodes, and the base station (BS) much more efficient. An example is given in Fig. 19.2(a), which shows that there are nine wireless sensors and one BS in the WSN, and the remaining energy of each sensor is quite different. If we connect all the sensors directly to the BS, it is very likely that some of them will run out of energy very quickly because they may be located far from the BS and the energy consumed for communication will be very high. For this reason, we can first divide these sensors into k groups (clusters), then select k CHs to receive messages from sensors in the same cluster, and finally forward these messages to the BS. Fig. 19.2(b) shows a possible way to select the CH for each of the two clusters. Different from the data clustering problem, in the CHS problem we need to consider not only the distances between the group members (sensors) and their CHs, but also the remaining energy, bandwidth, loading, and so forth, of the node that will be selected as the CH.

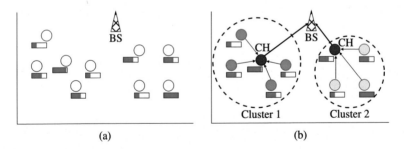

(a)   (b)

Figure 19.2 The cluster-head selection problem. (a) The distribution of wireless sensors. (b) The results of cluster-head selection of each cluster.

The CHS problem (also called the cluster-head election problem) can be defined as follows.

Definition 15 (The cluster-head selection problem $\mathbb{P}_{10}$). Given a set of sensors $w = \{w_1, w_2, \ldots, w_n\}$, where n is the number of sensors in the WSN, select k sensors $c = \{c_1, c_2, \ldots, c_k\}$ out of the set of sensors w as the CHs to minimize the value of the objective function $f(c)$ so as to maximize the lifetime of the WSN:

$$\min_{c \in 2^W \wedge |c|=k} f(c).$$

The objective function $f()$ is composed of a set of metrics that are of concern for the lifetime of the WSN. Here, we use the concerns described in (García-Nájera et al., 2021) for the CHS problem as an example to explain what kinds of objectives are usually considered in such a problem. Given Definition 15, all we have to do now is to define $f(c)$ in such a way that all the objectives that need to be taken into account for the CHS problem are included:

$$f(c) = \mathbb{N}\left(f_1(c)\right) + \mathbb{N}\left(f_2(c)\right) + \mathbb{N}\left(f_3(c)\right), \tag{19.2}$$

where $\mathbb{N}()$ is the normalization function for $f_1(c)$, $f_2(c)$, and $f_3(c)$; $f_1(c)$ represents the average distance from all CHs to the BS and is defined as

$$f_1(c) = \frac{1}{k}\sum_{j=1}^{k} \|c_j, b_s\|, \tag{19.3}$$

where c_j denotes the j-th CH, b_s denotes the base station, and $\|a, b\|$ denotes the distance between nodes a and b; $f_2()$ represents the average distance from the sensors to their respective CHs and is defined as

$$f_2(c) = \frac{1}{k}\sum_{j=1}^{k}\left(\frac{1}{m_j}\sum_{w_i \in \pi_j} \|w_i, c_j\|\right), \tag{19.4}$$

where π_j denotes the j-th cluster, m_j denotes the number of sensors in π_j, and w_i denotes the i-th sensor; and $f_3()$ represents the average residual energy of all CHs and is defined as

$$-f_3(c) = \frac{1}{k}\sum_{j=1}^{k} E_I - E_w(c_i), \tag{19.5}$$

where E_I denotes the initial energy of all sensors in the WSN and $E_w(c_i)$ denotes the energy consumed by the CH c_i. Eventually, many more factors of a WSN can also be taken into consideration for the CHS problem, such as the energy consumption of all the sensors (Heinzelman et al., 2000) for measuring the energy expended in transmitting and receiving an l-bit data over a distance d, denoted $E_t(l, d)$ and $E_r(l, d)$, respectively, and defined by

$$E_t(l, d) = T_e \times l + \epsilon \times l \times d^2, \qquad (19.6)$$

$$E_r(l, d) = T_e \times l, \qquad (19.7)$$

where T_e denotes the energy dissipated per bit to run the transmitter or receiver circuitry and ϵ denotes the energy dissipation of the amplifier of the transmitter.

19.2.2 Solution encoding

An intuitive way to encode a solution is to encode it as an n-bit string $s_i = (s_{i,1}, s_{i,2}, \ldots, s_{i,n})$, $s_{i,j} \in \{0, 1\}$, and $k = \sum_{j=1}^{n} s_{i,j}$, where n is the number of sensors and k is the number of clusters. Besides, $s_{i,j} = 1$ indicates that the j-th sensor is selected as a CH; $s_{i,j} = 0$ otherwise. This solution encoding schema makes it easy for the search algorithm to identify which sensors are CHs and which are not so that the sensors that are not CHs can be easily assigned to an appropriate cluster based on, say, the distance (or communication cost) between the sensors that are not CHs and the sensors that are CHs. The clustering results can be denoted by $\pi = \{\pi_1, \pi_2, \ldots, \pi_k\}$, $\pi_i \cap \pi_j = \emptyset$, that is, π_i denotes a set of sensors associated with the i-th CH. By using this solution encoding schema, the metaheuristic algorithm for the one-max problem can also be used for the CHS problem with a minor modification, that is, the number of 1's in a solution has to be equal to k. Based on Definition 15, another possible solution encoding schema is to encode a solution as $s_i = \{s_{i,1}, s_{i,2}, \ldots, s_{i,k}\}$, $s_{i,j} \in \{1, 2, \ldots, n\}$, and $s_{i,x} \neq s_{i,y}$ for $x \neq y$. With this solution encoding schema, a subsolution $s_{i,j}$ now encodes the j-th CH. Of course, we need to determine to which clusters all the sensors belong when using this kind of solution encoding schema.

19.2.3 Metaheuristic algorithm for cluster-head selection

The low-energy adaptive clustering hierarchy (LEACH) (Heinzelman et al., 2000) is apparently the most well-known method for solving the CHS problem, which elects the CHs every round by using the following threshold:

$$T(w_i) = \begin{cases} \dfrac{P}{1 - P \times (r \mod (1/P))}, & \text{if } w_i \in G, \\ 0, & \text{otherwise,} \end{cases} \qquad (19.8)$$

where w_i represents a sensor node, P represents the desired percentage of CH, r represents the current round, and G represents a set of nodes that have not been selected (elected) as the CHs in $1/P$ rounds. It means that Eq. (19.8) will be used to determine whether the node w_i will be a CH or not. Each sensor node will first choose a random number r_i uniformly distributed in the range $[0, 1]$ and then check to see if r_i is less than the threshold $T(w_i)$. If it is, then the sensor w_i will become a CH for the current round.

Algorithm 24 Metaheuristic algorithm for cluster-head selection.

1 Initialize the WSN environment E	I
2 Randomly create the initial population s	I
3 **While** the termination criterion is not met	
4 $v = \text{Transition}(s, E)$	T
5 $f_v = \text{Evaluation}(v', E)$	E
6 $s = \text{Determination}(v', s, f_v, f_s, E)$	D
7 **End**	
8 **Output** s and E	O

Algorithm 24 gives a simple example to explain how to apply a meta-heuristic algorithm to the CHS problem. The example shows that it will first take into account all the related information of a WSN environment, as shown in line 1, such as the energy of each sensor node, and then randomly create the initial solution. The transition and determination operators, of course, have to match the solution encoding schema. When using a bit string to encode a solution to indicate if a sensor is used as a CH or a set of integers to represent a set of CHs, the transition and determination operators for combinatorial optimizations can then be used by a metaheuristic algorithm for the CHS problem. Since the results of Algorithm 24 can be used to update the CHs of a WSN environment, this design makes it possible to use a metaheuristic algorithm to generate a static solution for the WSN environment once in a period or to generate solutions adaptively for the WSN environment multiple times at different periods.

19.3. Traffic light control

19.3.1 Problem description and definition

Modern cars have been around for more than 100 years, that is, they have come into global use since the early 20th century. Compared with ancient travel modalities, which mainly relied on human and animal power, the invention of the car made it possible to reduce not only the traveling costs but also the traveling time. Hence, the number of vehicles on the road has increased significantly in the late 20th century or even earlier. But when we enjoy the convenience of cars, this results in carbon emissions. How to reduce the number and duration of traffic jams has become a critical problem, because traffic jams lead to unnecessary carbon emissions. Unlike the addition of more infrastructure to mitigate the traffic jams and improve traffic, a "good" traffic light control plan (TLCP) is also useful for providing a convenient traffic environment for people. Importantly, it is an inexpensive solution compared with the addition of infrastructure.

Fig. 19.3 gives a simple example to explain the main concerns of the TLCP. In this example, there are two intersections, I_1 and I_2, each of which has four traffic lights. It can be easily seen that direction D_4 at intersection I_1 and direction D_8 at intersection I_2 have the pressure of vehicles. If the duration of green lights of D_4 and D_8 can be increased, this may reduce the length of queues of these two directions. However, simply increasing the duration of green lights for the vehicles moving in directions D_4 and D_8 without taking the other directions of these two intersections into consideration may generate traffic congestions in the other directions, such as D_1, D_5, and D_7. Because it is very expensive and difficult to verify a traffic control plan in the real world, several simulation tools—such as Saturn (https://saturnsoftware2.co.uk/), Paramics (https://www.paramics.co.uk/en/), Aimsun (http://www.aimsun.com/), and SUMO (https://sumo.dlr.de/index.html)—have been presented for the development of traffic control plans.

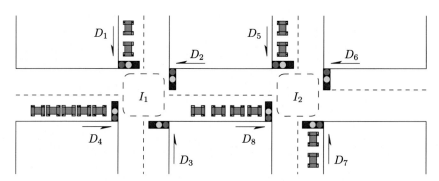

Figure 19.3 The traffic light control problem.

Fig. 19.4 shows the four phases of four traffic lights at an intersection that can also be used by a traffic simulation tool. In these phases, "G," "r," and "y" represent, respectively, green, red, and yellow light. When

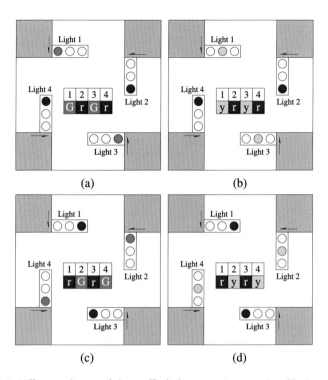

Figure 19.4 Different phases of the traffic lights at an intersection (Tsai et al., 2021). (a) Phase 1. (b) Phase 2. (c) Phase 3. (d) Phase 4.

submitting this traffic control plan to SUMO, the duration of the traffic light state for each phase will also be needed. We can set the duration of phase 1 (GrGr) to, say, 40 seconds. This means that lights 1 and 3 will be green and lights 2 and 4 will be red for 40 seconds, as shown in Fig. 19.4(a). The duration of phase 2 (yryr) can then be set to, say, 5 seconds, so that lights 1 and 3 will be yellow and lights 2 and 4 will be red for 5 seconds, as shown in Fig. 19.4(b). If we further set the duration of phase 3 to 40 seconds and the duration of phase 4 to 5 seconds, these four phases can be used as a TLCP. To develop a good TLCP, we will first define the traffic light cycle optimization problem (TLCOP) (Tsai et al., 2021) based on a well-known definition given in (García-Nieto et al., 2013).

Definition 16 (The traffic light control problem $\mathbb{P}_{11}$). Given a map that contains a set of intersections, a set of traffic lights, a set of vehicles, and simulation time, in the TLCP the aim is to (1) maximize the number of vehicles arriving at their destination, (2) minimize the number of vehicles not arriving at their destination, and (3) minimize the duration of all vehicle journeys. The three objectives can be integrated into an objective function $f()$ so that the goal of this optimization problem can be defined as follows:

$$\min_{s \in A} f(s) = \frac{d_v(s) + w_v(s) + c(s) \cdot s_t}{d(s)^2 + c_r},$$

where:

- A denotes the set of all possible solutions,
- s denotes a possible solution,
- $d_v(s)$ denotes the total journey time of vehicles that reach their destination,
- $w_v(s)$ denotes the total stop and waiting time of all vehicles,
- $c(s)$ denotes the number of vehicles that do not reach their destination,
- s_t denotes the simulation time,
- $d(s)$ denotes the number of vehicles that reach their destination, and
- c_r denotes a ratio defined by

$$c_r = \sum_{i=1}^{i_n} \sum_{j=1}^{p_n} d_{i,j}^r \cdot \frac{g_{i,j}}{r_{i,j}},$$

where $g_{i,j}$ denotes the number of "green" traffic lights for the j-th phase at the i-th intersection, $r_{i,j}$ denotes the number of "red" traffic lights for the j-th phase at the i-th intersection, $d_{i,j}^r$ denotes the duration of the j-th phase at the i-th intersection, i_n denotes the number of intersections, and p_n denotes the number of phases of each intersection. Note that the minimum value of $r_{i,j}$ is set to 1 to avoid division by zero.

19.3.2 Solution encoding

A simple way to encode a solution of the TLCP is to encode it as a tuple of four values representing the durations of the four phases of traffic lights at each intersection. That is, assuming that there are m intersections in the TLCP, the length of a solution will be $4 \times m$. Each solution of a metaheuristic algorithm can be encoded as $s_i = \{s_{i,1}^1, s_{i,1}^2, s_{i,1}^3, s_{i,1}^4, \ldots, s_{i,m}^1, s_{i,m}^2, s_{i,m}^3, s_{i,m}^4\}$, where $s_{i,j}^k$ represents the duration of the k-th phase at the j-th intersection of the i-th solution. An alternative solution encoding schema can be found in (Tsai et al., 2021), where the durations of the yellow and red lights are set equal to 5 seconds for all the intersections. In this way, the solution encoding can be simplified, that is, the length of the solution and thus the solution and search space of this problem can be reduced. Based on this assumption, a solution will no longer need to encode the durations of the second and fourth phases. A solution of the TLCP can then be encoded as $s_i = \{s_{i,1}^1, s_{i,1}^2, \ldots, s_{i,m}^1, s_{i,m}^2\}$, where $s_{i,j}^k$ again represents the duration of the k-th phase at the j-th intersection of the i-th solution.

19.3.3 Metaheuristic algorithm for traffic light control

Except for the problem definition, solution encoding, and objective function, most of a metaheuristic algorithm can be applied directly to the TLCP without many modifications. However, this application requires that a traffic simulation tool be used to evaluate the candidate solution generated by a metaheuristic algorithm. Fig. 19.5 gives a simple example to illustrate how to use a metaheuristic algorithm and SUMO to develop a good TLCP. In this figure, the left-hand side represents a general process of a metaheuristic algorithm. It will first initialize the parameters and create the initial populations. Each solution will then encode the durations of the four phases of the traffic lights at all the intersections we are interested in. After transition, the evaluation operator will create a file named .add.xml to pass to SUMO each candidate solution s_i, which encodes a TLCP for all the traffic lights at all the intersections we are interested in. In this example, for the solution s_i, the durations of the four phases of the traffic lights at the first intersection I_1 are set equal to 40, 5, 45, and 5 seconds, which will be used to generate the file .add.xml, as shown in the right-hand side of Fig. 19.5. In addition to the file .add.xml, the files .net.xml and .rou.xml, which are used to provide the network map and vehicle route information, are also required by SUMO to establish a traffic simulation. As the output of the simulation, SUMO will generate a file named "tripinfos.xml" that contains the information of the traffic state. The evaluation operator of the metaheuristic algorithm can then use the information in the file tripinfos.xml to calculate the objective value of the candidate solution s_i using the objective function defined in Definition 16. This evaluation process will be used by the determination operator of the metaheuristic algorithm for the evaluation of each candidate solution.

19.4. Hyperparameter optimization

19.4.1 Problem description and definition

Although the so-called hyperparameter optimization (HPO) for DNNs is a promising research topic, it is not a brand-new problem. In fact, most,

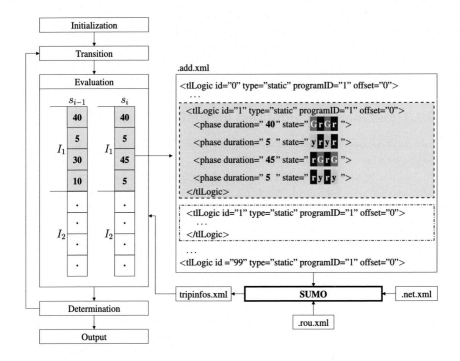

Figure 19.5 Traffic light control problem.

if not all, machine learning algorithms, such as support vector machine (SVM), decision trees, and neural networks (NNs), face the problem of determining a set of good parameters to enhance their performance and learning results. If the ultimate goal of a search algorithm $\mathcal{A}_s$ is to find the best hyperparameters x and y in the solution space that minimize the value of the function $f(x, y)$ for the learning algorithm $\mathcal{A}_L$ to find the best results, we typically assume that these two hyperparameters are of equal importance. An intuitive way to search for the best hyperparameters is to use the grid search algorithm to check a set of combinations of values of these two hyperparameters. However, it is unlikely to reflect the ground truth; otherwise, the grid search will easily find good values of x and y. It may happen that $f(x, y) = g(x) + h(y) \approx g(x)$, where $h(y)$ represents an unimportant hyperparameter while $g(x)$ represents an important hyperparameter. This means that x has a stronger impact than y; thus, we should invest much more computation resource to search for different values of x instead of y to find a better setting of x; therefore, the discussions in (Bergstra and Bengio, 2012; Bergstra et al., 2011; Li et al., 2018; Ozaki et al., 2017; Wang et al.,

2018b) conclude that random search or heuristic algorithms may be able to find better settings of hyperparameters for a learning algorithm. With these concerns, the HPO problem can be defined as follows.

Definition 17 (The hyperparameter optimization problem $\mathbb{P}_{12}$). Given a dataset $\mathcal{D}$ that can be divided into a training set $\mathcal{D}^t$ and a testing set $\mathcal{D}^v$, where $\mathcal{D}^t \cup \mathcal{D}^v = \mathcal{D}$ and $\mathcal{D}^t \cap \mathcal{D}^v = \emptyset$, the ultimate goal of this problem is to find a set of hyperparameters λ for the learning algorithm $\mathcal{A}_L^\lambda$ that minimizes the loss function $\mathcal{L}()$ by using a search algorithm $\mathcal{A}_s$ and the training set $\mathcal{D}^t$:

$$\min_{\lambda \in \Lambda} \mathcal{L}(\mathcal{M}_\lambda, \mathcal{D}^v),$$

where:

- Λ denotes the set of all possible solutions,
- λ denotes a possible solution, that is, a set of hyperparameters,
- $\mathcal{M}_\lambda = \mathcal{A}_L^\lambda(\mathcal{D}^t)$ denotes the model generated by the learning algorithm $\mathcal{A}_L$ with λ and $\mathcal{D}^t$, and
- $\lambda = \mathcal{A}_s(\mathcal{A}_L(\mathcal{D}^t))$ denotes a set of good hyperparameters found by the search algorithm $\mathcal{A}_s$ based on the results of the learning algorithm $\mathcal{A}_L$ with the training set $\mathcal{D}^t$.

As far as this problem definition is concerned, the model $\mathcal{M}_\lambda$ can be a set of classifiers, a set of classification rules, or even a set of hyperplanes that can be used to classify unseen data. The learning algorithm $\mathcal{A}_L$ can be either a traditional classification algorithm, such as the k-nearest neighbor (KNN), or a DNN algorithm, such as the convolutional neural network (CNN). The search algorithm $\mathcal{A}_S$ can be either a heuristic algorithm, such as random search, or a metaheuristic algorithm, such as GA.

19.4.2 Solution encoding

A solution of the HPO problem depends on the learning algorithm that we will use to generate the learning model because each learning algorithm has a different set of hyperparameters. For KNN, the hyperparameter will be the number of neighbors k. For SVM, the hyperparameters will be the kernel function (such as radial basis, polynomial, and sigmoid functions), C, and γ. For DNN, the hyperparameters will be the batch size, the number of hidden layers, the number of neurons for each hidden layer, and the learning rate (Tsai et al., 2021). Because the types of hyperparameters may be different, the solution encoding schema of HPO typically needs to be a hybrid vector. For example, a solution for the four hyperparameters h_{p1}, h_{p2}, h_{p3}, and h_{p4} can be encoded as $s_i = \{s_{i,1}, s_{i,2}, \ldots, s_{i,6}\}$. The subsolutions $s_{i,1}$ and $s_{i,2}$ represent, respectively, the batch size (the first hyperparameter h_{p1}) and the number of hidden layers (the second hyperparameter h_{p2}). The subsolutions $s_{i,3}$, $s_{i,4}$, and $s_{i,5}$ represent, respectively, the number of neurons in the first, second, and third hidden layers that can be regarded as the third hyperparameter h_{p3}. The last subsolution $s_{i,6}$ is the fourth hyperparameter h_{p4} that represents the learning rate. Because the first five subsolutions can be encoded as a set of integers but the last subsolution has to be encoded as a real number, this kind of solution encoding schema is a hybrid vector; thus, the transition operator of a metaheuristic algorithm has to take into account how to exchange such different types of subsolutions. Unlike encoding a solution as a hybrid vector, we can of course encode all the subsolutions of a solution as integer. A possible way is to use a set of indices to represent a set of real numbers, such as using 1, 2, 3, 4, and 5 to represent the learning rates 0.01, 0.02, 0.03, 0.1, and 0.2 of a DNN. Although this may result in loss of precision in search of the hyperparameter, it is a possible solution to make a metaheuristic algorithm suitable for the combinatorial optimization problem (COP) work for this problem.

19.4.3 Metaheuristic algorithm for hyperparameter optimization

With Definition 17, it is easy to find or design a metaheuristic algorithm for solving this problem. According to our observations, most metaheuristic algorithms, with a few modifications or even without any modifications, can be applied to this problem; as a result, simulated annealing, GA, PSO, and even search economics (Camero et al., 2019; Kim and Cho, 2019; Lee et al., 2018; Lorenzo et al., 2017; Qolomany et al., 2017; da Silva et al., 2018; Srivastava et al., 2019; Tsai et al., 2020, 2021; Ye, 2017; Young et al., 2015) have been used for solving the HPO problem of DNN.

Fig. 19.6 shows a simple example to explain how to apply a metaheuristic algorithm to the HPO problem of DNN. In the example, all the operators of the metaheuristic algorithm except for the evaluation operator need not be changed if the metaheuristic algorithm is capable of solving the COP. Similar to metaheuristic algorithms for the TLCOP that require a traffic simulation tool to evaluate a candidate solution, here, we need to use the candidate solution (a set of hyperparameters) for DNN to train a model to get the objective or fitness value from the accuracy rate or loss

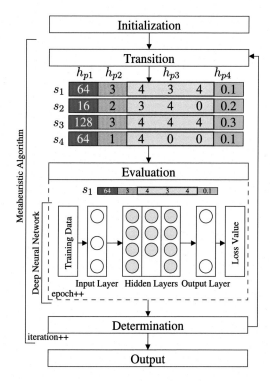

Figure 19.6 An illustration of metaheuristic algorithms for hyperparameter optimization problems.

value. This implies that a DNN must be embedded in such a design for evaluating a candidate solution of the metaheuristic algorithm. This application also shows that it is the evaluation operator that may take a much longer computation time than all the other operators of a metaheuristic algorithm because a training process of DNN is required to evaluate each candidate solution that would normally take a lot of computation time even though it is realized with GPUs.

19.5. Convolutional neural network filter pruning

19.5.1 Problem description and definition

The so-called pruning is another promising research topic of DNN that has attracted the attention of researchers from different research domains in recent years because pruning technologies are useful for reducing the model size of DNNs, thus making it possible to increase the recognition speed. A pruned model has a better generalization capability than the original model (LeCun et al., 1989), which means that a pruned model will in general perform much better than the original model for unseen datasets. In early studies (Hassibi et al., 1993; He et al., 2017; LeCun et al., 1989; Li et al., 2017; Liu et al., 2017; Reed, 1993), the pruning methods were divided into two categories: unstructured (Hassibi et al., 1993; LeCun et al., 1989) and structured methods (He et al., 2017; Li et al., 2017; Liu et al., 2017). Unstructured pruning methods aim to reduce the size of a model by removing unimportant weights based on the importance of all the weights (also called parameters) for the loss value in terms of a predefined measurement, such as the second-order Hessian matrix of the loss function that is used to determine the saliency (importance) of each weight to remove the unimportant weights in Optimal Brain Damage (OBD) (LeCun et al., 1989) and Optimal Brain Surgeon (OBS) (Hassibi et al., 1993). Since the pruned model is obtained by removing a certain number of unimportant weights from the original model, or, to be more specific, by setting them to zero, the weight matrix may become sparse and produce irregular accesses to memory during the propagation; thus, specialized hardware may be required to speed up a pruned model. To deal with this problem, structured pruning methods (He et al., 2017; Li et al., 2017; Liu et al., 2017) have been developed to remove a certain number of weight groups from the model, such as relevant weights of the same convolutional filter. Since the focuses of these studies (Hassibi et al., 1993; He et al., 2017; LeCun et al., 1989; Li et al., 2017; Liu et al., 2017; Reed, 1993) are on ways to determine whether weights, filters, and channels are important or not, most of them use only the greedy or other heuristic search algorithms to search for the weights, filters, and channels that need to be pruned. In addition to the greedy or heuristic algorithms for reducing the model size of a DNN, some early studies showed that metaheuristic algorithms, such as GA (Schaffer et al., 1992; Whitley et al., 1990) and DE (Deng et al., 2020; Ilonen et al., 2003), are able to reduce the sizes of architectures and models or improve the accuracy of NNs. Some recent studies (Hu et al., 2018; Lin et al., 2020; Wang et al., 2018c; Wu et al., 2019; Zhou et al., 2021) showed that GA and PSO can be used to prune the model of a DNN.

Fig. 19.7 shows a simple example to explain how to prune the connections of an NN or DNN to reduce its model size. In this example, the connections $\textcircled{1} \rightarrow \textcircled{3}$, $\textcircled{2} \rightarrow \textcircled{4}$, $\textcircled{3} \rightarrow \textcircled{6}$, $\textcircled{4} \rightarrow \textcircled{7}$, $\textcircled{5} \rightarrow \textcircled{8}$, $\textcircled{6} \rightarrow \textcircled{9}$, and $\textcircled{8} \rightarrow \textcircled{10}$ will be pruned so that the computations of this model will be decreased.

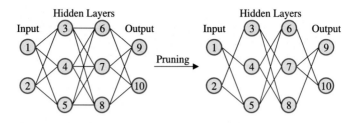

Figure 19.7 Connection pruning.

Fig. 19.8 shows another example to explain how to prune the neurons of an NN or DNN to reduce its model size. In this example, neurons ④ and ⑦ will first be pruned, followed by the relevant connections ① → ④, ② → ④, ④ → ⑥, ④ → ⑦, ④ → ⑧, ③ → ⑦, ⑤ → ⑦, ⑦ → ⑨, and ⑦ → ⑩. As a result, the total number of connections of this model is decreased from 21 to 12.

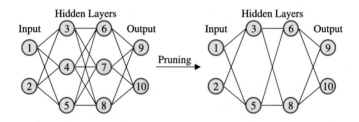

Figure 19.8 Weight pruning.

Fig. 19.9 gives an example to explain how filter pruning works on the model of a CNN. In this example, there are three input feature maps and four filters, namely, $f_{i,1}$, $f_{i,2}$, $f_{i,3}$, and $f_{i,4}$, in the i-th layer. It will generate four feature maps for the next layer of the CNN. If we remove the fourth filter $f_{i,4}$ in the i-th layer, then the number of output feature maps will be decreased to three, so that the computation cost for filter $f_{i,4}$ will be eliminated at later layers. Here, it is defined as a slightly modified version of the well-known definition given in (Lin et al., 2020) for filter pruning of a CNN model.

Definition 18 (The model pruning problem of CNN $\mathbb{P}_{13}$). Given a dataset $\mathcal{D}$ that can be divided into a training set $\mathcal{D}^t$ and a testing set $\mathcal{D}^v$ in such a way that $\mathcal{D}^t \cup \mathcal{D}^v = \mathcal{D}$ and $\mathcal{D}^t \cap \mathcal{D}^v = \emptyset$, as well as a pretrained model $\mathcal{M}$ with n convolutional layers $c = \{c_1, c_2, c_3, \ldots, c_n\}$ and a set of filters

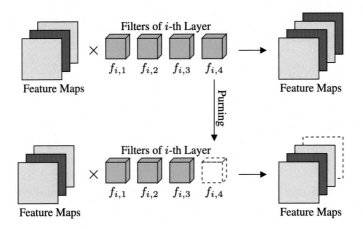

Figure 19.9 Filter pruning.

$f = \{f_{1,1}, f_{1,2}, \ldots, f_{1,|f_1|}, \ldots, f_{n,1}, f_{n,2}, \ldots, f_{1,|f_n|}\}$, where $f_{i,j}$ represents the j-th filter in the i-th layer and $|f_i|$ represents the number of filters in the i-th layer, the goal of this problem is to use a search algorithm $\mathcal{A}_s$ to find a pruned model to minimize its model size and computation cost while maximizing the accuracy rate for the given testing set $\mathcal{D}^v$, that is,

$$\min_{f' \in \mathbf{F}} \mathbb{A}(\mathcal{M}'', \mathcal{D}^v),$$

where:

- $\mathbf{F}$ is the set of all possible combinations of pruned filter sets, which can be regarded as the set of all possible candidate solutions,
- f' is the best pruned filter set found by the search algorithm $\mathcal{A}_s$,
- $\mathcal{M}'$ is the model generated by f',
- $\mathcal{M}''$ is a fine-tuned model of $\mathcal{M}'$, and
- $\mathbb{A}()$ is a measurement used to quantify the compression capacity of a search algorithm. It is typically composed of many factors, such as the number of floating-point operations per second (FLOPs), the model size (e.g., the total number of filters or weight parameters), and the accuracy rate of the pruned model.

19.5.2 Solution encoding

Given Definition 18, a solution can be encoded intuitively as $s_i = \{s_{i,1}^1, s_{i,1}^2, \ldots, s_{i,1}^{|f_1|}, \ldots, s_{i,n}^1, s_{i,n}^2, \ldots, s_{i,n}^{|f_n|}\}$, where $s_{i,j}^k \in \{0, 1\}$ indicates whether or not to prune the k-th filter in the j-th layer of the i-th solution, with $s_{i,j}^k = 0$ indicating pruning and $s_{i,j}^k = 1$ indicating no pruning. The length of s_i, denoted h, is equal to the total number of filters in all the convolutional layers, that is, $h = \sum_{i=1}^n |f_i|$. By using this solution encoding schema, the model pruning problem of CNN $\mathbb{P}_{13}$ can then be regarded as an extended version of the one-max problem; thus, metaheuristic algorithms for solving the one-max problem can be easily adapted for solving $\mathbb{P}_{13}$. Another possible way is to encode a solution of this problem as $s_i = \{s_{i,1}, s_{i,2}, \ldots, s_{i,\ddot{n}}\}$, where $s_{i,j} \neq s_{i,k}$ for $j \neq k$, $\ddot{n}$ represents how many filters need to be pruned, and $s_{i,j} \in [1, h]$ indicates the index of the filter to be pruned from the model.

19.5.3 Metaheuristic algorithm for convolutional neural network pruning

Similar to using a metaheuristic algorithm to find a set of good hyperparameters for a DNN, all the operators of a metaheuristic algorithm except for the evaluation operator need not be changed for solving $\mathbb{P}_{13}$ if the metaheuristic algorithm can be applied to a COP. As shown in the right-hand side of Fig. 19.10, the evaluation of each solution (how to prune the filters of a CNN) has to be based on a pretrained model $\mathcal{M}$. After generating a pruned model $\mathcal{M}'$ by removing some filters from $\mathcal{M}$ based on the solution s_i, the pruned model $\mathcal{M}'$ has to be fine-tuned and a verification process is carried out to get a fine-tuned model $\mathcal{M}''$ and its loss value $\mathbb{A}(\mathcal{M}'', \mathcal{D}')$ as the objective or fitness value of the solution s_i. In this way, the metaheuristic algorithm can now be used for solving $\mathbb{P}_{13}$. When a different pruning method is used to downsize the CNN model, the major parts that need to be redesigned are the solution encoding schema and the way to reconstruct the pruned model.

19.6. Discussion

In this chapter, we give five examples to show how to apply metaheuristic algorithms to different kinds of problems. Each of these five examples can be regarded as a standalone intelligent application or as a module of such an information system. Fig. 19.11 shows an information system for the internet of things (IoT) as an example to show which part of such a system can use a metaheuristic algorithm to improve its performance or make it more intelligent. In this example, we emphasize metaheuristic algorithms can be used as modules of (1) sensors and end devices, (2) edge

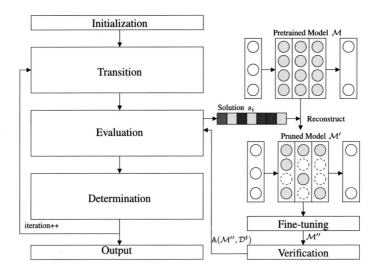

Figure 19.10 The metaheuristic algorithm for CNN pruning (Tsai et al., 2022).

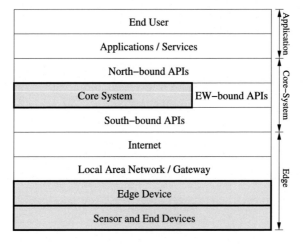

Figure 19.11 The relationship between modules using and not using metaheuristic algorithms in an information system. Note that EW-bound APIs represents East- and West-bound APIs.

devices, and (3) core systems. As a module of the sensors and end devices, a metaheuristic algorithm can be used to improve the performance of sensors or end devices, such as using a metaheuristic algorithm for CHS in WSNs or to improve the performance of data preprocessing of sensors or end devices. A possible application of a metaheuristic algorithm as a module of an edge device is to use it to realize a lightweight intrusion detection system (IDS) to classify normal and abnormal behaviors for a set of sensors or end devices. A simple way to use a metaheuristic algorithm as a module of a core system is to use it as a data analysis or machine learning algorithm, which has been discussed in many studies on IoT systems and other modern information systems (Tsai et al., 2018).

Although these five applications cannot represent all possible applications of metaheuristic algorithms, we can see from these examples that the possibilities of metaheuristic algorithms are almost infinite. Our experience in using metaheuristic algorithms for real-world applications and systems shows that the learning steps can be summarized as follows: first learn how to use a metaheuristic algorithm to find the solution of an optimization problem, then use it as a standalone system, and finally use it as a module of an information system.

CHAPTER TWENTY

Conclusion and future research directions

20.1. Conclusion

From the discussions in the previous 19 chapters, especially Chapter 19, we know that nowadays metaheuristics have become a research domain—just like a basic discipline, such as statistics, economics, or physics—that will strongly impact our daily lives in the foreseeable future. Even so, metaheuristic algorithms, just like different countries in the world, are developed independently; meanwhile, useful strategies are also adopted from the others. Most metaheuristic algorithms have their own distinguishing features, but the notations, definitions, and descriptions are "quite different" from each other—just like different languages and development strategies are used in different countries. For example, a solution is referred to as a chromosome or individual in the genetic algorithm (GA), as a particle in particle swarm optimization (PSO), as a routing path of an ant in ant colony optimization (ACO), and so on. A similar situation exists for operators of metaheuristic algorithms. This is why we presented a unified framework as a language translator for metaheuristic algorithms described in this book. The discussions in this book started with the problem definitions of several optimization problems, followed first by a unified framework for metaheuristic algorithms and then by exhaustive search (ES), hill climbing (HC), and six well-known metaheuristic algorithms, namely, simulated annealing (SA), tabu search (TS), GA, ACO, PSO, and differential evolution (DE), based on the unified framework. The advanced issues of metaheuristic algorithms—encoding schema, operators for transition, evaluation, and determination, parallelization, hybrid metaheuristic algorithms, hyperheuristics, and metaheuristic algorithms with local search, were also discussed. After that, pattern reduction (PR) and search economics (SE) were used as examples to show how the performance of existing metaheuristic algorithms can be improved and how a new metaheuristic algorithm can be developed. All of them can be regarded as essential background and domain knowledge of metaheuristic algorithms. They are not all and there is simply no way to cover all in this book; however, these examples provide insight into metaheuristic algorithms, and we

Handbook of Metaheuristic Algorithms
https://doi.org/10.1016/B978-0-44-319108-4.00034-4

hope that this will lead the audience of this book to come up with valuable ideas and solutions in using metaheuristic algorithms.

In addition to improved versions of SA, TS, GA, DE, ACO, and PSO, a lot of new metaheuristic algorithms have also been presented to solve different optimization problems. These "new" metaheuristic algorithms are inspired by the behavior of insects, social behavior of humans, physics, or even astronomy. The essentials of some new metaheuristic algorithms are similar to ACO, PSO, or DE (i.e., without any unique features) because these three metaheuristic algorithms provide us very useful ways to generate candidate solutions. How to develop more efficient and effective metaheuristic algorithms and to what kind of optimization problems they can be applied are two important questions we need to often ask ourselves. We should try looking at these issues from a positive perspective although some of the new metaheuristic algorithms are not really new, some of them do provide us with new ideas to solve optimization problems. We should not throw away the apple because of the core. We must think outside the box to explore more possibilities. Finding distinguishing features of new information technologies and systems is a very useful way to find answers to the aforementioned questions because new technologies can lead to revolutions. By adopting strong points of existing metaheuristic algorithms and developing more efficient and effective mechanisms, we can develop better metaheuristic algorithms for solving optimization problems.

Fig. 20.1 shows a family tree of metaheuristic algorithms, which shows that this research domain is still thriving and prosperous. In this family tree, we divide metaheuristic algorithms into six groups based on their abbreviations, names, and inspirations; namely, (1) neighborhood heuristic, (2) evolutionary computation, (3) swarm intelligence, (4) human intelligence, (5) astronomy and physics, and (6) natural phenomena. The swarm intelligence group can be further divided into insects and animals. We prefer classifying metaheuristic algorithms this way because such a classification makes it easy for us to differentiate and remember them. An alternative way is to differentiate metaheuristic algorithms by the transition operator so that they can be classified by how solutions are transited, say, perturbation or recombination vs. velocity update. This classification provides a better way to remind ourselves how new candidate solutions are generated during the convergence process. From this perspective, it is not hard to imagine that metaheuristic algorithms can also be classified based on their determination operator and other operators.

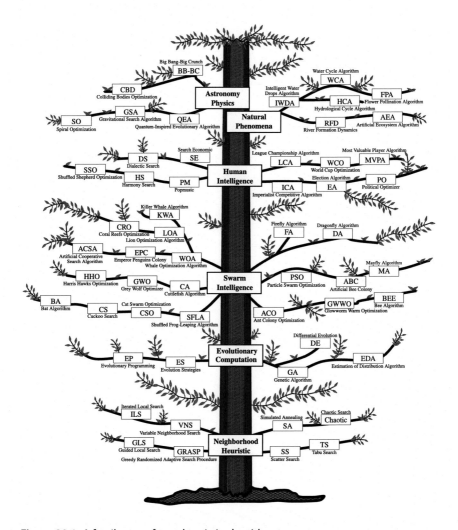

Figure 20.1 A family tree of metaheuristic algorithms.

Table 20.1 provides more detailed information of these metaheuristic algorithms; namely, name, abbreviation, year of publication, and the first major reference. Below we discuss the main waves of metaheuristic algorithms since the 1960s.

- **Local optimum:** The first wave was from 1960 to 1990. The major focus was on finding a way for metaheuristic algorithms to escape from local optima or to avoid falling into local optima at early iterations. TS and SA are good representatives of this wave.

Table 20.1　A list of metaheuristic algorithms.

Neighborhood heuristic

Scatter search	SS	1977	(Glover, 1977)
Simulated annealing	SA	1983	(Černý, 1985; Kirkpatrick et al., 1983)
Tabu search	TS	1986	(Glover, 1986, 1989, 1990a)
Chaotic search	Chaotic	1992	(Chen and Aihara, 1995; Nozawa, 1992)
Greedy randomized adaptive search procedure	GRASP	1995	(Feo and Resende, 1995)
Variable neighborhood search	VNS	1997	(Mladenović and Hansen, 1997)
Guided local search	GLS	1996	(Voudouris and Tsang, 1999, 1996)
Iterated local search	ILS	2002	(Lourenço et al., 2003)

Evolutionary computation

Genetic algorithm	GA	1962	(Holland, 1962)
Evolution strategies	ES	1965	(Rechenberg, 1965)
Evolutionary programming	EP	1966	(Fogel et al., 1966)
Differential evolution	DE	1995	(Storn and Price, 1995)
Estimation of distribution algorithm	EDA	1996	(Baluja, 1994; Pelikan et al., 2000)

Swarm intelligence by insects

Ant colony optimization	ACO	1991	(Dorigo et al., 1991)
Particle swarm optimization	PSO	1995	(Kennedy and Eberhart, 1995)
Artificial bee colony	ABC	2005	(Karaboga, 2005)
Bee algorithm	BEE	2005	(Pham et al., 2005)
Glowworm swarm optimization	GWWO	2005	(Krishnanand and Ghose, 2005)
Firefly algorithm	FA	2008	(Yang, 2009)
Dragonfly algorithm	DA	2015	(Mirjalili, 2016)
Mayfly algorithm	MA	2020	(Zervoudakis and Tsafarakis, 2020)

continued on next page

Table 20.1 (*continued*)

Swarm intelligence by animals

Shuffled frog-leaping algorithm	SFLA	2006	(Eusuff et al., 2006)
Cat swarm optimization	CSO	2006	(Chu et al., 2006)
Cuckoo search	CS	2009	(Yang and Deb, 2009)
Bat algorithm	BA	2010	(Yang, 2010)
Coral reefs optimization	CRO	2013	(Salcedo-Sanz et al., 2013)
Cuttlefish algorithm	CA	2013	(Eesa et al., 2013)
Artificial cooperative search algorithm	ACSA	2013	(Civicioglu, 2013)
Gray wolf optimizer	GWO	2014	(Mirjalili et al., 2014)
Lion optimization algorithm	LOA	2016	(Yazdani and Jolai, 2016)
Whale optimization algorithm	WOA	2016	(Mirjalili and Lewis, 2016)
Killer whale algorithm	KWA	2017	(Biyanto et al., 2017)
Harris hawks optimization	HHO	2019	(Heidari et al., 2019)
Emperor penguins colony	EPC	2019	(Harifi et al., 2019)

Human intelligence

Popmusic	PM	1999	(Taillard and Voss, 2002)
Harmony search	HS	2001	(Geem, 2009; Geem et al., 2001)
Imperialist competitive algorithm	ICA	2007	(Atashpaz-Gargari and Lucas, 2007)
League championship algorithm	LCA	2009	(Kashan, 2009)
Dialectic search	DS	2009	(Kadioglu and Sellmann, 2009)
Election algorithm	EA	2015	(Emami and Derakhshan, 2015)
Search economics	SE	2015	(Tsai, 2015)
World cup optimization	WCO	2016	(Razmjooy et al., 2016)
Most valuable player algorithm	MVPA	2017	(Bouchekara, 2017)
Shuffled shepherd optimization	SSO	2020	(Kaveh et al., 2020)
Political optimizer	PO	2020	(Askari et al., 2020)

continued on next page

Table 20.1 (*continued*)

Astronomy, physics, and natural phenomena

Quantum-inspired evolutionary algorithm	QEA	2002	(Han and Kim, 2002)
River formation dynamics	RFD	2007	(Rabanal et al., 2007)
Intelligent water drops algorithm	IWDA	2008	(Shah-Hosseini, 2008)
Gravitational search algorithm	GSA	2009	(Rashedi et al., 2009)
Big Bang–Big Crunch	BB–BC	2010	(Jaradat and Ayob, 2010)
Spiral optimization	SO	2010	(Kenichi and Keiichiro, 2010)
Water cycle algorithm	WCA	2012	(Eskandar et al., 2012)
Flower pollination algorithm	FPA	2012	(Yang, 2012)
Colliding bodies optimization	CBD	2014	(Kaveh and Mahdavi, 2014)
Artificial ecosystem algorithm	AEA	2016	(Baczyński, 2016)
Hydrological cycle algorithm	HCA	2017	(Wedyan et al., 2017)

- **Multiple search directions:** The second wave was from 1990 to 2000. The main focus was on using multiple search directions in each iteration. Representative methods can be found in the studies of evolutionary computation, e.g., GA. Since a large proportion of studies of metaheuristic algorithms were on evolutionary computation in this period, focuses of these studies were also on the combinatorial optimization problem.

- **Balance of diversification and intensification:** The third wave was focused on swarm intelligence, e.g., ACO and PSO. The importance of diversification and intensification was emphasized by Glover and Kochenberger (Glover and Kochenberger, 2003), and both ACO and PSO were designed to embed mechanisms to balance the diversification and intensification of searches during the convergence process. Since a good mechanism to balance the diversification and intensification may be able to improve the effectiveness and efficiency of a metaheuristic algorithm, this research issue has attracted the attention of researchers from different disciplines.

- **From discrete to continuous:** The fourth wave was from 2010 or even earlier to the present. As mentioned before, if we encode a solution as a vector while transiting the current solution to another candidate solution via a velocity update operator, the metaheuristic algorithm will be able to find a good solution in solving continuous optimization problems, such as PSO and DE. This is an important reason why some metaheuristic algorithms presented in recent years used similar ways to encode the solutions and design the transition operators. Other critical research trends have been discovered from 2000 onward, such as (1) using multiple strategies for search operators of metaheuristic algorithms, (2) dynamically, adaptively, or self-adaptively adjusting the parameters of a metaheuristic algorithm, (3) combining a metaheuristic algorithm with other (meta)heuristic algorithm(s), (4) using a historical vector to keep a set of successful parameters or partial solutions, or even (5) using the fitness landscape description to determine the search directions or regions. It is clear that the possibilities are infinite.

20.2. Future research directions

Although most metaheuristic algorithms were presented years ago, there is plenty of room to improve their performance. With this book, we can set up a goal to enhance the performance of existing metaheuristic algorithms or develop a brand-new metaheuristic algorithm to satisfy the needs of new environments and new requirements we confront today. This implies that when designing a high-performance metaheuristic algorithm, one needs to consider not only the current open issues, but also new computer systems, platforms, and environments. To help the audience of the book find possible ways to develop a "good metaheuristic algorithm," possible high-impact research trends are given below.

- **Quality of the end result:** The basic idea of metaheuristic algorithms is to use a strategic guess to find an approximate solution or the optimal solution in the search space; however, most of them cannot guarantee to find the optimal solution(s) for complex optimization problems in a reasonable time. How to improve the end results and how to reduce the computation time are two very important open issues and will remain so. To improve the quality of the end results, several useful methods have been presented. Adding a better initialization operator to generate refined initial seeds is an intuitive way to kick off a metaheuristic algorithm with a good starting point that may allow it to find a better end result than a similar metaheuristic algorithm with random initial seeds. Since a better initialization operator may be able to improve the quality of the end results, a better transition operator may be used to improve the end results as well. For example, for solving the traveling salesman problem, the OX crossover operator of GA, but not the one-point crossover operator, is guaranteed not to generate infeasible candidate solutions. Several recent studies focused on combining a metaheuristic algorithm and deep learning, which has become a research trend that provides an alternative way for finding better results in solving optimization problems.

- **Computation time and response time:** No matter what kinds of high-performance machines we have today, accelerating metaheuristic algorithms has always been a critical goal. Acceleration methods can be divided into two types: algorithm-oriented and hardware-oriented methods. Since algorithm-oriented acceleration methods may be able to reduce computations or searches that are essentially redundant, this can be regarded as computation time reduction. Using a short-term or long-term memory to keep track of recent searches (e.g., TS or SE) and using a detection operator to detect and remove computations that are redundant (e.g., pattern reduction) are two possible ways for a metaheuristic algorithm to avoid redundant searches and computations during the convergence process. Both of them may be able to enhance the search performance of a metaheuristic algorithm; therefore, avoiding redundant searches is a research trend.

Since most hardware-oriented acceleration methods simply use more computers to complete the computation tasks of a metaheuristic algorithm, this can be regarded as response time reduction. Using parallel computing platforms is a useful way to reduce the response time of a metaheuristic algorithm. However, with the development of parallel and distributed computing systems, there are many ways by which this can be done. But such solutions may not always be useful for small optimization problems. Also, parallel metaheuristic algorithms are designed in different ways, and different parallel environments have different requirements. All these issues need to be considered when developing a high-performance metaheuristic algorithm. When using metaheuristic algorithms in a multi-thread environment, one may employ shared memory to reduce the communication costs between different threads, but when using metaheuristic algorithms in a cloud computing environment (e.g., MapReduce), one needs to take into account the communication costs between server and workers (i.e., clients). Among these parallel environments, parallel computing will not be useful when the communication overhead is higher than the response time reduction. How much communication and additional overheard will be incurred when using a metaheuristic algorithm on parallel computing systems is thus a key issue that we need to consider before we go for such a solution. Although hardware-oriented acceleration methods may not be able to reduce the computation time of a metaheuristic algorithm, an ultimate way to speed up a metaheuristic algorithm is to leverage the strength of both algorithm-oriented and hardware-oriented methods to fully utilize the computing power of a computer, just like installing a couple of performance packs to boost the engine of, say, a Mustang GT.

- **Balance of diversification and intensification:** Predicting convergence of most metaheuristic algorithms is difficult, because there are many uncertain and stochastic factors. However, if we could predict the convergence of a metaheuristic algorithm more accurately, we could easily enhance its search performance with some applicable modifications. A compromise approach is to use the so-called convergence curves of a metaheuristic algorithm, which may provide insight into how it works. Although the convergence analysis, prediction, and proof of a metaheuristic algorithm are still difficult, the information thus obtained is very useful in the development of new metaheuristic algorithms or in the application of a metaheuristic algorithm to real systems because the information tells us about the possibilities and limitations of a metaheuristic algorithm.

  How to balance diversification and intensification in the search of a solution is another open issue. One of the possible solutions is to automatically adjust the tendency between these two factors based on the search results during the convergence process or design a set of predefined thresholds and/or rules before the convergence process to adjust the importance of these two factors. Most of the solutions can certainly provide better search performance than fixing the proportion of these two factors; thus, it is a useful way to enhance the search performance of a metaheuristic algorithm. In addition to automatically adjusting the proportion of diversification and intensification in the search of a solution, how to automatically adjust parameter settings is another open issue because dynamic parameter settings are likely to provide better results than fixed parameter settings. Several recent studies attempted to develop an automatic adjustment strategy for parameter settings.

- **Knowledge-based vs. knowledge-free:** Since many successful cases show that domain knowledge is very useful to enhance the performance of a metaheuristic algorithm when solving a specific optimization problem, how to use the domain knowledge is a critical issue. If the design of a metaheuristic algorithm is based on domain knowledge, it can be regarded as a knowledge-based method. An intuitive approach is to allow such a metaheuristic algorithm to adopt a specific heuristic algorithm for solving a particular optimization problem. For example, we can use GA as the global search mechanism and k-means as the local search mechanism for solving the clustering problem. Such a combination eventually outperforms GA alone. It can be easily seen that domain knowledge is useful to enhance the performance of a metaheuristic algorithm; therefore, how to avoid incurring additional cost and how to avoid unnecessary integration are two critical issues that need to be considered.

Developing a metaheuristic algorithm that is capable of solving many kinds of optimization problems without domain knowledge is the ultimate goal. Hybrid heuristics and hyperheuristics provide possible approaches that can be regarded as knowledge-free. Although several hybrid heuristics and hyperheuristics have been presented, open issues that need to be addressed still exist, such as how to transit information between two different (meta)heuristic algorithms. An important research issue is that most hybridization methods incur additional computation costs. How to avoid these additional computation costs and how to avoid conflict integration are two potential research trends.

In addition to adding search operators with domain knowledge and combining two or more (meta)heuristic algorithms, another research trend is to keep some searched solutions to depict roughly the fitness landscape, which may in turn provide abundant information for a metaheuristic algorithm to determine later searches more accurately than a metaheuristic algorithm without fitness landscape. Using the fitness landscape is just like using a roadmap and GPS to a city we have never visited before; it will be much easier for us to explore the city or get to a special region we are interested in, e.g., the global optimum.

- **Open sources:** There are several open sources that can be used to realize a metaheuristic algorithm, which makes it possible to rapidly develop a metaheuristic algorithm or even the proof of concept of a new idea. Some open sources which can simplify the development process of a metaheuristic algorithm because they can be used as a function, to be invoked just like how the function `printf()` is called in C, can be found on GitHub (https://github.com/). This, however, does not mean that we can simply ignore the details of a metaheuristic algorithm and use them as a black box. In fact, researchers in this field need to pay much attention to the details and deeply understand every single part of a metaheuristic algorithm to develop a good metaheuristic algorithm; otherwise, chances are small that we can provide a better solution than others because almost every software engineer can use the open source to realize a metaheuristic algorithm, just like calling the function `printf()` in C.

The advance of computer systems will certainly be very useful to reduce the computation time of metaheuristic algorithms. Now the question arises, what does "reasonable time" for a metaheuristic algorithm actually mean? In fact, the reasonable time depends on the computer system. A computer today may be 100 times faster than a computer 20 years ago. Optimization problems that could not be solved efficiently before can now be solved efficiently. Although computer systems are still not able to allow a metaheuristic algorithm to completely solve an NP-hard or NP-complete problem in a reasonable time, they do allow a metaheuristic algorithm to find an approximate solution, or even the optimal solution, in a reasonable time if the problem size is "not too large." When we persist in reducing the computation time of a metaheuristic algorithm, the size of an optimization problem to be solved is another concern that we have to take into account. This means that making it possible for a metaheuristic algorithm to solve a larger optimization problem is not the only important thing; rather, it is also important to take into account the size of a problem to be solved. One good reason is that the size of an optimization problem in a real system may be fixed; therefore, the kind of optimization problem that we want to solve is the most important thing that we need to consider. In summary, after reading this book, the reader should understand most of the metaheuristic algorithms presented in recent years, be able to improve the search performance of a metaheuristic algorithm, and perhaps even develop a brand-new metaheuristic algorithm. Improving the `Initialization()`, `Transition()`, `Evaluation()`, and `Determination()` operators and developing a new search operator are all possible ways to enhance the search performance of a metaheuristic algorithm. In the end, we hope that this book serves as a good starting point for the reader to develop good metaheuristic algorithms for solving optimization problems, because everything is possible in this research field.

Interpretations and analyses of simulation results

A.1. Interpretations of metaheuristics

After realization of a metaheuristic algorithm, testing it to further understand it is a critical step. The test results will be very useful for improving the performance of the metaheuristic algorithm in question or for developing a new method. To understand the performance of a newly implemented metaheuristic algorithm, an intuitive way is to record their final results. But if we want to better understand the metaheuristic algorithm, it is better to keep more information because more details can be obtained from more complete testing data. The following results are the average objective values of each iteration of 100 runs of SA for the one-max problem of size $n = 100$ shown in Section 5.3.1.

```
50.010
50.450
50.990
51.460
51.930
52.320
.
.
.
100.000
100.000
100.000
```

Most studies on metaheuristic algorithms do not use the results of a single run to claim anything because different runs may give different results. For this reason, most studies on metaheuristic algorithms are usually based on the simulation results of several runs. Experienced researchers will also pay attention to the impact of "different parameters" of a metaheuristic algorithm on "different data distributions" to better understand the performance of the metaheuristic algorithm. How to display the simulation results then has become an important issue because without a proper way to display simulation results, they are just numbers, that is, it will be very hard to interpret their meaning, not to mention extracting useful information out of them. No matter what tools are used, be it Microsoft Excel, OpenOffice Calc, gnuplot, or even the other display libraries in Java, Python, and MATLAB®, they are all useful tools for displaying simulation results, providing some simple ways to help us understand the performance of metaheuristic algorithms. In this appendix, we use `gnuplot` as an example to show how to display simulation results.

A.1.1 Quality of the end result

Now the question arises, how close is the end result of a metaheuristic algorithm for an optimization problem to the optimal solution of the optimization problem? This is because we may already know that metaheuristic algorithms provide an alternative way to find the optimal, or an approximate, solution much faster than ES algorithms. To evaluate the performance of different search algorithms, most studies provide simulation results to show the performance of metaheuristics for the "optimization problem," with different "search operators" and different "parameters." We can tabulate the results, as shown in Table A.1, to compare the performance of HC and SA, which is a simple but common way to show the performance of each algorithm in most research papers.

Table A.1 Number of evaluations needed by HC and SA to find the optimal solution.

Problem size	HC	SA
$n = 10$	59	79
$n = 20$	145	148
$n = 50$	432	403
$n = 100$	965	826

As shown in Table A.2, compared with Table A.1, although six more comparisons between HC and SA for solving the one-max problem are provided, it is still very difficult to understand the meaning of the simulation results. This shows that some of the details may not be found when we confront too many numbers.

By looking only at the table, it may be very difficult to figure these situations out. As shown in Figs. A.1 and A.2, if figures are used to display

Table A.2 Number of evaluations needed by HC and SA to find the optimal solution for problems of sizes from $n = 10$ to $n = 100$.

Problem size	HC	SA
$n = 10$	59	79
$n = 20$	145	148
$n = 30$	218	217
$n = 40$	277	306
$n = 50$	432	403
$n = 60$	486	460
$n = 70$	854	574
$n = 80$	786	576
$n = 90$	856	895
$n = 100$	965	826

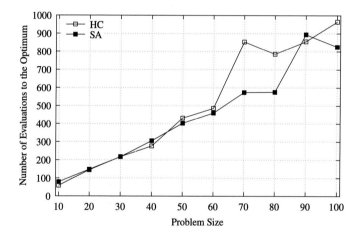

Figure A.1 Simulation results of HC and SA.

the same simulation result, it may be easier to examine the results from different angles. In addition to HC, which can find the optimal solution using a lower number of iterations than SA for solving the one-max problem in some cases, it can be easily seen that the number of evaluations taken by HC and SA will grow linearly with the problem size, as shown in Fig. A.1. Now we know that there exists no perfect way to display the simulation results of metaheuristics; we can, however, try different ways to display the results so that we may be able to find more hidden information. Table A.2

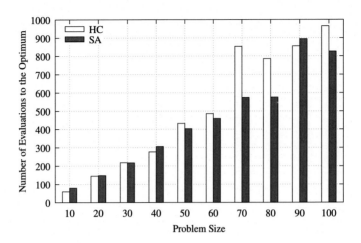

Figure A.2 Simulation results of HC and SA.

and Figs. A.1 and A.2 present three different ways to show the simulation results; however, these are not the only ways to show the results.

An example to show how Fig. A.1 is created using gnuplot is given in Listing A.1. For more details, the reader is referred to the official website of gnuplot (http://gnuplot.info/) or the book entitled "Gnuplot in Action, Second Edition" (Janert, 2015).

Listing A.1: src/gnuplot/app01/hcsa-10-100-o.gp: Gnuplot for Fig. A.1.

```
 1 reset
 2 set terminal pdfcairo enhanced color font "Times,14"
 3 set title font "Times-Bold,14"
 4 set size ratio 0.618
 5
 6 set style line 1 lc rgb 'black'  lw 1 lt 1 pt 4 pi -1 ps 0.6
 7 set style line 2 lc rgb 'black'  lw 1 lt 1 pt 5 pi -1 ps 0.6
 8
 9 set grid xtics ytics
10 set pointintervalbox 0.01
11
12 set key top left vertical inside
13
14 set ylabel offset character 1,0,0
15 set ylabel "Number of Evaluations to the Optimum"
16 set xlabel "Problem Size\n"
17
18 set xrange [-0.1:9.1]
19 set xtics ("10" 0, "20" 1, "30" 2, "40" 3, "50" 4, "60" 5, "70" 6, "80" 7, "90" 8, "100" 9)
20
21 set output "hcsa-10-100.pdf"
22
23 plot "hc10-100.txt" u 1 t "HC" w lp ls 1, "sa10-100.txt" u 1 t "SA" w lp ls 2
```

As shown in lines 1–4, this program first restores the default settings of the graph-related options before it moves on to setting the output terminal (e.g., file), the default font, and the scale factors for the size of the plot. This program will then set not only different line styles to show the results of different search algorithms, as shown in lines 6–7, but it will also have grid lines drawn on the plot to help us visualize the results, as shown in line 9, and control the size gap between point symbols and lines, as shown in line 10. Line 12 sets a key or legend containing a title and a sample for each plot in the graph to be shown in the top-left corner of the output figure. Line 14 adjusts the position of the y-label to move it closer to the y-axis while lines 15–16 label the y-axis as "Number of Evaluations to the Optimum" and the x-axis as "Problem Size." Then gnuplot will automatically determine the scope of the x-axis and the y-axis based on the input data, if not specified. In this example, this program will set the scope of the x-axis from −0.1 to 9.1, as shown in line 18. Line 19 sets the major tics on the x-axis as "10," "20," "30," "40," "50," "60," "70," "80," "90," and "100" to represent the problem sizes. Line 21 sets the output filename to "hcsa-10-100.pdf." Line 23 invokes the plot command to actually draw the lines for the results of HC and SA saved in the files "hc-10-100.txt" and "sa-10-100.txt," the contents of which are repeated below for easy reference. Moreover, "u 1" in the plot command means that the plot of the line is for the first column of the data saved in the file specified.

HC-10-100.TXT

```
59, 10.000
145, 20.000
218, 30.000
277, 40.000
432, 50.000
486, 60.000
854, 70.000
786, 80.000
856, 90.000
965, 100.000
```

SA-10-100.TXT

```
79, 10.000
148, 20.000
217, 30.000
306, 40.000
403, 50.000
460, 60.000
574, 70.000
576, 80.000
895, 90.000
826, 100.000
```

A different way to draw the plots can be created by making just a few changes to the code given in Listing A.1. An example is given in Listing A.2, where a histogram (Fig. A.2) instead of lines is used to display the same simulation results, by modifying only lines 6–7, 18, 21, and 23 of Listing A.1.

Listing A.2: src/gnuplot/app01/hcsa-10-100-1-o.gp: Gnuplot for Fig. A.2.

```
1 reset
2 set terminal pdfcairo enhanced color font "Times,14"
3 set title font "Times-Bold,14"
4 set size ratio 0.618
5
6 set style data histograms
7 set style fill solid .5 border -1
8
9 set grid xtics ytics
10 set pointintervalbox 0.01
11
12 set key top left vertical inside
13
14 set ylabel offset character 1,0,0
15 set ylabel "Number of Evaluations to the Optimum"
16 set xlabel "Problem Size\n"
17
18 set xrange [-0.5:9.5]
19 set xtics ("10" 0, "20" 1, "30" 2, "40" 3, "50" 4, "60" 5, "70" 6, "80" 7, "90" 8, "100" 9)
20
21 set output "hcsa-10-100-1.pdf"
22
23 plot "hc10-100.txt" u 1 t "HC" fill solid .0 border -1, "sa10-100.txt" u 1 t "SA" fill solid .7
       border -1 lc rgb "black"
```

A.1.2 Convergence curves

In addition to the quality of the results, the convergence curve is another key tool for the studies on metaheuristic algorithms because we may analyze the convergence characteristics (e.g., convergence speed, stability, limitation) of a metaheuristic algorithm for an optimization problem by looking at the details of the convergence curve. Without the convergence results, it is very difficult to understand the results by merely looking at the numbers. As shown in Fig. A.3, it can be easily seen that ACS is able to find the tour with a distance less than 440 in no more than 5000 evaluations on average, which is equivalent to 250 iterations on average because the number of ants is set equal to 20 in this case, i.e., 250 iterations = 5000 evaluations/20 ants per iteration = 5000 evaluations/20 evaluations per iteration. The results after 15,000 evaluations (i.e., 750 iterations) are very close to the results after 25,000 evaluations (i.e., 1250 iterations). If we look at this figure very carefully, we see that ACS is able to find better results by increasing the number of iterations. Compared to Fig. A.3, Fig. A.4 further shows that the convergence curve of ACS for the period between iteration 400 and iteration 1200 is not flat. In fact, the convergence trend of ACS shows that it can still improve the results after 1000 iterations. These two figures explain that many details are hidden in simulation results, and we will not see the problem if we do not interpret the simulation results carefully. The "right things" to do before you claim the research results of metaheuristics is therefore to *hypothesize boldly, and prove it carefully*.[1] We will then not be misled by the simulation results.

As shown in Listing A.3, gnuplot can be used to create Fig. A.3 to show the convergence curves of ACS for eil51. Unlike that depicted in Listing A.1, the *x*-axis and *y*-axis, as shown in lines 14–15 of Listing A.3, are labeled as "Number of Evaluations" and "Objective Value," respectively. The output of this program, as shown in line 17, will be written to the file named "result-acs-tsp-eil51-1.pdf." This example implies that the scope of the *x*-axis will be automatically determined by gnuplot based on the input data because the set xrange command is not given as in the previous examples. The pathname of the file to which the result of ACS for the TSP benchmark eil51 will be written is represented by the alias "file_tsp_acs_eil51" because it is sort of long, as shown in line 19. With

---

[1] This slogan, proposed by Shih Hu (https://en.wikipedia.org/wiki/Hu_Shih), is inspired by John Dewey and Thomas Henry Huxley.

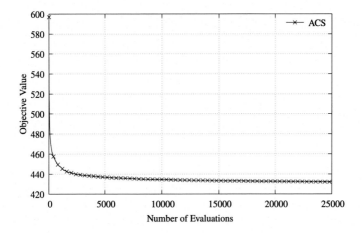

Figure A.3 Simulation results of ACS for the TSP benchmark eil51.

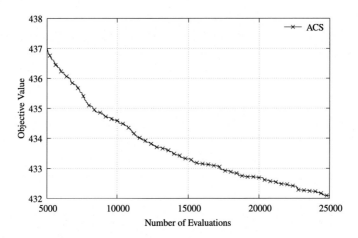

Figure A.4 Detailed simulation results of Fig. A.3.

alias, the plot instruction can be made much shorter and clearer, as shown in line 21. In general, alias is a useful way to deal with repeated use of long pathnames. Finally, this program will invoke the `plot` command to draw all the plots for the results residing in "result-acs-tsp-eil51.txt" using line style 1. As a result, it provides insight into the convergence curves of ACS for eil51, as shown in Fig. A.3. Even better, we can use the same program to create Fig. A.4. All we have to do is to add "set yrange [432:438]" to Listing A.3.

Listing A.3: src/gnuplot/app01/result-aco-tsp-eil51-1-o.gp: Gnuplot for Fig. A.3.

```
1 reset
2 set terminal pdfcairo enhanced color font "Times,14"
3 set title font "Times-Bold,14"
4 set size ratio 0.618
5
6 set style line 1 lc rgb 'black' lw 1 lt 1 pt 2 pi -30 ps 0.6
7
8 set grid xtics ytics
9 set pointintervalbox 0.01
10
11 set key top right vertical inside
12
13 set ylabel offset character 1,0,0
14 set ylabel "Objective Value"
15 set xlabel "Number of Evaluations"
16
17 set output "result-aco-tsp-eil51-1.pdf"
18
19 file_tsp_acs_eil51 = '"../../src/3-tsp/result/result-aco-tsp-eil51.txt"'
20
21 plot @file_tsp_acs_eil51 u 1 t "ACS" with lp ls 1
```

A.1.3 Number of evaluations and computation time

Using only the convergence curve of the number of iterations to evaluate the performance of different metaheuristics may not be fair because different algorithms may use a different number of search directions per iteration to search for the solution of an optimization problem. Another way to compare the performance of different metaheuristics is to observe the convergence with the same number of evaluations because the evaluation of the objective values of all the candidate solutions is computationally much more expensive than the evaluation of other operators in most cases. If all the candidate solutions are regarded as a deck of cards, each evaluation of the solution during the convergence process of a metaheuristic algorithm can be regarded as the action of drawing a card (i.e., generating a new candidate solution) and then checking it (i.e., evaluating this new candidate solution). For this reason, checking the same number of candidate solutions is widely accepted as a fair comparison between different search algorithms for most scientific communities of metaheuristic algorithms. This means that by using the number of evaluations for measurement, we know exactly how many candidate solutions have been checked by a metaheuristic algorithm. The stop criterion of different metaheuristic algorithms can then be set equal to the number of candidate solutions to make the comparison as fair as possible.

The computation time is another promising way for evaluating the performance of metaheuristics. This means that the stop criterion can be set to the same "running time" for different metaheuristics in solving the same optimization problem. As shown in Listing 3.4 (Section 3.2.2), we use the following code,

```
63    const clock_t begin = clock();
64    const int rc = get<0>(t)(argc, argv);
65    const clock_t end = clock();
66
67    const double cpu_time = static_cast<double>(end - begin) / CLOCKS_PER_SEC;
68    cerr << "# CPU time used: " << cpu_time << " seconds." << endl; //
```

to time each metaheuristic algorithm, such as HC, to know how long a metaheuristic algorithm takes to converge instead of just using the number of iterations to evaluate the performance of a metaheuristic algorithm. Even if we realize that a metaheuristic algorithm will take a lot of computation time, we may still not know which part of the search process takes most computation time. In some cases, we may integrate several operators or concepts to a metaheuristic algorithm so that they will add additional computation costs. Because of this, it is useful to show the computation time of each operator so that we know which one needs to be improved to accelerate the metaheuristic algorithm under development.

A.2. Analyses of metaheuristics

After the performance of metaheuristics is evaluated by using some sorts of measurements, the question that arises is how we can better understand their performance. In this section, we will discuss three different ways to analyze metaheuristics, namely, (1) the impact of different operators and parameters, (2) the time complexity, and (3) statistical analysis.

A.2.1 Impact of parameters and operators

Since most metaheuristics have more than one parameter, they can be flexibly adjusted for different optimization problems. In this section, we will use GA for the one-max problem as an example to show the impact of different parameters and operators. The population size of GA is set equal to 10, the crossover rate is set equal to 80% (i.e., 0.8), and the mutation rate is set equal to from 1% (i.e., 0.01) up to 10% (i.e., 0.10). All results are the average of 100 runs, and 100,000 evaluations (i.e., 10,000 generations) were performed each run for the one-max problem.

As shown in Table A.3, GA with a mutation rate of 0.01 (i.e., 1%) is able to find the optimal solution faster than all the other settings on

Table A.3 Simulation results of GA for the one-max problem of size $n = 100$ with different mutation rates.

Mutation rate	0.01	0.02	0.03	0.04	0.05	0.06	0.07	0.08	0.09	0.10
Best solution	100.00	94.72	90.18	87.12	84.82	83.31	81.89	80.55	79.77	79.30

average. However, the numbers themselves do not show the details of the performance of GA.

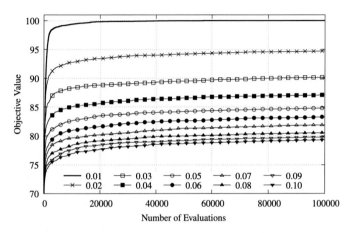

Figure A.5 Convergence trend of GA for the one-max problem of size $n = 100$ with different mutation rates.

Fig. A.5 shows a different way to present these results to see what happens exactly when different mutation rates are used by GA for the one–max problem. The 10 lines in this figure represent the convergence curves of GA with different mutation rates. GA is able to find the near optimal solution at the early stage during the convergence process when the mutation rate is set equal to 0.01. Needless to say, it is difficult to imagine the convergence trend of GA in solving the one-max problem from Table A.3, but it can be easily seen from Fig. A.5 because Table A.3 does not provide the convergence trends. One of the important things that can also be observed from this figure is that the convergence will slow down as the mutation rate increases. These situations are quite obvious when the mutation rate is set equal to 0.02 or higher. By analyzing the impact of the mutation rate in this way, we obtain useful information to apply GA to other complex optimization problems.

As shown in Fig. A.6, it is easy to see the convergence curves with these three different crossover operators of GA, i.e., the one-point, two-

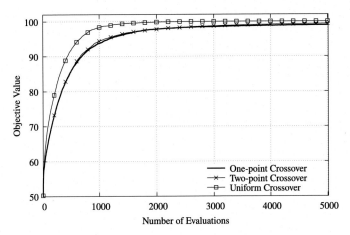

Figure A.6 Simulation results of GA for the one-max problem of size $n = 100$ with different crossover operators.

point, and uniform crossover operators. In addition to showing that the uniform crossover operator finds better results than the other two crossover operators, this figure also shows that the convergence of GA with uniform crossover is much faster than that of GA with the other two crossover operators.

As shown in Listings A.5 and A.6, only a few modifications to the one-point crossover of Listing A.4 are needed to get the so-called two-point and uniform crossovers. As shown in lines 17–32 of Listing A.5, a two-point crossover operator, as the name suggests, will use two points to split a chromosome into three parts, and then GA will exchange the middle part of a pair of chromosomes to perform the crossover procedure. As shown in lines 17–28 of Listing A.6, for the uniform crossover operator, each gene has a 50% chance of being exchanged with the gene in another chromosome.

Listing A.4: src/c++/5-eval/main/ga.h: An implementation of one-point crossover.

```
183 // one-point crossover
184 inline ga::population ga::crossover(population& curr_pop, double cr)
185 {
186     const size_t mid = curr_pop.size()/2;
187     for (size_t i = 0; i < mid; i++) {
188         const double f = static_cast<double>(rand()) / RAND_MAX;
189         if (f <= cr) {
190             int xp = rand() % curr_pop[0].size();        // crossover point
191             for (size_t j = xp; j < curr_pop[0].size(); j++)
192                 swap(curr_pop[i][j], curr_pop[mid+i][j]);
```

```
193          }
194     }
195     return curr_pop;
196 } //
```

Listing A.5: src/c++/5-eval/main/gax2.h: An implementation of two-point crossover.

```
16 // two-point crossover
17 inline gax2::population gax2::crossover(population& curr_pop, double cr)
18 {
19     const int mid = curr_pop.size()/2;
20     for (int i = 0; i < mid; i++) {
21         const double f = static_cast<double>(rand()) / RAND_MAX;
22         if (f <= cr) {
23             int xp1 = rand() % curr_pop[0].size(); // crossover point 1
24             int xp2 = rand() % curr_pop[0].size(); // crossover point 2
25             if (xp2 < xp1)
26                 swap(xp1, xp2);
27             for (int j = xp1; j < xp2; j++)
28                 swap(curr_pop[i][j], curr_pop[mid+i][j]);
29         }
30     }
31     return curr_pop;
32 } //
```

Listing A.6: src/c++/5-eval/main/gaxu.h: An implementation of uniform crossover.

```
16 // uniform crossover
17 inline gaxu::population gaxu::crossover(population& curr_pop, double cr)
18 {
19     const int mid = curr_pop.size()/2;
20     for (int i = 0; i < mid; i++) {
21         for (size_t j = 0; j < curr_pop[0].size(); j++) {
22             const double f = static_cast<double>(rand()) / RAND_MAX;
23             if (f <= 0.5) // yes, 0.5 instead of cr
24                 swap(curr_pop[i][j], curr_pop[mid+i][j]);
25         }
26     }
27     return curr_pop;
28 } //
```

A.2.2 Complexity and statistical analyses

Although we can use convergence, computation time, and the number of evaluations to measure the performance of any metaheuristic algorithm, complexity and convergence analysis can provide us with more accurate information. Because most metaheuristics consist of too many random factors, analysis of metaheuristics is difficult. Of course, we can use a Markov model to better understand the convergence of metaheuristics, but how close analysis and simulation results are is still an open issue. In fact, we can conduct some easy complexity analyses by using traditional time complexity analysis technologies to estimate the computation costs of the operators of metaheuristics. The easiest way is to analyze the number of loops or the number of steps each operator takes. We typically (though not always) say that a loop nested in another will take $O(n^2)$ steps. After all the operators for a metaheuristic algorithm are analyzed, we may conduct a rough time complexity analysis for the metaheuristic algorithm. Although this kind of analysis is not good enough, it provides us with additional information to understand the performance of metaheuristics. Convergence and time complexity analyses are still open issues, and several groups focus on this research topic.

Statistical analysis is another promising analysis tool. The Wilcoxon test and other statistical analysis methods (Derrac et al., 2011) can be used to understand whether the difference of the end results between metaheuristics is significant or not. This is because it is hard to discriminate the results of different metaheuristics when they are very close to each other. For example, for the function optimization problem (e.g., the Ackley function), we may want to know if algorithm $\mathcal{A}$ is better than algorithm $\mathcal{B}$ when the result of $\mathcal{A}$ is 0.00001 and the result of $\mathcal{B}$ is 0.000011. It is hard to say which one is better because the difference is very small and it may be caused by random factors. As a consequence, statistical analysis technologies are used to see if the difference between these two simulation results is significant and thus help us understand if algorithm $\mathcal{A}$ can find better results than algorithm $\mathcal{B}$. In brief, if the result of $\mathcal{A}$ is significantly better than the result of $\mathcal{B}$, then we can claim that $\mathcal{A}$ outperforms $\mathcal{B}$ for the optimization problem in question with the particular parameter settings.

A.3. Discussion

The "quality of the end result" and "computation speed" are two critical measures that can be used to evaluate the performance of a metaheuristic algorithm. However, the final results alone cannot help us understand the distinguishing features of a metaheuristic algorithm. This appendix shows that the information provided by the convergence curve may be useful for understanding the performance of a metaheuristic algorithm in detail. The reason is that the convergence process of most metaheuristic algorithms may undergo the first three stages of a product life cycle.[2] Without information on convergence, it is very difficult to fully understand the capabilities of a metaheuristic algorithm. As shown in Fig. A.7, the convergence curve of a metaheuristic algorithm for solving an optimization problem consists of the initial, growth, and stable stages. A metaheuristic algorithm has high potential to find better results at the initial and growth stages, but not at the stable stage. This is why the information of convergence is very important to understand the performance of a metaheuristic algorithm. In addition to the convergence curve, using the computation time to replace the number of iterations has attracted attention because the computation time can more accurately reflect the computation speed of a metaheuristic algorithm in practice.

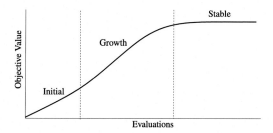

Figure A.7 Convergence curve of a metaheuristic algorithm.

The impact of parameters, time complexity, and statistical analyses are useful to analyze the performance of metaheuristic algorithms. We have to bear in mind that no single metaheuristic algorithm is able to outperform all other metaheuristic algorithms for all optimization problems. To be more specific, a more detailed analysis is required to help us avoid drawing wrong conclusions. That is why we cannot use just the final results of two algorithms $\mathcal{A}$ and $\mathcal{B}$ to claim that one outperforms the other. Rather, we typically use the information provided by the convergence curve to make sure that both algorithms are at the stable stage for particular datasets of an optimization problem $\mathbb{P}$ so that we can conclude that $\mathcal{A}$ can find better results than $\mathcal{B}$ for solving the optimization problem $\mathbb{P}$ for these datasets.

---

[2] The market introduction, growth, maturity, and saturation and decline stages are the four representative stages to explain the product life cycle. See also https://en.wikipedia.org/wiki/Product_life-cycle_management_(marketing).

Supplementary source code

Here are the hyperlinks to the directories where the source code described in this chapter resides.

https://github.com/cwtsaiai/metaheuristics_2023/src/gnuplot/app01/
https://github.com/cwtsaiai/metaheuristics_2023/src/c++/5-eval/

Implementation in Python

In this book, C++ is used as the programming language of choice because it is one of the most widely used programming languages when speed is a concern. This appendix presents the Python implementation of the C++ programs described in the main text of this book, because Python has emerged as one of the most popular programming languages, especially in the area of data science and machine learning. We hope that this appendix will make it easier for the readers who are familiar with Python but not with C++ to learn how to realize all the metaheuristic algorithms described in this book. Since the Python implementation described in this appendix is essentially the same as the C++ implementation described in the main text, no explanations will be given.

On Linux, the command to run the programs described in this appendix is either

```
$ python <progname> [-h | <arg1> <arg2> ... <argN>]
```

or

```
$ ./<progname> [-h | <arg1> <arg2> ... <argN>]
```

where `<progname>` is the name of the program and nothing, `-h`, or `<arg1>` `<arg2>` ... `<argN>` are the arguments to the program. In other words, you can enter `-h` as the argument to the program to see the usage, you can enter `<arg1>` `<arg2>` ... `<argN>` as the arguments to the program, or you can even enter nothing as the argument to the program so that the default setting of the arguments will be used.[1]

For example, the program described in Listing B.9 can be invoked by either of the following commands:

---

[1] Note that the sole purpose of the default setting of the arguments is to get you started with a quick test.

```
$ python sa.py 100 1000 10 "" 0.00001 1.0
$ python sa.py
$ python sa.py -h
$ ./sa.py 100 1000 10 "" 0.00001 1.0
$ ./sa.py
$ ./sa.py -h
```

and the output of the commands with -h as its sole argument is
"Usage: python sa.py <#runs> <#evals> <#patterns> <filename_ini>
<min_temp> <max_temp>," which acts as a gentle reminder in case you
do not know or forget what parameters are required by sa.py. Of course,
you can always create a script and use it to invoke the program so that you
do not have to retype all the arguments every time you want to run the
program, just like ./search.sh described in Section 3.2.2.

Now that we have learned how to run all these programs, Table B.1
gives a complete list of the Python implementation of the C++ programs
described in the main text by the order of problems. The column labeled
"Python" shows the listing of the Python code. The column labeled "Path
to the Python code" gives the path to the Python code. The column labeled
"C++" shows where the C++ counterparts are described in the main text.
Note that in this column, "–" indicates that the C++ counterpart is not
listed in the main text and "Ditto to 'a listing'" implies that the program is
eventually a duplicate of the listing (a program residing in another directory)
that was written for another problem but can be reused for this problem.

Table B.1   Python implementation of the C++ programs by the order of problems.

Python	Path to the Python code	C++
B.1	src/python/0-es/1-onemax/es.py	3.1, 3.2, and 3.3
B.2	src/python/0-es/2-deception/es.py	–
B.3	src/python/1-onemax/ga.py	7.1
B.4	src/python/1-onemax/gar.py	7.2
B.5	src/python/1-onemax/gax2.py	A.5
B.6	src/python/1-onemax/gaxu.py	A.6
B.7	src/python/1-onemax/hc.py	3.4 and 3.5
B.8	src/python/1-onemax/hc_r.py	–
B.9	src/python/1-onemax/sa.py	5.1
B.10	src/python/1-onemax/sa_refinit.py	11.1
B.11	src/python/1-onemax/se.py	18.1

continued on next page

Table B.1 (*continued*)

Python	Path to the Python code	C++
B.12	src/python/1-onemax/ts.py	6.1
B.13	src/python/2-deception/gadp.py	7.3
B.14	src/python/2-deception/ga.py	Ditto to B.3
B.15	src/python/2-deception/gardp.py	–
B.16	src/python/2-deception/gar.py	Ditto to B.4
B.17	src/python/2-deception/hcdp.py	3.7
B.18	src/python/2-deception/hcdp_rm.py	–
B.19	src/python/2-deception/hcdp_r.py	–
B.20	src/python/2-deception/hc.py	Ditto to B.7
B.21	src/python/2-deception/sadp.py	5.2
B.22	src/python/2-deception/sa.py	Ditto to B.9
B.23	src/python/2-deception/tsdp.py	6.2
B.24	src/python/2-deception/ts.py	Ditto to B.12
B.25	src/python/3-tsp/aco.py	8.1
B.26	src/python/3-tsp/gacx.py	12.3
B.30	src/python/3-tsp/ga.py	12.1
B.27	src/python/3-tsp/gals.py	16.1
B.28	src/python/3-tsp/gaox.py	12.4
B.29	src/python/3-tsp/gapmx.py	12.2
B.31	src/python/3-tsp/hga.py	15.1
B.32	src/python/3-tsp/pga.py	14.1
B.33	src/python/3-tsp/sa.py	15.2
B.34	src/python/4-function/deb1.py	10.2
B.35	src/python/4-function/deb2.py	10.4
B.36	src/python/4-function/decb1.py	10.3
B.37	src/python/4-function/de.py	10.1
B.38	src/python/4-function/der2.py	10.5
B.39	src/python/4-function/functions.py	9.2
B.40	src/python/4-function/pso.py	9.1
B.41	src/python/5-eval/ga.py	Ditto to B.3
B.42	src/python/5-eval/gar.py	Ditto to B.4
B.43	src/python/5-eval/garx2.py	–
B.44	src/python/5-eval/garxu.py	–
B.45	src/python/5-eval/gax2.py	Ditto to B.5
B.46	src/python/5-eval/gaxu.py	Ditto to B.6
B.47	src/python/6-clustering/ga.py	–
B.48	src/python/6-clustering/prega.py	17.1
B.49	src/python/a-lib/lib.py	3.6

Table B.2 gives a complete list of the C++ programs and the Python counterparts by the order they are described in the main text.

Table B.2 Python implementation of the C++ programs by the order of discussions in the main text.

C++	Path to the C++ code	Python
3.1	src/c++/0-es/1-onemax/es_main.cpp	B.1
3.2	src/c++/0-es/1-onemax/es_search.h	B.1
3.3	src/c++/0-es/1-onemax/es_search.cpp	B.1
3.4	src/c++/1-onemax/main/main.cpp	B.7
3.5	src/c++/1-onemax/main/hc.h	B.7
3.6	src/c++/1-onemax/main/lib.h	B.49
3.7	src/c++/2-deception/main/hcdp.h	B.17
5.1	src/c++/1-onemax/main/sa.h	B.9
5.2	src/c++/2-deception/main/sadp.h	B.21
6.1	src/c++/1-onemax/main/ts.h	B.12
6.2	src/c++/2-deception/main/tsdp.h	B.23
7.1	src/c++/1-onemax/main/ga.h	B.3
7.2	src/c++/1-onemax/main/gar.h	B.4
7.3	src/c++/2-deception/main/gadp.h	B.13
8.1	src/c++/3-tsp/main/aco.h	B.25
9.1	src/c++/4-function/main/pso.h	B.40
9.2	src/c++/4-function/main/functions.h	B.39
10.1	src/c++/4-function/main/de.h	B.37
10.2	src/c++/4-function/main/deb1.h	B.34
10.3	src/c++/4-function/main/decb1.h	B.36
10.4	src/c++/4-function/main/deb2.h	B.35
10.5	src/c++/4-function/main/der2.h	B.38
11.1	src/c++/1-onemax/main/sa_refinit.h	B.10
11.2	src/c++/3-tsp/main/ga.h	B.30
11.3	src/c++/3-tsp/main/ga.h	B.30
12.1	src/c++/3-tsp/main/ga.h	B.30
12.2	src/c++/3-tsp/main/gapmx.h	B.29
12.3	src/c++/3-tsp/main/gacx.h	B.26
12.4	src/c++/3-tsp/main/gaox.h	B.28
14.1	src/c++/3-tsp/main/pga.h	B.32
15.1	src/c++/3-tsp/main/hga.h	B.31
15.2	src/c++/3-tsp/main/sa.h	B.33
16.1	src/c++/3-tsp/main/gals.h	B.27
17.1	src/c++/6-clustering/main/prega.h	B.48
18.1	src/c++/1-onemax/main/se.h	B.11
A.4	src/c++/5-eval/main/ga.h	B.41 and B.3
A.5	src/c++/5-eval/main/gax2.h	B.45 and B.5
A.6	src/c++/5-eval/main/gaxu.h	B.46 and B.6

Below we provide a complete list of the Python implementation of the C++ programs, be it described in the main text or not.

Listing B.1: src/python/0-es/1-onemax/es.py: The implementation of ES in Python.

```
1  #!/usr/bin/env python
2  import numpy as np; import time; import sys; import os
3  from lib import *
4
5  class es:
6      def __init__(self, num_bits = 10):
7          self.num_bits = num_bits
8
9      def run(self):
10          max_sol = 2**self.num_bits - 1
11          s = 0
12          fs = self.evaluate(s)
13          print('{0} # {1:0{2}b}'.format(fs, s, self.num_bits))
14          v = s
15          while v < max_sol:
16              v = self.transit(v)
17              fv = self.evaluate(v)
18              fs, s = self.determine(fv, v, fs, s)
19              print('{0} # {1:0{2}b}'.format(fs, v, self.num_bits))
20
21      def transit(self, s):
22          s += 1
23          return s
24
25      def evaluate(self, s):
26          return bin(s).count("1")
27
28      def determine(self, fv, v, fs, s):
29          if fv > fs:
30              fs, s = fv, v
31          return fs, s
32
33      def print_parameters(self, algname):
34          print("# name of the search algorithm: '%s'" % algname)
35          print("# number of bits: %d" % self.num_bits)
36
37  def usage():
38      lib.usage("<num_bits>", common_params = "")
39
40  if __name__ == "__main__":
41      if ac() == 2 and ss(1) == "-h": usage()
42      elif ac() == 1: search = es()
43      elif ac() == 2: search = es(si(1))
44      else: usage()
45      lib.run(search, None)
```

Listing B.2: src/python/0-es/2-deception/es.py: The implementation of ES in Python.

```
1  #!/usr/bin/env python
2  import numpy as np; import time; import sys; import os
3  from lib import *
4
```

```
 5 class es:
 6     def __init__(self, num_bits = 10):
 7         self.num_bits = num_bits
 8
 9     def run(self):
10         max_sol = 2**self.num_bits - 1
11         s = 0
12         fs = self.evaluate(s)
13         print('{0} # {1:0{2}b}'.format(fs, s, self.num_bits))
14         v = s
15         while v < max_sol:
16             v = self.transit(v)
17             fv = self.evaluate(v)
18             fs, s = self.determine(fv, v, fs, s)
19             print('{0} # {1:0{2}b}'.format(fs, v, self.num_bits))
20
21     def transit(self, s):
22         s += 1
23         return s
24
25     def evaluate(self, s):
26         return abs(s - (1 << (self.num_bits - 2)))
27
28     def determine(self, fv, v, fs, s):
29         if fv > fs:
30             fs, s = fv, v
31         return fs, s
32
33     def print_parameters(self, algname):
34         print("# name of the search algorithm: '%s'" % algname)
35         print("# number of bits: %d" % self.num_bits)
36
37 def usage():
38     lib.usage("<num_bits>", common_params = "")
39
40 if __name__ == "__main__":
41     if ac() == 2 and ss(1) == "-h": usage()
42     elif ac() == 1: search = es()
43     elif ac() == 2: search = es(si(1))
44     else: usage()
45     lib.run(search, None)
```

Listing B.3: src/python/1-onemax/ga.py: The implementation of GA in Python.

```
 1 #!/usr/bin/env python
 2 import numpy as np; import time; import sys; import os
 3 from lib import *
 4
 5 class ga:
 6     def __init__(self, num_runs = 3, num_evals = 1000, num_patterns_sol = 10, filename_ini = "",
        pop_size = 10, crossover_rate = 0.6, mutation_rate = 0.01, num_players = 3):
 7         self.num_runs = num_runs
 8         self.num_evals = num_evals
 9         self.num_patterns_sol = num_patterns_sol
10         self.filename_ini = filename_ini
11         self.pop_size = pop_size
12         self.crossover_rate = crossover_rate
13         self.mutation_rate = mutation_rate
14         self.num_players = num_players
15         self.best_obj_val = 0.0
16         self.best_sol = None
```

```
17        self.seed = int(time.time())
18        np.random.seed(self.seed)
19
20    def run(self):
21        avg_obj_val = 0.0
22        avg_obj_val_eval = np.zeros(self.num_evals)
23        for r in range(self.num_runs):
24            eval_count = 0
25            best_so_far = 0.0
26            # 0. initialization
27            curr_pop = self.init()
28            while eval_count < self.num_evals:
29                # 1. evaluation
30                curr_obj_vals = self.evaluate(curr_pop)
31                self.update_best_sol(curr_pop, curr_obj_vals)
32                for i in range(self.pop_size):
33                    if best_so_far < curr_obj_vals[i]:
34                        best_so_far = curr_obj_vals[i]
35                    if eval_count < self.num_evals:
36                        avg_obj_val_eval[eval_count] += best_so_far
37                        eval_count += 1
38                # 2. determination
39                curr_pop = self.select(curr_pop, curr_obj_vals, self.num_players)
40                # 3. transition
41                curr_pop = self.crossover(curr_pop, self.crossover_rate)
42                curr_pop = self.mutation(curr_pop, self.mutation_rate)
43            avg_obj_val += self.best_obj_val
44        # 4. output
45        avg_obj_val /= self.num_runs
46        for i in range(self.num_evals):
47            avg_obj_val_eval[i] /= self.num_runs
48            print("%.3f" % avg_obj_val_eval[i])
49        return curr_pop
50
51    def init(self):
52        if len(self.filename_ini) == 0:
53            curr_pop = np.array([np.array([1 if x >= 0.5 else 0 for x in np.random.rand(self.
    num_patterns_sol)]) for i in range(self.pop_size)])
54        else:
55            with open(self.filename_ini, "r") as f:
56                curr_pop = np.array(f.read().strip().split(), dtype=int).reshape(self.pop_size,
    self.num_patterns_sol)
57        self.best_obj_val = 0
58        return curr_pop
59
60    def evaluate(self, curr_pop):
61        return np.array(list(map(sum, curr_pop)))
62
63    def select(self, curr_pop, curr_obj_vals, num_players):
64        tmp_pop = []
65        for i in range(self.pop_size):
66            k = np.random.randint(self.pop_size)
67            f = curr_obj_vals[k]
68            for j in range(1, num_players):
69                n = np.random.randint(self.pop_size)
70                if curr_obj_vals[n] > f:
71                    k = n
72                    f = curr_obj_vals[k]
73            tmp_pop.append(curr_pop[k].copy())
74        return tmp_pop
75
76    def update_best_sol(self, curr_pop, curr_obj_vals):
77        for i in range(self.pop_size):
78            if curr_obj_vals[i] > self.best_obj_val:
79                self.best_obj_val = curr_obj_vals[i]
```

```
80                  self.best_sol = curr_pop[i]
81
82      def crossover(self, curr_pop, cr):
83          mid = int(self.pop_size / 2)
84          for i in range(mid):
85              f = np.random.rand()
86              if f <= cr:
87                  xp = np.random.randint(self.num_patterns_sol)
88                  for j in range(xp, self.num_patterns_sol):
89                      curr_pop[i][j], curr_pop[mid+i][j] = curr_pop[mid+i][j], curr_pop[i][j]
90          return curr_pop
91
92      def mutation(self, curr_pop, mr):
93          for i in range(self.pop_size):
94              for j in range(self.num_patterns_sol):
95                  f = np.random.rand()
96                  if f <= mr:
97                      curr_pop[i][j] ^= 1
98          return curr_pop
99
100     def print_parameters(self, algname):
101         print("# name of the search algorithm: '%s'" % algname)
102         print("# number of runs: %s" % self.num_runs)
103         print("# number of evaluations: %s" % self.num_evals)
104         print("# number of patterns: %s" % self.num_patterns_sol)
105         print("# filename of the initial seeds: '%s'" % self.filename_ini)
106         print("# population size: %s" % self.pop_size)
107         print("# crossover rate: %s" % self.crossover_rate)
108         print("# mutation rate: %s" % self.mutation_rate)
109         print("# number of players: %s" % self.num_players)
110         print("# seed: %s" % self.seed)
111
112 def usage():
113     lib.usage("<pop_size> <cr> <mr> <#players>")
114
115 if __name__ == "__main__":
116     if ac() == 2 and ss(1) == "-h": usage()
117     elif ac() == 1: search = ga()
118     elif ac() == 9: search = ga(si(1), si(2), si(3), ss(4), si(5), sf(6), sf(7), si(8))
119     else: usage()
120     lib.run(search, lib.print_population)
```

Listing B.4: src/python/1-onemax/gar.py: The implementation of GAR in Python.

```
1 #!/usr/bin/env python
2 import numpy as np; import time; import sys; import os
3 from lib import *
4 from ga import ga
5
6 class gar(ga):
7     def __init__(self, num_runs = 3, num_evals = 1000, num_patterns_sol = 10, filename_ini = "",
        pop_size = 10, crossover_rate = 0.6, mutation_rate = 0.01, num_players = 0):
8         super().__init__(num_runs, num_evals, num_patterns_sol, filename_ini, pop_size,
        crossover_rate, mutation_rate, num_players)
9
10    def select(self, curr_pop, curr_obj_vals, num_players):
11        # 1. compute the probabilities of the roulette wheel
12        total = sum(curr_obj_vals)
13        tmp_pop = []
14        for i in range(self.pop_size):
15            # 2. select the individuals for the next generation
```

```
16              f = total * np.random.rand()
17              for j in range(self.pop_size):
18                  f -= curr_obj_vals[j]
19                  if f <= 0.0:
20                      tmp_pop.append(curr_pop[j].copy())
21                      break
22          return tmp_pop
23
24 def usage():
25      lib.usage("<pop_size> <cr> <mr> <#players>")
26
27 if __name__ == "__main__":
28      if ac() == 2 and ss(1) == "-h": usage()
29      elif ac() == 1: search = gar()
30      elif ac() == 9: search = gar(si(1), si(2), si(3), ss(4), si(5), sf(6), sf(7), si(8))
31      else: usage()
32      lib.run(search, lib.print_population)
```

Listing B.5: src/python/1-onemax/gax2.py: The implementation of GAX2 in Python.

```
 1 #!/usr/bin/env python
 2 import numpy as np; import time; import sys; import os
 3 from lib import *
 4 from ga import ga
 5
 6 class gax2(ga):
 7      def __init__(self, num_runs = 3, num_evals = 1000, num_patterns_sol = 10, filename_ini = "",
         pop_size = 10, crossover_rate = 0.6, mutation_rate = 0.01, num_players = 0):
 8          super().__init__(num_runs, num_evals, num_patterns_sol, filename_ini, pop_size,
         crossover_rate, mutation_rate, num_players)
 9
10      def crossover(self, curr_pop, cr):
11          # two-point crossover
12          mid = self.pop_size // 2
13          for i in range(mid):
14              f = np.random.rand()
15              if f <= cr:
16                  xp1 = np.random.randint(self.num_patterns_sol)
17                  xp2 = np.random.randint(self.num_patterns_sol)
18                  if xp2 < xp1:
19                      xp1, xp2 = xp2, xp1
20                  for j in range(xp1, xp2):
21                      curr_pop[i][j], curr_pop[mid+i][j] = curr_pop[mid+i][j], curr_pop[i][j]
22          return curr_pop
23
24 def usage():
25      lib.usage("<pop_size> <cr> <mr> <#players>")
26
27 if __name__ == "__main__":
28      if ac() == 2 and ss(1) == "-h": usage()
29      elif ac() == 1: search = gax2()
30      elif ac() == 9: search = gax2(si(1), si(2), si(3), ss(4), si(5), sf(6), sf(7), si(8))
31      else: usage()
32      lib.run(search, lib.print_population)
```

Listing B.6: src/python/1-onemax/gaxu.py: The implementation of GAXU in Python.

```
1  #!/usr/bin/env python
2  import numpy as np; import time; import sys; import os
3  from lib import *
4  from ga import ga
5
6  class gaxu(ga):
7      def __init__(self, num_runs = 3, num_evals = 1000, num_patterns_sol = 10, filename_ini = "",
         pop_size = 10, crossover_rate = 0.6, mutation_rate = 0.01, num_players = 0):
8          super().__init__(num_runs, num_evals, num_patterns_sol, filename_ini, pop_size,
         crossover_rate, mutation_rate, num_players)
9
10     def crossover(self, curr_pop, cr):
11         # uniform crossover
12         mid = self.pop_size // 2
13         for i in range(mid):
14             for j in range(self.num_patterns_sol):
15                 f = np.random.rand()
16                 if f <= 0.5: # yes, 0.5 instead of cr
17                     curr_pop[i][j], curr_pop[mid+i][j] = curr_pop[mid+i][j], curr_pop[i][j]
18         return curr_pop
19
20 def usage():
21     lib.usage("<pop_size> <cr> <mr> <#players>")
22
23 if __name__ == "__main__":
24     if ac() == 2 and ss(1) == "-h": usage()
25     elif ac() == 1: search = gaxu()
26     elif ac() == 9: search = gaxu(si(1), si(2), si(3), ss(4), si(5), sf(6), sf(7), si(8))
27     else: usage()
28     lib.run(search, lib.print_population)
```

Listing B.7: src/python/1-onemax/hc.py: The implementation of HC in Python.

```
1  #!/usr/bin/env python
2  import numpy as np; import time; import sys; import os
3  from lib import *
4
5  class hc:
6      def __init__(self, num_runs = 3, num_evals = 1000, num_patterns_sol = 10, filename_ini = ""):
7          self.num_runs = num_runs
8          self.num_evals = num_evals
9          self.num_patterns_sol = num_patterns_sol
10         self.filename_ini = filename_ini
11         self.seed = int(time.time())
12         np.random.seed(self.seed)
13
14     def run(self):
15         avg_obj_val = 0.0
16         avg_obj_val_eval = np.zeros(self.num_evals)
17         for r in range(self.num_runs):
18             # 0. Initialization
19             sol, obj_val = self.init()
20             eval_count = 0
21             avg_obj_val_eval[eval_count] += obj_val
22             eval_count += 1
23             while eval_count < self.num_evals:
```

```
24              # 1. Transition
25              tmp_sol = self.transit(sol)
26              # 2. Evaluation
27              tmp_obj_val = self.evaluate(tmp_sol)
28              # 3. Determination
29              sol, obj_val = self.determine(tmp_sol, tmp_obj_val, sol, obj_val)
30              avg_obj_val_eval[eval_count] += obj_val
31              eval_count += 1
32           avg_obj_val += obj_val
33        # 4. Output
34        avg_obj_val /= self.num_runs
35        for i in range(self.num_evals):
36           avg_obj_val_eval[i] /= self.num_runs
37           print("%.3f" % avg_obj_val_eval[i])
38        return '{0:0{1}b}'.format(sol, self.num_patterns_sol)
39
40    def init(self):
41        if len(self.filename_ini) == 0:
42           sol = np.random.randint(2, size = self.num_patterns_sol, dtype=int)
43        else:
44           with open(self.filename_ini, "r") as f:
45               sol = np.array(f.read().strip().split(), dtype=int)
46           sol = int(sol.dot(2**np.arange(sol.size)[::-1]))
47        obj_val = self.evaluate(sol)
48        return sol, obj_val
49
50    def transit(self, s):
51        if s == 0 or s == (1 << self.num_patterns_sol) - 1:
52           return s
53        s += 1 if np.random.rand() < 0.5 else -1
54        return s
55
56    def evaluate(self, s):
57        return bin(s).count("1")
58
59    def determine(self, v, fv, s, fs):
60        return (v, fv) if fv > fs else (s, fs)
61
62    def print_parameters(self, algname):
63        print("# name of the search algorithm: '%s'" % algname)
64        print("# number of runs: %s" % self.num_runs)
65        print("# number of evaluations: %s" % self.num_evals)
66        print("# number of patterns: %s" % self.num_patterns_sol)
67        print("# filename of the initial seeds: '%s'" % self.filename_ini)
68        print("# seed: %s" % self.seed)
69
70 def usage():
71    lib.usage("")
72
73 if __name__ == "__main__":
74    if ac() == 2 and ss(1) == "-h": usage()
75    elif ac() == 1: search = hc()
76    elif ac() == 5: search = hc(si(1), si(2), si(3), ss(4))
77    else: usage()
78    lib.run(search, lib.print_solution)
```

Listing B.8: src/python/1-onemax/hc_r.py: The implementation of HC_R in Python.

```
1 #!/usr/bin/env python
2 import numpy as np; import time; import sys; import os
3 from lib import *
```

```
 4 from hc import hc
 5
 6 class hc_r(hc):
 7     def __init__(self, num_runs = 3, num_evals = 1000, num_patterns_sol = 10, filename_ini = "")
       :
 8         super().__init__(num_runs, num_evals, num_patterns_sol, filename_ini)
 9
10     def transit(self, s):
11         i = np.random.randint(self.num_patterns_sol)
12         s ^= 1 << self.num_patterns_sol - 1 - i
13         return s
14
15 def usage():
16     lib.usage("")
17
18 if __name__ == "__main__":
19     if ac() == 2 and ss(1) == "-h": usage()
20     elif ac() == 1: search = hc_r()
21     elif ac() == 5: search = hc_r(si(1), si(2), si(3), ss(4))
22     else: usage()
23     lib.run(search, lib.print_solution)
```

Listing B.9: src/python/1-onemax/sa.py: The implementation of SA in Python.

```
 1 #!/usr/bin/env python
 2 import numpy as np; import time; import sys; import os
 3 from lib import *
 4
 5 class sa:
 6     def __init__(self, num_runs = 3, num_evals = 1000, num_patterns_sol = 10, filename_ini = "",
       min_temp = 0.00001, max_temp = 1.0):
 7         self.num_runs = num_runs
 8         self.num_evals = num_evals
 9         self.num_patterns_sol = num_patterns_sol
10         self.filename_ini = filename_ini
11         self.min_temp = min_temp
12         self.max_temp = max_temp
13         self.seed = int(time.time())
14         np.random.seed(self.seed)
15
16     def run(self):
17         avg_obj_val = 0.0
18         avg_obj_val_eval = np.zeros(self.num_evals)
19         for r in range(self.num_runs):
20             # 0. Initialization
21             obj_val, sol, best_obj_val, best_sol, curr_temp = self.init()
22             eval_count = 0
23             avg_obj_val_eval[eval_count] += obj_val
24             eval_count += 1
25             while eval_count < self.num_evals:
26                 # 1. Transition
27                 tmp_sol = self.transit(sol)
28                 # 2. Evaluation
29                 tmp_obj_val = self.evaluate(tmp_sol)
30                 # 3. Determination
31                 if self.determine(tmp_obj_val, obj_val, curr_temp):
32                     obj_val, sol = tmp_obj_val, tmp_sol
33                 if obj_val > best_obj_val:
34                     best_obj_val, best_sol = obj_val, sol
35                 curr_temp = self.annealing(curr_temp)
36                 avg_obj_val_eval[eval_count] += best_obj_val
```

```
37                      eval_count += 1
38                      avg_obj_val += best_obj_val
39              # 4. Output
40              avg_obj_val /= self.num_runs
41              for i in range(self.num_evals):
42                  avg_obj_val_eval[i] /= self.num_runs
43                  print("%.3f" % avg_obj_val_eval[i])
44              return best_sol
45
46      def init(self):
47          if len(self.filename_ini) == 0:
48              sol = np.array([1 if x >= 0.5 else 0 for x in np.random.rand(self.num_patterns_sol)
            ])
49          else:
50              with open(self.filename_ini, "r") as f:
51                  sol = np.array(f.read().strip().split(), dtype=int)
52          obj_val = self.evaluate(sol)
53          return obj_val, sol, obj_val, sol, self.max_temp
54
55      def transit(self, sol):
56          t = sol.copy()
57          i = np.random.randint(len(sol))
58          t[i] ^= 1
59          return t
60
61      def evaluate(self, sol):
62          return sum(sol)
63
64      def determine(self, tmp_obj_val, obj_val, temperature):
65          r = np.random.rand()
66          p = np.exp((tmp_obj_val - obj_val) / temperature)
67          return r < p
68
69      def annealing(self, temperature):
70          return 0.9 * temperature
71
72      def print_parameters(self, algname):
73          print("# name of the search algorithm: '%s'" % algname)
74          print("# number of runs: %s" % self.num_runs)
75          print("# number of evaluations: %s" % self.num_evals)
76          print("# number of patterns: %s" % self.num_patterns_sol)
77          print("# filename of the initial seeds: '%s'" % self.filename_ini)
78          print("# minimum temperature: %s" % self.min_temp)
79          print("# maximum temperature: %s" % self.max_temp)
80          print("# seed: %s" % self.seed)
81
82 def usage():
83      lib.usage("<min_temp> <max_temp>")
84
85 if __name__ == "__main__":
86      if ac() == 2 and ss(1) == "-h": usage()
87      elif ac() == 1: search = sa()
88      elif ac() == 7: search = sa(si(1), si(2), si(3), ss(4), sf(5), sf(6))
89      else: usage()
90      lib.run(search, lib.print_solution)
```

Listing B.10: src/python/1-onemax/sa_refinit.py: The implementation of SA_REFINIT in Python.

```
1 #!/usr/bin/env python
2 import numpy as np; import time; import sys; import os
3 from lib import *
```

```
4
5  class sa_refinit:
6      def __init__(self, num_runs = 3, num_evals = 1000, num_patterns_sol = 10, filename_ini = "",
       min_temp = 0.00001, max_temp = 1.0, num_samplings = 10, num_same_bits = 2):
7          self.num_runs = num_runs
8          self.num_evals = num_evals
9          self.num_patterns_sol = num_patterns_sol
10         self.filename_ini = filename_ini
11         self.min_temp = min_temp
12         self.max_temp = max_temp
13         self.num_samplings = num_samplings
14         self.num_same_bits = num_same_bits
15         self.seed = int(time.time())
16         np.random.seed(self.seed)
17
18     def run(self):
19         avg_obj_val = 0.0
20         avg_obj_val_eval = np.zeros(self.num_evals)
21         for r in range(self.num_runs):
22             # 0. Initialization
23             obj_val, sol, best_obj_val, best_sol, curr_temp = self.init(avg_obj_val_eval)
24             eval_count = self.num_samplings
25             avg_obj_val_eval[eval_count] += obj_val
26             eval_count += 1
27             while eval_count < self.num_evals:
28                 # 1. Transition
29                 tmp_sol = self.transit(sol)
30                 # 2. Evaluation
31                 tmp_obj_val = self.evaluate(tmp_sol)
32                 # 3. Determination
33                 if self.determine(tmp_obj_val, obj_val, curr_temp):
34                     obj_val, sol = tmp_obj_val, tmp_sol
35                 if obj_val > best_obj_val:
36                     best_obj_val, best_sol = obj_val, sol
37                 curr_temp = self.annealing(curr_temp)
38                 avg_obj_val_eval[eval_count] += best_obj_val
39                 eval_count += 1
40             avg_obj_val += best_obj_val
41         # 4. Output
42         avg_obj_val /= self.num_runs
43         for i in range(self.num_evals):
44             avg_obj_val_eval[i] /= self.num_runs
45             print("%.3f" % avg_obj_val_eval[i])
46         return best_sol
47
48     def init(self, avg_obj_val_eval):
49         if len(self.filename_ini) > 0:
50             with open(self.filename_ini, "r") as f:
51                 sol = np.array(f.read().strip().split(), dtype=int)
52         else:
53             tmp_sol = np.zeros(self.num_patterns_sol, dtype=int)
54             for j in range(self.num_samplings):
55                 for i in range(0, self.num_patterns_sol, self.num_same_bits):
56                     tmp_sol[i:i+self.num_same_bits] = np.random.randint(2, size = self.
       num_same_bits)
57                 tmp_obj_val = self.evaluate(tmp_sol)
58                 if j == 0 or tmp_obj_val > obj_val:
59                     best_obj_val = obj_val = tmp_obj_val
60                     sol = tmp_sol
61                 avg_obj_val_eval[j] += obj_val
62         obj_val = self.evaluate(sol)
63         return obj_val, sol, obj_val, sol, self.max_temp
64
65     def transit(self, sol):
66         t = sol.copy()
```

```
67          i = np.random.randint(len(sol))
68          t[i] ^= 1
69          return t
70
71      def evaluate(self, sol):
72          return sum(sol)
73
74      def determine(self, tmp_obj_val, obj_val, temperature):
75          r = np.random.rand()
76          p = np.exp((tmp_obj_val - obj_val) / temperature)
77          return r < p
78
79      def annealing(self, temperature):
80          return 0.9 * temperature
81
82      def print_parameters(self, algname):
83          print("# name of the search algorithm: 'sa_refinit'")
84          print("# number of runs: %s" % self.num_runs)
85          print("# number of evaluations: %s" % self.num_evals)
86          print("# number of patterns: %s" % self.num_patterns_sol)
87          print("# filename of the initial seeds: '%s'" % self.filename_ini)
88          print("# minimum temperature: %s" % self.min_temp)
89          print("# maximum temperature: %s" % self.max_temp)
90          print("# number of samples: %s" % self.num_samplings)
91          print("# number of same bits: %s" % self.num_same_bits)
92          print("# seed: %s" % self.seed)
93
94  def usage():
95      lib.usage("<min_temp> <max_temp> <#samples> <#same_bits>")
96
97  if __name__ == "__main__":
98      if ac() == 2 and ss(1) == "-h": usage()
99      elif ac() == 1: search = sa_refinit()
100     elif ac() == 9: search = sa_refinit(si(1), si(2), si(3), ss(4), sf(5), sf(6), si(7), si(8))
101     else: usage()
102     lib.run(search, lib.print_solution)
```

Listing B.11: src/python/1-onemax/se.py: The implementation of SE in Python.

```
1  #!/usr/bin/env python
2  import numpy as np; import time; import sys; import os; import math
3  from lib import *
4
5  class se:
6      def __init__(self, num_runs = 3, num_evals = 10000, num_bits_sol = 100, filename_ini = "",
          num_searchers = 1, num_regions = 1, num_samples = 1, num_players = 1, scatter_plot = 0):
7          self.num_runs = num_runs
8          self.num_evals = num_evals
9          self.num_bits_sol = num_bits_sol
10         self.filename_ini = filename_ini
11         self.num_searchers = num_searchers
12         self.num_regions = num_regions
13         self.num_samples = num_samples
14         self.num_players = num_players
15         self.scatter_plot = scatter_plot != 0
16         self.num_id_bits = int(math.log2(num_regions))
17         self.id_bits = np.empty((num_regions, self.num_id_bits), dtype=int)
18         self.seed = int(time.time())
19         np.random.seed(self.seed)
20
21     def run(self):
```

```
22          avg_obj_val_eval = np.zeros(self.num_evals)
23          for r in range(self.num_runs):
24              self.eval_count = 0
25              self.init()                          # 1. initialization
26              self.resource_arrangement()          # 2. resource arrangement
27              while self.eval_count < self.num_evals:
28                  eval_cc = self.eval_count
29                  self.vision_search(eval_cc)      # 3. vision search
30                  self.marketing_survey()          # 4. marketing survey
31                  avg_obj_val_eval[eval_cc:min(self.eval_count, self.num_evals)] += self.
        best_so_far
32          avg_obj_val_eval /= self.num_runs
33          for i in range(self.num_evals):
34              print("%.3f" % avg_obj_val_eval[i])
35
36      # 1. initialization
37      def init(self):
38          n_searchers = self.num_searchers
39          n_regions = self.num_regions
40          n_samples = self.num_samples
41          n_bits_sol = self.num_bits_sol
42          # set aside arrays for searchers, samples, and sampleV
43          self.searcher_sol = np.array([np.array([0 if _ < 0.5 else 1 for _ in np.random.rand(
        n_bits_sol)], dtype=int) for _ in range(n_searchers)], dtype=int)
44          self.sample_sol = np.zeros((n_regions, n_samples, n_bits_sol), dtype=int)
45          self.sample_sol_best = np.zeros((n_regions, n_bits_sol), dtype=int)
46          self.sampleV_sol = np.zeros((n_searchers, n_regions, 2*n_samples, n_bits_sol), dtype=int
        )
47          self.searcher_sol_fitness = np.zeros(n_searchers, dtype=float)
48          self.sample_sol_fitness = np.zeros((n_regions, n_samples), dtype=float)
49          self.sample_sol_best_fitness = np.zeros(n_regions, dtype=float)
50          self.sampleV_sol_fitness = np.zeros((n_searchers, n_regions, 2*n_samples), dtype=float)
51          self.best_sol = np.zeros(n_bits_sol, dtype=int)
52          self.best_so_far = 0
53
54      # 2. resource arrangement
55      def resource_arrangement(self):
56          n_searchers = self.num_searchers
57          n_regions = self.num_regions
58          n_samples = self.num_samples
59          n_bits_sol = self.num_bits_sol
60          n_id_bits = self.num_id_bits
61          # 2.1. initialize searchers and regions
62          ## regions_id_bits()
63          self.id_bits = np.array([np.array(list('{0:0{1}b}'.format(j, n_id_bits))) for j in range
        (n_regions)], dtype=int)
64          # 2.1.1 assign searchers to regions
65          self.searcher_region_id = np.array([i % n_regions for i in range(n_searchers)], dtype=
        int)
66          ## assign_searcher to region (i, i % num_regions)
67          self.searcher_sol[:, :n_id_bits] = np.array([self.id_bits[i % n_regions] for i in range(
        n_searchers)], dtype=int)
68          # 2.1.2 initialize sample solutions
69          for j in range(n_regions):
70              for k in range(n_samples):
71                  self.sample_sol[j][k][:n_id_bits] = self.id_bits[j]
72                  self.sample_sol[j][k][n_id_bits:] = np.array([0 if _ < 0.5 else 1 for _ in np.
        random.rand(n_bits_sol-n_id_bits)])
73          # 2.2. initialize investment and how long regions have not been searched
74          self.ta = np.zeros(n_regions)
75          self.tb = np.ones(n_regions)
76          for i in range(n_searchers):
77              j = self.searcher_region_id[i]
78              self.ta[j] += 1
79              self.tb[j] = 1
```

```
80      # 2.3. initialize expected values (ev)
81      self.expected_value = np.zeros((n_searchers, n_regions))
82      self.T_j = np.zeros(n_regions)
83      self.M_j = np.zeros(n_regions)
84
85  # 3. vision search
86  def vision_search(self, eval):
87      # 3.1 construct V (searcher X sample)
88      if eval > 0:
89          self.transit()
90      # 3.2 compute the expected value of all regions of searchers
91      self.compute_expected_value(eval)
92      # 3.3 select region to which a searcher will be assigned
93      self.vision_selection(self.num_players, eval)
94
95  # 3.1 construct V (searcher X sample)
96  def transit(self):
97      n_searchers = self.num_searchers
98      n_regions = self.num_regions
99      n_samples = self.num_samples
100     n_bits_sol = self.num_bits_sol
101     n_id_bits = self.num_id_bits
102     # 3.1.1 exchange information between searchers and samples
103     for i in range(n_searchers):
104         for j in range(n_regions):
105             for k in range(n_samples):
106                 x = np.random.randint(n_bits_sol + 1)
107                 m = k << 1
108                 n = m + 1
109                 self.sampleV_sol[i][j][m][:n_id_bits] = self.id_bits[j]
110                 self.sampleV_sol[i][j][n][:n_id_bits] = self.id_bits[j]
111                 self.sampleV_sol[i][j][m][n_id_bits:x] = self.searcher_sol[i][n_id_bits:x]
112                 self.sampleV_sol[i][j][n][n_id_bits:x] = self.sample_sol[j][k][n_id_bits:x]
113                 self.sampleV_sol[i][j][m][x:n_bits_sol] = self.sample_sol[j][k][x:n_bits_sol
]
114                 self.sampleV_sol[i][j][n][x:n_bits_sol] = self.searcher_sol[i][x:n_bits_sol]
115     # 3.1.2 randomly change one bit of each solution in sampleV_sol
116     for i in range(n_searchers):
117         for j in range(n_regions):
118             for k in range(2*n_samples):
119                 m = np.random.randint(n_bits_sol) # bit to mutate
120                 if m >= n_id_bits:
121                     self.sampleV_sol[i][j][k][m] ^= 1
122
123  # 3.2 expected value for onemax problem
124  def compute_expected_value(self, eval):
125     n_searchers = self.num_searchers
126     n_regions = self.num_regions
127     n_samples = self.num_samples
128     n_bits_sol = self.num_bits_sol
129     # 3.2.1 fitness value of searchers and sampleV_sol (new candidate solutions)
130     if eval == 0:
131         # 3.2.1a fitness value of searchers
132         self.searcher_sol_fitness = np.array([self.evaluate_fitness(self.searcher_sol[i])
for i in range(n_searchers)])
133     else:
134         # 3.2.1b fitness value of sampleV_sol (new candidate solutions)
135         for i in range(n_searchers):
136             j = self.searcher_region_id[i]
137             if self.scatter_plot == 1:
138                 print("%d %.6f %d " % (self.eval_count, j-0.1*i+0.1*n_regions/2, i), end='')
139             for k in range(n_samples):
140                 n = np.random.randint(2*n_samples)
141                 f = self.evaluate_fitness(self.sampleV_sol[i][j][n])
142                 if f > self.searcher_sol_fitness[i]:
```

```
143                        self.searcher_sol[i] = self.sampleV_sol[i][j][n]
144                        self.searcher_sol_fitness[i] = f
145                    if f > self.sample_sol_fitness[j][k]:
146                        self.sample_sol[j][k] = self.sampleV_sol[i][j][n]
147                        self.sample_sol_fitness[j][k] = f
148            if self.scatter_plot == 1:
149                print()
150        # 3.2.2 fitness value of samples
151        if eval == 0:
152            self.sample_sol_fitness = np.array([np.array([self.evaluate_fitness(self.sample_sol[
     j][k]) for k in range(n_samples)]) for j in range(n_regions)])
153        total_sample_fitness = 0.0 # f(m_j)
154        for j in range(n_regions):
155            rbj = 0.0
156            b = -1
157            for k in range(n_samples):
158                total_sample_fitness += self.sample_sol_fitness[j][k]
159                # update fbj
160                if self.sample_sol_fitness[j][k] > rbj:
161                    b = k
162                    rbj = self.sample_sol_fitness[j][k]
163            if b >= 0:
164                self.sample_sol_best_fitness[j] = rbj
165                self.sample_sol_best[j] = self.sample_sol[j][b]
166        # 3.2.3 M_j
167        self.M_j = self.sample_sol_best_fitness / total_sample_fitness
168        # 3.2.4 T_j
169        self.T_j = self.ta / self.tb
170        # 3.2.5 compute the expected_value
171        self.expected_value = np.array([self.T_j * self.M_j for _ in range(n_searchers)])
172
173    # subfunction: 3.2.1 fitness value
174    def evaluate_fitness(self, sol):
175        self.eval_count += 1
176        return sum(sol)
177
178    # 3.3 select region to which a searcher will be assigned
179    def vision_selection(self, player, eval):
180        n_searchers = self.num_searchers
181        n_regions = self.num_regions
182        n_samples = self.num_samples
183        n_bits_sol = self.num_bits_sol
184        n_players = self.num_players
185        self.tb += 1
186        # find index of the best vij
187        for i in range(n_searchers):
188            j = np.random.randint(n_regions)
189            ev = self.expected_value[i][j]
190            for _ in range(n_players - 1):
191                c = np.random.randint(n_regions)
192                if self.expected_value[i][c] > ev:
193                    j = c
194                    ev = self.expected_value[i][c]
195            # assign searcher i to region j
196            self.searcher_region_id[i] = j
197            # update ta[j] and tb[j]
198            self.ta[j] += 1
199            self.tb[j] = 1
200
201    # 4. marketing survey
202    def marketing_survey(self):
203        for j in range(self.num_regions):
204            if self.tb[j] > 1:
205                self.ta[j] = 1
206        for i in range(self.num_searchers):
```

```
207              if self.searcher_sol_fitness[i] > self.best_so_far:
208                  self.best_so_far = self.searcher_sol_fitness[i]
209                  self.best_sol = self.searcher_sol[i]
210          for j in range(self.num_regions):
211              for k in range(self.num_samples):
212                  if self.sample_sol_fitness[j][k] > self.best_so_far:
213                      self.best_so_far = self.sample_sol_fitness[j][k]
214                      self.best_sol = self.sample_sol[j][k]
215
216      def print_parameters(self, algname):
217          print("# name of the search algorithm: '%s'" % algname)
218          print("# number of runs: %s" % self.num_runs)
219          print("# number of evaluations: %s" % self.num_evals)
220          print("# number of patterns: %s" % self.num_bits_sol)
221          print("# filename of the initial seeds: '%s'" % self.filename_ini)
222          print("# number of searchers: %s" % self.num_searchers)
223          print("# number of regions: %s" % self.num_regions)
224          print("# number of samples: %s" % self.num_samples)
225          print("# number of players: %s" % self.num_players)
226          print("# scatter plot?: %s" % self.scatter_plot)
227          print("# seed: %s" % self.seed)
228
229 def usage():
230     lib.usage("<#searchers> <#regions> <#samples> <#players> <scatter_plot?>")
231
232 if __name__ == "__main__":
233     if ac() == 2 and ss(1) == "-h": usage()
234     elif ac() == 1: search = se()
235     elif ac() == 10: search = se(si(1), si(2), si(3), ss(4), si(5), si(6), si(7), si(8), si(9))
236     else: usage()
237     lib.run(search, lib.print_population)
```

Listing B.12: src/python/1-onemax/ts.py: The implementation of TS in Python.

```
1 #!/usr/bin/env python
2 import numpy as np; import time; import sys; import os
3 from lib import *
4
5 class ts:
6     def __init__(self, num_runs = 3, num_evals = 1000, num_patterns_sol = 10, filename_ini = "",
        num_neighbors = 3, siz_tabulist = 7):
7         self.num_runs = num_runs
8         self.num_evals = num_evals
9         self.num_patterns_sol = num_patterns_sol
10        self.filename_ini = filename_ini
11        self.num_neighbors = num_neighbors
12        self.siz_tabulist = siz_tabulist
13        self.eval_count = 0
14        self.tabulist = []
15        self.seed = int(time.time())
16        np.random.seed(self.seed)
17
18    def run(self):
19        avg_obj_val = 0.0
20        avg_obj_val_eval = np.zeros(self.num_evals)
21        for r in range(self.num_runs):
22            # 0. initialization
23            obj_val, sol, best_obj_val, best_sol = self.init()
24            self.eval_count = 0
25            avg_obj_val_eval[self.eval_count] += obj_val
26            while self.eval_count < self.num_evals-1:
```

```
27                     # 1. transition and evaluation
28                     tmp_sol, tmp_obj_val, n_evals = self.select_neighbor_not_in_tabu(sol)
29                     if n_evals > 0:
30                         # 2. determination
31                         if tmp_obj_val > obj_val:
32                             obj_val, sol = tmp_obj_val, tmp_sol
33                         if obj_val > best_obj_val:
34                             best_obj_val, best_sol = obj_val, sol
35                         avg_obj_val_eval[self.eval_count] += best_obj_val
36                         for i in range(1, n_evals):
37                             avg_obj_val_eval[self.eval_count - i] = avg_obj_val_eval[self.eval_count
      ]
38             avg_obj_val += best_obj_val
39         # 4. output
40         avg_obj_val /= self.num_runs
41         for i in range(self.num_evals):
42             avg_obj_val_eval[i] /= self.num_runs
43             print("%.3f" % avg_obj_val_eval[i])
44         return best_sol
45
46     def init(self):
47         if len(self.filename_ini) == 0:
48             sol = np.array([1 if x >= 0.5 else 0 for x in np.random.rand(self.num_patterns_sol)
      ])
49         else:
50             with open(self.filename_ini, "r") as f:
51                 sol = np.array(f.read().strip().split(), dtype=int)
52         obj_val = self.evaluate(sol)
53         self.tabulist.clear()
54         return obj_val, sol, obj_val, sol
55
56     def transit(self, sol):
57         t = sol.copy()
58         i = np.random.randint(len(sol))
59         t[i] ^= 1
60         return t
61
62     def select_neighbor_not_in_tabu(self, sol):
63         t, f_t, n_evals = None, 0, 0
64         for i in range(self.num_neighbors):
65             v = self.transit(sol)
66             if not self.in_tabu(v):
67                 n_evals += 1
68                 f_v = self.evaluate(v)
69                 if f_v > f_t:
70                     t, f_t = v, f_v
71                 if self.eval_count >= self.num_evals-1:
72                     break
73         if n_evals > 0:
74             self.append_to_tabu_list(t)
75         return t, f_t, n_evals
76
77     def in_tabu(self, sol):
78         for s in self.tabulist:
79             if np.array_equal(s, sol):
80                 return True
81         return False
82
83     def evaluate(self, sol):
84         self.eval_count += 1
85         return sum(sol)
86
87     def append_to_tabu_list(self, sol):
88         self.tabulist.append(sol)
89         if len(self.tabulist) > self.siz_tabulist:
```

```
90              self.tabulist.pop(0)
91
92     def print_parameters(self, algname):
93         print("# name of the search algorithm: '%s'" % algname)
94         print("# number of runs: %s" % self.num_runs)
95         print("# number of evaluations: %s" % self.num_evals)
96         print("# number of patterns: %s" % self.num_patterns_sol)
97         print("# filename of the initial seeds: '%s'" % self.filename_ini)
98         print("# number of neighbors: %s" % self.num_neighbors)
99         print("# size of the tabu list: %s" % self.siz_tabulist)
100        print("# seed: %s" % self.seed)
101
102 def usage():
103     lib.usage("<#neighbors> <tabulist_size>")
104
105 if __name__ == "__main__":
106     if ac() == 2 and ss(1) == "-h": usage()
107     elif ac() == 1: search = ts()
108     elif ac() == 7: search = ts(si(1), si(2), si(3), ss(4), si(5), si(6))
109     else: usage()
110     lib.run(search, lib.print_solution)
```

Listing B.13: src/python/2-deception/gadp.py: The implementation of
GADP in Python.

```
1 #!/usr/bin/env python
2 import numpy as np; import time; import sys; import os
3 from lib import *
4 from ga import ga
5
6 class gadp(ga):
7     def __init__(self, num_runs = 3, num_evals = 1000, num_patterns_sol = 10, filename_ini = "",
        pop_size = 10, crossover_rate = 0.6, mutation_rate = 0.01, num_players = 0):
8         super().__init__(num_runs, num_evals, num_patterns_sol, filename_ini, pop_size,
        crossover_rate, mutation_rate, num_players)
9
10    def evaluate(self, curr_pop):
11        n = np.zeros(self.pop_size)
12        for i in range(self.pop_size):
13            sol = curr_pop[i]
14            n[i] = abs(int(sol.dot(2**np.arange(sol.size)[::-1])) - (1 << (sol.size - 2)))
15        return n
16
17 def usage():
18     lib.usage("<pop_size> <cr> <mr> <#players>")
19
20 if __name__ == "__main__":
21     if ac() == 2 and ss(1) == "-h": usage()
22     elif ac() == 1: search = gadp()
23     elif ac() == 9: search = gadp(si(1), si(2), si(3), ss(4), si(5), sf(6), sf(7), si(8))
24     else: usage()
25     lib.run(search, lib.print_population)
```

Listing B.14: src/python/2-deception/ga.py: The implementation of GA
in Python.

```
1 Ditto to B.3.
```

Listing B.15: src/python/2-deception/gardp.py: The implementation of GARDP in Python.

```
1  #!/usr/bin/env python
2  import numpy as np; import time; import sys; import os
3  from lib import *
4  from gar import gar
5
6  class gardp(gar):
7      def __init__(self, num_runs = 3, num_evals = 1000, num_patterns_sol = 10, filename_ini = "",
         pop_size = 10, crossover_rate = 0.6, mutation_rate = 0.01, num_players = 0):
8          super().__init__(num_runs, num_evals, num_patterns_sol, filename_ini, pop_size,
         crossover_rate, mutation_rate, num_players)
9
10     def evaluate(self, curr_pop):
11         n = np.zeros(self.pop_size)
12         for i in range(self.pop_size):
13             sol = curr_pop[i]
14             n[i] = abs(int(sol.dot(2**np.arange(sol.size)[::-1])) - (1 << (sol.size - 2)))
15         return n
16
17 def usage():
18     lib.usage("<pop_size> <cr> <mr> <#players>")
19
20 if __name__ == "__main__":
21     if ac() == 2 and ss(1) == "-h": usage()
22     elif ac() == 1: search = gardp()
23     elif ac() == 9: search = gardp(si(1), si(2), si(3), ss(4), si(5), sf(6), sf(7), si(8))
24     else: usage()
25     lib.run(search, lib.print_population)
```

Listing B.16: src/python/2-deception/gar.py: The implementation of GAR in Python.

```
1  Ditto to B.4.
```

Listing B.17: src/python/2-deception/hcdp.py: The implementation of HCDP in Python.

```
1  #!/usr/bin/env python
2  import numpy as np; import time; import sys; import os
3  from lib import *
4  from hc import hc
5
6  class hcdp(hc):
7      def __init__(self, num_runs = 3, num_evals = 1000, num_patterns_sol = 10, filename_ini = "")
         :
8          super().__init__(num_runs, num_evals, num_patterns_sol, filename_ini)
9
10     def evaluate(self, s):
11         return abs(s - (1 << (self.num_patterns_sol - 2)))
12
13 def usage():
14     lib.usage("")
15
16 if __name__ == "__main__":
17     if ac() == 2 and ss(1) == "-h": usage()
18     elif ac() == 1: search = hcdp()
```

```
19    elif ac() == 5: search = hcdp(si(1), si(2), si(3), ss(4))
20    else: usage()
21    lib.run(search, lib.print_solution)
```

Listing B.18: src/python/2-deception/hcdp_rm.py: The implementation of HCDP_RM in Python.

```
1  #!/usr/bin/env python
2  import numpy as np; import time; import sys; import os
3  from lib import *
4  from hcdp_r import hcdp_r
5
6  class hcdp_rm(hcdp_r):
7      def __init__(self, num_runs = 3, num_evals = 1000, num_patterns_sol = 10, filename_ini = ""):
          :
8          super().__init__(num_runs, num_evals, num_patterns_sol, filename_ini)
9
10     def determine(self, v, fv, s, fs):
11         return (v, fv) if fv >= fs else (s, fs)
12
13 def usage():
14     lib.usage("")
15
16 if __name__ == "__main__":
17     if ac() == 2 and ss(1) == "-h": usage()
18     elif ac() == 1: search = hcdp_rm()
19     elif ac() == 5: search = hcdp_rm(si(1), si(2), si(3), ss(4))
20     else: usage()
21     lib.run(search, lib.print_solution)
```

Listing B.19: src/python/2-deception/hcdp_r.py: The implementation of HCDP_R in Python.

```
1  #!/usr/bin/env python
2  import numpy as np; import time; import sys; import os
3  from lib import *
4  from hcdp import hcdp
5
6  class hcdp_r(hcdp):
7      def __init__(self, num_runs = 3, num_evals = 1000, num_patterns_sol = 10, filename_ini = ""):
          :
8          super().__init__(num_runs, num_evals, num_patterns_sol, filename_ini)
9
10     def transit(self, s):
11         i = np.random.randint(self.num_patterns_sol)
12         s ^= 1 << self.num_patterns_sol - 1 - i
13         return s
14
15 def usage():
16     lib.usage("")
17
18 if __name__ == "__main__":
19     if ac() == 2 and ss(1) == "-h": usage()
20     elif ac() == 1: search = hcdp_r()
21     elif ac() == 5: search = hcdp_r(si(1), si(2), si(3), ss(4))
22     else: usage()
23     lib.run(search, lib.print_solution)
```

510

Implementation in Python

Listing B.20: src/python/2-deception/hc.py: The implementation of HC in Python.

```
1 Ditto to B.7.
```

Listing B.21: src/python/2-deception/sadp.py: The implementation of SADP in Python.

```
1 #!/usr/bin/env python
2 import numpy as np; import time; import sys; import os
3 from lib import *
4 from sa import sa
5
6 class sadp(sa):
7     def __init__(self, num_runs = 3, num_evals = 20000, num_patterns_sol = 4, filename_ini = "",
        min_temp = 0.00001, max_temp = 1.0):
8         super().__init__(num_runs, num_evals, num_patterns_sol, filename_ini, min_temp, max_temp
        )
9
10    def evaluate(self, sol):
11        return abs(int(sol.dot(2**np.arange(sol.size)[::-1])) - (1 << (sol.size - 2)))
12
13 def usage():
14     lib.usage("<min_temp> <max_temp>")
15
16 if __name__ == "__main__":
17     if ac() == 2 and ss(1) == "-h": usage()
18     elif ac() == 1: search = sadp()
19     elif ac() == 7: search = sadp(si(1), si(2), si(3), ss(4), sf(5), sf(6))
20     else: usage()
21     lib.run(search, lib.print_solution)
```

Listing B.22: src/python/2-deception/sa.py: The implementation of SA in Python.

```
1 Ditto to B.9.
```

Listing B.23: src/python/2-deception/tsdp.py: The implementation of TSDP in Python.

```
1 #!/usr/bin/env python
2 import numpy as np; import time; import sys; import os
3 from lib import *
4 from ts import ts
5
6 class tsdp(ts):
7     def __init__(self, num_runs = 3, num_evals = 1000, num_patterns_sol = 10, filename_ini = "",
        num_neighbors = 3, siz_tabulist = 7):
8         super().__init__(num_runs, num_evals, num_patterns_sol, filename_ini, num_neighbors,
        siz_tabulist)
9
10    def evaluate(self, sol):
11        self.eval_count += 1
12        return abs(int(sol.dot(2**np.arange(sol.size)[::-1])) - (1 << (sol.size - 2)))
13
```

```
14      def select_neighbor_not_in_tabu(self, sol):
15          t, f_t, n_evals = super().select_neighbor_not_in_tabu(sol)
16          if n_evals == 0:
17              while (True):
18                  t = np.random.randint(2, size = len(sol))
19                  if not self.in_tabu(t):
20                      break
21              self.append_to_tabu_list(t)
22              f_t = self.evaluate(t)
23              n_evals = 1
24          return t, f_t, n_evals
25
26 def usage():
27     lib.usage("<#neighbors> <tabulist_size>")
28
29 if __name__ == "__main__":
30     if ac() == 2 and ss(1) == "-h": usage()
31     elif ac() == 1: search = tsdp()
32     elif ac() == 7: search = tsdp(si(1), si(2), si(3), ss(4), si(5), si(6))
33     else: usage()
34     lib.run(search, lib.print_solution)
```

Listing B.24: src/python/2-deception/ts.py: The implementation of TS in Python.

```
1 Ditto to B.12.
```

Listing B.25: src/python/3-tsp/aco.py: The implementation of ACO in Python.

```
1 #!/usr/bin/env python
2 import numpy as np; import time; import sys; import os; import math
3 from lib import *
4
5 class aco:
6     def __init__(self, num_runs = 3, num_evals = 25000, num_patterns_sol = 51, filename_ini = ""
    , filename_ins = "eil51.tsp", pop_size = 20, alpha = 0.1, beta = 2.0, rho = 0.1, q0 = 0.9):
7         self.num_runs = num_runs
8         self.num_evals = num_evals
9         self.num_patterns_sol = num_patterns_sol
10        self.filename_ini = filename_ini
11        self.filename_ins = filename_ins
12        self.pop_size = pop_size
13        self.alpha = alpha
14        self.beta = beta
15        self.rho = rho
16        self.q0 = q0
17        self.seed = int(time.time())
18        np.random.seed(self.seed)
19
20    def run(self):
21        avg_obj_val = 0.0
22        avg_obj_val_eval = np.zeros(self.num_evals)
23        for r in range(self.num_runs):
24            eval_count = 0
25            best_so_far = sys.float_info.max
26            # 0. initialization
27            curr_pop, curr_obj_vals = self.init()
```

```
28              while eval_count < self.num_evals:
29                  # 1. transition
30                  self.construct_solutions(curr_pop, self.dist, self.ph, self.beta, self.q0)
31                  # 2. evaluation
32                  curr_obj_vals = self.evaluate(curr_pop)
33                  for i in range(self.pop_size):
34                      if best_so_far > curr_obj_vals[i]:
35                          best_so_far = curr_obj_vals[i]
36                      if eval_count < self.num_evals:
37                          avg_obj_val_eval[eval_count] += best_so_far
38                          eval_count += 1
39                  # 3. determination
40                  self.update_best_sol(curr_pop, curr_obj_vals)
41                  self.update_global_ph(self.ph, self.best_sol, self.best_obj_val)
42              avg_obj_val += self.best_obj_val
43          # 4. output
44          avg_obj_val /= self.num_runs
45          for i in range(self.num_evals):
46              avg_obj_val_eval[i] /= self.num_runs
47              print("%.3f" % avg_obj_val_eval[i])
48          self.show_optimum()
49          # return curr_pop
50          return self.best_sol
51
52      def init(self):
53          # 1. initialization
54          curr_obj_vals = np.zeros(self.pop_size)
55          self.best_sol = np.zeros(self.num_patterns_sol, dtype=int)
56          self.best_obj_val = sys.float_info.max
57          # 2. input the TSP benchmark
58          with open(self.filename_ins, "r") as f:
59              self.ins_tsp = np.array(f.read().strip().split(), dtype=float).reshape(self.
    num_patterns_sol, 2)
60          # 3. input the optimal solution
61          with open(self.filename_ins + ".opt", "r") as f:
62              self.opt_sol = np.array(f.read().strip().split(), dtype=int) - 1
63          # 4. initial solutions
64          if len(self.filename_ini) > 0:
65              with open(self.filename_ini, "r") as f:
66                  curr_pop = np.array(f.read().strip().split(), dtype=int).reshape(self.pop_size,
    self.num_patterns_sol)
67          else:
68              curr_pop = np.array([np.arange(self.num_patterns_sol) for i in range(self.pop_size)
    ]).reshape(self.pop_size, self.num_patterns_sol)
69              for i in range(self.pop_size):
70                  np.random.shuffle(curr_pop[i])
71          # 5. construct the distance matrix
72          self.ph = np.zeros((self.num_patterns_sol, self.num_patterns_sol))
73          self.dist = np.zeros((self.num_patterns_sol, self.num_patterns_sol))
74          for i in range(self.num_patterns_sol):
75              for j in range(self.num_patterns_sol):
76                  self.dist[i][j] = math.dist(self.ins_tsp[i], self.ins_tsp[j])
77          # 6. create the pheromone table, by first constructing a solution using the nearest
    neighbor method
78          n = np.random.randint(self.pop_size)
79          cities = list(curr_pop[n])
80          self.best_sol[0] = cities[0]
81          cities.pop(0)
82          Lnn = 0.0
83          for i in range(1, self.num_patterns_sol):
84              r = self.best_sol[i-1]
85              min_s = cities[0]      # simply choose a city
86              min_dist = sys.float_info.max
87              for s in cities:
88                  if self.dist[r][s] < min_dist:
```

```
89                          min_dist = self.dist[r][s]
90                          min_s = s
91                  s = min_s
92                  self.best_sol[i] = s
93                  cities.remove(min_s)
94                  Lnn += self.dist[r][s]
95              r = self.best_sol[self.num_patterns_sol-1]
96              s = self.best_sol[0]
97              Lnn += self.dist[r][s]
98              self.tau0 = 1 / (self.num_patterns_sol * Lnn)
99              for r in range(self.num_patterns_sol):
100                 for s in range(self.num_patterns_sol):
101                     self.ph[r][s] = 0.0 if r == s else self.tau0
102             return curr_pop, curr_obj_vals
103
104         def construct_solutions(self, curr_pop, dist, ph, beta, q0):
105             # 1. tau eta
106             tau_eta = np.zeros((self.num_patterns_sol, self.num_patterns_sol))
107             for r in range(self.num_patterns_sol):
108                 for s in range(self.num_patterns_sol):
109                     tau_eta[r][s] = 0.0 if r == s else ph[r][s] * math.pow(1.0 / dist[r][s], beta)
110             # 2. construct the solution: first, the first city of the tour
111             tmp_pop = curr_pop.tolist()
112             for j in range(self.pop_size):
113                 asol = tmp_pop[j]
114                 n = np.random.randint(self.num_patterns_sol)
115                 curr_pop[j][0] = asol[n]
116                 asol.remove(asol[n])
117             # 3. then, the remaining cities of the tour
118             for i in range(1, self.num_patterns_sol):
119                 # for each step
120                 for j in range(self.pop_size):
121                     # for each ant
122                     asol = tmp_pop[j]
123                     r = curr_pop[j][i-1]
124                     s = asol[0]
125                     x = 0
126                     q = np.random.rand()
127                     if q <= q0:
128                         # 3.1. exploitation
129                         max_tau_eta = tau_eta[r][s]
130                         for k in range(len(asol)):
131                             if tau_eta[r][asol[k]] > max_tau_eta:
132                                 s = asol[k]
133                                 x = k
134                                 max_tau_eta = tau_eta[r][s]
135                     else:
136                         # 3.2. biased exploration
137                         total = 0.0
138                         for k in range(len(asol)):
139                             total += tau_eta[r][asol[k]]
140                         # 3.3. choose the next city based on the probability
141                         f = total * np.random.rand()
142                         for k in range(len(asol)):
143                             f -= tau_eta[r][asol[k]]
144                             if f <= 0:
145                                 s = asol[k]
146                                 x = k
147                                 break
148                     curr_pop[j][i] = s
149                     asol.remove(asol[x])
150                     # 4.1. update local pheromones, 0 to n-1
151                     self.update_local_ph(ph, r, s)
152             # 4.2. update local pheromones, n-1 to 0
153             for k in range(self.pop_size):
```

```
154              # for each ant
155              r = curr_pop[k][self.num_patterns_sol-1]
156              s = curr_pop[k][0]
157              self.update_local_ph(ph, r, s)
158
159     def evaluate(self, curr_pop):
160         tour_dist = np.zeros(self.pop_size)
161         for k in range(self.pop_size):
162             for i in range(self.num_patterns_sol):
163                 r = curr_pop[k][i]
164                 s = curr_pop[k][(i+1) % self.num_patterns_sol]
165                 tour_dist[k] += self.dist[r][s]
166         return tour_dist
167
168     def update_best_sol(self, curr_pop, curr_obj_vals):
169         for i in range(self.pop_size):
170             if curr_obj_vals[i] < self.best_obj_val:
171                 self.best_obj_val = curr_obj_vals[i].copy()
172                 self.best_sol = curr_pop[i].copy()
173
174     def update_global_ph(self, ph, best_sol, best_obj_val):
175         for i in range(self.num_patterns_sol):
176             r = best_sol[i]
177             s = best_sol[(i+1) % self.num_patterns_sol]
178             ph[s][r] = ph[r][s] = (1 - self.alpha) * ph[r][s] + self.alpha * (1 / self.
        best_obj_val)
179
180     def update_local_ph(self, ph, r, s):
181         ph[s][r] = ph[r][s] = (1 - self.rho) * ph[r][s] + self.rho * self.tau0
182
183     def show_optimum(self):
184         opt_dist = 0.0
185         for i in range(self.num_patterns_sol):
186             r = self.opt_sol[i]
187             s = self.opt_sol[(i+1) % self.num_patterns_sol]
188             opt_dist += self.dist[r][s]
189         lib.print_solution(self.best_sol, prefix = "# Current route: ")
190         lib.printf("# Optimum distance: %.3f", opt_dist)
191         lib.print_solution(self.opt_sol, prefix = "# Optimum route: ")
192
193     def print_parameters(self, algname):
194         print("# name of the search algorithm: '%s'" % algname)
195         print("# number of runs: %s" % self.num_runs)
196         print("# number of evaluations: %s" % self.num_evals)
197         print("# number of patterns (subsolutions) each solution: %s" % self.num_patterns_sol)
198         print("# filename of the initial seeds: '%s'" % self.filename_ini)
199         print("# filename of the benchmark: '%s'" % self.filename_ins)
200         print("# population size (i.e., number of ants): %s" % self.pop_size)
201         print("# alpha: %s" % self.alpha)
202         print("# beta: %s" % self.beta)
203         print("# rho: %s" % self.rho)
204         print("# Q: %s" % self.q0)
205         print("# seed: %s" % self.seed)
206
207 def usage():
208     lib.usage("<filename_ins> <pop_size> <alpha> <beta> <rho> <Q>")
209
210 if __name__ == "__main__":
211     if ac() == 2 and ss(1) == "-h": usage()
212     elif ac() == 1: search = aco()
213     elif ac() == 11: search = aco(si(1), si(2), si(3), ss(4), ss(5), si(6), sf(7), sf(8), sf(9),
        sf(10))
214     else: usage()
215     lib.run(search, None)
```

Listing B.26: src/python/3-tsp/gacx.py: The implementation of GACX in Python.

```
1 #!/usr/bin/env python
2 import numpy as np; import time; import sys; import os; import math
3 from lib import *
4 from ga import ga
5
6 class gacx(ga):
7     def __init__(self, num_runs = 3, num_evals = 120000, num_patterns_sol = 1002, filename_ini =
        "", filename_ins = "pr1002.tsp", pop_size = 120, crossover_rate = 0.4, mutation_rate = 0.1,
        num_players = 3):
8         super().__init__(num_runs, num_evals, num_patterns_sol, filename_ini, filename_ins,
        pop_size, crossover_rate, mutation_rate, num_players)
9
10    def crossover(self, curr_pop, cr):
11        tmp_pop = curr_pop.copy()
12        mid = self.pop_size // 2
13        ssz = self.num_patterns_sol
14        for i in range(mid):
15            f = np.random.rand()
16            if f <= cr:
17                mask = np.zeros(ssz, dtype=bool)
18                c1 = i
19                c2 = i+mid
20                c = tmp_pop[c1][0]
21                n = tmp_pop[c2][0]
22                mask[0] = True
23                try:
24                    while c != n:
25                        for j in range(ssz):
26                            if n == tmp_pop[c1][j]:
27                                if mask[j]:
28                                    raise StopIteration
29                                c = n
30                                n = tmp_pop[c2][j]
31                                mask[j] = True
32                                break
33                except StopIteration:
34                    for j in range(ssz):
35                        if not mask[j]:
36                            tmp_pop[c1][j], tmp_pop[c2][j] = tmp_pop[c2][j], tmp_pop[c1][j]
37        return tmp_pop
38
39 def usage():
40     lib.usage("<filename_ins> <pop_size> <cr> <mr> <#players>")
41
42 if __name__ == "__main__":
43     if ac() == 2 and ss(1) == "-h": usage()
44     elif ac() == 1: search = gacx()
45     elif ac() == 10: search = gacx(si(1), si(2), si(3), ss(4), ss(5), si(6), sf(7), sf(8), si(9)
        )
46     else: usage()
47     lib.run(search, None)
```

Listing B.27: src/python/3-tsp/gals.py: The implementation of GALS in Python.

```
1 #!/usr/bin/env python
2 import numpy as np; import time; import sys; import os; import math
3 from lib import *
```

```python
4
5  class gals:
6      def __init__(self, num_runs = 3, num_evals = 120000, num_patterns_sol = 51, filename_ini = "
       ", filename_ins = "eil51.tsp", pop_size = 100, crossover_rate = 0.4, mutation_rate = 0.1,
       num_players = 3, num_ls = 100, ls_flag = 0):
7          self.num_runs = num_runs
8          self.num_evals = num_evals
9          self.num_patterns_sol = num_patterns_sol
10         self.filename_ini = filename_ini
11         self.filename_ins = filename_ins
12         self.pop_size = pop_size
13         self.crossover_rate = crossover_rate
14         self.mutation_rate = mutation_rate
15         self.num_players = num_players
16         self.num_ls = num_ls
17         self.ls_flag = ls_flag
18         self.seed = int(time.time())
19         np.random.seed(self.seed)
20
21     def run(self):
22         self.avg_obj_val = 0.0
23         self.avg_obj_val_eval = np.zeros(self.num_evals)
24         for r in range(self.num_runs):
25             self.eval_count = 0
26             self.best_so_far = sys.float_info.max
27             curr_pop, curr_obj_vals = self.init()
28             while self.eval_count < self.num_evals:
29                 # 1. evaluation
30                 curr_obj_vals = self.evaluate(curr_pop, self.dist)
31                 self.update_best_sol(curr_pop, curr_obj_vals)
32                 if 0:
33                     for i in range(self.pop_size):
34                         if self.best_so_far > curr_obj_vals[i]:
35                             self.best_so_far = curr_obj_vals[i]
36                         if self.eval_count < self.num_evals:
37                             self.avg_obj_val_eval[self.eval_count] += self.best_so_far
38                             self.eval_count += 1
39                 # 2. determination
40                 curr_pop = self.tournament_select(curr_pop, curr_obj_vals, self.num_players)
41                 # 3. transition
42                 curr_pop = self.crossover_ox(curr_pop, self.crossover_rate)
43                 curr_pop = self.mutate(curr_pop, self.mutation_rate)
44                 if self.ls_flag == 0:
45                     curr_pop = self.two_opt(curr_pop, self.dist, self.num_ls)
46                 else:
47                     curr_obj_vals = self.evaluate(curr_pop, self.dist)
48                     n = 0
49                     v = curr_obj_vals[0]
50                     for i in range(1, self.pop_size):
51                         if curr_obj_vals[i] < v:
52                             n = i
53                             v = curr_obj_vals[i]
54                     curr_pop[n] = self.two_opt_sol(curr_pop[n], self.dist, self.num_ls)
55                 self.update_best_sol(curr_pop, curr_obj_vals)
56             self.avg_obj_val += self.best_so_far
57         # 4. output
58         self.avg_obj_val /= self.num_runs
59         self.avg_obj_val_eval /= self.num_runs
60         for i in range(self.num_evals):
61             print("%.3f" % self.avg_obj_val_eval[i])
62         self.show_optimum()
63         return curr_pop
64
65     def init(self):
66         # 1. initialization
```

```
67          self.best_sol = np.zeros(self.num_patterns_sol, dtype=int)
68          self.best_obj_val = sys.float_info.max
69          # 2. input the TSP benchmark
70          with open(self.filename_ins, "r") as f:
71              self.ins_tsp = np.array(f.read().strip().split(), dtype=float).reshape(self.
    num_patterns_sol, 2)
72          # 3. input the optimal solution
73          with open(self.filename_ins + ".opt", "r") as f:
74              self.opt_sol = np.array(f.read().strip().split(), dtype=int) - 1
75          # 4. initial solutions
76          if len(self.filename_ini) > 0:
77              with open(self.filename_ini, "r") as f:
78                  curr_pop = np.array(f.read().strip().split(), dtype=int).reshape(self.pop_size,
    self.num_patterns_sol)
79          else:
80              curr_pop = np.array([np.arange(self.num_patterns_sol) for i in range(self.pop_size)
    ])
81              for i in range(self.pop_size):
82                  np.random.shuffle(curr_pop[i])
83          # 5. construct the distance matrix
84          self.dist = np.zeros((self.num_patterns_sol, self.num_patterns_sol))
85          for i in range(self.num_patterns_sol):
86              for j in range(self.num_patterns_sol):
87                  self.dist[i][j] = math.dist(self.ins_tsp[i], self.ins_tsp[j])
88          return curr_pop, np.zeros(self.pop_size)
89
90      def evaluate(self, curr_pop, dist):
91          tour_dist = np.zeros(self.pop_size)
92          for p in range(self.pop_size):
93              tour_dist[p] = self.evaluate_sol(curr_pop[p], dist)
94          return tour_dist
95
96      def evaluate_sol(self, curr_sol, dist):
97          tour_dist = 0.0
98          for i in range(self.num_patterns_sol):
99              r = curr_sol[i]
100             s = curr_sol[(i+1) % self.num_patterns_sol]
101             tour_dist += self.dist[r][s]
102         if self.best_so_far > tour_dist:
103             self.best_so_far = tour_dist
104         if self.eval_count < self.num_evals:
105             self.avg_obj_val_eval[self.eval_count] += self.best_so_far
106             self.eval_count += 1
107         return tour_dist
108
109     def update_best_sol(self, curr_pop, curr_obj_vals):
110         for i in range(self.pop_size):
111             if curr_obj_vals[i] < self.best_obj_val:
112                 self.best_obj_val = curr_obj_vals[i]
113                 self.best_sol = curr_pop[i]
114
115     def tournament_select(self, curr_pop, curr_obj_vals, num_players):
116         tmp_pop = np.zeros((self.pop_size, self.num_patterns_sol), dtype=int)
117         for i in range(self.pop_size):
118             k = np.random.randint(self.pop_size)
119             f = curr_obj_vals[k]
120             for j in range(1, num_players):
121                 n = np.random.randint(self.pop_size)
122                 if curr_obj_vals[n] < f:
123                     k = n
124                     f = curr_obj_vals[k]
125             tmp_pop[i] = curr_pop[k]
126         return tmp_pop
127
128     def crossover_ox(self, curr_pop, cr):
```

```
129        tmp_pop = curr_pop.copy()
130        mid = self.pop_size // 2
131        ssz = self.num_patterns_sol
132        for i in range(mid):
133            f = np.random.rand()
134            if f <= cr:
135                # 1. create the mapping sections
136                xp1 = np.random.randint(ssz + 1)
137                xp2 = np.random.randint(ssz + 1)
138                if xp1 > xp2:
139                    xp1, xp2 = xp2, xp1
140                # 2. indices to the two parents and offspring
141                p = [ i, i+mid ]
142                # 3. the main process of ox
143                for k in range(2):
144                    c1 = p[k]
145                    c2 = p[1-k]
146                    # 4. mask the genes between xp1 and xp2
147                    s1 = curr_pop[c1]
148                    s2 = curr_pop[c2]
149                    msk1 = np.zeros(ssz, dtype=bool)
150                    for j in range(xp1, xp2):
151                        msk1[s1[j]] = True
152                    msk2 = np.zeros(ssz, dtype=bool)
153                    for j in range(0, ssz):
154                        msk2[j] = msk1[s2[j]]
155                    # 5. replace the genes that are not masked
156                    j = xp2 % ssz
157                    for z in range(ssz):
158                        if not msk2[z]:
159                            tmp_pop[c1][j] = s2[z]
160                            j = (j+1) % ssz
161        return tmp_pop
162
163    def mutate(self, curr_pop, mr):
164        tmp_pop = curr_pop
165        ssz = self.num_patterns_sol
166        for i in range(self.pop_size):
167            f = np.random.rand()
168            if f <= mr:
169                m1 = np.random.randint(ssz)      # mutation point
170                m2 = np.random.randint(ssz)      # mutation point
171                tmp_pop[i][m1], tmp_pop[i][m2] = tmp_pop[i][m2], tmp_pop[i][m1]
172        return tmp_pop
173
174    def two_opt(self, curr_pop, dist, num_ls):
175        tmp_pop = np.zeros((self.pop_size, self.num_patterns_sol), dtype=int)
176        for i in range(self.pop_size):
177            tmp_pop[i] = self.two_opt_sol(curr_pop[i], dist, num_ls)
178        return tmp_pop
179
180    def two_opt_sol(self, curr_sol, dist, num_ls):
181        r_start = np.random.randint(self.pop_size)
182        ls_count = 0
183        tmp_sol = curr_sol
184        tmp_sol_dist = self.evaluate_sol(tmp_sol, dist)
185        for i in range(r_start, self.num_patterns_sol - 1):
186            for k in range(i+1, self.num_patterns_sol):
187                tmp_sol_2opt = tmp_sol.copy()
188                e = k
189                for b in range(i, k+1):
190                    tmp_sol_2opt[b] = tmp_sol[e]
191                    e -= 1
192                tmp_sol_2opt_dist = self.evaluate_sol(tmp_sol_2opt, dist)
193                if tmp_sol_dist > tmp_sol_2opt_dist:
```

```
194                        tmp_sol = tmp_sol_2opt
195                        tmp_sol_dist = tmp_sol_2opt_dist
196                    ls_count += 1
197                    if ls_count == num_ls:
198                        return tmp_sol
199         return tmp_sol
200
201     def show_optimum(self):
202         opt_dist = 0.0
203         for i in range(self.num_patterns_sol):
204             r = self.opt_sol[i]
205             s = self.opt_sol[(i+1) % self.num_patterns_sol]
206             opt_dist += self.dist[r][s]
207         lib.print_solution(self.best_sol, prefix = "# Current route: ")
208         lib.printf("# Optimum distance: %.3f", opt_dist)
209         lib.print_solution(self.opt_sol, prefix = "# Optimum route: ")
210
211     def print_parameters(self, algname):
212         print("# name of the search algorithm: '%s'" % algname)
213         print("# number of runs: %s" % self.num_runs)
214         print("# number of evaluations: %s" % self.num_evals)
215         print("# number of patterns (subsolutions) each solution: %s" % self.num_patterns_sol)
216         print("# filename of the initial seeds: '%s'" % self.filename_ini)
217         print("# filename of the benchmark: '%s'" % self.filename_ins)
218         print("# population size: %s" % self.pop_size)
219         print("# crossover rate: %s" % self.crossover_rate)
220         print("# mutation rate: %s" % self.mutation_rate)
221         print("# number of players: %s" % self.num_players)
222         print("# number of local searches per evaluation: %s" % self.num_ls)
223         print("# fine-tune? 0:population; 1:solution: %s" % self.ls_flag)
224         print("# seed: %s" % self.seed)
225
226 def usage():
227     lib.usage("<filename_ins> <pop_size> <cr> <mr> <#players> <#ls_per_eval> <ls sol?>")
228
229 if __name__ == "__main__":
230     if ac() == 2 and ss(1) == "-h": usage()
231     elif ac() == 1: search = gals()
232     elif ac() == 12: search = gals(si(1), si(2), si(3), ss(4), ss(5), si(6), sf(7), sf(8), si(9)
        , si(10), si(11))
233     else: usage()
234     lib.run(search, None)
```

Listing B.28: src/python/3-tsp/gaox.py: The implementation of GAOX in Python.

```
1  #!/usr/bin/env python
2  import numpy as np; import time; import sys; import os; import math
3  from lib import *
4  from ga import ga
5
6  class gaox(ga):
7      def __init__(self, num_runs = 3, num_evals = 120000, num_patterns_sol = 1002, filename_ini =
        "", filename_ins = "pr1002.tsp", pop_size = 120, crossover_rate = 0.4, mutation_rate = 0.1,
        num_players = 3):
8          super().__init__(num_runs, num_evals, num_patterns_sol, filename_ini, filename_ins,
        pop_size, crossover_rate, mutation_rate, num_players)
9
10     def crossover(self, curr_pop, cr):
11         # originally crossover_ox, renamed for override
12         tmp_pop = curr_pop.copy()
13         mid = self.pop_size // 2
```

```
14        ssz = self.num_patterns_sol
15        for i in range(mid):
16            f = np.random.rand()
17            if f <= cr:
18                # 1. create the mapping sections
19                xp1 = np.random.randint(ssz + 1)
20                xp2 = np.random.randint(ssz + 1)
21                if xp1 > xp2:
22                    xp1, xp2 = xp2, xp1
23                # 2. indices to the two parents and offspring
24                p = [ i, i+mid ]
25                # 3. the main process of ox
26                for k in range(2):
27                    c1 = p[k]
28                    c2 = p[1-k]
29                    # 4. mask the genes between xp1 and xp2
30                    s1 = curr_pop[c1]
31                    s2 = curr_pop[c2]
32                    msk1 = np.zeros(ssz, dtype=bool)
33                    for j in range(xp1, xp2):
34                        msk1[s1[j]] = True
35                    msk2 = np.zeros(ssz, dtype=bool)
36                    for j in range(0, ssz):
37                        msk2[j] = msk1[s2[j]]
38                    # 5. replace the genes that are not masked
39                    j = xp2 % ssz
40                    for z in range(ssz):
41                        if not msk2[z]:
42                            tmp_pop[c1][j] = s2[z]
43                            j = (j+1) % ssz
44        return tmp_pop
45
46 def usage():
47     lib.usage("<filename_ins> <pop_size> <cr> <mr> <#players>")
48
49 if __name__ == "__main__":
50     if ac() == 2 and ss(1) == "-h": usage()
51     elif ac() == 1: search = gaox()
52     elif ac() == 10: search = gaox(si(1), si(2), si(3), ss(4), ss(5), si(6), sf(7), sf(8), si(9)
        )
53     else: usage()
54     lib.run(search, None)
```

Listing B.29: src/python/3-tsp/gapmx.py: The implementation of GAPMX in Python.

```
1 #!/usr/bin/env python
2 import numpy as np; import time; import sys; import os; import math
3 from lib import *
4 from ga import ga
5
6 class gapmx(ga):
7     def __init__(self, num_runs = 3, num_evals = 120000, num_patterns_sol = 1002, filename_ini =
        "", filename_ins = "pr1002.tsp", pop_size = 120, crossover_rate = 0.4, mutation_rate = 0.1,
        num_players = 3):
8         super().__init__(num_runs, num_evals, num_patterns_sol, filename_ini, filename_ins,
        pop_size, crossover_rate, mutation_rate, num_players)
9
10    def crossover(self, curr_pop, cr):
11        # originally crossover_ox, renamed for override
12        tmp_pop = curr_pop.copy()
13        mid = self.pop_size // 2
```

```
14            ssz = self.num_patterns_sol
15            for i in range(mid):
16                # 1. select the two parents and offspring
17                f = np.random.rand()
18                if f <= cr:
19                    # 1. select the two parents and offspring
20                    p = [ i, i+mid ]
21                    # 2. select the mapping sections
22                    xp1 = np.random.randint(ssz + 1)
23                    xp2 = np.random.randint(ssz + 1)
24                    if xp1 > xp2:
25                        xp1, xp2 = xp2, xp1
26                    # 3. swap the mapping sections
27                    for j in range(xp1, xp2):
28                        tmp_pop[p[0]][j], tmp_pop[p[1]][j] = tmp_pop[p[1]][j], tmp_pop[p[0]][j]
29                    # 4. fix the duplicates
30                    for k in range(2):
31                        z = p[k]
32                        for j in range(ssz):
33                            if j < xp1 or j >= xp2:
34                                c = curr_pop[z][j]
35                                m = xp1
36                                while m < xp2:
37                                    if c == tmp_pop[z][m]:
38                                        c = curr_pop[z][m]
39                                        m = xp1      # restart the while loop
40                                    else:
41                                        m += 1       # move on to the next
42                                if c != curr_pop[z][j]:
43                                    tmp_pop[z][j] = c
44            return tmp_pop
45
46 def usage():
47     lib.usage("<filename_ins> <pop_size> <cr> <mr> <#players>")
48
49 if __name__ == "__main__":
50     if ac() == 2 and ss(1) == "-h": usage()
51     elif ac() == 1: search = gapmx()
52     elif ac() == 10: search = gapmx(si(1), si(2), si(3), ss(4), ss(5), si(6), sf(7), sf(8), si
       (9))
53     else: usage()
54     lib.run(search, None)
```

Listing B.30: src/python/3-tsp/ga.py: The implementation of GA in Python.

```
1 #!/usr/bin/env python
2 import numpy as np; import time; import sys; import os; import math
3 from lib import *
4
5 class ga:
6     def __init__(self, num_runs = 3, num_evals = 120000, num_patterns_sol = 1002, filename_ini =
       "", filename_ins = "pr1002.tsp", pop_size = 120, crossover_rate = 0.4, mutation_rate = 0.1,
       num_players = 3):
7         self.num_runs = num_runs
8         self.num_evals = num_evals
9         self.num_patterns_sol = num_patterns_sol
10        self.filename_ini = filename_ini
11        self.filename_ins = filename_ins
12        self.pop_size = pop_size
13        self.crossover_rate = crossover_rate
14        self.mutation_rate = mutation_rate
```

```
15          self.num_players = num_players
16          self.seed = int(time.time())
17          np.random.seed(self.seed)
18
19      def run(self):
20          self.avg_obj_val = 0.0
21          self.avg_obj_val_eval = np.zeros(self.num_evals)
22          for r in range(self.num_runs):
23              eval_count = 0
24              best_so_far = sys.float_info.max
25              curr_pop, curr_obj_vals = self.init()
26              while eval_count < self.num_evals:
27                  # 1. evaluation
28                  curr_obj_vals = self.evaluate(curr_pop, self.dist)
29                  self.update_best_sol(curr_pop, curr_obj_vals)
30                  for i in range(self.pop_size):
31                      if best_so_far > curr_obj_vals[i]:
32                          best_so_far = curr_obj_vals[i]
33                      if eval_count < self.num_evals:
34                          self.avg_obj_val_eval[eval_count] += best_so_far
35                          eval_count += 1
36                  # 2. determination
37                  curr_pop = self.tournament_select(curr_pop, curr_obj_vals, self.num_players)
38                  # 3. transition
39                  curr_pop = self.crossover(curr_pop, self.crossover_rate)
40                  curr_pop = self.mutate(curr_pop, self.mutation_rate)
41              self.avg_obj_val += best_so_far
42          # 4. output
43          self.avg_obj_val /= self.num_runs
44          self.avg_obj_val_eval /= self.num_runs
45          for i in range(self.num_evals):
46              print("%.3f" % self.avg_obj_val_eval[i])
47          self.show_optimum()
48          return curr_pop
49
50      def init(self):
51          # 1. initialization
52          self.best_sol = np.zeros(self.num_patterns_sol, dtype=int)
53          self.best_obj_val = sys.float_info.max
54          # 2. input the TSP benchmark
55          with open(self.filename_ins, "r") as f:
56              self.ins_tsp = np.array(f.read().strip().split(), dtype=float).reshape(self.
    num_patterns_sol, 2)
57          # 3. input the optimal solution
58          with open(self.filename_ins + ".opt", "r") as f:
59              self.opt_sol = np.array(f.read().strip().split(), dtype=int) - 1
60          # 4. initial solutions
61          if len(self.filename_ini) > 0:
62              with open(self.filename_ini, "r") as f:
63                  curr_pop = np.array(f.read().strip().split(), dtype=int).reshape(self.pop_size,
    self.num_patterns_sol)
64          else:
65              curr_pop = np.array([np.arange(self.num_patterns_sol) for i in range(self.pop_size)
    ])
66              for i in range(self.pop_size):
67                  np.random.shuffle(curr_pop[i])
68          # 5. construct the distance matrix
69          self.dist = np.zeros((self.num_patterns_sol, self.num_patterns_sol))
70          for i in range(self.num_patterns_sol):
71              for j in range(self.num_patterns_sol):
72                  self.dist[i][j] = math.dist(self.ins_tsp[i], self.ins_tsp[j])
73          return curr_pop, np.zeros(self.pop_size)
74
75      def evaluate(self, curr_pop, dist):
76          tour_dist = np.zeros(self.pop_size)
```

```
77          for p in range(self.pop_size):
78              for i in range(self.num_patterns_sol):
79                  r = curr_pop[p][i]
80                  s = curr_pop[p][(i+1) % self.num_patterns_sol]
81                  tour_dist[p] += self.dist[r][s]
82          return tour_dist
83
84      def update_best_sol(self, curr_pop, curr_obj_vals):
85          for i in range(self.pop_size):
86              if curr_obj_vals[i] < self.best_obj_val:
87                  self.best_obj_val = curr_obj_vals[i]
88                  self.best_sol = curr_pop[i]
89
90      def tournament_select(self, curr_pop, curr_obj_vals, num_players):
91          tmp_pop = np.zeros((self.pop_size, self.num_patterns_sol), dtype=int)
92          for i in range(self.pop_size):
93              k = np.random.randint(self.pop_size)
94              f = curr_obj_vals[k]
95              for j in range(1, num_players):
96                  n = np.random.randint(self.pop_size)
97                  if curr_obj_vals[n] < f:
98                      k = n
99                      f = curr_obj_vals[k]
100             tmp_pop[i] = curr_pop[k]
101         return tmp_pop
102
103     def crossover(self, curr_pop, cr):
104         tmp_pop = curr_pop.copy()
105         mid = self.pop_size // 2
106         ssz = self.num_patterns_sol
107         for i in range(mid):
108             f = np.random.rand()
109             if f <= cr:
110                 # 1. one-point crossover
111                 xp = np.random.randint(ssz)
112                 s1 = tmp_pop[i]
113                 s2 = tmp_pop[mid+i]
114                 s1[xp:], s2[xp:] = s2[xp:], s1[xp:]
115                 # 2. correct the solutions
116                 for j in range(i, self.pop_size, mid):
117                     s = tmp_pop[j]
118                     # find the number of times each city was visited,
119                     # which can be either 0, 1, or 2, i.e., not visited,
120                     # visited once, and visited twice
121                     visit = np.zeros(ssz, dtype=int)
122                     for k in range(ssz):
123                         visit[s[k]] += 1
124                     # put cities not visited in bag
125                     bag = []
126                     for k in range(ssz):
127                         if visit[k] == 0:
128                             bag.append(k)
129                     # correct cities visited twice, by replacing one
130                     # of which by one randomly chosen from the bag, if
131                     # necessary
132                     if len(bag) > 0:
133                         bag = np.array(bag)
134                         np.random.shuffle(bag)
135                         n = 0
136                         for k in range(xp, ssz):
137                             if visit[s[k]] == 2:
138                                 s[k] = bag[n]
139                                 n += 1
140         return tmp_pop
141
```

```
142     def mutate(self, curr_pop, mr):
143         tmp_pop = curr_pop
144         ssz = self.num_patterns_sol
145         for i in range(self.pop_size):
146             f = np.random.rand()
147             if f <= mr:
148                 m1 = np.random.randint(ssz)      # mutation point
149                 m2 = np.random.randint(ssz)      # mutation point
150                 tmp_pop[i][m1], tmp_pop[i][m2] = tmp_pop[i][m2], tmp_pop[i][m1]
151         return tmp_pop
152
153     def show_optimum(self):
154         opt_dist = 0.0
155         for i in range(self.num_patterns_sol):
156             r = self.opt_sol[i]
157             s = self.opt_sol[(i+1) % self.num_patterns_sol]
158             opt_dist += self.dist[r][s]
159         lib.print_solution(self.best_sol, prefix = "# Current route: ")
160         lib.printf("# Optimum distance: %.3f", opt_dist)
161         lib.print_solution(self.opt_sol, prefix = "# Optimum route: ")
162
163     def print_parameters(self, algname):
164         print("# name of the search algorithm: '%s'" % algname)
165         print("# number of runs: %s" % self.num_runs)
166         print("# number of evaluations: %s" % self.num_evals)
167         print("# number of patterns (subsolutions) each solution: %s" % self.num_patterns_sol)
168         print("# filename of the initial seeds: '%s'" % self.filename_ini)
169         print("# filename of the benchmark: '%s'" % self.filename_ins)
170         print("# population size: %s" % self.pop_size)
171         print("# crossover rate: %s" % self.crossover_rate)
172         print("# mutation rate: %s" % self.mutation_rate)
173         print("# number of players: %s" % self.num_players)
174         print("# seed: %s" % self.seed)
175
176 def usage():
177     lib.usage("<filename_ins> <pop_size> <cr> <mr> <#players>")
178
179 if __name__ == "__main__":
180     if ac() == 2 and ss(1) == "-h": usage()
181     elif ac() == 1: search = ga()
182     elif ac() == 10: search = ga(si(1), si(2), si(3), ss(4), ss(5), si(6), sf(7), sf(8), si(9))
183     else: usage()
184     lib.run(search, None)
```

Listing B.31: src/python/3-tsp/hga.py: The implementation of HGA in Python.

```
1 #!/usr/bin/env python
2 import numpy as np; import time; import sys; import os; import math
3 from lib import *
4 from sa import sa
5
6 class hga:
7     def __init__(self, num_runs = 3, num_evals = 120000, num_patterns_sol = 51, filename_ini = "
         ", filename_ins = "eil51.tsp", pop_size = 120, crossover_rate = 0.4, mutation_rate = 0.1,
         num_players = 3, sa_num_evals = 10, sa_time2run = 0.1, max_temp = 1.0, min_temp = 0.00001):
8         self.num_runs = num_runs
9         self.num_evals = num_evals
10        self.num_patterns_sol = num_patterns_sol
11        self.filename_ini = filename_ini
12        self.filename_ins = filename_ins
13        self.pop_size = pop_size
```

```
14         self.crossover_rate = crossover_rate
15         self.mutation_rate = mutation_rate
16         self.num_players = num_players
17         self.sa_num_evals = sa_num_evals
18         self.sa_time2run = sa_time2run
19         self.max_temp = max_temp
20         self.min_temp = min_temp
21         self.seed = int(time.time())
22         np.random.seed(self.seed)
23
24    def run(self):
25        self.avg_obj_val = 0.0
26        self.avg_obj_val_eval = np.zeros(self.num_evals)
27        for r in range(self.num_runs):
28            self.eval_count = 0
29            self.best_so_far = sys.float_info.max
30            curr_pop, curr_obj_vals = self.init()
31            while self.eval_count < self.num_evals:
32                # 1. evaluation
33                curr_obj_vals = self.evaluate(curr_pop, self.dist)
34                self.update_best_sol(curr_pop, curr_obj_vals)
35                # 2. determination
36                curr_pop = self.tournament_select(curr_pop, curr_obj_vals, self.num_players)
37                # 3. transition
38                curr_pop = self.crossover_ox(curr_pop, self.crossover_rate)
39                curr_pop = self.mutate(curr_pop, self.mutation_rate)
40                if self.eval_count > self.num_evals * self.sa_time2run:
41                    curr_pop = self.saf(curr_pop, curr_obj_vals, self.num_evals)
42            self.avg_obj_val += self.best_obj_val
43        # 4. output
44        self.avg_obj_val /= self.num_runs
45        self.avg_obj_val_eval /= self.num_runs
46        for i in range(self.num_evals):
47            print("%.3f" % self.avg_obj_val_eval[i])
48        self.show_optimum()
49        return curr_pop
50
51    def saf(self, curr_pop, curr_obj_vals, total_evals):
52        tmp_pop = curr_pop
53        search = sa(1,
54                    self.sa_num_evals,
55                    self.num_patterns_sol,
56                    "",
57                    self.max_temp,
58                    self.min_temp,
59                    self.dist,
60                    True,
61                    total_evals)
62        n = curr_obj_vals.argmax()
63        tmp_pop[n], self.eval_count, self.avg_obj_val_eval, self.best_obj_val, self.best_sol = search.run(curr_pop[n], self.eval_count, self.avg_obj_val_eval, self.best_obj_val, self.best_sol)
64        return tmp_pop
65
66    def init(self):
67        # 1. initialization
68        self.best_sol = np.zeros(self.num_patterns_sol, dtype=int)
69        self.best_obj_val = sys.float_info.max
70        # 2. input the TSP benchmark
71        with open(self.filename_ins, "r") as f:
72            self.ins_tsp = np.array(f.read().strip().split(), dtype=float).reshape(self.num_patterns_sol, 2)
73        # 3. input the optimal solution
74        with open(self.filename_ins + ".opt", "r") as f:
75            self.opt_sol = np.array(f.read().strip().split(), dtype=int) - 1
```

```
76          # 4. initial solutions
77          if len(self.filename_ini) > 0:
78              with open(self.filename_ini, "r") as f:
79                  curr_pop = np.array(f.read().strip().split(), dtype=int).reshape(self.pop_size,
        self.num_patterns_sol)
80          else:
81              curr_pop = np.array([np.arange(self.num_patterns_sol) for i in range(self.pop_size)
        ])
82              for i in range(self.pop_size):
83                  np.random.shuffle(curr_pop[i])
84          # 5. construct the distance matrix
85          self.dist = np.zeros((self.num_patterns_sol, self.num_patterns_sol))
86          for i in range(self.num_patterns_sol):
87              for j in range(self.num_patterns_sol):
88                  self.dist[i][j] = math.dist(self.ins_tsp[i], self.ins_tsp[j])
89          return curr_pop, np.zeros(self.pop_size)
90
91      def evaluate(self, curr_pop, dist):
92          tour_dist = np.zeros(self.pop_size)
93          for p in range(self.pop_size):
94              tour_dist[p] = self.evaluate_sol(curr_pop[p], dist)
95          return tour_dist
96
97      def evaluate_sol(self, curr_sol, dist):
98          tour_dist = 0.0
99          for i in range(self.num_patterns_sol):
100             r = curr_sol[i]
101             s = curr_sol[(i+1) % self.num_patterns_sol]
102             tour_dist += self.dist[r][s]
103         if self.best_so_far > tour_dist:
104             self.best_so_far = tour_dist
105         if self.eval_count < self.num_evals:
106             self.avg_obj_val_eval[self.eval_count] += self.best_so_far
107             self.eval_count += 1
108         return tour_dist
109
110     def update_best_sol(self, curr_pop, curr_obj_vals):
111         for i in range(self.pop_size):
112             if curr_obj_vals[i] < self.best_obj_val:
113                 self.best_obj_val = curr_obj_vals[i]
114                 self.best_sol = curr_pop[i]
115
116     def tournament_select(self, curr_pop, curr_obj_vals, num_players):
117         tmp_pop = np.zeros((self.pop_size, self.num_patterns_sol), dtype=int)
118         for i in range(self.pop_size):
119             k = np.random.randint(self.pop_size)
120             f = curr_obj_vals[k]
121             for j in range(1, num_players):
122                 n = np.random.randint(self.pop_size)
123                 if curr_obj_vals[n] < f:
124                     k = n
125                     f = curr_obj_vals[k]
126             tmp_pop[i] = curr_pop[k]
127         return tmp_pop
128
129     def crossover_ox(self, curr_pop, cr):
130         tmp_pop = curr_pop.copy()
131         mid = self.pop_size // 2
132         ssz = self.num_patterns_sol
133         for i in range(mid):
134             f = np.random.rand()
135             if f <= cr:
136                 # 1. create the mapping sections
137                 xp1 = np.random.randint(ssz + 1)
138                 xp2 = np.random.randint(ssz + 1)
```

```
139                    if xp1 > xp2:
140                        xp1, xp2 = xp2, xp1
141                    # 2. indices to the two parents and offspring
142                    p = [ i, i+mid ]
143                    # 3. the main process of ox
144                    for k in range(2):
145                        c1 = p[k]
146                        c2 = p[1-k]
147                        # 4. mask the genes between xp1 and xp2
148                        s1 = curr_pop[c1]
149                        s2 = curr_pop[c2]
150                        msk1 = np.zeros(ssz, dtype=bool)
151                        for j in range(xp1, xp2):
152                            msk1[s1[j]] = True
153                        msk2 = np.zeros(ssz, dtype=bool)
154                        for j in range(0, ssz):
155                            msk2[j] = msk1[s2[j]]
156                        # 5. replace the genes that are not masked
157                        j = xp2 % ssz
158                        for z in range(ssz):
159                            if not msk2[z]:
160                                tmp_pop[c1][j] = s2[z]
161                                j = (j+1) % ssz
162            return tmp_pop
163
164        def mutate(self, curr_pop, mr):
165            tmp_pop = curr_pop
166            ssz = self.num_patterns_sol
167            for i in range(self.pop_size):
168                f = np.random.rand()
169                if f <= mr:
170                    m1 = np.random.randint(ssz)      # mutation point
171                    m2 = np.random.randint(ssz)      # mutation point
172                    tmp_pop[i][m1], tmp_pop[i][m2] = tmp_pop[i][m2], tmp_pop[i][m1]
173            return tmp_pop
174
175        def show_optimum(self):
176            opt_dist = 0.0
177            for i in range(self.num_patterns_sol):
178                r = self.opt_sol[i]
179                s = self.opt_sol[(i+1) % self.num_patterns_sol]
180                opt_dist += self.dist[r][s]
181            lib.print_solution(self.best_sol, prefix = "# Current route: ")
182            lib.printf("# Optimum distance: %.3f", opt_dist)
183            lib.print_solution(self.opt_sol, prefix = "# Optimum route: ")
184
185        def print_parameters(self, algname):
186            print("# name of the search algorithm: '%s'" % algname)
187            print("# number of runs: %s" % self.num_runs)
188            print("# number of evaluations: %s" % self.num_evals)
189            print("# number of patterns (subsolutions) each solution: %s" % self.num_patterns_sol)
190            print("# filename of the initial seeds: '%s'" % self.filename_ini)
191            print("# filename of the benchmark: '%s'" % self.filename_ins)
192            print("# population size: %s" % self.pop_size)
193            print("# crossover rate: %s" % self.crossover_rate)
194            print("# mutation rate: %s" % self.mutation_rate)
195            print("# number of players: %s" % self.num_players)
196            print("# number of evaluations for SA: %s" % self.sa_num_evals)
197            print("# run SA after n%% of evaluations: %s" % self.sa_time2run)
198            print("# maximum temperature for SA: %s" % self.max_temp)
199            print("# minimum temperature for SA: %s" % self.min_temp)
200            print("# seed: %s" % self.seed)
201
202 def usage():
```

```
203    lib.usage("<filename_ins> <pop_size> <cr> <mr> <#players> <#evals_sa> <time2run> <
       max_temp_sa> <min_temp_sa>")
204
205 if __name__ == "__main__":
206    if ac() == 2 and ss(1) == "-h": usage()
207    elif ac() == 1: search = hga()
208    elif ac() == 14: search = hga(si(1), si(2), si(3), ss(4), ss(5), si(6), sf(7), sf(8), si(9),
       si(10), sf(11), sf(12), sf(13))
209    else: usage()
210    lib.run(search, None)
```

Listing B.32: src/python/3-tsp/pga.py: The implementation of PGA in Python.

```
1 #!/usr/bin/env python
2 import numpy as np; import time; import sys; import os; import math; import threading
3 from lib import *
4
5 class atomic_int:
6     def __init__(self, v):
7         self.count = v
8         self.lock = threading.Lock()
9
10    def set(self, v):
11        with self.lock:
12            self.count = v
13            return self.count
14
15    def dec(self):
16        with self.lock:
17            self.count -= 1
18            return self.count
19
20 class pga:
21    def __init__(self, num_runs = 3, num_evals = 120000, num_patterns_sol = 1002, filename_ini =
       "", filename_ins = "pr1002.tsp", pop_size = 120, crossover_rate = 0.4, mutation_rate = 0.1,
       num_players = 3, num_threads = 6):
22        self.num_runs = num_runs
23        self.num_evals = num_evals
24        self.num_patterns_sol = num_patterns_sol
25        self.filename_ini = filename_ini
26        self.filename_ins = filename_ins
27        self.pop_size = pop_size
28        self.crossover_rate = crossover_rate
29        self.mutation_rate = mutation_rate
30        self.num_players = num_players
31        self.num_threads = num_threads
32        self.seed = int(time.time())
33        np.random.seed(self.seed)
34
35    def run(self):
36        self.avg_obj_val = 0.0
37        self.avg_obj_val_eval = np.zeros(self.num_evals)
38        cv = threading.Condition()
39        icnt = atomic_int(self.num_threads)
40        for r in range(self.num_runs):
41            curr_pop, curr_obj_vals = self.init()
42            tmp_curr_pop = curr_pop
43            threads = []
44            for i in range(self.num_threads):
45                thread = threading.Thread(name = "Thread-" + str(i),
46                                          target = self.thread_run,
```

```
47                                args = (curr_pop,            # r/w (s_idx - e_idx)
48                                        tmp_curr_pop,         # r/w (s_idx - e_idx)
49                                        curr_obj_vals,        # r/w (s_idx - e_idx)
50                                        i, # thread_idx,      # r/o
51                                        self.num_threads,     # r/o
52                                        self.num_evals,       # r/o
53                                        self.pop_size // self.num_threads,   # r/o
54                                        self.crossover_rate,  # r/o
55                                        self.mutation_rate,   # r/o
56                                        self.num_players,     # r/o
57                                        cv,                   # r/w
58                                        icnt                  # r/w
59                                        )
60                                )
61                thread.start()
62                threads.append(thread)
63            for thread in threads:
64                thread.join()
65        # 4. output
66        self.avg_obj_val /= self.num_runs
67        for i in range(self.num_evals):
68            self.avg_obj_val_eval[i] /= self.num_runs
69            print("%.3f" % self.avg_obj_val_eval[i])
70        self.show_optimum()
71        return curr_pop
72
73    def init(self):
74        # 1. initialization
75        self.best_sol = np.zeros(self.num_patterns_sol, dtype=int)
76        self.best_obj_val = sys.float_info.max
77        # 2. input the TSP benchmark
78        with open(self.filename_ins, "r") as f:
79            self.ins_tsp = np.array(f.read().strip().split(), dtype=float).reshape(self.
num_patterns_sol, 2)
80        # 3. input the optimal solution
81        with open(self.filename_ins + ".opt", "r") as f:
82            self.opt_sol = np.array(f.read().strip().split(), dtype=int) - 1
83        # 4. initial solutions
84        if len(self.filename_ini) > 0:
85            with open(self.filename_ini, "r") as f:
86                curr_pop = np.array(f.read().strip().split(), dtype=int).reshape(self.pop_size,
self.num_patterns_sol)
87        else:
88            curr_pop = np.array([np.arange(self.num_patterns_sol) for i in range(self.pop_size)
])
89            for i in range(self.pop_size):
90                np.random.shuffle(curr_pop[i])
91        # 5. construct the distance matrix
92        self.dist = np.zeros((self.num_patterns_sol, self.num_patterns_sol))
93        for i in range(self.num_patterns_sol):
94            for j in range(self.num_patterns_sol):
95                self.dist[i][j] = math.dist(self.ins_tsp[i], self.ins_tsp[j])
96        return curr_pop, np.zeros(self.pop_size)
97
98    def thread_run(self,
99                   curr_pop,
100                  tmp_curr_pop,
101                  curr_obj_vals,
102                  thread_idx,
103                  num_threads,
104                  num_evals,
105                  thread_pop_size,
106                  crossover_rate,
107                  mutation_rate,
108                  num_players,
```

```
109                     cv,
110                     icnt
111                     ):
112             # 0. declaration
113             num_gens = num_evals // self.pop_size
114             s_idx = thread_idx * thread_pop_size
115             e_idx = s_idx + thread_pop_size
116             gen_count = 0
117             while gen_count < num_gens:
118                 # 1. evaluation
119                 for i in range(s_idx, e_idx):
120                     curr_obj_vals[i] = self.evaluate(curr_pop[i])
121                 cv.acquire()
122                 if icnt.dec() > 0:
123                     cv.wait()
124                 else:
125                     # 1.1. sync
126                     z = 0
127                     avg = self.avg_obj_val_eval[gen_count * self.pop_size : ]       # a slice
128                     for i in range(self.pop_size):
129                         if curr_obj_vals[i] < self.best_obj_val:
130                             self.best_obj_val = curr_obj_vals[i]
131                             z = i
132                         avg[i] += self.best_obj_val
133                     self.best_sol = curr_pop[z]
134                     # 1.2. continue
135                     icnt.set(num_threads)
136                     cv.notify_all()
137                 cv.release()
138                 # 2. determination
139                 self.tournament_select(curr_pop, curr_obj_vals, tmp_curr_pop, num_players,
            thread_idx, thread_pop_size)
140                 cv.acquire()
141                 if icnt.dec() > 0:
142                     cv.wait()
143                 else:
144                     # 2.1. sync
145                     curr_pop = tmp_curr_pop
146                     # 2.2. continue
147                     icnt.set(num_threads)
148                     cv.notify_all()
149                 cv.release()
150                 # 3. transition
151                 self.crossover_ox(curr_pop, tmp_curr_pop, crossover_rate, thread_idx,
            thread_pop_size)
152                 self.mutate(curr_pop, mutation_rate, thread_idx, thread_pop_size)
153                 gen_count += 1
154             self.avg_obj_val += self.best_obj_val
155
156     def evaluate(self, sol):
157         tour_dist = 0.0
158         for i in range(self.num_patterns_sol):
159             r = sol[i]
160             s = sol[(i+1) % self.num_patterns_sol]
161             tour_dist += self.dist[r][s]
162         return tour_dist
163
164     def tournament_select(self, curr_pop, curr_obj_vals, tmp_curr_pop, num_players, thread_idx,
            thread_pop_size):
165         s_idx = thread_idx * thread_pop_size
166         e_idx = s_idx + thread_pop_size
167         for i in range(s_idx, e_idx):
168             k = np.random.randint(self.pop_size)
169             f = curr_obj_vals[k]
170             for j in range(1, num_players):
```

```
171                    n = np.random.randint(self.pop_size)
172                    if curr_obj_vals[n] < f:
173                        k = n
174                        f = curr_obj_vals[k]
175                tmp_curr_pop[i] = curr_pop[k]
176
177    def crossover_ox(self, curr_pop, tmp_curr_pop, cr, thread_idx, thread_pop_size):
178        mid = thread_pop_size // 2
179        s_idx = thread_idx * thread_pop_size
180        m_idx = s_idx + mid
181        e_idx = s_idx + thread_pop_size
182        ssz = self.num_patterns_sol
183        for i in range(s_idx, m_idx):
184            f = np.random.rand()
185            if f <= cr:
186                # 1. create the mapping sections
187                xp1 = np.random.randint(ssz + 1)
188                xp2 = np.random.randint(ssz + 1)
189                if xp1 > xp2:
190                    xp1, xp2 = xp2, xp1
191                # 2. indices to the two parents and offspring
192                p = [ i, i+mid ]
193                # 3. the main process of ox
194                for k in range(2):
195                    c1 = p[k]
196                    c2 = p[1-k]
197                    # 4. mask the genes between xp1 and xp2
198                    s1 = curr_pop[c1]
199                    s2 = curr_pop[c2]
200                    msk1 = np.zeros(ssz, dtype=bool)
201                    for j in range(xp1, xp2):
202                        msk1[s1[j]] = True
203                    msk2 = np.zeros(ssz, dtype=bool)
204                    for j in range(0, ssz):
205                        msk2[j] = msk1[s2[j]]
206                    # 5. replace the genes that are not masked
207                    j = xp2 % ssz
208                    for z in range(ssz):
209                        if not msk2[z]:
210                            tmp_curr_pop[c1][j] = s2[z]
211                            j = (j+1) % ssz
212        for i in range(s_idx, e_idx):
213            curr_pop[i] = tmp_curr_pop[i]
214
215    def mutate(self, curr_pop, mr, thread_idx, thread_pop_size):
216        s_idx = thread_idx * thread_pop_size
217        e_idx = s_idx + thread_pop_size
218        for i in range(s_idx, e_idx):
219            f = np.random.rand()
220            if f <= mr:
221                m1 = np.random.randint(self.num_patterns_sol)   # mutation point
222                m2 = np.random.randint(self.num_patterns_sol)   # mutation point
223                curr_pop[i][m1], curr_pop[i][m2] = curr_pop[i][m2], curr_pop[i][m1]
224
225    def show_optimum(self):
226        opt_dist = 0.0
227        for i in range(self.num_patterns_sol):
228            r = self.opt_sol[i]
229            s = self.opt_sol[(i+1) % self.num_patterns_sol]
230            opt_dist += self.dist[r][s]
231        lib.print_solution(self.best_sol, prefix = "# Current route: ")
232        lib.printf("# Optimum distance: %.3f", opt_dist)
233        lib.print_solution(self.opt_sol, prefix = "# Optimum route: ")
234
235    def print_parameters(self, algname):
```

```
236        print("# name of the search algorithm: '%s'" % algname)
237        print("# number of runs: %s" % self.num_runs)
238        print("# number of evaluations: %s" % self.num_evals)
239        print("# number of patterns (subsolutions) each solution: %s" % self.num_patterns_sol)
240        print("# filename of the initial seeds: '%s'" % self.filename_ini)
241        print("# filename of the benchmark: '%s'" % self.filename_ins)
242        print("# population size: %s" % self.pop_size)
243        print("# crossover rate: %s" % self.crossover_rate)
244        print("# mutation rate: %s" % self.mutation_rate)
245        print("# number of players: %s" % self.num_players)
246        print("# number of threads: %s" % self.num_threads)
247        print("# seed: %s" % self.seed)
248
249 def usage():
250        lib.usage("<filename_ins> <pop_size> <cr> <mr> <#players> <#threads>")
251
252 if __name__ == "__main__":
253        if ac() == 2 and ss(1) == "-h": usage()
254        elif ac() == 1: search = pga()
255        elif ac() == 11: search = pga(si(1), si(2), si(3), ss(4), ss(5), si(6), sf(7), sf(8), si(9),
           si(10))
256        else: usage()
257        lib.run(search, None)
```

Listing B.33: src/python/3-tsp/sa.py: The implementation of SA in Python.

```
1 #!/usr/bin/env python
2 import numpy as np; import time; import sys; import os
3 from lib import *
4
5 class sa:
6     def __init__(self, num_runs, num_evals, num_patterns_sol, filename_ini, max_temp, min_temp,
          dist, continue_flag, total_evals):
7         self.num_runs = num_runs
8         self.num_evals = num_evals
9         self.num_patterns_sol = num_patterns_sol
10        self.filename_ini = filename_ini
11        self.max_temp = max_temp
12        self.min_temp = min_temp
13        self.dist = dist
14        self.continue_flag = continue_flag
15        self.total_evals = total_evals
16        self.seed = int(time.time())
17        # np.random.seed(self.seed)
18
19    def run(self, sol_orig, eval_count_orig, avg_obj_val_eval_orig, best_obj_val_orig,
          best_sol_orig):
20        eval_count_saved = eval_count_orig
21        for r in range(self.num_runs):
22            eval_count = eval_count_saved
23            # 0. Initialization
24            curr_sol = self.init(sol_orig)
25            curr_temp = self.max_temp
26            best_obj_val = best_obj_val_orig
27            best_sol = best_sol_orig
28            curr_sol = sol_orig
29            obj_val = self.evaluate(curr_sol, self.dist)
30            while eval_count < min(self.total_evals, eval_count_saved + self.num_evals):
31                # 1. Transition
32                tmp_sol = self.transit(curr_sol)
33                # 2. Evaluation
```

```
34                      tmp_obj_val = self.evaluate(tmp_sol, self.dist)
35                      # 3. Determination
36                      if self.determine(tmp_obj_val, obj_val, curr_temp):
37                          obj_val, curr_sol = tmp_obj_val, tmp_sol
38                      if obj_val < best_obj_val:
39                          best_obj_val, best_sol = obj_val, curr_sol
40                      curr_temp = self.annealing(curr_temp)
41                      avg_obj_val_eval_orig[eval_count] += best_obj_val
42                      eval_count += 1
43                  eval_count_orig = eval_count
44                  if best_obj_val < best_obj_val_orig:
45                      best_obj_val_orig = best_obj_val
46                      best_sol_orig = best_sol
47                      sol_orig = best_sol
48                  else:
49                      sol_orig = curr_sol
50          return sol_orig, eval_count_orig, avg_obj_val_eval_orig, best_obj_val_orig,
       best_sol_orig
51
52      def init(self, sol):
53          if not self.continue_flag:
54              sol = np.arange(len(sol))
55              np.random.shuffle(sol)
56          return sol
57
58      def transit(self, sol):
59          t = sol.copy()
60          ssz = len(sol)
61          i = np.random.randint(ssz)
62          j = np.random.randint(ssz)
63          t[i], t[j] = t[j], t[i]
64          return t
65
66      def evaluate(self, sol, dist):
67          tour_dist = 0.0
68          ssz = len(sol)
69          for i in range(ssz):
70              r = sol[i]
71              s = sol[(i+1) % ssz]
72              tour_dist += dist[r][s]
73          return tour_dist
74
75      def determine(self, tmp_obj_val, obj_val, temperature):
76          r = np.random.rand()
77          p = np.exp((tmp_obj_val - obj_val) / temperature)
78          return r > p
79
80      def annealing(self, temperature):
81          return 0.9 * temperature
82
83      def print_parameters(self, algname):
84          print("# name of the search algorithm: '%s'" % algname)
85          print("# number of runs: %s" % self.num_runs)
86          print("# number of evaluations: %s" % self.num_evals)
87          print("# number of patterns: %s" % self.num_patterns_sol)
88          print("# filename of the initial seeds: '%s'" % self.filename_ini)
89          print("# maximum temperature: %s" % self.max_temp)
90          print("# minimum temperature: %s" % self.min_temp)
91          print("# dist: %s" % self.dist)
92          print("# continue_flag: %s" % self.continue_flag)
93          print("# seed: %s" % self.seed)
94
95  if __name__ == "__main__":
96      pass
```

Listing B.34: src/python/4-function/deb1.py: The implementation of DEB1 in Python.

```python
#!/usr/bin/env python
import numpy as np; import time; import sys; import os; import functions
from lib import *
from de import de

class deb1(de):
    def __init__(self, num_runs = 3, num_evals = 1000, num_dims = 2, filename_ini = "",
    filename_ins = "mvfAckley", pop_size = 10, F = 0.7, cr = 0.5, v_min = -32.0, v_max = 32.0):
        super().__init__(num_runs, num_evals, num_dims, filename_ini, filename_ins, pop_size, F,
    cr, v_min, v_max)

    def mutate(self, curr_pop):
        new_pop = np.zeros((self.pop_size, self.num_dims))
        for i in range(self.pop_size):
            s1 = curr_pop[np.random.randint(self.pop_size)]
            s2 = curr_pop[np.random.randint(self.pop_size)]
            new_pop[i] = np.clip(self.best_sol + self.F * (s1 - s2), self.v_min, self.v_max)
        return new_pop

def usage():
    lib.usage("<#runs> <#evals> <#dims> <filename_ini> <filename_ins> <pop_size> <F> <cr> <v_min
    > <v_max>", common_params = "")

if __name__ == "__main__":
    if ac() == 2 and ss(1) == "-h": usage()
    elif ac() == 1: search = deb1()
    elif ac() == 11: search = deb1(si(1), si(2), si(3), ss(4), ss(5), si(6), sf(7), sf(8), sf(9)
    , sf(10))
    else: usage()
    lib.run(search, lib.print_solution)
```

Listing B.35: src/python/4-function/deb2.py: The implementation of DEB2 in Python.

```python
#!/usr/bin/env python
import numpy as np; import time; import sys; import os; import functions
from lib import *
from de import de

class deb2(de):
    def __init__(self, num_runs = 3, num_evals = 1000, num_dims = 2, filename_ini = "",
    filename_ins = "mvfAckley", pop_size = 10, F = 0.7, cr = 0.5, v_min = -32.0, v_max = 32.0):
        super().__init__(num_runs, num_evals, num_dims, filename_ini, filename_ins, pop_size, F,
    cr, v_min, v_max)

    def mutate(self, curr_pop):
        new_pop = np.zeros((self.pop_size, self.num_dims))
        for i in range(self.pop_size):
            s1 = curr_pop[np.random.randint(self.pop_size)]
            s2 = curr_pop[np.random.randint(self.pop_size)]
            s3 = curr_pop[np.random.randint(self.pop_size)]
            s4 = curr_pop[np.random.randint(self.pop_size)]
            new_pop[i] = np.clip(self.best_sol + self.F * (s1 - s2) + self.F * (s3 - s4), self.
    v_min, self.v_max)
        return new_pop

def usage():
    lib.usage("<#runs> <#evals> <#dims> <filename_ini> <filename_ins> <pop_size> <F> <cr> <v_min
    > <v_max>", common_params = "")
```

```
22
23 if __name__ == "__main__":
24     if ac() == 2 and ss(1) == "-h": usage()
25     elif ac() == 1: search = deb2()
26     elif ac() == 11: search = deb2(si(1), si(2), si(3), ss(4), ss(5), si(6), sf(7), sf(8), sf(9)
       , sf(10))
27     else: usage()
28     lib.run(search, lib.print_solution)
```

Listing B.36: src/python/4-function/decb1.py: The implementation of DECB1 in Python.

```
1 #!/usr/bin/env python
2 import numpy as np; import time; import sys; import os; import functions
3 from lib import *
4 from de import de
5
6 class decb1(de):
7     def __init__(self, num_runs = 3, num_evals = 1000, num_dims = 2, filename_ini = "",
       filename_ins = "mvfAckley", pop_size = 10, F = 0.7, cr = 0.5, v_min = -32.0, v_max = 32.0):
8         super().__init__(num_runs, num_evals, num_dims, filename_ini, filename_ins, pop_size, F,
       cr, v_min, v_max)
9
10     def mutate(self, curr_pop):
11         new_pop = np.zeros((self.pop_size, self.num_dims))
12         for i in range(self.pop_size):
13             s1 = curr_pop[np.random.randint(self.pop_size)]
14             s2 = curr_pop[np.random.randint(self.pop_size)]
15             new_pop[i] = np.clip(curr_pop[i] + self.F * (self.best_sol - curr_pop[i]) + self.F *
       (s1 - s2), self.v_min, self.v_max)
16         return new_pop
17
18 def usage():
19     lib.usage("<#runs> <#evals> <#dims> <filename_ini> <filename_ins> <pop_size> <F> <cr> <v_min
       > <v_max>", common_params = "")
20
21 if __name__ == "__main__":
22     if ac() == 2 and ss(1) == "-h": usage()
23     elif ac() == 1: search = decb1()
24     elif ac() == 11: search = decb1(si(1), si(2), si(3), ss(4), ss(5), si(6), sf(7), sf(8), sf
       (9), sf(10))
25     else: usage()
26     lib.run(search, lib.print_solution)
```

Listing B.37: src/python/4-function/de.py: The implementation of DE in Python.

```
1 #!/usr/bin/env python
2 import numpy as np; import time; import sys; import os; import functions
3 from lib import *
4
5 class de:
6     def __init__(self, num_runs = 3, num_evals = 1000, num_dims = 2, filename_ini = "",
       filename_ins = "mvfAckley", pop_size = 10, F = 0.7, cr = 0.5, v_min = -32.0, v_max = 32.0):
7         self.num_runs = num_runs
8         self.num_evals = num_evals
9         self.num_dims = num_dims
10         self.filename_ini = filename_ini
```

```
11          self.filename_ins = filename_ins
12          self.pop_size = pop_size
13          self.F = F
14          self.cr = cr
15          self.v_min = v_min
16          self.v_max = v_max
17          self.seed = int(time.time())
18          np.random.seed(self.seed)
19
20      def run(self):
21          avg_obj_val = 0.0
22          avg_obj_val_eval = np.zeros(self.num_evals)
23          for r in range(self.num_runs):
24              eval_count = 0
25              # 0. initialization
26              curr_pop, curr_obj_vals = self.init()
27              while eval_count < self.num_evals:
28                  # 1. mutation
29                  curr_pop_v = self.mutate(curr_pop)
30                  # 2. recombination
31                  curr_pop_u = self.recombine(curr_pop, curr_pop_v)
32                  # 3. evaluation
33                  curr_obj_vals_u = self.evaluate(curr_pop_u)
34                  # 4. selection
35                  curr_pop = self.select(curr_pop, curr_pop_u, curr_obj_vals, curr_obj_vals_u)
36                  for i in range(self.pop_size):
37                      if self.best_obj_val > curr_obj_vals[i]:
38                          self.best_obj_val = curr_obj_vals[i]
39                          self.best_sol = curr_pop[i]
40                      if eval_count < self.num_evals:
41                          avg_obj_val_eval[eval_count] += self.best_obj_val
42                          eval_count +=1
43              avg_obj_val += self.best_obj_val
44          # 4. output
45          avg_obj_val /= self.num_runs
46          for i in range(self.num_evals):
47              avg_obj_val_eval[i] /= self.num_runs
48              print("%.15f" % avg_obj_val_eval[i])
49          # return curr_pop
50          return self.best_sol
51
52      def init(self):
53          # 1. initialize all best solution and its objective value
54          self.best_sol = self.v_max * np.ones(self.num_dims)
55          self.best_obj_val = sys.float_info.max
56          # 2. initialize the positions and velocities of particles
57          if len(self.filename_ini) == 0:
58              curr_pop = self.v_min + (self.v_max - self.v_min) * np.random.rand(self.pop_size,
    self.num_dims)
59          else:
60              with open(self.filename_ini, "r") as f:
61                  curr_pop = np.array(f.read().strip().split(), dtype=float).reshape(self.pop_size
    , self.num_dims)
62          # 3. evaluate the initial population
63          curr_obj_vals = self.evaluate(curr_pop)
64          return curr_pop, curr_obj_vals
65
66      def evaluate(self, curr_pop):
67          return np.array([functions.function_table[self.filename_ins](self.num_dims, x) for x in
    curr_pop])
68
69      def mutate(self, curr_pop):
70          new_pop = np.zeros((self.pop_size, self.num_dims))
71          for i in range(self.pop_size):
72              s1 = curr_pop[np.random.randint(self.pop_size)]
```

```
73              s2 = curr_pop[np.random.randint(self.pop_size)]
74              s3 = curr_pop[np.random.randint(self.pop_size)]
75              new_pop[i] = np.clip(s1 + self.F * (s2 - s3), self.v_min, self.v_max)
76          return new_pop
77
78      def recombine(self, curr_pop, curr_pop_v):
79          new_pop = np.zeros((self.pop_size, self.num_dims))
80          for i in range(self.pop_size):
81              s = np.random.randint(self.num_dims)
82              for j in range(self.num_dims):
83                  r = np.random.rand()
84                  new_pop[i][j] = curr_pop_v[i][j] if (r < self.cr or s == j) else curr_pop[i][j]
85          return new_pop
86
87      def select(self, curr_pop, curr_pop_u, curr_obj_vals, curr_obj_vals_u):
88          new_pop = np.zeros((self.pop_size, self.num_dims))
89          for i in range(self.pop_size):
90              if curr_obj_vals_u[i] < curr_obj_vals[i]:
91                  new_pop[i] = curr_pop_u[i]
92                  curr_obj_vals[i] = curr_obj_vals_u[i]
93              else:
94                  new_pop[i] = curr_pop[i]
95          return new_pop
96
97      def print_parameters(self, algname):
98          print("# name of the search algorithm: '%s'" % algname)
99          print("# number of runs: %s" % self.num_runs)
100         print("# number of evaluations: %s" % self.num_evals)
101         print("# number of dimensions: %s" % self.num_dims)
102         print("# filename of the initial seeds: '%s'" % self.filename_ini)
103         print("# filename of the benchmark: '%s'" % self.filename_ins)
104         print("# population size: %s" % self.pop_size)
105         print("# F: %s" % self.F)
106         print("# crossover rate: %s" % self.cr)
107         print("# v_min: %s" % self.v_min)
108         print("# v_max: %s" % self.v_max)
109         print("# seed: %s" % self.seed)
110
111 def usage():
112     lib.usage("<#runs> <#evals> <#dims> <filename_ini> <filename_ins> <pop_size> <F> <cr> <v_min
        > <v_max>", common_params = "")
113
114 if __name__ == "__main__":
115     if ac() == 2 and ss(1) == "-h": usage()
116     elif ac() == 1: search = de()
117     elif ac() == 11: search = de(si(1), si(2), si(3), ss(4), ss(5), si(6), sf(7), sf(8), sf(9),
        sf(10))
118     else: usage()
119     lib.run(search, lib.print_solution)
```

Listing B.38: src/python/4-function/der2.py: The implementation of DER2 in Python.

```
1 #!/usr/bin/env python
2 import numpy as np; import time; import sys; import os; import functions
3 from lib import *
4 from de import de
5
6 class der2(de):
7     def __init__(self, num_runs = 3, num_evals = 1000, num_dims = 2, filename_ini = "",
        filename_ins = "mvfAckley", pop_size = 10, F = 0.7, cr = 0.5, v_min = -32.0, v_max = 32.0):
```

```
 8        super().__init__(num_runs, num_evals, num_dims, filename_ini, filename_ins, pop_size, F,
       cr, v_min, v_max)
 9
10    def mutate(self, curr_pop):
11        new_pop = np.zeros((self.pop_size, self.num_dims))
12        for i in range(self.pop_size):
13            s1 = curr_pop[np.random.randint(self.pop_size)]
14            s2 = curr_pop[np.random.randint(self.pop_size)]
15            s3 = curr_pop[np.random.randint(self.pop_size)]
16            s4 = curr_pop[np.random.randint(self.pop_size)]
17            s5 = curr_pop[np.random.randint(self.pop_size)]
18            new_pop[i] = np.clip(s1 + self.F * (s2 - s3) + self.F * (s4 - s5), self.v_min, self.
       v_max)
19        return new_pop
20
21 def usage():
22    lib.usage("<#runs> <#evals> <#dims> <filename_ini> <filename_ins> <pop_size> <F> <cr> <v_min
       > <v_max>", common_params = "")
23
24 if __name__ == "__main__":
25    if ac() == 2 and ss(1) == "-h": usage()
26    elif ac() == 1: search = der2()
27    elif ac() == 11: search = der2(si(1), si(2), si(3), ss(4), ss(5), si(6), sf(7), sf(8), sf(9)
       , sf(10))
28    else: usage()
29    lib.run(search, lib.print_solution)
```

Listing B.39: src/python/4-function/functions.py: The implementation of PSO-FUN in Python.

```
 1 #!/usr/bin/env python
 2 import math
 3
 4 sqrt = math.sqrt
 5 sin = math.sin
 6 cos = math.cos
 7 exp = math.exp
 8 pi = math.pi
 9
10 def mvfAckley(n, x):
11     s1 = s2 = 0.0
12     for i in range(n):
13         s1 += x[i] * x[i]
14         s2 += cos(2.0 * pi * x[i])
15     return -20.0 * exp(-0.2 * sqrt(s1/n)) + 20.0 - exp(s2/n) + exp(1.0)
16
17 function_table = {
18     "mvfAckley": mvfAckley
19 }
20
21 if __name__ == "__main__":
22     pass
```

Listing B.40: src/python/4-function/pso.py: The implementation of PSO in Python.

```
 1 #!/usr/bin/env python
 2 import numpy as np; import time; import sys; import os; import functions
```

```
 3 from lib import *
 4
 5 class pso:
 6     def __init__(self, num_runs = 3, num_evals = 1000, num_dims = 2, filename_ini = "",
       filename_ins = "mvfAckley", pop_size = 10, omega = 0.5, c1 = 1.0, c2 = 1.5, v_min = -32.0,
       v_max = 32.0):
 7         self.num_runs = num_runs
 8         self.num_evals = num_evals
 9         self.num_dims = num_dims
10         self.filename_ini = filename_ini
11         self.filename_ins = filename_ins
12         self.pop_size = pop_size
13         self.omega = omega
14         self.c1 = c1
15         self.c2 = c2
16         self.v_min = v_min
17         self.v_max = v_max
18         self.seed = int(time.time())
19         np.random.seed(self.seed)
20
21     def run(self):
22         avg_obj_val = 0.0
23         avg_obj_val_eval = np.zeros(self.num_evals)
24         for r in range(self.num_runs):
25             eval_count = 0
26             best_so_far = sys.float_info.max
27             # 0. initialization
28             curr_pop, curr_obj_vals, velocity = self.init()
29             while eval_count < self.num_evals:
30                 # 1. compute new velocity of each particle
31                 velocity = self.new_velocity(curr_pop, velocity)
32                 # 2. adjust all the particles to their new positions
33                 curr_pop = self.new_position(curr_pop, velocity)
34                 # 3. evaluation
35                 curr_obj_vals = self.evaluate(curr_pop)
36                 for i in range(self.pop_size):
37                     if best_so_far > curr_obj_vals[i]:
38                         best_so_far = curr_obj_vals[i]
39                     if eval_count < self.num_evals:
40                         avg_obj_val_eval[eval_count] += best_so_far
41                         eval_count += 1
42                 # 4. update
43                 self.update_pb(curr_pop, curr_obj_vals)
44                 self.update_gb(curr_pop, curr_obj_vals)
45                 self.update_best_sol()
46             avg_obj_val += self.best_obj_val
47         # 4. output
48         avg_obj_val /= self.num_runs
49         for i in range(self.num_evals):
50             avg_obj_val_eval[i] /= self.num_runs
51             print("%.15f" % avg_obj_val_eval[i])
52         # return curr_pop
53         return self.best_sol
54
55     def init(self):
56         # 1. initialize all best solutions and their objective values
57         self.pbest = sys.float_info.max * np.ones((self.pop_size, self.num_dims))
58         self.pbest_obj_vals = sys.float_info.max * np.ones(self.pop_size)
59         self.gbest = self.v_max * np.ones(self.num_dims)
60         self.gbest_obj_val = sys.float_info.max
61         self.best_sol = self.gbest
62         self.best_obj_val = self.gbest_obj_val
63         # 2. initialize the positions and velocities of particles
64         if len(self.filename_ini) == 0:
```

```
65              curr_pop = self.v_min + (self.v_max - self.v_min) * np.random.rand(self.pop_size,
        self.num_dims)
66          else:
67              with open(self.filename_ini, "r") as f:
68                  curr_pop = np.array(f.read().strip().split(), dtype=float).reshape(self.pop_size
        , self.num_dims)
69          velocity = curr_pop / float(self.num_evals)
70          # 3. evaluation
71          curr_obj_vals = self.evaluate(curr_pop)
72          # 4. update
73          self.update_pb(curr_pop, curr_obj_vals)
74          self.update_gb(curr_pop, curr_obj_vals)
75          self.update_best_sol()
76          return curr_pop, curr_obj_vals, velocity
77
78      def evaluate(self, curr_pop):
79          return np.array([functions.function_table[self.filename_ins](self.num_dims, x) for x in
        curr_pop])
80
81      def update_pb(self, curr_pop, curr_obj_vals):
82          for i in range(self.pop_size):
83              if curr_obj_vals[i] < self.pbest_obj_vals[i]:
84                  self.pbest_obj_vals[i] = curr_obj_vals[i]
85                  self.pbest[i] = curr_pop[i]
86
87      def update_gb(self, curr_pop, curr_obj_vals):
88          for i in range(self.pop_size):
89              if curr_obj_vals[i] < self.gbest_obj_val:
90                  self.gbest_obj_val = curr_obj_vals[i]
91                  self.gbest = curr_pop[i]
92
93      def update_best_sol(self):
94          if self.gbest_obj_val < self.best_obj_val:
95              self.best_obj_val = self.gbest_obj_val
96              self.best_sol = self.gbest
97
98      def new_velocity(self, curr_pop, velocity):
99          r1 = np.random.rand(self.pop_size, self.num_dims)
100         r2 = np.random.rand(self.pop_size, self.num_dims)
101         new_v = np.zeros((self.pop_size, self.num_dims)) # FIXME, why this line is needed to
        make it right?!
102         new_v = self.omega * velocity + self.c1 * r1 * (self.pbest - curr_pop) + self.c2 * r2 *
        (self.gbest - curr_pop)
103         return np.clip(new_v, self.v_min, self.v_max)
104
105     def new_position(self, curr_pop, velocity):
106         return np.clip(curr_pop + velocity, self.v_min, self.v_max)
107
108     def print_parameters(self, algname):
109         print("# name of the search algorithm: '%s'" % algname)
110         print("# number of runs: %s" % self.num_runs)
111         print("# number of evaluations: %s" % self.num_evals)
112         print("# number of dimensions: %s" % self.num_dims)
113         print("# filename of the initial seeds: '%s'" % self.filename_ini)
114         print("# filename of the benchmark: '%s'" % self.filename_ins)
115         print("# population size: %s" % self.pop_size)
116         print("# omega: %s" % self.omega)
117         print("# c1: %s" % self.c1)
118         print("# c2: %s" % self.c2)
119         print("# v_min: %s" % self.v_min)
120         print("# v_max: %s" % self.v_max)
121
122 def usage():
123     lib.usage("<#runs> <#evals> <#dims> <filename_ini> <filename_ins> <pop_size> <omega> <c1> <
        c2> <v_min> <v_max>", common_params = "")
```

```
124
125 if __name__ == "__main__":
126     if ac() == 2 and ss(1) == "-h": usage()
127     elif ac() == 1: search = pso()
128     elif ac() == 12: search = pso(si(1), si(2), si(3), ss(4), ss(5), si(6), sf(7), sf(8), sf(9),
        sf(10), sf(11))
129     else: usage()
130     lib.run(search, lib.print_solution)
```

Listing B.41: src/python/5-eval/ga.py: The implementation of GA in Python.

```
1 Ditto to B.3.
```

Listing B.42: src/python/5-eval/gar.py: The implementation of GAR in Python.

```
1 Ditto to B.4.
```

Listing B.43: src/python/5-eval/garx2.py: The implementation of GARX2 in Python.

```
1 #!/usr/bin/env python
2 import numpy as np; import time; import sys; import os
3 from lib import *
4 from gar import gar
5
6 class garx2(gar):
7     def __init__(self, num_runs = 3, num_evals = 1000, num_patterns_sol = 10, filename_ini = "",
        pop_size = 10, crossover_rate = 0.6, mutation_rate = 0.01, num_players = 0):
8         super().__init__(num_runs, num_evals, num_patterns_sol, filename_ini, pop_size,
        crossover_rate, mutation_rate, num_players)
9
10     def crossover(self, curr_pop, cr):
11         # two-point crossover
12         mid = int(self.pop_size / 2)
13         for i in range(mid):
14             f = np.random.rand()
15             if f <= cr:
16                 xp1 = np.random.randint(self.num_patterns_sol)
17                 xp2 = np.random.randint(self.num_patterns_sol)
18                 if xp2 < xp1:
19                     xp1, xp2 = xp2, xp1
20                 for j in range(xp1, xp2):
21                     curr_pop[i][xp1:xp2], curr_pop[mid+i][j] = curr_pop[mid+i][j], curr_pop[i][j
        ]
22         return curr_pop
23
24 def usage():
25     lib.usage("<pop_size> <cr> <mr> <#players>")
26
27 if __name__ == "__main__":
28     if ac() == 2 and ss(1) == "-h": usage()
29     elif ac() == 1: search = garx2()
30     elif ac() == 9: search = garx2(si(1), si(2), si(3), ss(4), si(5), sf(6), sf(7), si(8))
31     else: usage()
32     lib.run(search, lib.print_population)
```

Listing B.44: src/python/5-eval/garxu.py: The implementation of GARXU in Python.

```python
1 #!/usr/bin/env python
2 import numpy as np; import time; import sys; import os
3 from lib import *
4 from gar import gar
5
6 class garxu(gar):
7     def __init__(self, num_runs = 3, num_evals = 1000, num_patterns_sol = 10, filename_ini = "",
        pop_size = 10, crossover_rate = 0.6, mutation_rate = 0.01, num_players = 0):
8         super().__init__(num_runs, num_evals, num_patterns_sol, filename_ini, pop_size,
        crossover_rate, mutation_rate, num_players)
9
10    def crossover(self, curr_pop, cr):
11        # uniform crossover
12        mid = self.pop_size // 2
13        for i in range(mid):
14            for j in range(self.num_patterns_sol):
15                f = np.random.rand()
16                if f <= 0.5: # yes, 0.5 instead of cr
17                    curr_pop[i][j], curr_pop[mid+i][j] = curr_pop[mid+i][j], curr_pop[i][j]
18        return curr_pop
19
20 def usage():
21     lib.usage("<pop_size> <cr> <mr> <#players>")
22
23 if __name__ == "__main__":
24     if ac() == 2 and ss(1) == "-h": usage()
25     elif ac() == 1: search = garxu()
26     elif ac() == 9: search = garxu(si(1), si(2), si(3), ss(4), si(5), sf(6), sf(7), si(8))
27     else: usage()
28     lib.run(search, lib.print_population)
```

Listing B.45: src/python/5-eval/gax2.py: The implementation of GAX2 in Python.

```python
1 Ditto to B.5.
```

Listing B.46: src/python/5-eval/gaxu.py: The implementation of GAXU in Python.

```python
1 Ditto to B.6.
```

Listing B.47: src/python/6-clustering/ga.py: The implementation of GA in Python.

```python
1 #!/usr/bin/env python
2 import numpy as np; import time; import sys; import os; import math
3 from lib import *
4
5 class ga:
6     def __init__(self, num_runs = 3, num_evals = 1000, num_patterns_sol = 150, num_dims = 4,
        num_clusters = 3, filename_ini = "", filename_ins = "iris.data", pop_size = 20, cr = 0.6, mr
        = 0.1, num_players = 3, pra = 1):
```

```
7          self.num_runs = num_runs
8          self.num_evals = num_evals
9          self.num_patterns_sol = num_patterns_sol
10         self.num_dims = num_dims
11         self.num_clusters = num_clusters
12         self.filename_ini = filename_ini
13         self.filename_ins = filename_ins
14         self.pop_size = pop_size
15         self.cr = cr
16         self.mr = mr
17         self.num_players = num_players
18         self.pra = pra != 0 # do redundant analysis if true
19         self.seed = int(time.time())
20         np.random.seed(self.seed)
21
22     def print_parameters(self, algname):
23         print("# name of the search algorithm: '%s'" % algname)
24         print("# number of runs: %s" % self.num_runs)
25         print("# number of evaluations: %s" % self.num_evals)
26         print("# number of patterns: %s" % self.num_patterns_sol)
27         print("# number of dimensions each pattern: %s" % self.num_dims)
28         print("# number of clusters: %s" % self.num_clusters)
29         print("# filename of the initial seeds: '%s'" % self.filename_ini)
30         print("# name of the data set to be clustered: '%s'" % self.filename_ins)
31         print("# population size: %s" % self.pop_size)
32         print("# crossover rate: %s" % self.cr)
33         print("# mutation rate: %s" % self.mr)
34         print("# number of players: %s" % self.num_players)
35         print("# pra?: %s" % self.pra)
36         print("# seed: %s" % self.seed)
37
38     def run(self):
39         avg_obj_val = 0.0
40         avg_obj_val_eval = np.zeros(self.num_evals)
41         num_gens = self.num_evals // self.pop_size
42         if self.pra:
43             avg_final_state = np.zeros((self.num_runs, num_gens))
44             all_final_state = np.zeros((num_gens,))
45         for r in range(self.num_runs):
46             eval_count = 0
47             best_so_far = sys.float_info.max
48             # 0. initialization
49             curr_pop = self.init()
50             if self.pra:
51                 pop_conv = np.zeros((num_gens, self.pop_size, self.num_patterns_sol), dtype=int)
52                 final_state = np.zeros((self.num_runs, self.pop_size, self.num_patterns_sol),
dtype=int)
53                 curr_gen = 0
54             while eval_count < self.num_evals:
55                 # 1. evaluation
56                 curr_obj_vals, centroids = self.evaluate(curr_pop)
57                 self.update_best_sol(curr_pop, curr_obj_vals)
58                 if self.pra:
59                     self.redundant_analysis_1(curr_gen, curr_pop, pop_conv)
60                     curr_gen += 1
61                 for i in range(self.pop_size):
62                     if best_so_far > curr_obj_vals[i]:
63                         best_so_far = curr_obj_vals[i]
64                     if eval_count < self.num_evals:
65                         avg_obj_val_eval[eval_count] += best_so_far
66                         eval_count += 1
67                 # 2. determination
68                 curr_pop = self.tournament_select(curr_pop, curr_obj_vals, self.num_players)
69                 # 3. transition
70                 curr_pop = self.crossover(curr_pop, self.cr)
```

```
71              curr_pop = self.mutate(curr_pop, self.mr)
72              curr_pop = self.okm(curr_pop, centroids)
73           avg_obj_val += self.best_obj_val
74           if self.pra:
75              self.redundant_analysis_2(curr_gen, r, pop_conv, final_state, avg_final_state,
    all_final_state)      # pra
76        avg_obj_val /= self.num_runs
77        avg_obj_val_eval /= self.num_runs
78        for i in range(self.num_evals):
79           print("%.3f" % avg_obj_val_eval[i])
80        return curr_pop
81
82    def init(self):
83        self.best_obj_val = sys.float_info.max
84        self.best_sol = None
85        # 1. input the dataset to be clustered
86        if len(self.filename_ins) > 0:
87           with open(self.filename_ins, "r") as f:
88              self.ins_clustering = np.array(f.read().strip().split(), dtype=float).reshape(
    self.num_patterns_sol, self.num_dims)
89        # 2. create the initial solutions
90        if len(self.filename_ini) > 0:
91           with open(self.filename_ini, "r") as f:
92              curr_pop = np.array(f.read().strip().split(), dtype=float).reshape(self.pop_size
    , self.num_patterns_sol)
93        else:
94           curr_pop = np.array([np.array([np.random.randint(self.num_clusters) for _ in range(
    self.num_patterns_sol)]) for _ in range(self.pop_size)])
95        return curr_pop
96
97    def evaluate(self, curr_pop):
98        pop_size = len(curr_pop)
99        centroids = np.zeros((self.pop_size, self.num_clusters, self.num_dims))
100       sse = np.zeros(pop_size)
101       for p in range(pop_size):
102          # 1. assign each pattern to its cluster
103          count_patterns = np.zeros(self.num_clusters, dtype=int)
104          for i in range(self.num_patterns_sol):
105             k = curr_pop[p][i]
106             centroids[p][k] += self.ins_clustering[i]
107             count_patterns[k] += 1
108          # 2. compute the centroids (means)
109          for k in range(self.num_clusters):
110             centroids[p][k] /= count_patterns[k]
111          # 3. compute sse
112          for i in range(self.num_patterns_sol):
113             k = curr_pop[p][i]
114             for d in range(self.num_dims):
115                sse[p] += math.pow(self.ins_clustering[i][d] - centroids[p][k][d], 2)
116       return sse, centroids
117
118    # tournament selection
119    def tournament_select(self, curr_pop, curr_obj_vals, num_players):
120       pop_size = len(curr_pop)
121       tmp_pop = np.empty((pop_size, self.num_patterns_sol), dtype=int)
122       for i in range(pop_size):
123          k = np.random.randint(pop_size)
124          f = curr_obj_vals[k]
125          for _ in range(1, num_players):
126             n = np.random.randint(pop_size)
127             if curr_obj_vals[n] < f:
128                k = n
129                f = curr_obj_vals[k]
130          tmp_pop[i] = curr_pop[k]
131       return tmp_pop
```

```
132
133    def update_best_sol(self, curr_pop, curr_obj_vals):
134        pop_size = len(curr_pop)
135        for i in range(pop_size):
136            if curr_obj_vals[i] < self.best_obj_val:
137                self.best_obj_val = curr_obj_vals[i]
138                self.best_sol = curr_pop[i]
139
140    # one-point crossover
141    def crossover(self, curr_pop, cr):
142        pop_size = len(curr_pop)
143        mid = pop_size // 2
144        for i in range(mid):
145            f = np.random.rand()
146            if f <= cr:
147                xp = np.random.randint(self.num_patterns_sol)    # crossover point
148                for j in range(xp, self.num_patterns_sol):
149                    curr_pop[i][j], curr_pop[mid+i][j] = curr_pop[mid+i][j], curr_pop[i][j]
150        return curr_pop
151
152    def mutate(self, curr_pop, mr):
153        pop_size = len(curr_pop)
154        for i in range(pop_size):
155            f = np.random.rand()
156            if f <= mr:
157                m = np.random.randint(self.num_patterns_sol)    # mutation point
158                curr_pop[i][m] = np.random.randint(self.num_clusters)
159        return curr_pop
160
161    # one-evaluation k-means
162    def okm(self, curr_pop, centroids):
163        pop_size = len(curr_pop)
164        for p in range(pop_size):
165            # 1. one iteration k-means
166            count = np.zeros(self.num_clusters, dtype=int)
167            for i in range(self.num_patterns_sol):
168                dist = sys.float_info.max
169                c = 0
170                for k in range(self.num_clusters):
171                    dist_tmp = math.dist(self.ins_clustering[i], centroids[p][k])
172                    if dist_tmp < dist:
173                        dist = dist_tmp
174                        c = k
175                curr_pop[p][i] = c
176                count[c] += 1
177            # 2. randomly assign a pattern to each empty cluster
178            for k in range(self.num_clusters):
179                if count[k] == 0:
180                    i = np.random.randint(self.num_patterns_sol)
181                    curr_pop[p][i] = k
182        return curr_pop
183
184    def redundant_analysis_1(self, curr_gen, curr_pop, pop_conv):
185        pop_conv[curr_gen] = curr_pop
186
187    def redundant_analysis_2(self, curr_gen, curr_run, pop_conv, final_state, avg_final_state,
       all_final_state):
188        # 1. find the generation at which each subsolution reaches the final state for each run
189        for i in range(self.pop_size):
190            for j in range(self.num_patterns_sol):
191                for k in range(curr_gen-1, 1, -1):
192                    if pop_conv[k][i][j] != pop_conv[k-1][i][j]:
193                        final_state[curr_run][i][j] = k
194                        break
195        # 2. accumulate the information obtained above for each run
```

```
196        for k in range(curr_gen):
197            final_state_count = 0
198            for i in range(self.pop_size):
199                for j in range(self.num_patterns_sol):
200                    if final_state[curr_run][i][j] <= k:
201                        final_state_count += 1
202            avg_final_state[curr_run][k] = final_state_count / self.pop_size
203            all_final_state[k] += avg_final_state[curr_run][k]
204        # 3. compute the percentage of subsolutions that reach the final state for all runs
205        if curr_run == self.num_runs-1:
206            print("# rd: ", end='')
207            for k in range(curr_gen):
208                print(("%.3f" if k == 0 else ", %.3f") % (all_final_state[k] / (self.num_runs *
        self.num_patterns_sol)), end='')
209            print()
210
211 def usage():
212     lib.usage("<#runs> <#evals> <#patterns> <#dims> <#clusters> <filename_ini> <filename_ins> <
        pop_size> <cr> <mr> <#players> <pra?>", common_params = "")
213
214 if __name__ == "__main__":
215     if ac() == 2 and ss(1) == "-h": usage()
216     if ac() == 1: search = ga()
217     elif ac() == 13: search = ga(si(1), si(2), si(3), si(4), si(5), ss(6), ss(7), si(8), sf(9),
        sf(10), si(11), si(12))
218     else: usage()
219     lib.run(search, lib.print_population)
```

Listing B.48: src/python/6-clustering/prega.py: The implementation of PREGA in Python.

```
1 #!/usr/bin/env python
2 import numpy as np; import time; import sys; import os; import math
3 from lib import *
4
5 class prega:
6     def __init__(self, num_runs = 3, num_evals = 1000, num_patterns_sol = 150, num_dims = 4,
        num_clusters = 3, filename_ini = "", filename_ins = "iris.data", pop_size = 20, cr = 0.6, mr
        = 0.1, num_players = 3, num_detections = 80, reduction_rate = 0.1, pra = 1):
7         self.num_runs = num_runs
8         self.num_evals = num_evals
9         self.num_patterns_sol = num_patterns_sol
10        self.num_dims = num_dims
11        self.num_clusters = num_clusters
12        self.filename_ini = filename_ini
13        self.filename_ins = filename_ins
14        self.pop_size = pop_size
15        self.cr = cr
16        self.mr = mr
17        self.num_players = num_players
18        self.num_detections = num_detections
19        self.reduction_rate = reduction_rate
20        self.pra = pra != 0 # do redundant analysis if true
21        self.seed = int(time.time())
22        np.random.seed(self.seed)
23
24    def print_parameters(self, algname):
25        print("# name of the search algorithm: '%s'" % algname)
26        print("# number of runs: %s" % self.num_runs)
27        print("# number of evaluations: %s" % self.num_evals)
28        print("# number of patterns: %s" % self.num_patterns_sol)
29        print("# number of dimensions each pattern: %s" % self.num_dims)
```

```
30          print("# number of clusters: %s" % self.num_clusters)
31          print("# filename of the initial seeds: '%s'" % self.filename_ini)
32          print("# name of the data set to be clustered: '%s'" % self.filename_ins)
33          print("# population size: %s" % self.pop_size)
34          print("# crossover rate: %s" % self.cr)
35          print("# mutation rate: %s" % self.mr)
36          print("# number of players: %s" % self.num_players)
37          print("# number of detections: %s" % self.num_detections)
38          print("# reduction rate: %s" % self.reduction_rate)
39          print("# pra?: %s" % self.pra)
40          print("# seed: %s" % self.seed)
41
42      def run(self):
43          avg_obj_val = 0.0
44          avg_obj_val_eval = np.zeros(self.num_evals)
45          if self.pra:
46              pr_kcounter = np.zeros(self.num_evals)
47              num_gens = self.num_evals // self.pop_size
48              avg_final_state = np.zeros((self.num_runs, num_gens))
49              all_final_state = np.zeros((num_gens,))
50          for r in range(self.num_runs):
51              eval_count = 0
52              best_so_far = sys.float_info.max
53              # 0. initialization
54              curr_pop = self.init()
55              centroid_dist = np.zeros((self.pop_size, self.num_clusters))
56              pr_mask = np.zeros((self.pop_size, self.num_patterns_sol), dtype=bool)
57              if self.pra:
58                  pop_conv = np.zeros((num_gens, self.pop_size, self.num_patterns_sol), dtype=int)
59                  final_state = np.zeros((self.num_runs, self.pop_size, self.num_patterns_sol),
            dtype=int)
60                  curr_gen = 0
61              while eval_count < self.num_evals:
62                  # 1. evaluation
63                  curr_obj_vals, centroids = self.evaluate(curr_pop)
64                  self.update_best_sol(curr_pop, curr_obj_vals)
65                  if self.pra:
66                      self.redundant_analysis_1(curr_gen, curr_pop, pop_conv)
67                      curr_gen += 1
68                      for p in range(self.pop_size):
69                          for i in range(self.num_patterns_sol):
70                              if pr_mask[p][i]:
71                                  pr_kcounter[p+eval_count] += 1
72                  for i in range(self.pop_size):
73                      if best_so_far > curr_obj_vals[i]:
74                          best_so_far = curr_obj_vals[i]
75                      if eval_count < self.num_evals:
76                          avg_obj_val_eval[eval_count] += best_so_far
77                          eval_count += 1
78                  # 2. determination
79                  curr_pop, pr_mask = self.tournament_select(curr_pop, curr_obj_vals, self.
            num_players, pr_mask)
80                  # 3. transition
81                  curr_pop = self.crossover(curr_pop, self.cr, pr_mask)
82                  curr_pop = self.mutate(curr_pop, self.mr, pr_mask)
83                  curr_pop = self.okm(curr_pop, pr_mask, eval_count, centroids, centroid_dist)
84                  if eval_count > self.num_detections:
85                      pr_mask = self.detection(curr_pop, pr_mask, (eval_count + self.pop_size) /
            self.pop_size, centroids, centroid_dist, self.reduction_rate)
86              avg_obj_val += self.best_obj_val
87              if self.pra:
88                  self.redundant_analysis_2(curr_gen, r, pop_conv, final_state, avg_final_state,
            all_final_state)
89          avg_obj_val /= self.num_runs
90          avg_obj_val_eval /= self.num_runs
```

```
91          for i in range(self.num_evals):
92              if self.pra:
93                  print("%.3f, %.3f" % (avg_obj_val_eval[i], pr_kcounter[i] / (self.num_runs *
       self.num_patterns_sol)))
94              else:
95                  print("%.3f" % avg_obj_val_eval[i])
96          return curr_pop
97
98      def init(self):
99          self.best_obj_val = sys.float_info.max
100         self.best_sol = None
101         # 1. input the dataset to be clustered
102         if len(self.filename_ins) > 0:
103             with open(self.filename_ins, "r") as f:
104                 self.ins_clustering = np.array(f.read().strip().split(), dtype=float).reshape(
       self.num_patterns_sol, self.num_dims)
105         # 2. create the initial solutions
106         if len(self.filename_ini) > 0:
107             with open(self.filename_ini, "r") as f:
108                 curr_pop = np.array(f.read().strip().split(), dtype=float).reshape(self.pop_size
       , self.num_patterns_sol)
109         else:
110             curr_pop = np.array([np.array([np.random.randint(self.num_clusters) for _ in range(
       self.num_patterns_sol)]) for _ in range(self.pop_size)])
111         return curr_pop
112
113     def evaluate(self, curr_pop):
114         pop_size = len(curr_pop)
115         centroids = np.zeros((self.pop_size, self.num_clusters, self.num_dims))
116         sse = np.zeros(pop_size)
117         for p in range(pop_size):
118             # 1. assign each pattern to its cluster
119             count_patterns = np.zeros(self.num_clusters, dtype=int)
120             for i in range(self.num_patterns_sol):
121                 k = curr_pop[p][i]
122                 centroids[p][k] += self.ins_clustering[i]
123                 count_patterns[k] += 1
124             # 2. compute the centroids (means)
125             for k in range(self.num_clusters):
126                 centroids[p][k] /= count_patterns[k]
127             # 3. compute sse
128             for i in range(self.num_patterns_sol):
129                 k = curr_pop[p][i]
130                 for d in range(self.num_dims):
131                     sse[p] += math.pow(self.ins_clustering[i][d] - centroids[p][k][d], 2)
132         return sse, centroids
133
134     # tournament selection
135     def tournament_select(self, curr_pop, curr_obj_vals, num_players, pr_mask):
136         pop_size = len(curr_pop)
137         tmp_pop = np.empty((pop_size, self.num_patterns_sol), dtype=int)
138         tmp_pr_mask = np.zeros((pop_size, self.num_patterns_sol), dtype=bool)
139         for i in range(pop_size):
140             k = np.random.randint(pop_size)
141             f = curr_obj_vals[k]
142             for _ in range(1, num_players):
143                 n = np.random.randint(pop_size)
144                 if curr_obj_vals[n] < f:
145                     k = n
146                     f = curr_obj_vals[k]
147             tmp_pop[i] = curr_pop[k]
148             tmp_pr_mask[i] = pr_mask[k]
149         return tmp_pop, tmp_pr_mask
150
151     def update_best_sol(self, curr_pop, curr_obj_vals):
```

```
152         pop_size = len(curr_pop)
153         for i in range(pop_size):
154             if curr_obj_vals[i] < self.best_obj_val:
155                 self.best_obj_val = curr_obj_vals[i]
156                 self.best_sol = curr_pop[i]
157
158     # one-point crossover
159     def crossover(self, curr_pop, cr, pr_mask):
160         pop_size = len(curr_pop)
161         mid = pop_size // 2
162         for i in range(mid):
163             f = np.random.rand()
164             if f <= cr:
165                 xp = np.random.randint(self.num_patterns_sol)   # crossover point
166                 for j in range(xp, self.num_patterns_sol):
167                     if not pr_mask[i][j]:
168                         curr_pop[i][j], curr_pop[mid+i][j] = curr_pop[mid+i][j], curr_pop[i][j]
169         return curr_pop
170
171     def mutate(self, curr_pop, mr, pr_mask):
172         pop_size = len(curr_pop)
173         for i in range(pop_size):
174             f = np.random.rand()
175             if f <= mr:
176                 m = np.random.randint(self.num_patterns_sol)      # mutation point
177                 if not pr_mask[i][m]:
178                     curr_pop[i][m] = np.random.randint(self.num_clusters)
179         return curr_pop
180
181     # one-evaluation k-means
182     def okm(self, curr_pop, pr_mask, eval, centroids, centroid_dist):
183         pop_size = len(curr_pop)
184         for p in range(pop_size):
185             # 1. one iteration k-means
186             count = np.zeros(self.num_clusters, dtype=int)
187             for i in range(self.num_patterns_sol):
188                 if not pr_mask[p][i] or eval > self.num_evals/2:
189                     dist = sys.float_info.max
190                     c = 0
191                     for k in range(self.num_clusters):
192                         dist_tmp = math.dist(self.ins_clustering[i], centroids[p][k])
193                         if dist_tmp < dist:
194                             dist = dist_tmp
195                             c = k
196                     curr_pop[p][i] = c
197                     count[c] += 1
198                     if centroid_dist[p][c] < dist:
199                         centroid_dist[p][c] = dist
200             # 2. randomly assign a pattern to each empty cluster
201             for k in range(self.num_clusters):
202                 if count[k] == 0:
203                     i = np.random.randint(self.num_patterns_sol)
204                     curr_pop[p][i] = k
205                     if pr_mask[p][i]:
206                         pr_mask[p][i] = False
207         return curr_pop
208
209     def detection(self, curr_pop, pr_mask, eval, centroids, centroid_dist, reduction_rate):
210         rr = eval * reduction_rate
211         for p in range(self.pop_size):
212             dist = np.empty(self.num_clusters, dtype=float)
213             for i in range(self.num_clusters):
214                 dist[i] = centroid_dist[p][i] * rr
215             for i in range(self.num_patterns_sol):
216                 if not pr_mask[p][i]:
```

```
217                        dist_tmp = math.dist(self.ins_clustering[i], centroids[p][curr_pop[p][i]])
218                        if dist_tmp < dist[curr_pop[p][i]]:
219                            pr_mask[p][i] = True
220            return pr_mask
221
222        def redundant_analysis_1(self, curr_gen, curr_pop, pop_conv):
223            pop_conv[curr_gen] = curr_pop
224
225        def redundant_analysis_2(self, curr_gen, curr_run, pop_conv, final_state, avg_final_state,
           all_final_state):
226            # 1. find the generation at which each subsolution reaches the final state for each run
227            for i in range(self.pop_size):
228                for j in range(self.num_patterns_sol):
229                    for k in range(curr_gen-1, 1, -1):
230                        if pop_conv[k][i][j] != pop_conv[k-1][i][j]:
231                            final_state[curr_run][i][j] = k
232                            break
233            # 2. accumulate the information obtained above for each run
234            for k in range(curr_gen):
235                final_state_count = 0
236                for i in range(self.pop_size):
237                    for j in range(self.num_patterns_sol):
238                        if final_state[curr_run][i][j] <= k:
239                            final_state_count += 1
240                avg_final_state[curr_run][k] = final_state_count / self.pop_size
241                all_final_state[k] += avg_final_state[curr_run][k]
242            # 3. compute the percentage of subsolutions that reach the final state for all runs
243            if curr_run == self.num_runs-1:
244                print("# rd: ", end='')
245                for k in range(curr_gen):
246                    print(("%.3f" if k == 0 else ", %.3f") % (all_final_state[k] / (self.num_runs *
           self.num_patterns_sol)), end='')
247                print()
248
249 def usage():
250     lib.usage("<#runs> <#evals> <#patterns> <#dims> <#clusters> <filename_ini> <filename_ins> <
        pop_size> <cr> <mr> <#players> <#detections> <rr> <pra?>", common_params = "")
251
252 if __name__ == "__main__":
253     if ac() == 2 and ss(1) == "-h": usage()
254     if ac() == 1: search = prega()
255     elif ac() == 15: search = prega(si(1), si(2), si(3), si(4), si(5), ss(6), ss(7), si(8), sf
        (9), sf(10), si(11), si(12), sf(13), si(14))
256     else: usage()
257     lib.run(search, lib.print_population)
```

Listing B.49: src/python/a-lib/lib.py: The implementation of LIB in Python.

```
1 #!/usr/bin/env python
2 import time; import sys; import os
3
4 ac = lambda: len(sys.argv)
5 si = lambda i: int(sys.argv[i])
6 sf = lambda i: float(sys.argv[i])
7 ss = lambda i: sys.argv[i]
8
9 _progname = os.path.basename(sys.argv[0])
10 _algname = _progname.split('.')[0]
11
12 class lib:
13     @staticmethod
```

```
14    def print_solution(sol, *, prefix = '# '):
15        print(prefix, end='')
16        print(*sol, sep=', ')
17
18    @staticmethod
19    def print_population(pop):
20        for sol in pop:
21            lib.print_solution(sol)
22
23    @staticmethod
24    def printf(fmt, *values):
25        print(fmt % values)
26
27    @staticmethod
28    def usage(extra_params, *, common_params = "<#runs> <#evals> <#patterns> <filename_ini>"):
29        if len(common_params) == 0:
30            print("Usage: python %s %s" % (_progname, extra_params))
31        else:
32            print("Usage: python %s %s %s" % (_progname, common_params, extra_params))
33        sys.exit(1)
34
35    @staticmethod
36    def run(search, print_result = print_solution):
37        search.print_parameters(_algname)
38        begin = time.time()
39        res = search.run()
40        if print_result is not None and res is not None:
41            print_result(res)
42        end = time.time()
43        print("# CPU time used: %.6f seconds." % (end - begin))
44
45 if __name__ == "__main__":
46    pass
```

Supplementary source code

Here are the hyperlinks to the directories where the source code described in this chapter resides.

https://github.com/cwtsaiai/metaheuristics_2023/src/python/0-es/1-onemax/
https://github.com/cwtsaiai/metaheuristics_2023/src/python/0-es/2-deception/
https://github.com/cwtsaiai/metaheuristics_2023/src/python/1-onemax/
https://github.com/cwtsaiai/metaheuristics_2023/src/python/2-deception/
https://github.com/cwtsaiai/metaheuristics_2023/src/python/3-tsp/
https://github.com/cwtsaiai/metaheuristics_2023/src/python/4-function/
https://github.com/cwtsaiai/metaheuristics_2023/src/python/5-eval/
https://github.com/cwtsaiai/metaheuristics_2023/src/python/6-clustering/
https://github.com/cwtsaiai/metaheuristics_2023/src/python/a-lib/

References

Abbasi, A.A., Younis, M., 2007. A survey on clustering algorithms for wireless sensor networks. Computer Communications 30, 2826–2841.

Abbass, H., Sarker, R., Newton, C., 2001. PDE: A Pareto-frontier differential evolution approach for multi-objective optimization problems. In: Proceedings of the Congress on Evolutionary Computation, pp. 971–978.

Abdel-Basset, M., Ding, W., El-Shahat, D., 2021. A hybrid Harris Hawks optimization algorithm with simulated annealing for feature selection. Artificial Intelligence Review 54, 593–637.

Ackley, D.H., 1987. A Connectionist Machine for Genetic Hillclimbing. Kluwer Academic Publishers, Norwell, MA.

Adorio, E.P., Diliman, U., 2005. MVF – Multivariate Test Functions Library in C for Unconstrained Global Optimization. Technical Report. University of the Philippines Diliman. www.geocities.ws/eadorio/mvf.pdf. (Accessed 1 August 2022).

Afsar, M.M., Tayarani-N, M.H., 2014. Clustering in sensor networks: A literature survey. Journal of Network and Computer Applications 46, 198–226.

Ahuja, R.K., Orlin, J.B., Tiwari, A., 2000. A greedy genetic algorithm for the quadratic assignment problem. Computers & Operations Research 27, 917–934.

Al-Sultan, K.S., 1995. A tabu search approach to the clustering problem. Pattern Recognition 28, 1443–1451.

Alba, E., 2005. Parallel Metaheuristics: A New Class of Algorithms. Wiley-Interscience, New York, NY.

Alba, E., Luque, G., Nesmachnow, S., 2013. Parallel metaheuristics: Recent advances and new trends. International Transactions in Operational Research 20, 1–48.

Ali, M., Pant, M., Abraham, A., 2009. Simplex differential evolution. Acta Polytechnica Hungarica 6, 95–115.

Ali, M.M., Törn, A., 2000. Optimization of carbon and silicon cluster geometry for Tersoff potential using differential evolution. In: Floudas, C.A., Pardalos, P.M. (Eds.), Optimization in Computational Chemistry and Molecular Biology: Local and Global Approaches. Springer US, Boston, MA, pp. 287–300.

Ali, W., Ahmed, A.A., 2019. Hybrid intelligent phishing website prediction using deep neural networks with genetic algorithm-based feature selection and weighting. IET Information Security 13, 659–669.

Alinaghian, M., Tirkolaee, E.B., Dezaki, Z.K., Hejazi, S.R., Ding, W., 2021. An augmented tabu search algorithm for the green inventory-routing problem with time windows. Swarm and Evolutionary Computation 60, 100802.

Alrashdi, Z., Sayyafzadeh, M., 2019. $(\mu - \lambda)$ evolution strategy algorithm in well placement, trajectory, control and joint optimisation. Journal of Petroleum Science & Engineering 177, 1042–1058.

Alvarez-Valdes, R., Crespo, E., Tamarit, J.M., 2002. Design and implementation of a course scheduling system using tabu search. European Journal of Operational Research 137, 512–523.

Angeline, P., 1998. Using selection to improve particle swarm optimization. In: Proceedings of the Congress on Evolutionary Computation, pp. 84–89.

Ankerst, M., Breunig, M.M., Kriegel, H.P., Sander, J., 1999. OPTICS: Ordering points to identify the clustering structure. In: Proceedings of the ACM SIGMOD International Conference on Management of Data, pp. 49–60.

Applegate, D.L., Bixby, R.E., Chvátal, V., Cook, W.J., 2007. The Traveling Salesman Problem: A Computational Study. Princeton University Press, Princeton, NJ.

Askari, Q., Younas, I., Saeed, M., 2020. Political optimizer: A novel socio-inspired meta-heuristic for global optimization. Knowledge-Based Systems 195, 105709.

Assad, A., Deep, K., 2018. A hybrid harmony search and simulated annealing algorithm for continuous optimization. Information Sciences 450, 246–266.

Atashpaz-Gargari, E., Lucas, C., 2007. Imperialist competitive algorithm: An algorithm for optimization inspired by imperialistic competition. In: Proceedings of the IEEE Congress on Evolutionary Computation, pp. 4661–4667.

Bäck, T., 1993. Optimal mutation rates in genetic search. In: Proceedings of the International Conference on Genetic Algorithms, pp. 2–8.

Bäck, T., 1996. Evolutionary Algorithms in Theory and Practice: Evolution Strategies, Evolutionary Programming, Genetic Algorithms. Oxford University Press, Inc., New York, NY.

Baczyński, D., 2016. A new concept of an artificial ecosystem algorithm for optimization problems. Control and Cybernetics 45, 5–36.

Bai, R., Kendall, G., 2005. An investigation of automated planograms using a simulated annealing based hyper-heuristic. In: Ibaraki, T., Nonobe, K., Yagiura, M. (Eds.), Meta-heuristics: Progress as Real Problem Solvers. Springer US, Boston, MA, pp. 87–108.

Baluja, S., 1994. Population-Based Incremental Learning: A Method for Integrating Genetic Search Based Function Optimization and Competitive Learning. Technical Report. Carnegie Mellon University, Pittsburgh, Pennsylvania. No. CMU-CS-94-163. Available at https://www.ri.cmu.edu/pub_files/pub1/baluja_shumeet_1994_2/baluja_shumeet_1994_2.pdf. (Accessed 1 August 2022).

Bandyopadhyay, S., Maulik, U., 2002. An evolutionary technique based on k-means algorithm for optimal clustering in R^N. Information Sciences 146, 221–237.

Battiti, R., Tecchiolli, G., 1994. The reactive tabu search. Annals of Operations Research 6, 107–220.

Beasley, D., Bull, D.R., Martin, R.R., 1993a. An overview of genetic algorithms: Part 1, fundamentals. University Computing 15, 58–69.

Beasley, D., Bull, D.R., Martin, R.R., 1993b. An overview of genetic algorithms: Part 2, research topics. University Computing 15, 170–181.

Bell, J.E., McMullen, P.R., 2004. Ant colony optimization techniques for the vehicle routing problem. Advanced Engineering Informatics 18, 41–48.

Bergstra, J., Bengio, Y., 2012. Random search for hyper-parameter optimization. Journal of Machine Learning Research 13, 281–305.

Bergstra, J.S., Bardenet, R., Bengio, Y., Kégl, B., 2011. Algorithms for hyper-parameter optimization. In: Proceedings of the International Conference on Neural Information Processing, pp. 2546–2554.

Bethke, A.D., 1976. Comparison of genetic algorithms and gradient-based optimizers on parallel processors: Efficiency of use of processing capacity. Technical Report. University of Michigan, College of Literature, Science, and the Arts, Computer and Communication Sciences Department. Available at https://deepblue.lib.umich.edu/bitstream/handle/2027.42/3571/bab2674.0001.001.pdf?sequence=5&isAllowed=y. (Accessed 1 August 2022).

Beyer, H.G., Schwefel, H.P., 2002. Evolution strategies – A comprehensive introduction. Natural Computing 1, 3–52.

Bi, J., Yuan, H., Duanmu, S., Zhou, M., Abusorrah, A., 2021. Energy-optimized partial computation offloading in mobile-edge computing with genetic simulated-annealing-based particle swarm optimization. IEEE Internet of Things Journal 8, 3774–3785.

Biyanto, T.R., Matradji, Irawan, S., Febrianto, H.Y., Afdanny, N., Rahman, A.H., Gunawan, K.S., Pratama, J.A., Bethiana, T.N., 2017. Killer whale algorithm: An algorithm inspired by the life of killer whale. In: Proceedings of the Information Systems International Conference, pp. 151–157.

Blum, C., 2005a. Ant colony optimization: Introduction and recent trends. Physics of Life Reviews 2, 353–373.

Blum, C., 2005b. Beam-ACO–hybridizing ant colony optimization with beam search: An application to open shop scheduling. Computers & Operations Research 32, 1565–1591.

Blum, C., Aguilera, M.J.B., Roli, A., Sampels, M., 2008. Hybrid Metaheuristics: An Emerging Approach to Optimization. Springer, Berlin, Heidelberg.

Blum, C., Dorigo, M., 2004. The hyper-cube framework for ant colony optimization. IEEE Transactions on Systems, Man and Cybernetics. Part B. Cybernetics 34, 1161–1172.

Blum, C., Puchinger, J., Raidl, G.R., Roli, A., 2010. A brief survey on hybrid metaheuristics. In: Proceedings of the International Conference on Bioinspired Optimization Methods and Their Applications, pp. 3–18.

Blum, C., Puchinger, J., Raidl, G.R., Roli, A., 2011. Hybrid metaheuristics in combinatorial optimization: A survey. Applied Soft Computing 11, 4135–4151.

Blum, C., Roli, A., 2003. Metaheuristics in combinatorial optimization: Overview and conceptual comparison. ACM Computing Surveys 35, 268–308.

Bonyadi, M.R., Michalewicz, Z., 2017. Particle swarm optimization for single objective continuous space problems: A review. Evolutionary Computation 25, 1–54.

Botee, H.M., Bonabeau, E., 1998. Evolving ant colony optimization. Advances in Complex Systems 01, 149–159.

Bouchekara, H.R.E.H., 2017. Most valuable player algorithm: A novel optimization algorithm inspired from sport. Information Sciences 20, 139–195.

Boyd, S., Vandenberghe, L., 2004. Convex Optimization. Cambridge University Press, Cambridge.

Bratton, D., Kennedy, J., 2007. Defining a standard for particle swarm optimization. In: Proceedings of the IEEE Swarm Intelligence Symposium, pp. 120–127.

Brest, J., Greiner, S., Boskovic, B., Mernik, M., Zumer, V., 2006. Self-adapting control parameters in differential evolution: A comparative study on numerical benchmark problems. IEEE Transactions on Evolutionary Computation 10, 646–657.

Brest, J., Maučec, M.S., Bošković, B., 2020. Differential evolution algorithm for single objective bound-constrained optimization: Algorithm j2020. In: Proceedings of the IEEE Congress on Evolutionary Computation, pp. 1–8.

Brest, J., Maučec, M.S., Bošković, B., 2017. Single objective real-parameter optimization: Algorithm jSO. In: Proceedings of the IEEE Congress on Evolutionary Computation, pp. 1311–1318.

Brest, J., Maučec, M.S., Bošković, B., 2019. The 100-digit challenge: Algorithm jDE100. In: Proceedings of the IEEE Congress on Evolutionary Computation, pp. 19–26.

Brindle, A., 1980. Genetic algorithms for function optimization. Ph.D. thesis. Edmonton, Minnesota.

Bujok, P., Kolenovsky, P., 2021. Differential evolution with distance-based mutation-selection applied to CEC 2021 single objective numerical optimisation. In: Proceedings of the IEEE Congress on Evolutionary Computation, pp. 849–856.

Bullnheimer, B., Hartl, R.F., Strauß, C., 1997. A new rank based version of the Ant System – A computational study. Technical Report. Institute of Management Science, University of Vienna. Available at https://epub.wu.ac.at/616/1/document.pdf. (Accessed 1 August 2022).

Burke, E., De Causmaecker, P., Vanden Berghe, G., 1999. A hybrid tabu search algorithm for the nurse rostering problem. In: Proceedings of the Asia-Pacific Conference on Simulated Evolution and Learning, pp. 187–194.

Burke, E., Kendall, G., Newall, J., Hart, E., Ross, P., Schulenburg, S., 2003. Hyperheuristics: An emerging direction in modern search technology. In: Glover, F., Kochenberger, G.A. (Eds.), Handbook of Metaheuristics. Kluwer Academic Publishers, Boston, MA, pp. 457–474.

Burke, E.K., Gendreau, M., Hyde, M., Kendall, G., Ochoa, G., Özcan, E., Qu, R., 2013. Hyper-heuristics: A survey of the state of the art. Journal of the Operational Research Society 64, 1695–1724.

Burke, E.K., Hyde, M., Kendall, G., Ochoa, G., Özcan, E., Woodward, J.R., 2010. A classification of hyper-heuristic approaches. In: Gendreau, M., Potvin, J.Y. (Eds.), Handbook of Metaheuristics. Springer US, Boston, MA, pp. 449–468.

Burke, E.K., McCollum, B., Meisels, A., Petrovic, S., Qu, R., 2007. A graph-based hyper-heuristic for educational timetabling problems. European Journal of Operational Research 176, 177–192.

Camero, A., Toutouh, J., Stolfi, D.H., Alba, E., 2019. Evolutionary deep learning for car park occupancy prediction in smart cities. In: Proceedings of the International Conference on Learning and Intelligent Optimization, pp. 386–401.

Cantú-Paz, E., 1998. A survey of parallel genetic algorithms. Calculateurs Paralleles, Reseaux et Systems Repartis 10, 141–171.

Casotto, A., Romeo, F., Sangiovanni-Vincentelli, A., 1987. A parallel simulated annealing algorithm for the placement of macro-cells. IEEE Transactions on Computer-Aided Design of Integrated Circuits and Systems 6, 838–847.

Černý, V., 1982. A thermodynamical approach to the traveling salesman problem: An efficient simulation algorithm. Technical Report. Comenius University, Bratislava, Czechoslovakia.

Černý, V., 1985. Thermodynamical approach to the traveling salesman problem: An efficient simulation algorithm. Journal of Optimization Theory and Applications 45, 41–51.

Chen, L., Aihara, K., 1995. Chaotic simulated annealing by a neural network model with transient chaos. Neural Networks 8, 915–930.

Chiang, M.C., Tsai, C.W., Yang, C.S., 2011. A time-efficient pattern reduction algorithm for k-means clustering. Information Sciences 181, 716–731.

Chu, S.C., Tsai, P.W., Pan, J.S., 2006. Cat swarm optimization. In: Proceedings of the Pacific Rim International Conference on Artificial Intelligence, pp. 854–858.

Chung, H., Shin, K.-s., 2020. Genetic algorithm-optimized multi-channel convolutional neural network for stock market prediction. Neural Computing & Applications 32, 7897–7914.

Civicioglu, P., 2013. Artificial cooperative search algorithm for numerical optimization problems. Information Sciences 229, 58–76.

Clerc, M., 1993. The swarm and the queen: Towards a deterministic and adaptive particle swarm optimization. In: Proceedings of the Congress on Evolutionary Computation, pp. 1951–1957.

Clerc, M., Kennedy, J., 2002. The particle swarm - explosion, stability, and convergence in a multidimensional complex space. IEEE Transactions on Evolutionary Computation 6, 58–73.

Coello, C.A.C., Lechuga, M.S., 2002. MOPSO: A proposal for multiple objective particle swarm optimization. In: Proceedings of the Congress on Evolutionary Computation, pp. 1051–1056.

Coello, C.A.C., Pulido, G.T., Lechuga, M.S., 2004. Handling multiple objectives with particle swarm optimization. IEEE Transactions on Evolutionary Computation 8, 256–279.

Colorni, A., Dorigo, M., Maniezzo, V., Varela, F., Bourgine, P., 1992. Distributed optimization by ant colonies. In: Proceedings of the European Conference on Artificial Life, pp. 134–152.

da Conceição, M., Ribeiro, L., 2004. Tabu search algorithms for water network optimization. European Journal of Operational Research 157, 746–758.

Cowling, P., Kendall, G., Soubeiga, E., 2001. A hyperheuristic approach to scheduling a sales summit. In: Proceedings of the International Conference on the Practice and Theory of Automated Timetabling, pp. 176–190.

Crainic, T.G., 2005. Parallel computation, co-operation, tabu search. In: Sharda, R., Voß, S., Rego, C., Alidaee, B. (Eds.), Metaheuristic Optimization via Memory and Evolution: Tabu Search and Scatter Search. Springer US, Boston, MA, pp. 283–302.

Crainic, T.G., Toulouse, M., 1998. Parallel metaheuristics. In: Crainic, G., Laporte, G. (Eds.), Fleet Management and Logistics. Kluwer Academic Publishers, Boston, MA, pp. 205–251.

Crainic, T.G., Toulouse, M., Gendreau, M., 1995. Synchronous tabu search parallelization strategies for multicommodity location-allocation with balancing requirements. Operations-Research-Spektrum 17, 113–123.

Croes, G.A., 1958. A method for solving traveling salesman problems. Operations Research 5, 791–812.

Cung, V.D., Martins, S.L., Ribeiro, C.C., Roucairol, C., 2002. Strategies for the parallel implementation of metaheuristics. In: Proceedings of the Essays and Surveys in Metaheuristics, pp. 263–308.

Dahal, K., Remde, S., Cowling, P., Colledge, N., 2008. Improving metaheuristic performance by evolving a variable fitness function. In: Proceedings of the Evolutionary Computation in Combinatorial Optimization, pp. 170–181.

Das, S., Abraham, A., Konar, A., 2008. Automatic clustering using an improved differential evolution algorithm. IEEE Transactions on Systems, Man and Cybernetics. Part A. Systems and Humans 38, 218–237.

Das, S., Mullick, S.S., Suganthan, P., 2016. Recent advances in differential evolution – An updated survey. Swarm and Evolutionary Computation 27, 1–30.

Das, S., Suganthan, P.N., 2011. Differential evolution: A survey of the state-of-the-art. IEEE Transactions on Evolutionary Computation 15, 4–31.

Davis, L., 1991. Handbook of Genetic Algorithms. Van Nostrand Reinhold, New York, NY.

De Jong, K.A., 1975. An Analysis of the Behavior of a Class of Genetic Adaptive Systems. Ph.D. thesis. Ann Arbor, Michigan.

De Jong, K.A., Spears, W.M., 1992. A formal analysis of the role of multi-point crossover in genetic algorithms. Annals of Mathematics and Artificial Intelligence 5, 1–26.

Deb, K., Kalyanmoy, D., 2001. Multi-Objective Optimization Using Evolutionary Algorithms. John Wiley & Sons, Inc., New York, NY.

Deb, K., Pratap, A., Agarwal, S., Meyarivan, T., 2002. A fast and elitist multiobjective genetic algorithm: NSGA-II. IEEE Transactions on Evolutionary Computation 6, 182–197.

Delahaye, D., Chaimatanan, S., Mongeau, M., 2019. Simulated annealing: From basics to applications. In: Gendreau, M., Potvin, J.Y. (Eds.), Handbook of Metaheuristics. Springer International Publishing, Cham, pp. 1–35.

Dell'Amico, M., Trubian, M., 1993. Applying tabu search to the job-shop scheduling problem. Annals of Operations Research 41, 231–252.

Deng, W., Liu, H., Xu, J., Zhao, H., Song, Y., 2020. An improved quantum-inspired differential evolution algorithm for deep belief network. IEEE Transactions on Instrumentation and Measurement 69, 7319–7327.

Derrac, J., García, S., Molina, D., Herrera, F., 2011. A practical tutorial on the use of nonparametric statistical tests as a methodology for comparing evolutionary and swarm intelligence algorithms. Swarm and Evolutionary Computation 1, 3–18.

Ding, S., Su, C., Yu, J., 2011. An optimizing BP neural network algorithm based on genetic algorithm. Artificial Intelligence Review 36, 153–162.

Doerner, K., Gutjahr, W.J., Hartl, R.F., Strauß, C., Stummer, C., 2004. Pareto ant colony optimization: A metaheuristic approach to multiobjective portfolio selection. Annals of Operations Research 131, 79–99.

Dorigo, M., Birattari, M., Stützle, T., 2006. Ant colony optimization. IEEE Computational Intelligence Magazine 1, 28–39.

Dorigo, M., Blum, C., 2005. Ant colony optimization theory: A survey. Theoretical Computer Science 344, 243–278.

Dorigo, M., Gambardella, L.M., 1997. Ant colony system: A cooperative learning approach to the traveling salesman problem. IEEE Transactions on Evolutionary Computation 1, 53–66.

Dorigo, M., Maniezzo, V., Colorni, A., 1991. Ant System: An Autocatalytic Optimizing Process. Technical Report. Dipartimento di Elettronica, Politecnico di Milano, Italy. Technical Report 91-016. Available at https://citeseerx.ist.psu.edu/viewdoc/download?rep=rep1&type=pdf&doi=10.1.1.51.4214. (Accessed 1 August 2022).

Dorigo, M., Maniezzo, V., Colorni, A., 1996. The Ant System: Optimization by a colony of cooperating agents. IEEE Transactions on Systems, Man and Cybernetics. Part B. Cybernetics 26, 29–41.

Dorigo, M., Stützle, T., 2004. Ant Colony Optimization. The MIT Press, Cambridge, MA.

Dorigo, M., Stützle, T., 2010. Ant colony optimization: Overview and recent advances. In: Gendreau, M., Potvin, J.Y. (Eds.), Handbook of Metaheuristics. Springer US, Boston, MA, pp. 227–263.

Dragoi, E.N., Dafinescu, V., 2016. Parameter control and hybridization techniques in differential evolution: A survey. Artificial Intelligence Review 45, 447–470.

Dua, D., Graff, C., 2017. UCI machine learning repository. http://archive.ics.uci.edu/ml. (Accessed 1 August 2022).

Duan, H., Yu, X., 2007. Hybrid ant colony optimization using memetic algorithm for traveling salesman problem. In: Proceedings of the IEEE International Symposium on Approximate Dynamic Programming and Reinforcement Learning, pp. 92–95.

Eberhart, R.C., Shi, Y., 1998. Comparison between genetic algorithms and particle swarm optimization. In: Proceedings of the International Conference on Evolutionary Programming, pp. 611–616.

Eberhart, R.C., Shi, Y., 2001. Particle swarm optimization: Developments, applications and resources. In: Proceedings of the Congress on Evolutionary Computation, pp. 81–86.

Eesa, A.S., Brifcani, A.M.A., Orman, Z., 2013. Cuttlefish algorithm – A novel bio-inspired optimization algorithm. International Journal of Scientific and Engineering Research 4, 1978–1986.

Eiben, A., Hinterding, R., Michalewicz, Z., 1999. Parameter control in evolutionary algorithms. IEEE Transactions on Evolutionary Computation 3, 124–141.

Elkan, C., 2003. Using the triangle inequality to accelerate k-means. In: Proceedings of the International Conference on International Conference on Machine Learning, pp. 147–153.

Emami, H., Derakhshan, F., 2015. Election algorithm: A new socio-politically inspired strategy. AI Communications 28, 591–603.

Engelbrecht, A.P., 2006. Fundamentals of Computational Swarm Intelligence. John Wiley & Sons, Hoboken, NJ.

Engin, O., Güçlü, A., 2018. A new hybrid ant colony optimization algorithm for solving the no-wait flow shop scheduling problems. Applied Soft Computing 72, 166–176.

Eskandar, H., Sadollah, A., Bahreininejad, A., Hamdi, M., 2012. Water cycle algorithm — A novel metaheuristic optimization method for solving constrained engineering optimization problems. Computers & Structures 110–111, 151–166.

Ester, M., Kriegel, H.P., Sander, J., Xu, X., 1996. A density-based algorithm for discovering clusters in large spatial databases with noise. In: Proceedings of the International Conference on Knowledge Discovery and Data Mining, pp. 226–231.

Eusuff, M., Lansey, K., Pasha, F., 2006. Shuffled frog-leaping algorithm: A memetic metaheuristic for discrete optimization. Engineering Optimization 38, 129–154.

Feo, T.A., Resende, M.G.C., 1995. Greedy randomized adaptive search procedures. Journal of Global Optimization 6, 109–133.

Ferreira, K.M., de Queiroz, T.A., 2018. Two effective simulated annealing algorithms for the location-routing problem. Applied Soft Computing 70, 389–422.

Fleetwood, K., 2004. An introduction to differential evolution. In: Proceedings of Mathematics and Statistics of Complex Systems (MASCOS) One Day Symposium, pp. 785–791.

Fogel, L., Owens, A., Walsh, M., 1966. Artificial Intelligence Through Simulated Evolution. Wiley, Chichester, WS.

Gambardella, L.M., Dorigo, M., 1995. Ant-Q: A reinforcement learning approach to the traveling salesman problem. In: Proceedings of the International Conference on Machine Learning, pp. 252–260.

García-Nájera, A., Zapotecas-Martínez, S., Miranda, K., 2021. Analysis of the multi-objective cluster head selection problem in WSNs. Applied Soft Computing 112, 107853.

García-Nieto, J., Olivera, A.C., Alba, E., 2013. Optimal cycle program of traffic lights with particle swarm optimization. IEEE Transactions on Evolutionary Computation 17, 823–839.

Geem, Z.W., 2009. Music-Inspired Harmony Search Algorithm: Theory and Applications. Springer, Berlin, Heidelberg.

Geem, Z.W., Kim, J.H., Loganathan, G., 2001. A new heuristic optimization algorithm: Harmony search. Simulation 76, 60–68.

Gendreau, M., Hertz, A., Laporte, G., 1994. A tabu search heuristic for the vehicle routing problem. Management Science 40, 1276–1290.

Gendreau, M., Potvin, J.Y., 2005. Tabu search. In: Burke, E.K., Kendall, G. (Eds.), Search Methodologies: Introductory Tutorials in Optimization and Decision Support Techniques. Springer US, Boston, MA, pp. 165–186.

Glover, F., 1977. Heuristics for integer programming using surrogate constraints. Decision Sciences 8, 156–166.

Glover, F., 1986. Future paths for integer programming and links to artificial intelligence. Computers & Operations Research 13, 533–549.

Glover, F., 1989. Tabu search—part I. ORSA Journal on Computing 1, 190–206.

Glover, F., 1990a. Tabu search—part II. ORSA Journal on Computing 2, 4–32.

Glover, F., 1990b. Tabu search: A tutorial. Interfaces 20, 74–94.

Glover, F., Kochenberger, G.A. (Eds.), 2003. Handbook of Metaheuristics. Kluwer Academic Publishers, Boston, MA.

Glover, F., Laguna, M., 1997. Tabu Search. Kluwer Academic Publishers, Boston, MA.

Glover, F., Laguna, M., Martí, R., 2007. Principles of tabu search. In: Gonzalez, T.F. (Ed.), Handbook of Approximation Algorithms and Metaheuristics. Chapman and Hall/CRC, Boca Raton, FL, pp. 23-1–23-12.

Glover, F., Taillard, E., Taillard, E., 1993. A user's guide to tabu search. Annals of Operations Research 41, 1–28.

Goldberg, D.E., 1989. Genetic Algorithms in Search, Optimization and Machine Learning. Addison-Wesley Longman Publishing Co., Inc., Boston, MA.

Goldberg, D.E., Deb, K., 1991. A comparative analysis of selection schemes used in genetic algorithms. In: Rawlins, G.J.E. (Ed.), Foundations of Genetic Algorithms, vol. 1. Morgan Kaufmann, San Mateo, CA, pp. 69–93.

Gonçalves, J.F., de Magalhães Mendes, J.J., Resende, M.G., 2005. A hybrid genetic algorithm for the job shop scheduling problem. European Journal of Operational Research 167, 77–95.

Gong, Y.J., Li, J.J., Zhou, Y., Li, Y., Chung, H.S.H., Shi, Y.H., Zhang, J., 2016. Genetic learning particle swarm optimization. IEEE Transactions on Cybernetics 46, 2277–2290.

Greening, D.R., 1990. Parallel simulated annealing techniques. Physica D: Nonlinear Phenomena 42, 293–306.

Grefenstette, J.J., 1999. Evolvability in dynamic fitness landscapes: A genetic algorithm approach. In: Proceedings of the Congress on Evolutionary Computation, pp. 2031–2038.

Grötschel, M., Lovász, L., Schrijver, A., 1981. The ellipsoid method and its consequences in combinatorial optimization. Combinatorica 1, 169–197.

Guo, S.M., Yang, C.C., 2015. Enhancing differential evolution utilizing eigenvector-based crossover operator. IEEE Transactions on Evolutionary Computation 19, 31–49.

Hadi, A.A., Mohamed, A.W., Jambi, K.M., 2021. Single-objective real-parameter optimization: Enhanced LSHADE-SPACMA algorithm. In: Yalaoui, F., Amodeo, L., Talbi, E.G. (Eds.), Heuristics for Optimization and Learning. Springer International Publishing, Cham, pp. 103–121.

Han, J., Kamber, M., Pei, J., 2012. Data Mining: Concepts and Techniques, third edition. Morgan Kaufmann, Boston, MA.

Han, K.H., Kim, J.H., 2002. Quantum-inspired evolutionary algorithm for a class of combinatorial optimization. IEEE Transactions on Evolutionary Computation 6, 580–593.

Han, L., Kendall, G., 2003. An investigation of a tabu assisted hyper-heuristic genetic algorithm. In: Proceedings of the Congress on Evolutionary Computation, pp. 2230–2237.

Hancock, P.J.B., 1994. An empirical comparison of selection methods in evolutionary algorithms. In: Proceedings of the Evolutionary Computing, pp. 80–94.

Harifi, S., Khalilian, M., Mohammadzadeh, J., Ebrahimnejad, S., 2019. Emperor penguins colony: A new metaheuristic algorithm for optimization. Evolutionary Intelligence 12, 211–226.

Harik, G.R., Lobo, F.G., 1999. A parameter-less genetic algorithm. In: Proceedings of the Genetic and Evolutionary Computation Conference, pp. 258–265.

Harrison, K.R., Engelbrecht, A.P., Ombuki-Berman, B.M., 2017. Self-adaptive particle swarm optimization: A review and analysis of convergence. Swarm Intelligence 12, 187–226.

Hassan, R., 2004. Particle Swarm Optimization: Method and Applications. Technical Report. Engineering Systems Division, Massachusetts Institute of Technology. Available at https://dspace.mit.edu/bitstream/handle/1721.1/68163/16-888-spring-2004/contents/lecture-notes/l13_msdo_pso.pdf. (Accessed 1 August 2022).

Hassibi, B., Stork, D.G., Wolff, G.J., 1993. Optimal brain surgeon and general network pruning. In: Proceedings of the IEEE International Conference on Neural Networks, pp. 293–299.

He, Y., Zhang, X., Sun, J., 2017. Channel pruning for accelerating very deep neural networks. In: Proceedings of the IEEE International Conference on Computer Vision, pp. 1398–1406.

Heidari, A.A., Mirjalili, S., Faris, H., Aljarah, I., Mafarja, M., Chen, H., 2019. Harris hawks optimization: Algorithm and applications. Future Generation Computer Systems 97, 849–872.

Heinzelman, W., Chandrakasan, A., Balakrishnan, H., 2000. Energy-efficient communication protocol for wireless microsensor networks. In: Proceedings of the Annual Hawaii International Conference on System Sciences, pp. 1–10.

Helsgaun, K., 2000. An effective implementation of the Lin–Kernighan traveling salesman heuristic. European Journal of Operational Research 126, 106–130.

Helsgaun, K., 2009. General k-opt submoves for the Lin—Kernighan TSP heuristic. Mathematical Programming Computation 1, 119–163.

Henderson, D., Jacobson, S.H., Johnson, A.W., 2003. The theory and practice of simulated annealing. In: Glover, F., Kochenberger, G.A. (Eds.), Handbook of Metaheuristics. Kluwer Academic Publishers, Boston, MA, pp. 287–319.

Hertz, A., de Werra, D., 1988. Using tabu search techniques for graph coloring. Computing 39, 345–351.

Higashi, N., Iba, H., 2003. Particle swarm optimization with Gaussian mutation. In: Proceedings of the Swarm Intelligence Symposium, pp. 72–79.

Hill, R.R., 1999. A Monte-Carlo study of genetic algorithm initial population generation methods. In: Proceedings of the Conference on Winter Simulation: Simulation—A Bridge to the Future, pp. 543–547.

Hinterding, R., Michalewicz, Z., Eiben, A., 1997. Adaptation in evolutionary computation: a survey. In: Proceedings of IEEE International Conference on Evolutionary Computation, pp. 65–69.

Holland, J.H., 1962. Outline for a logical theory of adaptive systems. Journal of the ACM 9, 297–314.

Holland, J.H., 1975. Adaptation in Natural and Artificial Systems. University of Michigan Press, Ann Arbor, MI.

Holland, J.H., 1992. Adaptation in Natural and Artificial Systems. MIT Press, Cambridge, MA.

Hoseini, P., Shayesteh, M.G., 2010. Hybrid ant colony optimization, genetic algorithm, and simulated annealing for image contrast enhancement. In: Proceedings of the IEEE Congress on Evolutionary Computation, pp. 1–6.

Hu, Y., Sun, S., Li, J., Wang, X., Gu, Q., 2018. A novel channel pruning method for deep neural network compression. https://doi.org/10.48550/ARXIV.1805.11394.

Huang, K.L., Liao, C.J., 2008. Ant colony optimization combined with taboo search for the job shop scheduling problem. Computers & Operations Research 35, 1030–1046.

Ilonen, J., Kamarainen, J.K., Lampinen, J., 2003. Differential evolution training algorithm for feed-forward neural networks. Neural Processing Letters 17, 93–105.

Ingber, L., 1989. Very fast simulated re-annealing. Mathematical and Computer Modelling 12, 967–973.

Ingber, L., 1993. Simulated annealing: Practice versus theory. Mathematical and Computer Modelling 18, 29–57.

Jadon, S.S., Tiwari, R., Sharma, H., Bansal, J.C., 2017. Hybrid artificial bee colony algorithm with differential evolution. Applied Soft Computing 58, 11–24.

Jain, A.K., Murty, M.N., Flynn, P.J., 1999. Data clustering: A review. ACM Computing Surveys 31, 264–323.

Jain, R.K., 1991. The Art of Computer Systems Performance Analysis: Techniques for Experimental Design, Measurement, Simulation, and Modeling. Wiley, New York, NY.

Jamil, M., Yang, X.S., 2013. A literature survey of benchmark functions for global optimisation problems. International Journal of Mathematical Modelling and Numerical Optimisation 4, 150–193.

Janert, P.K., 2015. Gnuplot in Action, 2nd ed. Manning Publications Co., Greenwich, CT.

Jaradat, G.M., Ayob, M., 2010. Big Bang-Big Crunch optimization algorithm to solve the course timetabling problem. In: Proceedings of the International Conference on Intelligent Systems Design and Applications, pp. 1448–1452.

Jiang, J., Han, F., Ling, Q., Wang, J., Li, T., Han, H., 2020. Efficient network architecture search via multiobjective particle swarm optimization based on decomposition. Neural Networks 123, 305–316.

Jin, Y., 2005. A comprehensive survey of fitness approximation in evolutionary computation. Soft Computing 9, 3–12.

Jin, Y., Olhofer, M., Sendhoff, B., 2001. Managing approximate models in evolutionary aerodynamic design optimization. In: Proceedings of the Congress on Evolutionary Computation, pp. 592–599.

Jones, T., 1995. Evolutionary Algorithms, Fitness Landscapes and Search. Ph.D. thesis. The University of New Mexico, Albuquerque, New Mexico.

Juang, C.F., 2004. A hybrid of genetic algorithm and particle swarm optimization for recurrent network design. IEEE Transactions on Systems, Man and Cybernetics. Part B. Cybernetics 34, 997–1006.

Kadioglu, S., Sellmann, M., 2009. Dialectic search. In: Proceedings of the International Conference on Principles and Practice of Constraint Programming, pp. 486–500.

Kao, Y.T., Zahara, E., 2008. A hybrid genetic algorithm and particle swarm optimization for multimodal functions. Applied Soft Computing 8, 849–857.

Karaboga, D., 2005. An idea based on honey bee swarm for numerical optimization. Technical Report. Department of Computer Engineering, Erciyes University. Available at https://abc.erciyes.edu.tr/pub/tr06_2005.pdf. (Accessed 1 August 2022).

Karamichailidou, D., Kaloutsa, V., Alexandridis, A., 2021. Wind turbine power curve modeling using radial basis function neural networks and tabu search. Renewable Energy 163, 2137–2152.

Kashan, A.H., 2009. League championship algorithm: A new algorithm for numerical function optimization. In: Proceedings of the International Conference of Soft Computing and Pattern Recognition, pp. 43–48.

Katoch, S., Chauhan, S.S., Kumar, V., 2021. A review on genetic algorithm: past, present, and future. Multimedia Tools and Applications 80, 8091–8126.

Katsigiannis, Y.A., Georgilakis, P.S., Karapidakis, E.S., 2012. Hybrid simulated annealing–tabu search method for optimal sizing of autonomous power systems with renewables. IEEE Transactions on Sustainable Energy 3, 330–338.

Kauffman, S.A., Weinberger, E.D., 1989. The NK model of rugged fitness landscapes and its application to maturation of the immune response. Journal of Theoretical Biology 141, 211–245.

Kaveh, A., Mahdavi, V., 2014. Colliding bodies optimization: A novel meta-heuristic method. Computers & Structures 139, 18–27.

Kaveh, A., Zaerreza, A., Hosseini, S.M., 2020. Shuffled shepherd optimization method simplified for reducing the parameter dependency. Iranian Journal of Science and Technology, Transactions of Civil Engineering, 1–15.

Kelly, J.D., Davis, L., 1991. A hybrid genetic algorithm for classification. In: Proceedings of the International Joint Conference on Artificial Intelligence, pp. 645–650.

Kenichi, T., Keiichiro, Y., 2010. Primary study of spiral dynamics inspired optimization. IEEJ Transactions on Electrical and Electronic Engineering 6, S98–S100.

Kennedy, J., 1999. Small worlds and mega-minds: Effects of neighborhood topology on particle swarm performance. In: Proceedings of the Congress on Evolutionary Computation, pp. 1931–1938.

Kennedy, J., Eberhart, R.C., 1995. Particle swarm optimization. In: Proceedings of the IEEE International Conference on Neural Networks, pp. 1942–1948.

Kennedy, J., Eberhart, R.C., 1997. A discrete binary version of the particle swarm algorithm. In: Proceedings of the IEEE International Conference on Systems, Man, and Cybernetics, pp. 4104–4108.

Khachaturyan, A., Semenovsovskaya, S., Vainshtein, B., 1981. The thermodynamic approach to the structure analysis of crystals. Acta Crystallographica. Section A 37, 742–754.

Kim, H.S., Cho, S.B., 2000. Application of interactive genetic algorithm to fashion design. Engineering Applications of Artificial Intelligence 13, 635–644.

Kim, T.Y., Cho, S.B., 2019. Particle swarm optimization-based CNN-LSTM networks for forecasting energy consumption. In: Proceedings of the IEEE Congress on Evolutionary Computation, pp. 1510–1516.

Kirkpatrick, S., Gelatt, C.D., Vecchi, M.P., 1983. Optimization by simulated annealing. Science 220, 671–680.

Krishna, K., Narasimha Murty, M., 1999. Genetic k-means algorithm. IEEE Transactions on Systems, Man and Cybernetics. Part B. Cybernetics 29, 433–439.

Krishnanand, K., Ghose, D., 2005. Detection of multiple source locations using a glowworm metaphor with applications to collective robotics. In: Proceedings of the IEEE Swarm Intelligence Symposium, pp. 84–91.

Kuo, T., Hwang, S.Y., 1996. A genetic algorithm with disruptive selection. IEEE Transactions on Systems, Man and Cybernetics. Part B. Cybernetics 26, 299–307.

Lai, J.Z., Liaw, Y.C., Liu, J., 2008. A fast VQ codebook generation algorithm using codeword displacement. Pattern Recognition 41, 315–319.

Lalwani, S., Sharma, H., Satapathy, S.C., Deep, K., Bansal, J.C., 2019. A survey on parallel particle swarm optimization algorithms. Arabian Journal for Science and Engineering 44, 2899–2923.

Lander, E.S., Green, P., Abrahamson, J., Barlow, A., Daly, M.J., Lincoln, S.E., Newburg, L., 1987. MAPMAKER: An interactive computer package for constructing primary genetic linkage maps of experimental and natural populations. Genomics 1, 174–181.

Larraanaga, P., Lozano, J.A., 2001. Estimation of Distribution Algorithms: A New Tool for Evolutionary Computation. Kluwer Academic Publishers, Norwell, MA.

Larrañaga, P., Kuijpers, C., Murga, R., Inza, I., Dizdarevic, S., 1999. Genetic algorithms for the travelling salesman problem: A review of representations and operators. Artificial Intelligence Review 13, 129–170.

LeCun, Y., Denker, J., Solla, S., 1989. Optimal brain damage. In: Proceedings of the International Conference on Neural Information Processing Systems, pp. 598–605.

Lee, W.Y., Park, S.M., Sim, K.B., 2018. Optimal hyperparameter tuning of convolutional neural networks based on the parameter-setting-free harmony search algorithm. Optik 172, 359–367.

Leung, F.H.F., Lam, H.K., Ling, S.H., Tam, P.K.S., 2003. Tuning of the structure and parameters of a neural network using an improved genetic algorithm. IEEE Transactions on Neural Networks 14, 79–88.

Li, H., Kadav, A., Durdanovic, I., Samet, H., Graf, H.P., 2017. Pruning filters for efficient ConvNets. In: Proceedings of the International Conference on Learning Representations.

Li, L., Jamieson, K., DeSalvo, G., Rostamizadeh, A., Talwalkar, A., 2018. Hyperband: A novel bandit-based approach to hyperparameter optimization. Journal of Machine Learning Research 18, 1–52.

Li, M., Hao, J.K., Wu, Q., 2022. Learning-driven feasible and infeasible tabu search for airport gate assignment. European Journal of Operational Research 302, 172–186.

Li, X., Gao, L., 2016. An effective hybrid genetic algorithm and tabu search for flexible job shop scheduling problem. International Journal of Production Economics 174, 93–110.

Li, Y., Jia, M., Han, X., Bai, X.S., 2021. Towards a comprehensive optimization of engine efficiency and emissions by coupling artificial neural network (ANN) with genetic algorithm (GA). Energy 225, 1–13.

Liang, J., Xu, W., Yue, C., Yu, K., Song, H., Crisalle, O.D., Qu, B., 2019. Multimodal multiobjective optimization with differential evolution. Swarm and Evolutionary Computation 44, 1028–1059.

Liang, J.J., Suganthan, P.N., 2005. Dynamic multi-swarm particle swarm optimizer. In: Proceedings IEEE Swarm Intelligence Symposium, pp. 124–129.

Liang, Y.C., Smith, A., 2004. An ant colony optimization algorithm for the redundancy allocation problem (RAP). IEEE Transactions on Reliability 53, 417–423.

Liaw, C.F., 2000. A hybrid genetic algorithm for the open shop scheduling problem. European Journal of Operational Research 124, 28–42.

Lin, F.T., Kao, C.Y., Hsu, C.C., 1993. Applying the genetic approach to simulated annealing in solving some NP-hard problems. IEEE Transactions on Systems, Man and Cybernetics 23, 1752–1767.

Lin, M., Ji, R., Zhang, Y., Zhang, B., Wu, Y., Tian, Y., 2020. Channel pruning via automatic structure search. In: Proceedings of the International Joint Conference on Artificial Intelligence, pp. 673–679.

Lin, S., 1965. Computer solutions of the traveling salesman problem. Bell System Technical Journal 44, 2245–2269.

Lin, S., Kernighan, B.W., 1973. An effective heuristic algorithm for the traveling-salesman problem. Operations Research 21, 2.

Lin, W.M., Cheng, F.S., Tsay, M.T., 2002. An improved tabu search for economic dispatch with multiple minima. IEEE Transactions on Power Systems 17, 108–112.

Liu, B., Wang, L., Jin, Y.H., Tang, F., Huang, D.X., 2005. Improved particle swarm optimization combined with chaos. Chaos, Solitons and Fractals 25, 1261–1271.

Liu, H., Motoda, H., 1998. Feature Extraction, Construction and Selection: A Data Mining Perspective. Kluwer Academic Publishers, Boston, MA.

Liu, J., Lampinen, J., 2002. A fuzzy adaptive differential evolution algorithm. In: Proceedings of the Conference on Computers, Communications, Control and Power Engineering, pp. 606–611.

Liu, Z., Li, J., Shen, Z., Huang, G., Yan, S., Zhang, C., 2017. Learning efficient convolutional networks through network slimming. In: Proceedings of the IEEE International Conference on Computer Vision, pp. 2755–2763.

Lorenzo, P.R., Nalepa, J., Kawulok, M., Ramos, L.S., Pastor, J.R., 2017. Particle swarm optimization for hyper-parameter selection in deep neural networks. In: Proceedings of the Genetic and Evolutionary Computation Conference, pp. 481–488.

Lourenço, H.R., Martin, O.C., Stützle, T., 2003. Iterated local search. In: Glover, F., Kochenberger, G.A. (Eds.), Handbook of Metaheuristics. Kluwer Academic Publishers, Boston, MA, pp. 321–353.

Loussaief, S., Abdelkrim, A., 2018. Convolutional neural network hyper-parameters optimization based on genetic algorithms. International Journal of Advanced Computer Science and Applications 9, 252–266.

Lozano, M., García-Martínez, C., 2010. Hybrid metaheuristics with evolutionary algorithms specializing in intensification and diversification: Overview and progress report. Computers & Operations Research 37, 481–497.

Luke, B.T., 2018. Simulated annealing cooling schedules. Available at http://web.archive. org/web/20190620071317/http://www.btluke.com/simanf1.html. (Accessed 1 August 2022).

Maaranen, H., Miettinen, K., Penttinen, A., 2006. On initial populations of a genetic algorithm for continuous optimization problems. Journal of Global Optimization 37, 405–436.

MacQueen, J.B., 1967. Some methods for classification and analysis of multivariate observations. In: Proceedings of the Berkeley Symposium on Mathematical Statistics and Probability, pp. 281–297.

Mahfoud, S.W., Goldberg, D.E., 1995. Parallel recombinative simulated annealing: A genetic algorithm. Parallel Computing 21, 1–28.

Mahi, M., Baykan, Ö.K., Kodaz, H., 2015. A new hybrid method based on particle swarm optimization, ant colony optimization and 3-opt algorithms for traveling salesman problem. Applied Soft Computing 30, 484–490.

Malan, K.M., Engelbrecht, A.P., 2013. A survey of techniques for characterising fitness landscapes and some possible ways forward. Information Sciences 241, 148–163.

Mantawy, A., Abdel-Magid, Y.L., Selim, S.Z., 1999. Integrating genetic algorithms, tabu search, and simulated annealing for the unit commitment problem. IEEE Transactions on Power Systems 14, 829–836.

Mavko, G., Mukerji, T., Dvorkin, J., 2009. The Rock Physics Handbook: Tools for Seismic Analysis of Porous Media. Cambridge University Press, Cambridge.

Merkle, D., Middendorf, M., Schmeck, H., 2002. Ant colony optimization for resource-constrained project scheduling. IEEE Transactions on Evolutionary Computation 6, 333–346.

Metropolis, N., Rosenbluth, A.W., Rosenbluth, M.N., Teller, A.H., 1953. Equation of state calculations by fast computing machines. Journal of Chemical Physics 21, 1087–1092.

Mezura-Montes, E., Reyes-Sierra, M., Coello, C.A.C., 2008. Multi-objective optimization using differential evolution: A survey of the state-of-the-art. In: Chakraborty, U.K. (Ed.), Advances in Differential Evolution. Springer, Berlin, Heidelberg, pp. 173–196.

Michalewicz, Z., 1996. Genetic Algorithms + Data Structures = Evolution Programs. Springer, Berlin, Heidelberg.

Michalski, R.S., 2000. Learnable evolution model: Evolutionary processes guided by machine learning. Machine Learning 38, 9–40.

Miller, G.F., Todd, P.M., Hegde, S.U., 1989. Designing neural networks using genetic algorithms. In: Proceedings of the International Conference on Genetic Algorithms, pp. 379–384.

Mininno, E., Neri, F., Cupertino, F., Naso, D., 2011. Compact differential evolution. IEEE Transactions on Evolutionary Computation 15, 32–54.

Mirjalili, S., 2016. Dragonfly algorithm: A new meta-heuristic optimization technique for solving single-objective, discrete, and multi-objective problems. Neural Computing & Applications 27, 1053–1073.

Mirjalili, S., Lewis, A., 2016. The whale optimization algorithm. Advances in Engineering Software 95, 51–67.

Mirjalili, S., Mirjalili, S.M., Lewis, A., 2014. Grey wolf optimizer. Advances in Engineering Software 69, 46–61.

Mitchell, M., 1996. An Introduction to Genetic Algorithms. MIT Press, Cambridge, MA.

Mladenović, N., Hansen, P., 1997. Variable neighborhood search. Computers & Operations Research 24, 1097–1100.

Mohamed, A.W., Mohamed, A.K., 2019. Adaptive guided differential evolution algorithm with novel mutation for numerical optimization. International Journal of Machine Learning and Cybernetics 10, 253–277.

Montana, D.J., Davis, L., 1989. Training feedforward neural networks using genetic algorithms. In: Proceedings of the International Joint Conference on Artificial Intelligence, pp. 762–767.

Moore, J., Chapman, R., Dozier, G., 2000. Multiobjective particle swarm optimization. In: Proceedings of the 38th Annual on Southeast Regional Conference, pp. 56–57.

Moscato, P., 1989. On Evolution, Search, Optimization, Genetic Algorithms and Martial Arts - Towards Memetic Algorithms. Technical Report. California Institute of Technology. Available at https://citeseerx.ist.psu.edu/viewdoc/download?doi=10.1.1. 27.9474&rep=rep1&type=pdf. (Accessed 1 August 2022).

Mostaghim, S., Teich, J., 2003. Strategies for finding good local guides in multi-objective particle swarm optimization (MOPSO). In: Proceedings of the IEEE Swarm Intelligence Symposium, pp. 26–33.

Mühlenbein, H., 1997. The equation for response to selection and its use for prediction. Evolutionary Computation 5, 303–346.

Mühlenbein, H., Paaß, G., 1996. From recombination of genes to the estimation of distributions I. Binary parameters. In: Proceedings of the Parallel Problem Solving from Nature — PPSN IV, pp. 178–187.

Nain, P., Deb, K., 2003. Computationally effective search and optimization procedure using coarse to fine approximations. In: Proceedings of the Congress on Evolutionary Computation, pp. 2081–2088.

Nara, K., Hayashi, Y., Ikeda, K., Ashizawa, T., 2001. Application of tabu search to optimal placement of distributed generators. In: Proceedings of the IEEE Power Engineering Society Winter Meeting, pp. 918–923.

Neri, F., Cotta, C., 2012. Memetic algorithms and memetic computing optimization: A literature review. Swarm and Evolutionary Computation 2, 1–14.

Neri, F., Tirronen, V., 2010. Recent advances in differential evolution: A survey and experimental analysis. Artificial Intelligence Review 33, 61–106.

Nikolaev, A.G., Jacobson, S.H., 2010. Simulated annealing. In: Gendreau, M., Potvin, J.Y. (Eds.), Handbook of Metaheuristics. Springer US, Boston, MA, pp. 1–39.

Niu, B., Zhu, Y., He, X., Wu, H., 2007. MCPSO: a multi-swarm cooperative particle swarm optimizer. Applied Mathematics and Computation 185, 1050–1062.

Nourani, Y., Andresen, B., 1998. A comparison of simulated annealing cooling strategies. Journal of Physics. A, Mathematical and General 31, 8373–8385.

Nowicki, E., Smutnicki, C., 1996. A fast taboo search algorithm for the job shop problem. Management Science 42, 797–813.

Nowostawski, M., Poli, R., 1999. Parallel genetic algorithm taxonomy. In: Proceedings of the International Conference on Knowledge-Based Intelligent Information Engineering Systems, pp. 88–92.

Nozawa, H., 1992. A neural network model as a globally coupled map and applications based on chaos. Chaos: An Interdisciplinary Journal of Nonlinear Science 2, 377–386.

Ochoa, G., Malan, K., 2019. Recent advances in fitness landscape analysis. In: Proceedings of the Genetic and Evolutionary Computation Conference, pp. 1077–1094.

Ochoa, G., Verel, S., Daolio, F., Tomassini, M., 2014. Local optima networks: A new model of combinatorial fitness landscapes. In: Recent Advances in the Theory and Application of Fitness Landscapes. Springer, Berlin, Heidelberg, pp. 233–262.

Omran, M., Engelbrecht, A.P., Salman, A., 2005. Particle swarm optimization method for image clustering. International Journal of Pattern Recognition and Artificial Intelligence 19, 297–321.

Omran, M.G., Salman, A.A., Engelbrecht, A.P., 2002. Image classification using particle swarm optimization. In: Proceedings of the Asia-Pacific Conference on Simulated Evolution and Learning, pp. 370–374.

Opara, K.R., Arabas, J., 2019. Differential evolution: A survey of theoretical analyses. Swarm and Evolutionary Computation 44, 546–558.

Ouelhadj, D., Petrovic, S., 2010. A cooperative hyper-heuristic search framework. Journal of Heuristics 16, 835–857.

Ow, P.S., Morton, T.E., 1988. Filtered beam search in scheduling†. International Journal of Production Research 26, 35–62.

Ozaki, Y., Yano, M., Onishi, M., 2017. Effective hyperparameter optimization using Nelder-Mead method in deep learning. IPSJ Transactions on Computer Vision and Applications 9, 1–12.

Özcan, E., Bilgin, B., Korkmaz, E.E., 2008. A comprehensive analysis of hyper-heuristics. Intelligent Data Analysis 12, 3–23.

Papadimitriou, C.H., Steiglitz, K., 1982. Combinatorial Optimization: Algorithms and Complexity. Dover Publications, Inc., Mineola, NY.

Parpinelli, R., Lopes, H., Freitas, A., 2002. Data mining with an ant colony optimization algorithm. IEEE Transactions on Evolutionary Computation 6, 321–332.

Pavai, G., Geetha, T.V., 2016. A survey on crossover operators. ACM Computing Surveys 49, 1–42.

Pedemonte, M., Nesmachnow, S., Cancela, H., 2011. A survey on parallel ant colony optimization. Applied Soft Computing 11, 5181–5197.

Pelikan, M., Goldberg, D.E., Cantú-Paz, E.E., 2000. Linkage problem, distribution estimation, and Bayesian networks. Evolutionary Computation 8, 311–340.

Pham, D., Ghanbarzadeh, A., Koc, E., Otri, S., Rahim, S., Zaidi, M., 2005. Bee Algorithm: A Novel Approach to Function Optimisation. Technical Report. The Manufacturing Engineering Centre, Cardiff University, Queen's University. No. MEC 0501. Available at https://www.researchgate.net/profile/Ebubekir-Koc/publication/260985621_The_Bees_Algorithm_Technical_Note/links/56c1bd8708aee5caccf911f1/The-Bees-Algorithm-Technical-Note.pdf. (Accessed 1 August 2022).

Pincus, M., 1970. A Monte Carlo method for the approximate solution of certain types of constrained optimization problems. Operations Research 18, 1225–1228.

Piotrowski, A.P., 2017. Review of differential evolution population size. Swarm and Evolutionary Computation 32, 1–24.

Poikolainen, I., Neri, F., Caraffini, F., 2015. Cluster-based population initialization for differential evolution frameworks. Information Sciences 297, 216–235.

Poli, R., Kennedy, J., Blackwell, T., 2007. Particle swarm optimization: An overview. Swarm Intelligence 1, 33–57.

Potts, J.C., Giddens, T.D., Yadav, S.B., 1994. The development and evaluation of an improved genetic algorithm based on migration and artificial selection. IEEE Transactions on Systems, Man and Cybernetics 24, 73–86.

Price, K.V., 2013. Differential evolution. In: Zelinka, I., Snášel, V., Abraham, A. (Eds.), Handbook of Optimization: From Classical to Modern Approach. Springer, Berlin, Heidelberg, pp. 187–214.

Punj, G., Stewart, D.W., 1983. Cluster analysis in marketing research: Review and suggestions for application. Journal of Marketing Research 20, 134–148.

Qin, A., Suganthan, P., 2005. Self-adaptive differential evolution algorithm for numerical optimization. In: Proceedings of the IEEE Congress on Evolutionary Computation, pp. 1785–1791.

Qin, A.K., Huang, V.L., Suganthan, P.N., 2009. Differential evolution algorithm with strategy adaptation for global numerical optimization. IEEE Transactions on Evolutionary Computation 13, 398–417.

Qolomany, B., Maabreh, M., Al-Fuqaha, A., Gupta, A., Benhaddou, D., 2017. Parameters optimization of deep learning models using particle swarm optimization. In: Proceedings of the International Wireless Communications and Mobile Computing Conference, pp. 1285–1290.

Rabanal, P., Rodríguez, I., Rubio, F., 2007. Using river formation dynamics to design heuristic algorithms. In: Proceedings of the International Conference on Unconventional Computation, pp. 163–177.

Rahnamayan, S., Tizhoosh, H.R., Salama, M.M.A., 2008. Opposition-based differential evolution. IEEE Transactions on Evolutionary Computation 12, 64–79.

Randall, M., Lewis, A., 2002. A parallel implementation of ant colony optimization. Journal of Parallel and Distributed Computing 62, 1421–1432.

Rashedi, E., Nezamabadi-pour, H., Saryazdi, S., 2009. GSA: A gravitational search algorithm. Information Sciences 179, 2232–2248.

Ratnaweera, A., Halgamuge, S.K., Watson, H.C., 2004. Self-organizing hierarchical particle swarm optimizer with time-varying acceleration coefficients. IEEE Transactions on Evolutionary Computation 8, 240–255.

Razmjooy, N., Khalilpour, M., Ramezani, M., 2016. A new meta-heuristic optimization algorithm inspired by FIFA world cup competitions: Theory and its application in PID designing for AVR system. Journal of Control, Automation and Electrical Systems 27.

Real, E., Moore, S., Selle, A., Saxena, S., Suematsu, Y.L., Tan, J., Le, Q.V., Kurakin, A., 2017. Large-scale evolution of image classifiers. In: Proceedings of the International Conference on Machine Learning, pp. 2902–2911.

Rechenberg, I., 1965. Cybernetic solution path of an experimental problem. Technical Report. Library Translation 1122, Royal Aircraft Establishment.

Reed, R., 1993. Pruning algorithms—A survey. IEEE Transactions on Neural Networks 4, 740–747.

Reeves, C.R. (Ed.), 1993. Modern Heuristic Techniques for Combinatorial Problems. John Wiley & Sons, Inc., New York, NY.

Reinelt, G., 1995. TSPLIB 95. Available at http://comopt.ifi.uni-heidelberg.de/software/TSPLIB95/tsp95.pdf. (Accessed 1 August 2022).

Reinelt, G., 1991. TSPLIB-A traveling salesman problem library. ORSA Journal on Computing 3, 376–384.

Reinelt, G., 2007. Optimal solutions for symmetric TSPs. Available at http://comopt.ifi.uni-heidelberg.de/software/TSPLIB95/STSP.html. (Accessed 1 August 2022).

Remde, S., Cowling, P., Dahal, K., Colledge, N., 2008. Evolution of fitness functions to improve heuristic performance. In: Proceedings of the Learning and Intelligent Optimization, pp. 206–219.

Reyes-Sierra, M., Coello, C.A.C., 2006. Multi-objective particle swarm optimizers: A survey of the state-of-the-art. International Journal of Computational Intelligence Research 2, 287–308.

Richter, H., Engelbrecht, A.P. (Eds.), 2014. Recent Advances in the Theory and Application of Fitness Landscapes. Springer, Berlin, Heidelberg.

Robič, T., Filipič, B., 2005. DEMO: Differential evolution for multiobjective optimization. In: Proceedings of the International Conference on Evolutionary Multi-Criterion Optimization, pp. 520–533.

Romeo, F., Sangiovanni-Vincentelli, A., 1991. A theoretical framework for simulated annealing. Algorithmica 6, 302–345.

Rostami, M., Berahmand, K., Forouzandeh, S., 2021. A novel community detection based genetic algorithm for feature selection. Journal of Big Data 8, 1–27.

Rutenbar, R.A., 1989. Simulated annealing algorithms: An overview. IEEE Circuits and Devices Magazine 5, 19–26.

Salcedo-Sanz, S., Ser, J.D., Gil-López, S., Landa-Torres, I., Portilla-Figueras, J.A., 2013. The coral reefs optimization algorithm: An efficient meta-heuristic for solving hard optimization problems. In: Proceedings of the Applied Stochastic Models and Data Analysis International Conference, pp. 751–758.

Salhi, S., 2002. Defining tabu list size and aspiration criterion within tabu search methods. Computers & Operations Research 29, 67–86.

Sallam, K.M., Elsayed, S.M., Sarker, R.A., Essam, D.L., 2017. Landscape-based adaptive operator selection mechanism for differential evolution. Information Sciences 418–419, 383–404.

Schaffer, J., Eshelman, L., 1991. On crossover as an evolutionary viable strategy. In: Proceedings of the International Conference on Genetic Algorithms, pp. 61–68.

Schaffer, J.D., Whitley, D., Eshelman, L.J., 1992. Combinations of genetic algorithms and neural networks: A survey of the state of the art. In: Proceedings of the International Workshop on Combinations of Genetic Algorithms and Neural Networks, pp. 1–37.

Schermer, D., Moeini, M., Wendt, O., 2019. A hybrid VNS/Tabu search algorithm for solving the vehicle routing problem with drones and en route operations. Computers & Operations Research 109, 134–158.

Schneider, J.J., Puchta, M., 2010. Investigation of acceptance simulated annealing — a simplified approach to adaptive cooling schedules. Physica A: Statistical Mechanics and its Applications 389, 5822–5831.

Shah-Hosseini, H., 2008. Intelligent water drops algorithm: A new optimization method for solving the multiple knapsack problem. International Journal of Intelligent Computing and Cybernetics 1, 193–212.

Shen, Q., Shi, W.M., Kong, W., 2008. Hybrid particle swarm optimization and tabu search approach for selecting genes for tumor classification using gene expression data. Computational Biology and Chemistry 32, 53–60.

Shi, Y., 2004. Particle swarm optimization. IEEE Connections 2, 8–13.

Shi, Y., Eberhart, R., 1998. A modified particle swarm optimizer. In: Proceedings of the IEEE International Conference on Evolutionary Computation Proceedings, pp. 69–73.

Shi, Y., Eberhart, R.C., 1999. Empirical study of particle swarm optimization. In: Proceedings of the Congress on Evolutionary Computation, pp. 1945–1950.

da Silva, G.L.F., Valente, T.L.A., Silva, A.C., de Paiva, A.C., Gattass, M., 2018. Convolutional neural network-based PSO for lung nodule false positive reduction on CT images. Computer Methods and Programs in Biomedicine 162, 109–118.

Silver, D., Huang, A., Maddison, C.J., Guez, A., Sifre, L., van den Driessche, G., Schrittwieser, J., Antonoglou, I., Panneershelvam, V., Lanctot, M., Dieleman, S., Grewe, D., Nham, J., Kalchbrenner, N., Sutskever, I., Lillicrap, T., Leach, M., Kavukcuoglu, K., Graepel, T., Hassabis, D., 2016. Mastering the game of Go with deep neural networks and tree search. Nature 529, 484–489.

Silver, N., 2012. The Signal and the Noise: Why So Many Predictions Fail–but Some Don't. Penguin Press, New York, NY.

Sinha, A., Goldberg, D.E., 2003. A Survey of Hybrid Genetic and Evolutionary Algorithms. Technical Report. IlliGAL Report No. 2003004. University of Illinois at Urbana-Champaign, Urbana, IL.

Socha, K., Dorigo, M., 2008. Ant colony optimization for continuous domains. European Journal of Operational Research 185, 1155–1173.

Srinivas, M., Patnaik, L., 1994. Adaptive probabilities of crossover and mutation in genetic algorithms. IEEE Transactions on Systems, Man and Cybernetics 24, 656–667.

Srinivas, N., Deb, K., 1994. Muiltiobjective optimization using nondominated sorting in genetic algorithms. Evolutionary Computation 2, 221–248.

Srivastava, D., Singh, Y., Sahoo, A., 2019. Auto tuning of RNN hyper-parameters using cuckoo search algorithm. In: Proceedings of the International Conference on Contemporary Computing, pp. 1–5.

Stacey, A., Jancic, M., Grundy, I., 2003. Particle swarm optimization with mutation. In: Proceedings of the Congress on Evolutionary Computation, pp. 1425–1430.

Stadler, P.F., 2002. Fitness landscapes. In: Lässig, M., Valleriani, A. (Eds.), Biological Evolution and Statistical Physics. Springer, Berlin, Heidelberg, pp. 183–204.

Stanovov, V., Akhmedova, S., Semenkin, E., 2018. LSHADE algorithm with rank-based selective pressure strategy for solving CEC 2017 benchmark problems. In: Proceedings of the IEEE Congress on Evolutionary Computation, pp. 1–8.

Stanovov, V., Akhmedova, S., Semenkin, E., 2021. NL-SHADE-RSP algorithm with adaptive archive and selective pressure for CEC 2021 numerical optimization. In: Proceedings of the IEEE Congress on Evolutionary Computation, pp. 809–816.

Storn, R., 1996. On the usage of differential evolution for function optimization. In: Proceedings of North American Fuzzy Information Processing, pp. 519–523.

Storn, R., Price, K., 1995. Differential evolution - A simple and efficient adaptive scheme for global optimization over continuous spaces. Technical Report. International Computer Science Institute (ICSI). Technical Report TR-95-012. Available at https://citeseerx.ist.psu.edu/viewdoc/download?doi=10.1.1.67.5398&rep=rep1&type=pdf. (Accessed 1 August 2022).

Storn, R., Price, K., 1997. Differential evolution – A simple and efficient heuristic for global optimization over continuous spaces. Journal of Global Optimization 11, 341–359.

Stützle, T., 1998. Parallelization strategies for ant colony optimization. In: Proceedings of the International Conference on Parallel Problem Solving from Nature, pp. 722–731.

Stützle, T., Hoos, H., 1996. Improving the ant system: A detailed report on the $\mathcal{MAX}$–$\mathcal{MIN}$ Ant System. Technical Report. FG Intellektik, FB Informatik, TU Darmstadt. Technical Report AIDA-96-12. Available at https://www.cs.ubc.ca/~hoos/Publ/aida-96-12r.ps. (Accessed 1 August 2022).

Stützle, T., Ruiz, R., 2018. Iterated local search. In: Martí, R., Pardalos, P.M., Resende, M.G.C. (Eds.), Handbook of Heuristics. Springer, Cham, pp. 579–605.

Suganthan, P.N., Hansen, N., Deb, J.J.L.K., Chen, Y., Auger, A., Tiwari, S., 2005. Problem Definitions and Evaluation Criteria for the CEC 2005 Special Session on Real-Parameter Optimization. Technical Report. Nanyang Technological University, Singapore and KanGAL Report Number 2005005 (Kanpur Genetic Algorithms Laboratory, IIT Kanpur). Available at https://www3.ntu.edu.sg/home/epnsugan/index_files/CEC-05/CEC05.htm. (Accessed 1 August 2022).

Suganthan, P.N., Hansen, N., Deb, J.J.L.K., Chen, Y., Auger, A., Tiwari, S., 2011. Problem Definitions and Evaluation Criteria for CEC 2011 Competition on Testing Evolutionary Algorithms on Real World Optimization Problems. Technical Report. Jadavpur University, India and Nanyang Technological University, Singapore. Available at https://www3.ntu.edu.sg/home/epnsugan/index_files/CEC11-RWP/CEC11-RWP.htm. (Accessed 1 August 2022).

Sultan, A.B.M., Mahmod, R., Sulaiman, M.N., Bakar, M.R.A., 2008. Selecting quality initial random seed for metaheuristic approaches: A case of timetabling problem. International Journal of the Computer, the Internet and Management 16, 38–45.

Suman, B., Kumar, P., 2006. A survey of simulated annealing as a tool for single and multi-objective optimization. Journal of the Operational Research Society 57, 1143–1160.

Sun, Y., Xue, B., Zhang, M., Yen, G.G., 2019. Evolving deep convolutional neural networks for image classification. IEEE Transactions on Evolutionary Computation 24, 394–407.

Sun, Y., Xue, B., Zhang, M., Yen, G.G., 2020. Completely automated CNN architecture design based on blocks. IEEE Transactions on Neural Networks and Learning Systems 31, 1242–1254.

Syswerda, G., 1989. Uniform crossover in genetic algorithms. In: Proceedings of the International Conference on Genetic Algorithms, pp. 2–9.

Syswerda, G., 1991. A study of reproduction in generational and steady-state genetic algorithms. In: Rawlins, G.J.E. (Ed.), Foundations of Genetic Algorithms, vol. 1. Morgan Kaufmann, San Mateo, CA, pp. 94–101.

Szu, H., Hartley, R., 1987. Fast simulated annealing. Physics Letters A 122, 157–162.

Taherkhani, M., Safabakhsh, R., 2016. A novel stability-based adaptive inertia weight for particle swarm optimization. Applied Soft Computing 38, 281–295.

Taillard, E., 2016. Tabu search. In: Siarry, P. (Ed.), Metaheuristics. Springer International Publishing, Cham, pp. 51–76.

Taillard, E.D., Voss, S., 2002. Popmusic — partial optimization metaheuristic under special intensification conditions. In: Essays and Surveys in Metaheuristics. Springer US, Boston, MA, pp. 613–629.

Talbi, E.G., 2009. Metaheuristics: From Design to Implementation. John Wiley & Sons, Hoboken, NJ.

Talbi, E.G., 2014. Hybrid Metaheuristics. Springer, Berlin, Heidelberg.

Talbi, E.G., Hafidi, Z., Geib, J.M., 1999. Parallel tabu search for large optimization problems. In: Voß, S., Martello, S., Osman, I.H., Roucairol, C. (Eds.), Meta-Heuristics: Advances and Trends in Local Search Paradigms for Optimization. Springer US, Boston, MA, pp. 345–358.

Tanabe, R., Fukunaga, A.S., 2013. Success-history based parameter adaptation for differential evolution. In: Proceedings of the IEEE Congress on Evolutionary Computation, pp. 71–78.

Tanabe, R., Fukunaga, A.S., 2014. Improving the search performance of SHADE using linear population size reduction. In: Proceedings of the IEEE Congress on Evolutionary Computation, pp. 1658–1665.

Tanese, R., 1987. Parallel genetic algorithms for a hypercube. In: Proceedings of the International Conference on Genetic Algorithms on Genetic Algorithms and Their Application, pp. 177–183.

Tasoulis, D., Pavlidis, N., Plagianakos, V., Vrahatis, M., 2004. Parallel differential evolution. In: Proceedings of the 2004 Congress on Evolutionary Computation, pp. 2023–2029.

Teijeiro, D., Pardo, X.C., González, P., Banga, J.R., Doallo, R., 2016. Implementing parallel differential evolution on spark. In: Proceedings of the European Conference on the Applications of Evolutionary Computation, pp. 75–90.

Teo, J., 2006. Exploring dynamic self-adaptive populations in differential evolution. Soft Computing 10, 673–686.

Thierens, D., Goldberg, D., 1994. Convergence models of genetic algorithm selection schemes. In: Proceedings of the International Conference on Parallel Problem Solving from Nature, pp. 119–129.

Ting, T.O., Yang, X.S., Cheng, S., Huang, K., 2015. Hybrid metaheuristic algorithms: Past, present, and future. In: Yang, X.S. (Ed.), Proceedings of the Recent Advances in Swarm Intelligence and Evolutionary Computation. Springer International Publishing, Cham, pp. 71–83.

Todorovski, M., Rajicic, D., 2006. An initialization procedure in solving optimal power flow by genetic algorithm. IEEE Transactions on Power Systems 21, 480–487.

Tong, D.L., Mintram, R., 2010. Genetic algorithm-neural network (GANN): A study of neural network activation functions and depth of genetic algorithm search applied to feature selection. International Journal of Machine Learning and Cybernetics 1, 75–87.

Toth, P., Vigo, D., 2003. The granular tabu search and its application to the vehicle-routing problem. European Journal of Operational Research 15, 333–346.

Tripathi, P.K., Bandyopadhyay, S., Pal, S.K., 2007. Multi-objective particle swarm optimization with time variant inertia and acceleration coefficients. Information Sciences 177, 5033–5049.

Tsai, C.W., 2009. On the Study of Efficient Metaheuristics via Pattern Reduction. Ph.D. thesis. National Sun Yat-sen University, Kaohsiung, Taiwan.

Tsai, C.W., 2015. Search economics: A solution space and computing resource aware search method. In: Proceedings of the IEEE International Conference on Systems, Man, and Cybernetics, pp. 2555–2560.

Tsai, C.W., 2016. An effective WSN deployment algorithm via search economics. Computer Networks 101, 178–191.

Tsai, C.W., Fang, Z.Y., 2021. An effective hyperparameter optimization algorithm for DNN to predict passengers at a metro station. ACM Transactions on Internet Technology 21, 1–24.

Tsai, C.W., Hong, T.P., Shiu, G.N., 2016. Metaheuristics for the lifetime of WSN: A review. IEEE Sensors Journal 16, 2812–2831.

Tsai, C.W., Hsia, C.H., Yang, S.J., Liu, S.J., Fang, Z.Y., 2020. Optimizing hyperparameters of deep learning in predicting bus passengers based on simulated annealing. Applied Soft Computing 88, 106068.

Tsai, C.W., Huang, K.W., Yang, C.S., Chiang, M.C., 2015. A fast particle swarm optimization for clustering. Soft Computing 19, 321–338.

Tsai, C.W., Huang, W.C., Chiang, M.H., Chiang, M.C., Yang, C.S., 2014. A hyper-heuristic scheduling algorithm for cloud. IEEE Transactions on Cloud Computing 2, 236–250.

Tsai, C.W., Lee, C.Y., Chiang, M.C., Yang, C.S., 2009. A fast VQ codebook generation algorithm via pattern reduction. Pattern Recognition Letters 30, 653–660.

Tsai, C.W., Li, Z.A., Chiang, M.C., Yang, C.S., 2017. A novel clustering algorithm for wireless sensor network based on search economics. In: Proceedings of the IEEE International Conference on Systems, Man, and Cybernetics, pp. 809–814.

Tsai, C.W., Lin, T.Y., Chiang, M.C., Yang, C.S., Hong, T.P., 2012. Continuous space pattern reduction for genetic clustering algorithm. In: Proceedings of the Genetic and Evolutionary Computation Conference, pp. 1475–1476.

Tsai, C.W., Liu, S.J., 2018. An effective IoT service-to-interface assignment algorithm via search economics. IEEE Internet of Things Journal 5, 1708–1718.

Tsai, C.W., Liu, S.J., 2019. SEIM: Search economics for influence maximization in online social networks. Future Generation Computer Systems 93, 1055–1064.

Tsai, C.W., Liu, S.J., 2020. Search economics for single-objective real-parameter optimization. In: Proceedings of the IEEE Congress on Evolutionary Computation, pp. 1–7.

Tsai, C.W., Teng, T.C., Liao, J.T., Chiang, M.C., 2021. An effective hybrid-heuristic algorithm for urban traffic light scheduling. Neural Computing & Applications 33, 17535–17549.

Tsai, C.W., Tsai, P.W., Chiang, M.C., Yang, C.S., 2018. Data analytics for internet of things: A review. WIREs Data Mining and Knowledge Discovery 8, e1261.

Tsai, C.W., Tseng, S.P., Chiang, M.C., Yang, C.S., 2010. A framework for accelerating metaheuristics via pattern reduction. In: Proceedings of the Genetic and Evolutionary Computation Conference, pp. 293–294.

Tsai, C.W., Tseng, S.P., Yang, C.S., Chiang, M.C., 2013. PREACO: A fast ant colony optimization for codebook generation. Applied Soft Computing 13, 3008–3020.

Tsai, C.W., Yang, C.S., Chiang, M.C., 2007. A novel pattern reduction algorithm for k-means based clustering. In: Proceedings of the IEEE International Conference on Systems, Man and Cybernetics, pp. 504–509.

Tsai, K.H., Tsai, C.W., Chiang, M.C., 2022. An effective metaheuristic-based pruning method for convolutional neural network. In: Proceedings of the Genetic and Evolutionary Computation Conference, pp. 1–4.

TSPLIB, 1995. Available at http://comopt.ifi.uni-heidelberg.de/software/TSPLIB95/index.html. (Accessed 1 August 2022).

Tsubakitani, S., Evans, J.R., 1998. Optimizing tabu list size for the traveling salesman problem. Computers & Operations Research 25, 91–97.

Tušar, T., Filipič, B., 2007. Differential evolution versus genetic algorithms in multiobjective optimization. In: Proceedings of the International Conference on Evolutionary Multi-Criterion Optimization, pp. 257–271.

van den Bergh, F., Engelbrecht, A., 2006. A study of particle swarm optimization particle trajectories. Information Sciences 176, 937–971.

van den Bergh, F., Engelbrecht, A.P., 2004. A cooperative approach to particle swarm optimization. IEEE Transactions on Evolutionary Computation 8, 225–239.

van der Merwe, D.W., Engelbrecht, A.P., 2003. Data clustering using particle swarm optimization. In: Proceedings of IEEE Congress on Evolutionary Computation, pp. 215–220.

Voudouris, C., Tsang, E., 1999. Guided local search and its application to the traveling salesman problem. European Journal of Operational Research 113, 469–499.

Voudouris, C., Tsang, E.P.K., 1996. Partial constraint satisfaction problems and guided local search. In: Proceedings of the International Conference on Practical Application of Constraint Technology, pp. 337–356.

Wang, D., Tan, D., Liu, L., 2018a. Particle swarm optimization algorithm: An overview. Soft Computing 22, 387–408.

Wang, J., Xu, J., Wang, X., 2018b. Combination of hyperband and Bayesian optimization for hyperparameter optimization in deep learning. https://doi.org/10.48550/ARXIV.1801.01596.

Wang, X.H., Li, J.J., 2004. Hybrid particle swarm optimization with simulated annealing. In: Proceedings of International Conference on Machine Learning and Cybernetics, pp. 2402–2405.

Wang, Y., Xu, C., Qiu, J., Xu, C., Tao, D., 2018c. Towards evolutionary compression. In: Proceedings of the ACM SIGKDD International Conference on Knowledge Discovery & Data Mining, pp. 2476–2485.

Wedyan, A., Whalley, J., Narayanan, A., 2017. Hydrological cycle algorithm for continuous optimization problems. Journal of Optimization 2017, 1–26.

Weile, D.S., Michielssen, E., 1997. Genetic algorithm optimization applied to electromagnetics: a review. IEEE Transactions on Antennas and Propagation 45, 343–353.

Whitley, D., 1989. The GENITOR algorithm and selection pressure: Why rank-based allocation of reproductive trials is best. In: Proceedings of the International Conference on Genetic Algorithms, pp. 111–121.

Whitley, D., 1994. A genetic algorithm tutorial. Statistics and Computing 4, 65–85.

Whitley, D., Starkweather, T., 1990. GENITOR II: A distributed genetic algorithm. Journal of Experimental and Theoretical Artificial Intelligence 2, 189–214.

Whitley, D., Starkweather, T., Bogart, C., 1990. Genetic algorithms and neural networks: Optimizing connections and connectivity. Parallel Computing 14, 347–361.

Wold, S., Esbensen, K., Geladi, P., 1987. Principal component analysis. Chemometrics and Intelligent Laboratory Systems 2, 37–52.

Wolfram Alpha, 2009. Available at http://www.wolframalpha.com/input/?i=graph+y%3D0.99%5Ex+and+y%3D0.99★(0.0001%2F0.99)%5E(x%2F1000)+from+x%3D1to1000. (Accessed 1 August 2022).

Wolpert, D.H., Macready, W.G., 1997. No free lunch theorems for optimization. IEEE Transactions on Evolutionary Computation 1, 67–82.

Wright, S., 1932. The roles of mutation, inbreeding, crossbreeding and selection in evolution. In: Proceedings of the International Congress of Genetics, pp. 356–366.

Wu, G., Mallipeddi, R., Suganthan, P., Wang, R., Chen, H., 2016. Differential evolution with multi-population based ensemble of mutation strategies. Information Sciences 329, 329–345.

Wu, G., Mallipeddi, R., Suganthan, P.N., 2017. Problem Definitions and Evaluation Criteria for the CEC 2017 Competition on Constrained Real-Parameter Optimization. Technical Report. National University of Defense Technology, Changsha, Hunan, P.R. China; Kyungpook National University, Daegu, South Korea; Nanyang Technological University, Singapore. Available at https://www3.ntu.edu.sg/home/epnsugan/index_files/CEC2017/CEC2017.htm. (Accessed 1 August 2022).

Wu, G., Shen, X., Li, H., Chen, H., Lin, A., Suganthan, P., 2018. Ensemble of differential evolution variants. Information Sciences 423, 172–186.

Wu, T., Shi, J., Zhou, D., Lei, Y., Gong, M., 2019. A multi-objective particle swarm optimization for neural networks pruning. In: Proceedings of the IEEE Congress on Evolutionary Computation, pp. 570–577.

Xie, L., Yuille, A., 2017. Genetic CNN. In: Proceedings of the IEEE International Conference on Computer Vision, pp. 1379–1388.

Xu, R., Wunsch-II, D.C., 2005. Survey of clustering algorithms. IEEE Transactions on Neural Networks 16, 645–678.

Xu, R., Wunsch-II, D.C., 2008. Clustering. Wiley, John & Sons, Inc., Hoboken, NJ.

Xue, B., Zhang, M., Browne, W.N., 2013. Particle swarm optimization for feature selection in classification: A multi-objective approach. IEEE Transactions on Cybernetics 43, 1656–1671.

Yang, Q., Chen, W.N., Yu, Z., Gu, T., Li, Y., Zhang, H., Zhang, J., 2017. Adaptive multimodal continuous ant colony optimization. IEEE Transactions on Evolutionary Computation 21, 191–205.

Yang, X.S., 2009. Firefly algorithms for multimodal optimization. In: Proceedings of the International Symposium on Stochastic Algorithms, pp. 169–178.

Yang, X.S., 2010. A new metaheuristic bat-inspired algorithm. In: Proceedings of the Nature Inspired Cooperative Strategies for Optimization, pp. 65–74.

Yang, X.S., 2012. Flower pollination algorithm for global optimization. In: Proceedings of the International Conference on Unconventional Computing and Natural Computation, pp. 240–249.

Yang, X.S., Deb, S., 2009. Cuckoo search via Lévy flights. In: Proceedings of the World Congress on Nature Biologically Inspired Computing, pp. 210–214.

Yazdani, M., Jolai, F., 2016. Lion optimization algorithm (LOA): A nature-inspired meta-heuristic algorithm. Journal of Computational Design and Engineering 3, 24–36.

Ye, F., 2017. Particle swarm optimization-based automatic parameter selection for deep neural networks and its applications in large-scale and high-dimensional data. PLoS ONE 12, 1–36.

Yetgin, H., Cheung, K.T.K., El-Hajjar, M., Hanzo, L.H., 2017. A survey of network lifetime maximization techniques in wireless sensor networks. IEEE Communications Surveys and Tutorials 19, 828–854.

Young, S.R., Rose, D.C., Karnowski, T.P., Lim, S.H., Patton, R.M., 2015. Optimizing deep learning hyper-parameters through an evolutionary algorithm. In: Proceedings of the Workshop on Machine Learning in High-Performance Computing Environments, pp. 1–5.

Zervoudakis, K., Tsafarakis, S., 2020. A mayfly optimization algorithm. Computers & Industrial Engineering 145, 106559.

Zhan, Z.H., Wang, Z.J., Jin, H., Zhang, J., 2020. Adaptive distributed differential evolution. IEEE Transactions on Cybernetics 50, 4633–4647.

Zhan, Z.H., Zhang, J., Li, Y., Chung, H.S.H., 2009. Adaptive particle swarm optimization. IEEE Transactions on Systems, Man and Cybernetics. Part B. Cybernetics 39, 1362–1381.

Zhang, H., Sun, G., 2002. Feature selection using tabu search method. Pattern Recognition 35, 701–711.

Zhang, J., Sanderson, A.C., 2009. JADE: Adaptive differential evolution with optional external archive. IEEE Transactions on Evolutionary Computation 13, 945–958.

Zhang, W., Maleki, A., Rosen, M.A., Liu, J., 2018. Optimization with a simulated annealing algorithm of a hybrid system for renewable energy including battery and hydrogen storage. Energy 163, 191–207.

Zhang, W.J., Xie, X.F., 2003. DEPSO: Hybrid particle swarm with differential evolution operator. In: Proceedings of the IEEE International Conference on Systems, Man and Cybernetics, pp. 3816–3821.

Zhou, Y., Yen, G.G., Yi, Z., 2021. A knee-guided evolutionary algorithm for compressing deep neural networks. IEEE Transactions on Cybernetics 51, 1626–1638.

Index

Printed in the United States
by Baker & Taylor Publisher Services